DISCARDED

BOWLING GREEN STATE UNIVERSITY

LIBRARY

MISMANAGED
TRADE?

KENNETH FLAMM

MISMANAGED TRADE?

*Strategic Policy
and the
Semiconductor Industry*

Bowling Green State University
Jerome Library

JUL 0 9 1996

SERIALS

BROOKINGS INSTITUTION PRESS
Washington, D.C.

1996

BOWLING GREEN STATE
UNIVERSITY LIBRARIES

Copyright © 1996 by
THE BROOKINGS INSTITUTION
1775 Massachusetts Avenue, N.W., Washington, D.C. 20036

All rights reserved

Library of Congress Cataloging-in-Publication data:

Flamm, Kenneth, 1951–
    Mismanaged trade : strategic policy and the semiconductor industry /
Kenneth Flamm.
        p. cm.
    Includes bibliographical references and index.
    ISBN 0-8157-2846-8 (cloth: alk. paper) — ISBN 0-8157-2847-6
    (pbk.: alk. paper)
        1. Semiconductor industry—Government policy—United States.
    2. Semiconductor industry—Government policy—Japan.   3. United
    States—Foreign economic relations—Japan.   4. Japan—Foreign
    economic relations—United States.   I. Title.
    HD9696.S43U466   1996
    382'.4562138152'0973—dc20   94-21324
                                                                  CIP

9 8 7 6 5 4 3 2 1

The paper used in this publication meets the minimum requirements of the
American National Standard for Information Sciences—Permanence of Paper for
Printed Library Materials, ANSI Z39.48-1984.

Set in Times Roman

Composition by Harlowe Typography, Inc.
Cottage City, Maryland

Printed by R.R. Donnelley and Sons Co.
Harrisonburg, Virginia

₿ THE BROOKINGS INSTITUTION

The Brookings Institution is an independent organization devoted to nonpartisan research, education, and publication in economics, government, foreign policy, and the social sciences generally. Its principal purposes are to aid in the development of sound public policies and to promote public understanding of issues of national importance.

The Institution was founded on December 8, 1927, to merge the activities of the Institute for Government Research, founded in 1916, the Institute of Economics, founded in 1922, and the Robert Brookings Graduate School of Economics and Government, founded in 1924.

The Board of Trustees is responsible for the general administration of the Institution, while the immediate direction of the policies, program, and staff is vested in the President, assisted by an advisory committee of the officers and staff. The by-laws of the Institution state: "It is the function of the Trustees to make possible the conduct of scientific research, and publication, under the most favorable conditions, and to safeguard the independence of the research staff in the pursuit of their studies and in the publication of the results of such studies. It is not a part of their function to determine, control, or influence the conduct of particular investigations or the conclusions reached."

The President bears final responsibility for the decision to publish a manuscript as a Brookings book. In reaching his judgment on the competence, accuracy, and objectivity of each study, the President is advised by the director of the appropriate research program and weighs the views of a panel of expert outside readers who report to him in confidence on the quality of the work. Publication of a work signifies that it is deemed a competent treatment worthy of public consideration but does not imply endorsement of conclusions or recommendations.

The Institution maintains its position of neutrality on issues of public policy in order to safeguard the intellectual freedom of the staff. Hence interpretations or conclusions in Brookings publications should be understood to be solely those of the authors and should not be attributed to the Institution, to its trustees, officers, or other staff members, or to the organizations that support its research.

Board of Trustees

James A. Johnson
Chairman

Leonard Abramson
Michael H. Armacost
Ronald J. Arnault
Elizabeth E. Bailey
Alan M. Dachs
Kenneth W. Dam
D. Ronald Daniel
Stephen Friedman

Vartan Gregorian
Bernadine P. Healy
Samuel Hellman
Warren Hellman
Robert A. Helman
Thomas W. Jones
Breene M. Kerr
Thomas G. Labrecque
Donald F. McHenry
Jessica Tuchman Mathews
David O. Maxwell

Constance Berry Newman
Maconda Brown O'Connor
Samuel Pisar
Rozanne L. Ridgway
Judith S. Rodin
Warren B. Rudman
Michael P. Schulhof
Robert H. Smith
Vincent J. Trosino
John D. Zeglis
Ezra K. Zilkha

Honorary Trustees

Vincent M. Barnett, Jr.
Rex J. Bates
Barton M. Biggs
Louis W. Cabot
Edward W. Carter
Frank T. Cary
A. W. Clausen
John L. Clendenin
William T. Coleman, Jr.
Lloyd N. Cutler
Bruce B. Dayton
Douglas Dillon
Charles W. Duncan, Jr.

Walter Y. Elisha
Robert F. Erburu
Robert D. Haas
Teresa Heinz
Andrew Heiskell
Roy M. Huffington
Vernon E. Jordan, Jr.
Nannerl O. Keohane
James T. Lynn
William McC. Martin, Jr.
Robert S. McNamara
Mary Patterson McPherson
Arjay Miller

Donald S. Perkins
J. Woodward Redmond
Charles W. Robinson
James D. Robinson III
David Rockefeller, Jr.
Howard D. Samuel
B. Francis Saul II
Ralph S. Saul
Henry B. Schacht
Robert Brookings Smith
Morris Tanenbaum
John C. Whitehead
James D. Wolfensohn

For
Dianne

Foreword

SINCE 1959 the United States and Japan have repeatedly negotiated with one another to resolve disputes involving trade in semiconductors. During these four decades the American and Japanese chip industries have been reluctant pioneers, cutting their way through a forest of issues created by the unique conditions of high-technology trade and investment. The new directions explored by the two countries' policymakers evolved into the 1986 U.S.-Japan Semiconductor Trade Arrangement. Today, as that experiment and its 1991 successor agreement are about to expire, the United States and Japan seem poised to again change course.

In this study Kenneth Flamm analyzes the public policy issues raised during these years of friction over semiconductor trade. Motivated by strategic economic and national security rationales, both the United States and Japan pursued policies designed to advance their industrial objectives. Policy conflicts occurred over trade, technology, development, investment, and antitrust enforcement as Japan's microelectronics industry caught up with its American competitors and pulled ahead in some areas. U.S. short-range tactics sometimes undercut long-term goals. The author concludes that over the long term improved trade rules and strengthened enforcement mechanisms must be developed to ensure open and competitive international markets for semiconductors. Effective solutions for trade in semiconductors will undoubtedly be applicable to other industries and ultimately may serve as the model for further strengthening the more general multilateral trading system codified in the General Agreement on Tariffs and Trade.

The author is grateful to Dan Hutcheson, F. M. Scherer, and Michael Smith for the detailed comments they made on the first draft of this study. Extensive comments received from Daryl Hatano, Thomas Howell, Matt

Rohde, John Steinbruner, Makoto Sumita, and Alan Wolff also greatly improved this book, as did comments from Bill Finan, Gary Hufbauer, Peter Reiss, Gary Saxonhouse, Carl Shapiro, Jack Triplett, and Philip Webre on articles based on this research. Executives in U.S., European, and Japanese electronics companies were generous with their time and extraordinarily helpful when they spoke with the author as he went about researching this book, as were many government officials in the United States, Japan, and the European Community. The author is also grateful to Japan's Ministry of International Trade and Industry and Ministry of Finance for their financial and logistical support of short stays at the Tokyo research institutes associated with those ministries and their help in arranging interviews with Japanese government and industry officials.

The author thanks the Rockefeller Brothers Fund and the John D. and Catherine T. MacArthur Foundation for their financial support of this study.

First-rate research assistance was provided by Yuko Iida Frost and Kaori Nakajima. Michael Treadway and James Schneider edited the manuscript. Melanie Allen, Ian Campbell, Cynthia Iglesias, and Andrew Solomon verified its factual content. Kris McDevitt and Ann Ziegler provided staff assistance. Ellen Garshick proofread the pages and Julia Petrakis compiled the index.

The views expressed in this book are those of the author and should not be ascribed to the persons or organizations whose assistance is acknowledged or to the trustees, officers, or other staff members of the Brookings Institution.

MICHAEL H. ARMACOST
President

May 1996
Washington, D.C

Contents

1. Introduction   1
   Challenges for Public Policy   *1*
   Technology and Industrial Structure in Semiconductors   *7*
   International Competition in Semiconductors   *18*
   Strategic Policy in the United States   *27*

2. New Competition: The Japanese Ascent in
   Semiconductors   39
   First Steps   *40*
   Early Exports in Electronics   *49*
   The Seal of Approval   *52*
   Protecting the Japanese Market   *55*
   The Integrated Circuit Arrives in Japan   *59*
   A Secret Truce in the "Patent War"   *68*
   Battles in the LSI Market   *70*
   First Steps toward Liberalization and "Dumping"   *73*
   The "Calculator War"   *75*
   Further Liberalization and Crisis   *77*
   Into the Japanese Market   *88*
   NTT Arrives on the VLSI Scene   *90*
   The VLSI Project   *94*
   The Continuing Role of Government   *113*
   Dependence in Silicon   *119*
   Summary   *124*

3. The Genesis of an American Trade Policy in
   Semiconductors, 1959–84   127
   A Threat to National Security, 1959   *128*
   Private versus Public Policy:
      Television Exports in the 1960s   *132*
   Competition in the 1970s   *136*

"Below Cost" Dumping *141*
Mounting Frictions: "Quality Dumping" *144*
Organizing a Response *147*
Competition, Collusion, or Predation?
   The 64K DRAM Wars *148*
Sectoral Negotiations *153*

4. **The Semiconductor Trade Arrangement and
     Its Aftermath**   159
A Thumbnail Historical Sketch *159*
Evolution of the Semiconductor Trade Regime *162*
Initial Implementation of the Arrangement,
   August 1986 to November 1987 *175*
The Privatization of Restraints,
   December 1987 to Mid-1989 *201*
"Coordination Structures" and "High Price Stability,"
   1989 to 1990 *212*
U.S. Memories to the Rescue *216*
Aftermath: The Second Semiconductor Trade
   Arrangement *223*

5. **Effects of the Semiconductor Trade Arrangement
     on Semiconductor Markets**   227
Impacts on Semiconductor Supply and Demand *231*
Impacts of Regional Price Differentials on Regional
   Welfare *272*
DRAMs versus EPROMs *278*
Import Promotion and the STA *279*
Conclusions *292*
Appendix 5A: The Economics of Contract Pricing *294*
Appendix 5B: Construction of Estimated FMVs *301*

6. **Dumping in DRAMs**   305
The Economic Rationality of Below-Marginal-Cost
   Pricing *307*
Cost Structures and Dumping in the Semiconductor
   Industry *311*
Modeling the Semiconductor Product Life Cycle *313*
Model Solution *334*
Conclusions *358*
Appendix 6A: A General Solution to the Problem of Optimal
   Capacity and Production Choice *359*

Appendix 6B: Detailed Solution with Specific Demand and
   Learning Curve Assumptions   365

7. Strategic Issues   372
    Conceptions of "Strategic"   372
    Semiconductor Dependency and Strategic Trade Policy   382
    Anecdotal Evidence on Private Collusion   396
    Empirical Tests for Competition   398
    The Costs of Facing a Cartel   405
    More Complex Stories about Cartels   415
    Conclusion   417
    Appendix 7A: Solution of the Model with Strategic
       Behavior   418

8. Conclusion: Mismanaged Trade?   425
    Why Semiconductors?   427
    Strategic Policy in Semiconductors   429
    Effects of the Semiconductor Trade Arrangement   433
    Where Do We Stand?   437
    Pricing and Dumping   442
    Where Do We Go from Here?   449
    The Global Semiconductor Conference   454
    From Here to There   455
    Mismanaged Trade?   457

Index   461

Tables

1-1. Share of Worldwide R&D Expenditures and Employment for
      U.S.-Based Companies and Their Majority-Owned Foreign
      Affiliates, by Parent's Industry, 1989   14

1-2. Capital Costs of a Typical Semiconductor Wafer Fabrication
      Facility, Mid-1970s–Early 1990s   16

1-3. Domestic R&D Expenditures and Employment by U.S.-Based
      Companies and Their Majority-Owned Foreign Affiliates as
      Shares of Worldwide Totals by Parent's Industry, 1989   19

1-4. World Market Shares of Merchant and Captive Semiconductor
      Manufacturers, by Region Where Headquartered, Selected
      Years, 1981–93   20

1-5. U.S. Government Shares of Total R&D Funding,
      by Industry, 1989   28

1-6. U.S. Industrial Consumption of Semiconductors,
      by Consuming Sector, Selected Years, 1963–87   35

1-7. Sources of R&D Funds in the U.S. Semiconductor Industry,
      Selected Periods, 1958–76   36

1-8. Federal Contracts as Share of Total U.S. Semiconductor
      Shipments, Selected Years, 1963–87   37

2-1. Transistor Output of U.S. and Japanese Companies during
      Transition to Overseas Production, 1957–68   47

2-2. FILP Loans to Japanese Machinery and Electronics Industries,
      1956–84   64

2-3. Sales by Japanese Semiconductor Companies to Affiliated and
      Nonaffiliated Companies, 1966–71   76

2-4. Japanese Measures to Liberalize Trade and Investment in the
      Electronics Industry, 1962–76   88

2-5. Japanese Expenditures on Integrated Circuit R&D,
      by Funding Source, 1963–79   97

2-6. Shares of Total VLSI Research Association Project Personnel,
      by Lab Group, 1976–86   108

2-7. MITI Expenditures on R&D Programs Related to
      Microelectronics, Fiscal Years 1983–92   114

2-8. Joint Patent Applications with NTT, by Number of
      Participating Companies, 1980–86   117

2-9. Government and NTT Semiconductor-Related R&D Funding
      in Relation to Private Japanese Semiconductor R&D Funding,
      1988   118

3-1. Factory Prices of Finished Integrated Circuits in Europe,
      Japan, and the United States, 1976   139

5-1. Market Concentration in DRAMs, 1979–89   256

5-2. Market Concentration in EPROMs, 1980–89   257

5-3. Shares of U.S. and Japanese EPROM Manufacturers in the
      Japanese Market, 1988–92   265

5-4. Estimated Welfare Effects of Removing Japanese Border
      Controls on DRAMs, 1987–88   274

5-5. Foreign Share of the Japanese Semiconductor Market Due to
      Changes in Foreign Penetration and Composition, Selected
      Periods, 1986–94   286

5-A1. Distribution of DRAM Contracts to Start (Lead) and Length,
      1984–89   296

5-B1. Foreign Market Value (FMV) Projected Cost Estimates for
      Japanese DRAM Manufacturers, by Company, Selected
      Quarters, 1986–89   302

6-1. Regression Results from Analysis of Firm-Level Learning
      Curves   340

6-2. Regression Results from Analysis of 1M DRAM Demand   344

6-3. Assumed Values for Empirical Parameters   346

6-4. Solution of the Optimal Control Problem, Nonstrategic
      Baseline Simulation   348

6-5. Semiconductor Industry Characteristics under Alternative
Symmetric Equilibrium Conditions   350
7-1. Solution of the Optimal Control Problem, Strategic Simulation   410
7-2. Simulated Welfare Effects of Cartelization on the DRAM
Market   413

Figures

1-1. Year-to-Year Changes in DRAM Prices, 1972–89   10
1-2. U.S. Semiconductor and Electronic Component Company
R&D as Share of Sales and Japanese Company IC R&D as
Share of Semiconductor Sales, 1976–94   12
1-3. U.S. Company Semiconductor Plant and Equipment
Investment and Japanese Company IC Plant and Equipment
Investment as Share of Semiconductor Sales, 1976–94   15
1-4. World Market Shares of Merchant Semiconductor Companies,
by Region, 1970–95   22
1-5. U.S. Share of Combined U.S. and Japanese Semiconductor
Industrial Employment, Value Added, and Sales, 1968–93   23
1-6. Federal Government Share of U.S. Electronic Components
R&D Expenditures, 1972–92   29
2-1. Output of Transistors and Diodes by Japanese Manufacturers,
September 1954–March 1956   46
2-2. Volume of U.S. Transistor Production, by Type, 1955–65   50
2-3. Value of U.S. Transistor Production, by Type, 1955–65   51
2-4. Import Shares of Japanese Semiconductor Consumption,
1966–94   62
2-5. Japanese Tax Incentives and Subsidized Lending for
Electronics Development, and Investment in Electronics by
Major Companies, 1961–71   65
2-6. Japanese R&D Subsidies for Electronics Development, and
R&D in Electronics by Major Companies, 1961–71   67
2-7. Japanese Consumption of Integrated Circuits and Import
Market Share, 1968–74   78
2-8. Major Results of the Japanese VLSI Project, 1975–90   101
2-9. Expenditures on VLSI Project and Subsequent Cooperative
Research, by Category, 1976–86   103
2-10. Linkages within VLSI Industrial System   104
2-11. Public and Private Funding of VLSI R&D Association,
1976–86   112
4-1. Errors in MITI Forecasts of DRAM Production, Quarterly,
1986–88, and Semiannually, 1987–94   198
4-2. Errors in MITI Forecasts of Japanese EPROM Production,
Quarterly, 1986–88, and Semiannually, 1987–94   199
4-3. Year-to-Year Changes in Fisher Ideal Price Indexes for DRAMs
and EPROMs, 1972–89   210

4-4.   Output of 1M DRAMs by Japanese and Non-Japanese
        Manufacturers, 1989   213

5-1.   Prices of 64K DRAMs in Contract, Distributor, and Spot
        Markets, 1983–85   235

5-2.   Year-to-Year Changes in DRAM Prices and Quantities
        Produced, 1972–89   238

5-3.   Year-to-Year Changes in EPROM Prices and Quantities
        Produced, 1972–89   239

5-4.   Prices for Four Generations of DRAMs, Pi Rule, 1974–89   241

5-5.   Prices for Three Generations of DRAMs, Bi Rule, 1987–95   242

5-6.   Quarterly Dataquest Monday Contract Prices and Average
        Selling Prices for 256K DRAMs, 1987–92   245

5-7.   Quarterly Dataquest Monday Contract Prices and Average
        Selling Prices for 1M DRAMs, 1987–92   246

5-8.   Monthly Contract Prices for 256K DRAMs in the United
        States and Japan, 1985–92   247

5-9.   Monthly Contract Prices for 1M DRAMs in the United States
        and Japan, 1987–92   248

5-10.   Quarterly Japanese and World Contract Prices for 64K
        DRAMs, 1985–89   249

5-11.   Monthly Spot Prices for 256K DRAMs in the United States
        and Japan, 1985–91   250

5-12.   Monthly Spot Prices for 1M DRAMs in the United States and
        Japan, 1986–92   251

5-13.   Quarterly Fisher Ideal Price Indexes for Large-User Purchases
        of DRAMs in the United States and Japan, 1986–89   253

5-14.   Quarterly Wholesale Prices for 128K and 256K EPROMs in
        the United States and Japan, Selected Years, 1985–92   255

5-15.   Hirschman-Herfindahl Indexes and Cumulative Output for
        Five Generations of DRAMs   258

5-16.   Hirschman-Herfindahl Indexes and Cumulative Output for
        Six Generations of EPROMs   259

5-17.   Yield Plot for IBM's 64K DRAM   261

5-18.   Japanese Output of DRAMs and EPROMs, 1986–95   263

5-19.   Japanese Exports of DRAMs and EPROMs as a Share of
        Total Output, 1986–95   264

5-20.   Average Foreign Market Value (FMV) across Companies for
        256K DRAMs, 1986–91   268

5-21.   Average Foreign Market Value (FMV) across Companies for
        1M DRAMs, 1986–91   270

5-22.   Prices and Estimated Foreign Market Value (FMV) for 512K
        and 1M EPROMs, 1987–90   271

5-23.   Alternative Measures of the Foreign Share of the Japanese
        Semiconductor Market, 1986–95   282

6-1.   Relationship between Semiconductor Price and Cost,
        Assuming "Small" Output-Sensitive Variable Costs   327

6-2. Empirical Approximation of the Learning Curve in
      Semiconductor Manufacture   330
6-3. Time Profile of 1M DRAM Costs and Prices in Simulated
      Nonstrategic Equilibrium with Gamma = 1   353
6-4. Time Profile of Semiconductor Costs and Prices in Simulated
      Nonstrategic Equilibrium with Gamma = 0   354
6-5. Historical Prices for 1M DRAMs Compared with Simulated
      Nonstrategic Equilibrium Time Profiles   355
6-6. Time Profile of Semiconductor Costs and Prices in Simulation
      with Very High Initial Yields   357
7-1. "Food Chain" Concept in Semiconductor Industry   375
7-2. Theoretical Case for Domestic Production of Semiconductors   384
7-3. Mitsubishi Competitive Marginal Revenue versus Cost of
      Manufacture (COM) Reported to Commerce Department,
      1989–91   405
7-4. NEC Competitive Marginal Revenue versus Cost of
      Manufacture (COM) Reported to Commerce Department,
      1989–91   406
7-5. Historical Prices for 1M DRAMs Compared with Simulated
      Strategic Equilibrium Time Profiles   412

MISMANAGED TRADE?

CHAPTER ONE

Introduction

IN JULY 1991 the U.S. government formally concluded negotiation of a new Semiconductor Trade Arrangement (STA) with Japan, and set out a framework for trade and investment in microelectronics between the two nations for the next five years. That agreement, replacing an earlier, pathbreaking 1986 pact, continued a controversial experiment in trade policy with the potential to alter the direction of the international trading system for high-technology goods. In 1996, as renewal of the 1991 agreement again loomed large, there were few signs that a consensus had yet emerged as to the lessons of the past decade's experience with semiconductor trade agreements. The rules of the game in international trade and investment continue to avoid very real challenges posed by both underlying economic fundamentals and real-world national policies in high-technology industries. This book is about how and why the semiconductor industry turned into the front line in a battle between national trade and industrial policies, and about the profound issues raised by this conflict.

Challenges for Public Policy

The 1991 semiconductor negotiations received surprisingly little notice as they were being concluded, probably because they addressed what was thought to be a troublesome little "sectoral" dispute between U.S. and Japanese companies. The U.S. semiconductor industry's global output, after all, was then only a tenth the size of the U.S. motor vehicle market.

1

U.S. firms' sales of semiconductor chips in 1990 were $19 billion (out of a worldwide industry total of $50 billion), compared with $200 billion in American motor vehicle shipments. And aside from the semiconductor industry itself, the only segments of U.S. industry to take an active interest in semiconductor trade policy were the computer and communications equipment sectors, the chief users of the products most affected by the negotiations.

Despite the lack of notice, the discussions carried out behind closed doors in government offices near the White House were anything but a minor sideshow. As in other high-technology sectors, where global competition had mainly pitted American and Japanese companies against one another, bilateral agreements between these two nations were historically the model for expanded arrangements as other countries' market presence grows.

By 1996, stakes that were already big had grown larger. Global semiconductor sales in 1995 had soared to $150 billion, now more than 20 percent of world vehicle sales. Forecasts predicted that by 2000, global chip sales would amount to 40 percent of global motor vehicle sales.[1] Increasingly, able new competitors in East Asia—particularly Korea and Taiwan—had emerged as major new players in the global chip industry, pushing past European companies in their share of the global market in 1994. As negotiators from Washington and Tokyo convened their trade talks, corporate and political leaders in Seoul, Taipei, and the capitals of Europe made little effort to disguise their intense interest in the proceedings. Today's bilateral precedents were clearly likely to shape tomorrow's multilateral system.

Precedent-setting features of the semiconductor pacts have already been mimicked in other sectoral trade negotiations. The semiconductor agreements' market share target for foreign sales in the Japanese chip market, widely hailed as a step toward "managed trade" by both supporters and opponents, was acknowledged by American government officials to be the inspiration for a 1992 bilateral agreement setting quantitative targets for increased sales of U.S.-made auto parts in the Japanese market.[2] More recently, apparent success in achieving market

1. See Dataquest, "Dataquest Predicts Worldwide Semiconductor Market to Double by 2000," press release, San Jose, Calif., October 2, 1995; and Department of Commerce, *U.S. Global Trade Outlook, 1995* (October 1995).

2. "[President George] Bush's aides . . . are philosophically opposed to 'managed trade'—the setting of quantifiable goals and market shares for each nation's industries. But

share targets in semiconductors led the Clinton administration's trade policymakers to focus on quantitative indicators of access to Japanese markets in their approach to a bilateral economic framework.[3] Japanese officials seemed just as determined to resist any further use of such targets or indicators.

Even more important, trade frictions in semiconductors were just the beginning of what promises to be a massive tangle of trade disputes in other high-technology sectors. In computers, telecommunications, satellites, pharmaceuticals, advanced materials, and aircraft, continuing industrial skirmishes have been taking place. Although the particulars vary, the central issues are almost always the same as those at the heart of the semiconductor negotiations.

There are two important issues. How can government subsidies and support for advanced technology—for decades an accepted and economically beneficial practice in the United States, Japan, and virtually all other industrialized countries—coexist with open international economic competition in global high-technology industries? The postwar trading system aspired to an ideal of free competition among firms from all nations operating in a single, open global market, with market outcomes determined only by the efficiency and effectiveness of individual companies. How can this vision be reconciled with the realities of a world in which national governments make large investments in new technologies, which may totally alter the industrial landscape, and inevitably favor those firms with the easiest access to the innovations created with these subsidies?[4]

---

they also realize that, as recent experience shows, that is the only way to deal with Japan. . . . 'We knew we had to pattern this agreement after the semiconductor agreement,' said an official in the Commerce Department. 'But we could not appear to be agreeing to specific numbers.'

The Japanese know that, too, and, eager to make Mr. Bush's trip a success, the Government here spent the last few weeks cajoling and arm-twisting to force Japanese automakers to once again revise plans to increase their imports. Because the announcements came from the companies, not the Government, the United States was able to argue that such 'unilateral' proposals from Japanese companies did not amount to managed trade." David E. Sanger, "Bush in Japan; A Trade Mission Ends in Tension as the 'Big Eight' of Autos Meet," New York Times, January 10, 1992, p. A1.

3. Bob Davis, "Economy: Clinton Team to Suggest Import Goals for Japan as Trade Talks Approach," Wall Street Journal, May 20, 1993, pp. A2, A11.

4. The 1993 GATT Agreement skirted the margins of this issue by explicitly permitting government subsidies of up to 75 percent of basic research conducted by industry and up to 50 percent of applied R&D costs through development of the first prototype. "World

Second, how must the world trading system be adapted to deal with the presence—or absence—of accepted norms for competitive behavior among high-technology firms, and for relations with their home governments? How much cooperation or collusion among nominally competitive firms, possibly mediated or coordinated by government, can be tolerated without making the idea of a competitive marketplace meaningless? Investments in high technology, if successful, create (at least temporarily) some element of monopoly power based on a firm's proprietary control over unique technological assets. What will be the ground rules for competition in global markets that must, virtually by definition, stray far from the textbook vision of perfect competition?

How these issues are settled will shape the world trading system for decades to come. Given its central importance, one might presume that thinking through the outlines of a coherent policy would be an overriding concern of U.S. trade policymaking. But this has not always been the case. The semiconductor industry has been simmering—or perhaps stewing—on the front burner of trade policy for over three decades. Instead of writing down a recipe, however, rapidly changing shifts of American trade policymakers have simply tossed additional ingredients into the pot, often ignoring the outcomes of their predecessors' experiments. Instead of drawing operational guidance from a long-term strategy and vision, American policy has largely been devised in terms of short-run tactics to fix the irritant of the moment on an industry-by-industry, case-by-case basis.

This study analyzes how thirty years of fierce international economic competition, and high-stakes political intervention, came to lock the American and Japanese chip industries onto their current course as reluctant pioneers of a new high-technology world order. For it is no accident that semiconductors have historically been at the cutting edge of trade friction in high-technology industries. There are three reasons why this has been so. First, semiconductors are an extreme example of a product undergoing rapid and sustained technical innovation, showcasing the distinctive features of high-technology industries that make them both economically important, and a challenge to the international trading system. Second, from the start, semiconductor companies fixed their gaze on a global market, producing a sharp and sustained international com-

---

Trade Talks: The Uruguay Round's Key Results," *Wall Street Journal*, December 15, 1993, p. A6.

petition. More rapidly than in other high-technology sectors, a broad array of companies from around the world—with various forms of support from their national governments—have entered into a mortal economic combat on this common, contested terrain. Setting the vague rules of the game for this battle inevitably involves nations, companies, and the very institutional fabric of the world trading system. Finally, governments and companies have often made decisions taking into explicit account the predicted reactions of others outside their own national, industrial, or firm boundaries: policies have been designed with *strategic* objectives in mind.

In short, semiconductors have been at the center stage of trade policy debate because they embody all the features of a high-technology industry most likely to test the limits of the existing trade regime. The rapid international diffusion of semiconductor technology created national companies from all corners of the industrialized world capable of competing in a single, global arena. Strategic concerns have linked the actions of companies and governments to one another in a way that clearly rebuts the usual presumption that the actions of individual, atomistic firms, disciplined only by the impersonal forces of a competitive market, are likely to determine industrial outcomes.

The bulk of this book analyzes how Japanese companies were able to successfully challenge American companies in semiconductors, the role that governments in the United States and Japan played in fostering their respective industries and regulating trade conflict when it periodically erupted, and the economic consequences of those government policies. The study covers a broad range of topics, and uses a variety of analytical tools. The chapters vary widely in their appeal to different sorts of readers.

Chapters 2 through 4 are historical and descriptive, covering the development of the Japanese semiconductor industry through the mid-1980s, the history of U.S.-Japan trade friction and trade policy over this period, and the negotiation and operation of the Semiconductor Trade Arrangement. In contrast with many earlier discussions of the development of the Japanese semiconductor industry, the account in this book highlights the recurring tendencies toward rivalry and competition among domestic chip producers. Japanese government policy basically focused on sheltering or assisting national firms against competitive threats from abroad. Successful cooperation among Japanese firms, and between government and industry, generally built on the common perception of a clear external threat (for example, a technological challenge, or U.S. trade actions). As the government's direct leverage over companies diminished with increasing economic liberalization in the 1970s and 1980s, the appearance of such an external threat became even more important in asserting government leadership over industry and organizing cooperative action within industry. The dilemma this has created for U.S. policy is one important subject discussed in this study. There is a recurring tension in U.S.-Japan trade policy between pursuit of a difficult, long-term struggle against anticompetitive patterns of behavior within Japanese government and industry that affect U.S. economic interests, and a quest for short-term success in narrower industrial trade objectives using policy tools that ultimately may distract from, rather than reinforce, the long-term objectives.

The next chapters are more technical and analytical. Chapter 5 provides a detailed empirical study of the economic impact of the STA. Chapter 6 develops a model of semiconductor production that incorporates key features of the industry—large, up-front R&D costs and significant learning curves in manufacturing—and through simulations, studies the relationship between price and cost over the life cycle of a memory chip under different assumptions about industrial structure. This relation between price and cost is at the heart of discussions of "dumping" of chips as an international trade problem. Chapter 7 explores the issue of strategic behavior—by governments or firms—and its implications for the national interest and policy. A concluding chapter summarizes the lessons of this study, takes a look at where U.S. policies are going, and proposes new policy initiatives in two major areas. The first is some revision of the rules of the game in international trade and investment that makes them more consistent with the special characteristics of high-technology products such as semiconductors. The second is a reformulation of our economic strategy with Japan that better supports our long-term goal of an open and competitive international marketplace, and is more deliberate in creating policies—and incentives—that reinforce movement by Japanese government and industry toward that goal.

The balance of this chapter builds some foundations. I next introduce the basics of technology and industrial structure in semiconductors. The next section presents a thumbnail sketch of the history of international competition in the industry. As a prelude to a study that is mainly about interaction with Japan, the last part of the chapter summarizes the evolution of the U.S. industry before Japan came on the scene and the role of the U.S. government in the development of semiconductor technology through the mid-1980s, when the current framework for competition was negotiated.

Technology and Industrial Structure in Semiconductors

In semiconductors, the ubiquitous silicon chips at the electronic heart of all manner of high-technology goods, an unprecedented number of the thorny economic issues that spawn trade conflicts in high-technology industries intersect. The chip industry's rate of technological progress places it at the dizzying forefront among high-tech industries, surpassing even the well-documented case of computers. (Memory chip prices have historically dropped at a rate of almost 35 percent a year, compared with a quality-adjusted price decline of roughly 20 percent in mainframe computers.)[5] With the high rate of innovation have come very short product lives, as newer generations of chips rapidly render older vintages obsolete. The relative size of investments in research and development marks the semiconductor business as one of the most technology-intensive of industries, again surpassing even computers, and is a source of significant economies of scale.

Other issues, less exclusively identified with high technology, further complicate matters. Semiconductor production is extraordinarily capital intensive, creating another potential source of economies of scale. Learning economies—the way that unit costs in some businesses fall with cumulative production experience—although well established in some other high-tech fields, like aircraft production (as well as in less technology-intensive areas, such as shipbuilding), and unimportant in others, are particularly critical to successful competition in semiconductors. Breakneck innovation, heavy investments in R&D and capital (and associated economies of scale), and learning economies make the semiconductor industry a virtual laboratory for high-tech trade friction. In a single arena, all the complications for the textbook model of perfectly competitive markets that are commonly found in one or another high-technology industry (but rarely are all simultaneously present in a single industry) have been combined.

5. See figure 1-1 and Jack E. Triplett, "Price and Technological Change in a Capital Good: A Survey of Research on Computers," in Dale W. Jorgenson and Ralph Landau, eds., *Technology and Capital Formation* (MIT Press, 1989), pp. 127–214.

Modern semiconductor electronics was originally developed during and after World War II. The first such solid-state components were diodes and rectifiers, used where previously certain types of vacuum tubes had been employed. In 1947 the transistor was invented at the Bell Telephone Laboratories, and proved to be the seed of a technological revolution. The transistor, with properties of both amplifier and switch, made it possible to replace hot, bulky, power-hungry, and fragile vacuum tubes with fast, cool, and small semiconductor components. From the start it was realized that these solid-state devices could revolutionize the just-emerging technology of the computer. The U.S. military invested heavily in both technologies, and the crucial link between computers and semiconductors continues to be the reason governments pay so much attention to the semiconductor industry.

Transistors and diodes are *discrete* devices, however, with just one circuit element per semiconductor device. The microelectronics revolution was to begin in earnest in 1959, when the *integrated circuit* (IC)—a semiconductor component with many devices, entire electronic circuits, constructed on a single silicon chip—was invented.

The primary force behind rapid technological improvement in semiconductors has been continuous advance in semiconductor manufacturing processes. Improvements in fabrication technology have steadily reduced the size of electronic circuit elements, making it possible to pack ever greater numbers of miniaturized circuit elements on a single chip, and stimulated the development of novel types of physical structures implementing standard electronic functions.

Memory chips are the largest single segment of the semiconductor market, estimated as 31 percent of global sales in 1995.[6] The dominant product (75 percent of memory sales in 1995) was the dynamic random access memory (DRAM), which by itself accounted for 23 percent of world semiconductor sales in 1995. The DRAM was to play a leading role in the unfolding U.S.-Japanese rivalry from the late 1970s to the present.

DRAMs have in many respects been the bellwether product of the semiconductor industry, since they have historically been the first product manufactured with each new vintage of processing technology.[7] By nature

6. Based on estimates found in Dataquest, "Worldwide Semiconductor Market Grew 40 percent in 1995," press release, San Jose, Calif., January 8, 1996; and Dataquest, "Worldwide Memory Market Is on Pace to Double by 1999; DRAM Key Driver," press release, November 6, 1995.

7. Although this is somewhat less true today. Increasingly, performance in micropro-

they are also well suited as a measure of advance in semiconductor manufacturing technology. The abstract logical design of a DRAM is simplicity itself, the endless repetition of a simple cell, and easily mastered; the details of its implementation and manufacture are everything. The principal and overwhelmingly important characteristic of a DRAM from the point of view of its consumers is its bit capacity, the amount of information it can hold. The impact of technical improvement is typically measured in cost per bit.

The first widely used commercial DRAM was the 1K memory (1 kilobit, or K, means 1,024 bits of information) introduced in 1970 by American semiconductor companies. New-generation DRAM chips (each with four times the capacity of the previous generation) have been introduced approximately every three years since the mid-1970s.

Innovation: The "Pi" Rule

Despite the more sophisticated processing required for higher density ICs, technological progress in semiconductor manufacturing is widely believed to have historically held the fabrication cost per area of silicon processed approximately constant. (Or equivalently, because the size of a mature chip product has been about constant, the manufacturing cost per chip as mass production has peaked has held roughly constant for successive generations of memory chips.) As a consequence, the cost per electronic function has fallen by three-quarters with the arrival of each new generation of DRAM with its fourfold greater capacity, implying a compound annual rate of decline of 37 percent in the cost of a bit of storage. This stylized, long-run relationship has even acquired a colorful label—the pi rule—since every generation of DRAM, it was claimed, approached a price of approximately pi (the mathematical constant 3.14159) dollars as production peaked.[8]

This, it turns out, was approximately true. Figure 1-1 shows two measures of quality-adjusted cost per DRAM and confirms that this extraor-

cessors has been limited by clock speed and the distances between transistors on a chip. High-performance computer processor chips are now often crammed as full of circuit elements as the densest memory chips, and the manufacturing technology used by the makers of the most advanced microprocessors is as demanding as that used in leading-edge memories.

8. See M. P. Lepselter and S. M. Sze, "DRAM Pricing Trends—The π Rule," *IEEE Circuits and Devices Magazine* (January 1985), p. 53.

Figure 1-1. *Year-to-Year Changes in DRAM Prices, 1972–89*

Rate of change (percent)

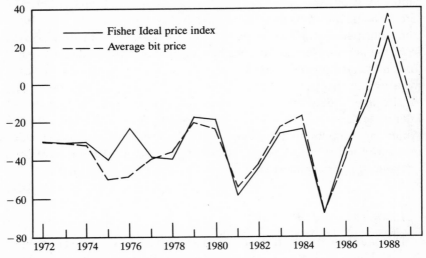

Source: Author's calculations. See chapter 5.

dinary decline in price was more than a fanciful phrase.[9] One DRAM price index in common use within the industry, the aggregate price per bit of memory sold, fell at a compound annual rate of 31 percent from 1977 to 1988. Over the entire 1974–89 period, the compound rate of quarterly decline in average cost per bit was 10.6 percent, equivalent to an annual rate of 36.1 percent. Another, technically superior index of DRAM price is a so-called Fisher Ideal price index. It more accurately measures the effects of price declines and shows a drop closer to 34 percent a year from 1977 to 1988.[10] A trend line fit to an annual Fisher Ideal price index, whose changes are depicted in figure 1-1, over the 1971–89 period, shows a long-term decline of about 42 percent a year.

9. Consulting firm Dataquest's quarterly estimates of average selling prices and quantities sold for all types of DRAMs were used to construct these measures.

10. Average price per bit is calculated by dividing aggregate sales revenues by total bits sold for all generations of chip. Even if prices remained absolutely constant for each type of chip from one period to the next, the average price per bit would change as a consequence of shifts in demand from one type of chip to another. In essence, price per bit is a weighted average of chip prices, where the weights vary across any two periods being compared.

The Fisher Ideal index, in contrast, uses identical weights in the two periods compared, thus avoiding confusion of the effects of price changes with shifts in usage from one generation to another. See chapter 5 for further details.

Other types of memory chips show only slightly less rapid declines in price. One study found a Fisher Ideal price index for all memory chips declining at an annual rate of almost 33 percent from 1977 to 1988.[11]

Interestingly, the long-term decline in DRAM prices is even more rapid if the period of extraordinary rise of chip prices after 1986, when the STA was signed, is excluded. Figure 1-1 shows how unusual the annual rates of change in DRAM prices after 1986 were: for the first time in the history of the industry, a substantial price increase was registered in 1988, and declines far smaller than historical norms were recorded in 1987 and 1989. How the STA might have been related to this slowing of the historical rate of price decline in DRAMs is an important subject explored in detail in this study.

A corollary to the rapid improvement in manufacturing technology that made these dramatic price declines possible has been an equally torrid pace for the introduction of new products. Even in the late 1950s, for example, more than half of all new types of transistors introduced were obsolete within two years; roughly the same life span applied to chips used in computer systems in the mid-1970s.[12] At IBM, memory products historically had about a one-year lifetime.[13] Nor does innovation appear to have slowed—one analysis shows new generations of DRAMs in the late 1980s being introduced at roughly the same rapid pace observed in the 1970s.[14]

Technological Intensity

Rapid technological change, of course, is the intended consequence of investment in research and development, and relative levels of resources plowed into R&D are the usual criteria used to define a high-

11. See Ellen R. Dulberger, "Sources of Price Decline in Computer Processors: Selected Electronic Components," in Murray F. Foss, Marilyn E. Manser, and Allan H. Young, eds., *Price Measurements and Their Uses* (University of Chicago Press and National Bureau of Economic Research, 1993), table 3.6.

12. See John E. Tilton, *International Diffusion of Technology: The Case of Semiconductors* (Brookings, 1971), p. 83; and Douglas W. Webbink, *Economic Report on the Semiconductor Industry: A Survey of Structure, Conduct and Performance* (Federal Trade Commission, 1977), p. 131.

13. William E. Harding, "Semiconductor Manufacturing in IBM, 1957 to the Present: A Perspective," *IBM Journal of Research and Development*, vol. 25 (September 1981), p. 653.

14. M. Therese Flaherty and Kathryn S. H. Huang, "The Myth of the Shortening Product Life Cycle," Harvard Business School, May 1988.

Figure 1-2. *U.S. Semiconductor and Electronic Component Company R&D as Share of Sales and Japanese Company IC R&D as Share of Semiconductor Sales, 1976-94*

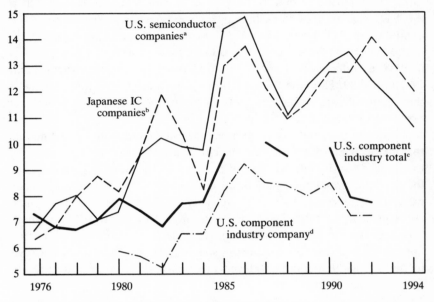

Source: MITI IC R&D as percent of semiconductor sales was estimated by taking the IC R&D-to-sales ratio from the MITI survey and adjusting to reflect aggregate value of semiconductor production versus IC production from MITI industrial production statistics. Semiconductor Industry Association, *SIA 1995 Annual Databook: Review of Global and U.S. Semiconductor Competitive Trends, 1978-1994* (San Jose, Calif.: Technicon Analytic Research, 1995), p. 40; Thomas R. Howell and others, *The Microelectronics Race: The Impact of Government Policy on International Competition* (Boulder, Colo.: Westview, 1988), pp. 218-19; National Science Foundation, *Research and Development in Industry: 1982*, NSF 84-325 (Arlington, Va., 1984), *1988*, NSF 90-319 (1990), *1991*, NSF 94-325 (1994) and unpublished MITI surveys of twelve Japanese IC companies.
    a. Members of the Semiconductor Industry Association (merchant producers only).
    b. Twelve Japanese semiconductor companies surveyed by the Ministry of International Trade and Industry (MITI).
    c. Total funds for R&D as a share of net domestic sales by U.S. electronic component companies.
    d. Company funds for R&D as a share of net domestic sales by U.S. electronic component companies.

technology industry. In the late 1980s American semiconductor companies funded R&D expenditures amounting to 11 to 13 percent of their global sales; available statistics suggest that Japanese companies invested equivalent percentages of sales in R&D over the past decade and a half. The size of the technological ante in this high-stakes global game has also clearly been rising. Figure 1-2 shows that, for both American and Japanese companies, R&D as a portion of sales roughly doubled from 1976 to 1986. Official National Science Foundation data, which include semiconductors in the broader industry category "electronic components" (in

which semiconductors accounted for almost 50 percent of electronic component value added, and a somewhat smaller share of sales in 1989), show
total, or company-funded, domestic R&D as a portion of domestic electronic component sales rose over time, just like semiconductors. But
although the technology intensity of semiconductors and the average for
all electronic components were roughly equal in 1976, by the early 1970s
semiconductors absorbed nearly double the R&D funds per dollar of
sales averaged in all components.

Bearing in mind that "electronic components" includes considerably
more than semiconductors, we can compare these figures with those for
other high-technology industries on a global basis (since high-technology
industries typically sell a substantial amount of output in foreign markets,
it is essential to include research performed and sales booked overseas
by American-based companies in presenting an accurate comparison).[15]
Table 1-1, based on data reported by U.S. multinational companies for
1989, shows some useful comparisons of electronic components with
other high-technology businesses.

One common definition of technological intensity is an R&D-to-sales
ratio exceeding the average for all manufacturing industry. By this definition electronic components rank near the top in a list of technology-
intensive industries, trailing only drugs and computers. Because semiconductors are considerably more research-intensive than electronic components as a whole, it seems likely that if more disaggregated data on
R&D share were available, semiconductors would be at the very top of
this list. Electronic components—and presumably semiconductors—rank
considerably lower when R&D employment as a share of global employment is the criterion, reflecting the fact that semiconductor manufacturing
is relatively labor intensive. Unlike other high-technology businesses that
produce *systems*—such as computers—U.S. chip companies have not for
the most part found it profitable to separate design and manufacture into
separate activities in their quest for competitive advantage.[16]

15. The National Science Foundation's data on industrial R&D spending show domestic
R&D as a share of domestic sales, which is a misleading basis for cross-industry comparisons when substantial research, or significant sales, take place within overseas affiliates and
the share of overseas affiliates varies substantially from industry to industry.

16. Although there are important examples of "fabless" chip companies, which do no
manufacturing and instead have their designs produced in outside fabrication facilities under
contract, these firms account for a relatively small share of industry output. The most
prominent examples are microprocessor and computer peripheral systems design specialists
such as Cyrix, Chips and Technologies, Weitek, and Cirrus Logic, and might arguably be

Table 1-1. *Share of Worldwide R&D Expenditures and Employment for U.S.-Based Companies and Their Majority-Owned Foreign Affiliates, by Parent's Industry, 1989*

Percent

Industry	R&D as proportion of sales to nonaffiliates	R&D employment as share of total employment
All industries	2.4	3.0
Petroleum	0.5	n.a.
Manufacturing	4.1	4.6
Chemicals and allied products	4.4	6.2
Industrial chemicals	3.6	6.2
Drugs	8.5	8.5
Other chemical products	2.9	5.6
Machinery, except electrical	6.3	6.9
Computers and office equipment	9.6	10.3
Electric and electronic equipment	6.1	5.3
Audio, video, and communications	7.7	5.9
Electronic components	7.8	5.8
Transportation equipment	6.2	6.3
Motor vehicles and equipment	n.a.	2.1
Other	n.a.	11.6
Other manufacturing	2.4	3.0
Instruments and related products	7.8	7.1
Wholesale trade	0.3	1.5
Services	0.8	0.7
Business services	1.9	1.2
Computer and data processing services	5.0	8.2

Source: Author's calculations based on Bureau of Economic Analysis, *U.S. Direct Investment Abroad, 1989 Benchmark Survey, Final Results* (Department of Commerce, 1992), tables 2K1, 2P1, 2Q1, 2Q4, 2R1, 3E8, 3F23a, 3G10, 3G11, 3I5. Data are classified by industry of parent and reflect operations by all parents of nonbank foreign affiliates and by majority-owned foreign affiliates only of these same parents.

n.a. Not available.

Scale Economies

The research-intensive nature of chip production has had significant consequences for the semiconductor industry. Perhaps the most noticeable has been important economies of scale in the production of chips. To a first approximation, whether one produces 10,000 chips or 10 million, the costs of developing the necessary technologies are roughly con-

---

designated as specialized computer peripheral producers (where the entire system just happens to be incorporated into a single chip or set of chips), rather than producers of standard semiconductor components. The blurring of lines between systems producers and semiconductor component manufacturers is an important topic addressed later.

Figure 1-3. *U.S. Company Semiconductor Plant and Equipment Investment and Japanese Company IC Plant and Equipment Investment as Share of Semiconductor Sales, 1976–94*[a]

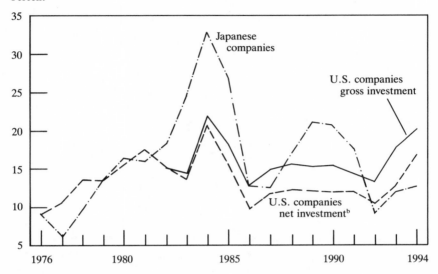

Percent

Source: See figure 1-2.
a. Japanese IC investment-to-semiconductor sales share estimated using MITI production data as in figure 1-2.
b. Gross investment less depreciation and retirements.

stant. Thus the larger the sales base, the lower the average unit cost of production.

These economies of scale in the use of the outcomes of R&D have created a constant pressure for manufacturers to seek out the widest possible market for their products to maximize the return on their relatively fixed R&D investments. As a consequence, the industry has been unrelentingly international in focus right from the start. Major producers quickly turned to foreign markets in their quest for a larger sales base. U.S. companies surged into Europe in the late 1960s and, after overcoming numerous obstacles, into Japan in the 1970s and 1980s.

Another important source of scale economies in semiconductor production has been the capital-intensive nature of the chip manufacturing process. Both American and Japanese companies have typically plowed from 10 to 20 percent of their sales into new plant and equipment investment, year after year. Figure 1-3 shows that in the early 1980s, one of the

Table 1-2. *Capital Costs of a Typical Semiconductor Wafer Fabrication Facility, Mid-1970s–Early 1990s*

	Mid-1970s	Mid-1980s	Early 1990s
Total cost of wafer fabrication facility (millions of dollars)[a]	30	100	300
Depreciation share of total wafer fabrication cost (percent)	15	49	61

Source: Integrated Circuit Engineering, cited in Reese M. Reynolds and Don R. Strom, "CEM: Process Latitude In a Bottle," *Semiconductor International* (October 1989), p. 123.
a. Assumes 4,000 wafer starts per week.

keys to the Japanese industry's rapid penetration of global markets was a share of sales invested in new plant that was substantially higher than that of U.S. firms.

Although investment as a share of company sales has been relatively stable for both American and Japanese semiconductor companies in recent years, the cost of a single full-scale manufacturing facility has been growing rapidly. Packing the maximum amount of circuitry onto a state-of-the-art chip requires increasingly expensive manufacturing equipment and facilities. The capital costs of a fabrication line for leading-edge chips rose from about 15 percent of the total fabrication cost in the mid-1970s, to about half of cost by the mid-1980s (table 1-2), and was projected to pass 60 percent of total cost in the early 1990s.[17] Today, the costs of building a leading-edge mass production facility for chips exceeds $1 billion.

Much of this equipment was highly specialized, with little or no scrap value outside of the semiconductor business; and because of the rapid pace of technological change, it had a short economic life. (Note the increasing share of investment absorbed by retirements and depreciation in table 1-2.) Investments in semiconductor manufacturing facilities, therefore, were often difficult to liquidate for more than a fraction of their acquisition cost. These investments thus took on the character of a sunk cost. The increasing magnitude of such sunk costs has made entry and exit from the industry more expensive and difficult. Moreover, since

17. See Reese M. Reynolds and Don R. Strom, "CEM: Process Latitude in a Bottle," *Semiconductor International* (October 1989), pp. 122–29. These figures refer to a state-of-the-art facility for manufacturing a high-volume, mass-produced product. Facilities investments relative to other costs would be substantially lower for a smaller facility used in producing smaller volumes of more specialized products. Unit manufacturing cost would also be a smaller share of product price for such more specialized, noncommodity products.

it typically takes a year or more to get a new plant up and running, a notoriously cyclical market adds another element of risk to such lengthy investment projects.[18]

Learning Economies

Given that processing costs are the major component of manufacturing cost, improving yields—the number of good chips that can be extracted from some area of processed silicon wafers—is the key to being profitable, especially in commoditylike semiconductors such as DRAMs, where proprietary design details count for little. Improved yields come from better control over the manufacturing process, which in turn comes with experience and learning by doing, as well as reductions in feature size, which permit more parts to be crammed onto a given surface area. Learning economies—decreases in unit cost that come from experience—are regarded as critical: typically, industry sources have reported that every doubling in cumulative output brings with it a 28 percent decline in unit costs.[19]

By definition all high-tech industries share the characteristic of research intensity with semiconductors. Available data, however, portray semiconductors as an extreme case in terms of both the importance of investments in R&D and the pace and speed of innovation produced by that investment in technology. Only a subset of high-tech industries displays the high capital intensity seen in semiconductors, and still another, perhaps smaller subset, exhibits learning economies as important as those observed in chip making. In short, the chip industry wraps up in a single package virtually every industrial attribute likely to spawn challenges to rules for international trade and investment based on textbook premises about competitive markets. The likelihood that the potential for conflict would deliver actual trouble was increased by a history of rapid diffusion of semiconductor technology across national boundaries, creating an intense international rivalry at a very early stage.

18. The world record for bringing a new chip plant on line appears to be held by NMB Semiconductor, a defunct Japanese company, which claimed that it took only nine months to go from initial groundbreaking on a new fabrication facility to initial production of 256K DRAMs in 1985. See Larry Waller, "DRAM Users and Makers: Shotgun Marriages Kick In," *Electronics* (November 1988), pp. 29–30.

19. Robert N. Noyce, "Microelectronics," *Scientific American*, September 1977, p. 67.

International Competition in Semiconductors

Solid-state electronics was largely developed in the United States, but overseas competitors soon came on the scene. Spurred by the generous terms on which American firms—particularly AT&T's Bell Telephone Laboratories and RCA—made the technology available to others, European and Japanese firms had achieved a significant presence by the late 1950s. By the 1960s Japanese solid-state radios and televisions were displacing a significant share of the American products sold in the U.S. market, and by the 1970s, Japanese IC-making skills honed on consumer electronics were being directed toward penetration of industrial electronic markets dominated by U.S. companies. The rapidity with which an enormous American technological advantage was whittled down to approximate parity was striking—and undoubtedly was an important reason semiconductors came to prominence as the prime battlefield for high-technology trade frictions.

The Global Spread of Semiconductor Manufacturing

Table 1-3 illustrates how, despite the international dispersion of technological competence, R&D in the U.S. semiconductor industry, as in other high-technology industries, has for the most part stuck close to home. (Published statistics suggest that semiconductor producers account for the vast bulk of sales by U.S. multinational electronic component firms, and that measures of the latter therefore are a good proxy for activities of the former.)[20] Nine percent of R&D funding by U.S. multi-

20. Bureau of Economic Analysis, *U.S. Direct Investment Abroad: 1989 Benchmark Survey, Final Results* (U.S. Department of Commerce, October 1992). Observe first that firms primarily producing semiconductors almost certainly account for the overwhelming share of output of *multinational* electronic component producers. Output of U.S. electronic component *factories* was about $60 billion in 1989; domestic and export sales by R&D-performing U.S. electronic component *companies* were about $53 billion. U.S. semiconductor sales accounted for 43 percent of electronic component output; sales of multinational electronic component home offices amounted to 42 percent of sales of R&D-performing electronic component companies. Since virtually all significant semiconductor producers operate foreign affiliates, one may infer that statistics on the populations of U.S. multinational electronic component firms and statistics on U.S. semiconductor companies are largely describing the same group.

Since parent-firm sales by U.S. electronic component makers operating as multinationals (in the Commerce Department benchmark survey) amounted to well under half of domestic sales by all R&D-performing U.S. electronic component makers (in the NSF

Table 1-3. *Domestic R&D Expenditures and Employment by U.S.-Based Companies and Their Majority-Owned Foreign Affiliates as Shares of Worldwide Totals, by Parent's Industry, 1989*

Industry	U.S. R&D expenditure as share of worldwide R&D expenditure	U.S. R&D employment as share of worldwide R&D employment
All industries	91.2	87.1
Petroleum	90.5	n.a.
Manufacturing	91.2	86.7
Chemicals and allied products	86.7	81.6
Industrial chemicals	89.7	83.7
Drugs	85.4	81.0
Other chemical products	89.4	83.3
Machinery, except electrical	86.8	84.2
Computers and office equipment	86.6	84.3
Electric and electronic equipment	95.8	92.2
Audio, video, and communications	97.8	96.1
Electronic components	90.9	88.2
Transportation equipment	94.1	88.2
Motors vehicles and equipment	n.a.	52.7
Other	n.a.	96.5
Other manufacturing	91.6	89.2
Instruments and related products	94.1	91.1
Wholesale trade	94.2	88.8
Services	86.8	88.6
Business services	93.7	95.0
Computer and data processing services	94.7	95.7

Source: See table 1-1.
n.a. Not available.

national companies and their foreign affiliates classified as primarily producing electronic components, and 12 percent of R&D employment, was funneled into facilities located outside the United States in 1989. This is similar to the pattern observed in other high-technology industries and contrasts markedly with the distribution of sales. In 1989 majority-owned foreign affiliates of U.S. electronic component companies accounted for 32 percent of worldwide sales to unaffiliated customers.

---

survey) in 1989, there were clearly a large number of American electronic component companies selling almost exclusively to the domestic market, including the large U.S. defense market. About half of electronic component output is active (vacuum tubes and semiconductors) and passive (resistors and capacitors) components. The balance is coils and transformers, connectors, printed circuit boards, miscellaneous parts, modules, and circuit assemblies.

Table 1-4. *World Market Shares of Merchant and Captive Semiconductor Manufacturers, by Region Where Headquartered, Selected Years, 1981–93*

Percent

Manufacturers	1981	1987	1990	1993[a]
North America				
Merchant	46.9	33.1	30.3	32.4
Captive	17.0	12.1	9.4	7.0
IBM	9.4	7.4	6.5	4.5
Europe	7.4	9.6	10.9	8.7
Japan	26.6	42.1	44.2	42.6
Rest of world	2.1	3.1	5.1	9.4
Memorandum:				
IBM share of North American				
captive production	55.2	60.9	68.9	65.5
IBM purchase/world market	n.a.	1.5	n.a.	0.9
IBM share of world production	9.4	7.4	6.5	4.5

Source: Integrated Circuit Engineering, *Status 1982, Mid-Term 1988, Status 1991, Status 1994: A Report on the Integrated Circuit Industry* (Scottsdale, Ariz.), pp. 1–10.
n.a. Not available.
a. Estimated.

Local production of components is widespread in the U.S. electronics industry for both political and economic reasons. Historically, for example, IBM had a well-known policy of sourcing most of the components used in its European and Japanese computer production from facilities in those regions.

However, lines around the "U.S. semiconductor industry" must be drawn carefully. Semiconductors are produced not only by firms labeled as semiconductor companies because that is their primary activity, but also by firms primarily engaged in producing other products, particularly electronic systems such as computers, using semiconductors as inputs. Indeed, some important producers of semiconductors sell virtually none of their output on the open market and instead consume it all; these are called captive producers. Producers who sell a significant amount of output externally are referred to as merchant producers.

All the major captive producers are located in North America. Indeed, one firm—IBM—historically accounted for about two-thirds of world-wide captive production. Table 1-4 shows one set of estimates of IBM's output relative to worldwide merchant chip production for selected years in the 1980s and 1990s.[21] In 1991 an estimated captive IC production of $4 billion ranked it as the largest American IC manufacturer and the third largest worldwide after Japan's Nippon Electric Corporation (NEC) and Toshiba.[22] In that same year IBM was estimated to account for about 6 percent of world semiconductor production, and purchased another 1 percent of world semiconductor output from outside suppliers. Within two years, however, IBM had slipped to third largest U.S. IC producer and sixth largest in the world.[23] Because most U.S. semiconductor industry data generally excluded IBM's production (in this book, I will note specifically when IBM's captive production is included), much of the empirical data fueling policy debates omitted this major American player. In fact, as is apparent in table 1-4, the historical trend toward reduced U.S. market share would have been even more pronounced if IBM's operations had been included.

The spread of competition for U.S. chip makers, in their global marketplace, is typically illustrated by figure 1-4, which shows a precipitous decline in the share of U.S.-based chip companies (excluding IBM) in world semiconductor sales, or some other measure of output, in the 1980s. Since the U.S. decline is largely the mirror image of a Japanese ascent, it is useful to focus on even more detailed comparisons of the U.S. and Japanese industries. Figure 1-5 shows that American companies accounted for about 40 percent of global sales by U.S.- and Japan-based firms in 1989, compared with almost 70 percent in 1970 when the consulting firm Dataquest first began producing these estimates. Since 1990 the U.S industry has regained global market share for the first time in fifteen years and is now at rough parity with its Japanese rivals in sales.

The same loaf may be sliced a bit differently if one compares activity by American and Japanese factories, that is, discriminating according to

21. The decline in the share of American captive production in world output during the 1980s reflects several factors: systems companies shutting down high-cost internal chip fabrication lines, rapid growth in merchant sales by companies such as Intel, and shifts in strategy leading captive producers to sell their chips on the open market as well. Until the late 1980s, when it began to sell its semiconductor products in significant volume, AT&T was the second largest captive producer in the United States; today it is classified as a merchant producer. Since 1991 IBM has aggressively begun to seek sales of its semiconductor output on the open market. See IBM, *Annual Report, 1991* (Armonk, N.Y., 1991).

22. See Integrated Circuit Engineering, *Status 1992: A Report on the Integrated Circuit Industry* (Scottsdale, Ariz., 1992), pp. 2-14 to 2-17, 2-32, 3-6.

23. Integrated Circuit Engineering, *Status 1994: A Report on the Integrated Circuit Industry* (Scottsdale, Ariz., 1994), pp. 1-14, 3-6.

Figure 1-4. *World Market Shares of Merchant Semiconductor Companies, by Region, 1970–95*

Percent

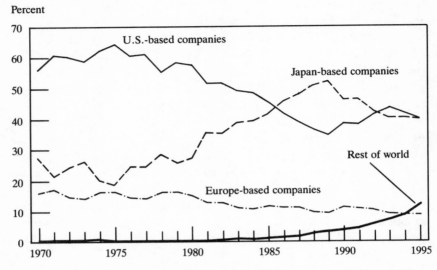

Sources: Dataquest unpublished historical estimates; Industry and Trade Strategies, *The U.S. Electronics Industry Complex*, report to the U.S. Congress, Office of Technology Analysis (Berkeley, Calif., October 1988), table 2; Dataquest, "Worldwide Semiconductor Market Grew 40 Percent in 1995," press release (January 1996); and "Worldwide Semiconductor Market Grew by 28 percent in 1994," press release (January 1995).

regions where the output is produced rather than the corporate nationality of the producer.[24] U.S. census data on factory shipments include Japanese firms' production in the United States, which, although still modest, increased significantly in the late 1980s.

More interestingly, the data show that the share of semiconductor value added within U.S. factories, despite a decrease through 1989, still greatly exceeded the share of U.S. factory sales. Since 1989 the U.S. share of value added has risen sharply and now exceeds that of Japanese factories, but U.S. factory shipments remained stagnant at under 40 percent of the binational total. Materials consumption in the United States, which probably more or less tracks wafer fabrication activity, declined even more steeply than the U.S. share of shipments. What these figures seem to show is that Japanese output is mainly concentrated in

24. I use official U.S. and Japanese manufacturing census data to construct these series. Note that the U.S. concept of value added corresponds to the Japanese concept of gross value added. (Japan also reports "plain" value added net of depreciation.)

Figure 1-5. *U.S. Share of Combined U.S. and Japanese Semiconductor Industrial Employment, Value Added, and Sales, 1968–93*

Percent

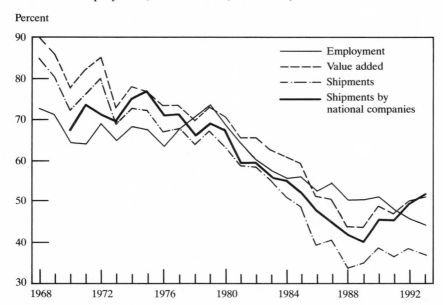

Sources: See figure 1-4; MITI, *Census of Manufacturers*, various years; and Bureau of the Census, *Annual Survey of Manufactures, 1993*.

manufacturing-intensive "commodity" products with a relatively small differential between price and materials costs, whereas U.S. output has increasingly emphasized relatively design-intensive products, where the spread between price and materials cost is much larger. In some cases, so-called fabless U.S. semiconductor companies entered the market over this period, shipping products designed internally but wholly manufactured outside, under contract, often in Japanese fabrication facilities. Thus although the data show a sharp decline in U.S. semiconductor *manufacturing* over the course of the 1980s, the U.S. semiconductor industry, particularly in the area of *design*, remained very much alive, even during the darkest days of the late 1980s.

American and Japanese firms today dominate global competition in semiconductors, with other Asian companies (mainly from Korea, but also in Taiwan) rapidly becoming significant players. For that reason, this book is mainly concerned with U.S.-Japanese competition in the industry. But it is worth considering for a moment the experience of the European companies, which were technically ahead of Japanese firms in the 1950s, and at rough parity in the 1960s, but which today trail far behind their Japanese competitors.[25]

The European Experience

The development of the European industry illustrates well how the organization and structure of the semiconductor industries of Japan and Europe have evolved quite differently from the way they did in the United States. In the United States, established electronic component producers—for the most part producers of vacuum tubes affiliated with electrical equipment manufacturers—became victims of technological change and fell by the wayside as entrepreneurial start-up companies were formed to push the development of new products.[26] As a consequence, by the mid-1960s, the American semiconductor industry was dominated by merchant producers, young companies that had specialized in the production of chips, then sold these chips at arm's length to an entirely different set of firms, the electronic equipment producers.

The development of this distinctive semiconductor industry structure in the United States was linked to a number of factors. On the demand side, much was owed, first, to the willingness of the military, the largest consumer of leading-edge components, to buy very expensive products from brand-new firms who offered the ultimate in performance in lieu of an established track record, and second, to the rise of a highly competitive commercial computer industry, which was also willing to buy the most advanced component technology from whomever offered it for sale.[27] Other factors at work included the high degree of labor mobility within American industry, which made it easy for engineers to leave established firms and start new ones if an existing company was slow to commercialize new developments; the ready availability of venture capi-

25. See, for example, Tilton, *International Diffusion of Technology*, p. 27, who shows Japanese firms trailing slightly behind most European firms in imitating major semiconductor innovations of the 1950s, but ahead of the Europeans (although still behind the United States) in the 1960s.

26. The standard source for this history is Tilton, *International Diffusion of Technology*.

27. See Kenneth Flamm, *Targeting the Computer: Government Support and International Competition* (Brookings, 1987), chap. 4, and Flamm, *Creating the Computer: Government, Industry, and High Technology* (Brookings, 1988), pp. 13–19. In the 1950s and early 1960s, military purchases also accounted for the bulk of sales of the most technologically advanced computers.

tal to fund such new spin-off companies; huge federal investments in R&D in the underlying technology base from which companies drew to develop their commercial products; and a first-class educational and scientific university infrastructure (likewise built with large doses of federal support and disposed to cooperate with industry as a consequence of the conditions tied to that funding).

The development of the semiconductor industry was quite different in Europe.[28] A more traditional, even hidebound, industrial structure prevailed: there was limited employee mobility among firms; scarce venture capital to fuel start-ups; little (until much later) government disposition to plow huge amounts via public procurement or R&D subsidy into leading-edge electronics or computers; and an academic sector with few links to, or interest in, industrial matters. Established electrical equipment manufacturers were the primary force driving investment in semiconductor electronics as they sought to produce cheaper components for use in their product lines. For the most part, semiconductors were developed and produced within existing, vertically integrated electrical equipment companies. Roughly the same (if vastly more successful) pattern of internalization of semiconductor production within larger systems producers emerged in Japan and is explored in the next chapter.

Driven largely by demand for use in consumer and industrial products, European production of discrete semiconductors (transistors and diodes) grew nicely in the 1950s. However, when the development of the integrated circuit first used in significant quantities in computers, particularly military computers, touched off an explosion in commercial demand for computers, European electrical equipment producers largely missed the shift in the market. As the hot new boxes produced by the American computer industry took over the European market in the early 1960s (the European computer industry lagged well behind), local electronics manufacturers began to realize that much of the American success was based on advanced semiconductors. When European countries embarked on crash programs to develop national computer industries in the mid- to late 1960s, the programs often contained some support for ICs.

But these national development programs for both computers and semiconductors were generally failures. Why they failed is an interesting

---

28. An extensive discussion of the comparative development of the European semiconductor industry is found in Franco Malerba, *The Semiconductor Business: The Economics of Rapid Growth and Decline* (University of Wisconsin Press, 1985).

and complex question that cannot be explored here.[29] In brief, the basic European strategy from the late 1960s on was to protect national markets with high tariff walls, then select "national champion" firms who were given favored treatment within the protected national market (generally receiving both direct subsidies and preferences in government procurement).[30] The major reasons for failure in the case of semiconductors were twofold: first, their sheltering from competition in the open market often meant that the European firms felt lessened pressure to stay technologically abreast in a rapidly changing marketplace; second, misguided and failed policies in the computer sector obstructed the development of a dynamic upstream market for chips used in computers like the one that was driving the IC industry in the United States.

Indeed, attempts to protect the European semiconductor and computer industries from imports created a vicious circle of sorts. High tariffs and high costs for imported semiconductors meant higher prices—and diminished sales—for European computer systems makers in both national and global markets. Diminished computer sales meant a smaller demand for locally produced semiconductors to be used in those computer systems. A weaker national semiconductor industry meant greater political pressure for protection, and so on. This apparent contradiction between protecting a chip industry and fostering a competitive, chip-using computer industry downstream is not unique to Europe, of course; it became a major source of division within the U.S. electronics industry in the late 1980s, after the Semiconductor Trade Arrangement was signed.

Because European chip manufacturers are largely vertically integrated divisions within electronics systems companies, this inherent policy con-

29. For a detailed analysis in computers, see Flamm, *Targeting the Computer*, chap. 5, and *Creating the Computer*, chap. 5.

30. See Giovanni Dosi, *Industrial Adjustment and Policy*, vol. 2: *Technical Change and Survival: Europe's Semiconductor Industry* (Sussex, U.K.: University of Sussex, 1981), pp. 26–41; and Malerba, *Semiconductor Business*, 1985, pp. 129–31, 188–200, for concise discussions of the pattern of government funding for European semiconductor companies in the late 1960s and 1970s. Although the United States, Japan, and Canada virtually eliminated all tariffs on semiconductors in 1985 on a most-favored-nation basis, the European Community continued to maintain a steep 14 percent duty on integrated circuits. In late 1995 the European Union agreed to slash semiconductor duties beginning in 1996. Tariffs greater than 7 percent were to be cut to 7 percent and those less than 7 percent were to be abolished. See "Europe to Cut Duties on Semiconductors," *Electronic Engineering Times* (December 18, 1995), p. 20.

flict was internalized within the European electronics firms. The solution chosen was to protect the domestic chip market, but to permit free investment within Europe by foreign producers. In that way, access to semiconductor technology developed abroad could be maintained and some degree of protection granted to the domestic industry, yet enough domestic competition preserved over the long run to ensure that prices eventually approached costs. Only a mildly negative effect on the computer industry would be felt, to the extent that foreign producers of the latest, most proprietary technology might be able to discriminate on price in the European market, behind the shelter of tariff and other barriers impeding chip imports. But for mature products, with multiple sources of supply, competition within the European market would eventually drive prices down.

The strategy contrasted markedly with that adopted in Japan, which built formidable walls around its domestic chip market, blocking both trade and investment, and strictly regulated the terms under which foreign technology could be imported. Another vital difference in Japanese policy seems to have been that Japan did not focus its promotional efforts on a single national champion, but instead chose to actively foster competition at home within the ranks of its sheltered domestic producers (while simultaneously promoting cooperation in R&D). As will be seen, this policy of maintaining competition within the domestic market was successfully defended by government bureaucrats against pressure from Japanese politicians to organize a national semiconductor champion. In both Japan and Europe—and increasingly in the United States—the driving force behind national support programs in semiconductors was the view that this sector was somehow strategic.

Strategic Policy in the United States

Finally, exploration of the role of U.S. policy in global competition in semiconductors must start by explicitly recognizing the enormously important role of government in developing the technology in the United States, a role that remains significant today. Table 1-5 clearly shows the important role that federal funding continues to play in electronic component research. In 1989 almost 20 percent of U.S. R&D performed by multinational electronic component makers was funded by sources outside the company itself (the vast bulk of this funding may be safely

Table 1-5. *U.S. Government Shares of Total R&D Funding,*
by Industry, 1989

Percent

Industry	U.S. federal and other noncompany funds for R&D as share of global R&D performed[a]
All industries	27.3
Petroleum	0.9
Manufacturing	29.3
Chemicals and allied products	1.6
Industrial chemicals	3.4
Drugs	0.5
Other chemical products	0.7
Machinery, except electrical	8.4
Computers and office equipment	8.7
Electric and electronic equipment	42.7
Audio, video, and communications	56.8
Electronic components	18.6
Transportation equipment	51.1
Motor vehicles and equipment	n.a.
Other	n.a.
Other manufacturing	28.3
Instruments and related products	39.3
Wholesale trade	19.8
Services	5.0
Business services	0.7
Computer and data processing services	0.7

Source: See table 1-1.
n.a. Not available.
a. These funds are overwhelmingly federal funds.

assumed to have come from the American government).[31] Although
smaller than that for some other high-tech industries (transportation
equipment, which is dominated by aerospace R&D; communications and
"other" electronic equipment; and instrumentation) with significant defense links, this figure is vastly greater than the federal share in other
commercial high-tech businesses such as computer hardware, drugs,
chemicals, and computer services.

31. The government share is considerably smaller (about 10 percent; see figure 1-6) if
all R&D-performing electronic component companies (that is, those with no foreign affiliates as well as those with them) are included. Virtually all significant U.S. semiconductor
makers have foreign affiliates, however, and the multinational segment of the electronic
components industry provides the more relevant comparison.

Figure 1-6. *Federal Government Share of U.S. Electronic Components R&D Expenditures, 1972–92*[a]

Percent

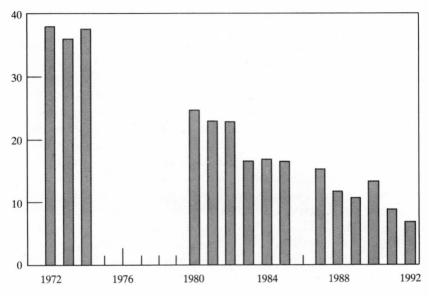

Source: Author's calculations based on National Science Foundation, *Selected Data on Research and Development in Industry: 1992* (Arlington, Va., 1994), tables SD-3, SD-4, SD-6, SD-7, SD-8, SD-9; National Science Foundation, *Research and Development in Industry: 1988* (1990), tables B-4, B-7, B-11, B-18, B-19.
a. Mid-1970s data are not available.

Furthermore, the trend has clearly been downward over time, so that semiconductors today benefit less from federal R&D funding than in the past. Figure 1-6 shows the decline of the government share in funding of the U.S. electronic component industry's domestic R&D from levels approaching 40 percent in the mid-1970s. The significance of these figures is that the semiconductor industry—in the United States, as well as overseas—has long been, and continues to be the focus of intense government involvement, with the power to shape competitive outcomes.

A brief consideration of the history of that involvement suggests that until fairly recently the motivation for U.S. government support in semiconductors was strategic only in the least economically interesting sense of the word, that is, driven primarily by national security concerns. Government's initial involvement with semiconductor technology had occurred during World War II as an effort to improve the reliability of silicon diodes used in radar. The military sponsored a huge research program investigating the fundamental properties of germanium and silicon that involved thirty to forty U.S. research labs.[32]

In the immediate postwar period the principal organization undertaking research into semiconductor electronics was the Bell Telephone Laboratories, which had resumed a prewar research effort into solid-state, semiconductor amplifiers. By the end of 1947, that effort had yielded the invention of the first crude transistor, a feat that later won Bell physicists William Shockley, Walter Brattain, and John Bardeen a Nobel Prize. Although Bell was not to receive a postwar semiconductor R&D contract from the military until after its invention of the transistor, defense users quickly grasped the transistor's potential value. It became clear that transistors, with their greater reliability, smaller size and weight, lower power consumption, and lower manufacturing cost, could potentially replace many types of vacuum tube amplifiers. Devices constructed from semiconductor materials could also perform most other electronic functions, including those of various types of diodes, resistors, and capacitors.

The U.S. Army's Signal Corps Engineering Laboratory at Fort Monmouth, New Jersey, quickly became the focal point for Defense Department efforts to apply the new technology for military ends.[33] Initial small-

32. See Ernest Braun and Stuart Macdonald, *Revolution in Miniature: The History and Impact of Semiconductor Electronics*, 2d ed. (Cambridge University Press), 1982, pp. 226–30. Some of the experiments undertaken at Purdue University as part of this effort were so similar to the later experiments undertaken at Bell Labs that resulted in the invention of the transistor that some have speculated that the transistor might have been invented years earlier if the Purdue researchers had been searching for amplification effects. See Richard R. Nelson, "The Link Between Science and Invention: the Case of the Transistor," in National Bureau of Economic Research, *The Rate and Direction of Inventive Activity: Economic and Social Factors* (Princeton University Press, 1962), pp. 549–86; and Thomas J. Misa, "Military Needs, Commercial Realities, and the Development of the Transistor, 1948–1958," in Merrit Roe Smith, ed., *Military Enterprise and Technological Change: Perspectives on the American Experience* (MIT Press, 1985), pp. 256–57.

33. After the war the Signal Corps also sponsored development of a photolithographic process for creating resistors and capacitors on ceramic substrates at the National Bureau of Standards and the Centralab division of the Globe Union Corporation for use in miniaturized proximity fuses. The basic photolithographic techniques later used in the manufacture of integrated circuits were derived from this technology. Jack S. Kilby, coinventor of the integrated circuit, had worked on this technology before moving on to Texas Instruments. See Kilby, "Invention of the Integrated Circuit," *IEEE Transactions on Electron Devices*, vol. ED-23 (July 1976), pp. 648–54.

The National Bureau of Standards and the Signal Corps also worked together on the research programs that led to the development of printed circuit boards and wave soldering techniques, which were to revolutionize electronics manufacture. As late as 1961, half of

scale Signal Corps funding of transistor research at Bell in 1950 had risen to 20 percent of total funding by 1952, and 50 percent of transistor work by 1953, and stayed at that level through 1955.[34] About 25 percent of Bell Labs' semiconductor research budget over the period 1949–58 was funded by defense contracts, and all of the early production of Western Electric, the Bell System's manufacturing affiliate, went to military shipments.[35]

After 1955 the Signal Corps began to fund semiconductor research and fundamental development at companies other than the Bell Labs (adding RCA and Pacific Semiconductor to the list in 1955), and in 1956 it doubled the size of the R&D effort to an average of $1 million a year. Even greater funding went into engineering development ("fundamental development" left off at the prototype stage, while "engineering development" created efficient and economic mass production technology). Army transistor engineering development contracts over the 1952–64 period amounted to $50 million, averaging $4 million annually.[36] The engineering development effort also jumped sharply in 1956—in that year alone, $15 million in contracts was appropriated, with funds flowing to virtually every semiconductor company in the United States.[37]

The occasion for the sharp increase in R&D funding in 1956 was a breakthrough in transistor manufacturing technology. In late 1955, influenced by an ongoing antitrust suit brought by the U.S. Justice Department, Bell Labs had permitted the transfer of its "diffusion process" technology to other companies. This technology, invented at Bell, with a

---

the value of U.S. shipments of printed circuit boards went to military users. See S. F. Danko, "Printed Circuits and Microelectronics," *Proceedings of the Institute of Radio and Electronic Engineers*, vol. 50 (May 1962), pp. 937–38, 941; and Misa, "Military Needs," pp. 263–64.

34. Misa, "Military Needs," p. 273.

35. Indeed, the first practical application of the transistor in laboratory equipment was undertaken by Bell for the navy in 1949. The first use of transistors in a computerlike digital circuit was in a gating matrix built at Bell Labs that same year as part of a "simulated warfare" computer. The first transistorized computer built in the United States and the early development of the power transistor were funded at Bell by the air force. C. A. Warren, B. McMillan, and B. D. Holbrook, "Military Systems Engineering and Research," in M. D. Fagen, ed., *A History of Engineering and Science in the Bell System: National Service in War and Peace (1925–1975)* (Murray Hill: Bell Telephone Laboratories, 1978), pp. 617–48; and W. S. Brown, B. D. Holbrook, and M. D. McIlroy, "Computer Science," in S. Millman, ed., *A History of Engineering and Science in the Bell System: Communications Sciences (1925–1980)* (Indianapolis: AT&T Bell Laboratories, 1984), pp. 351–98. See also Richard C. Levin, "The Semiconductor Industry," in Richard R. Nelson, ed., *Government and Technical Progress: A Cross-Industry Analysis* (Pergamon, 1982), p. 26.

36. Misa, "Military Needs," pp. 275–76.

37. Misa, "Military Needs," p. 282.

large share of the cost borne by the Signal Corps, made possible a great improvement in performance (in particular, the ability to deal with high-frequency signals) over the first crude point-contact transistors, and later generations of junction transistors.[38] With the rapid adoption of production technology for diffused base transistors by the U.S. industry, the Signal Corps had reason to spread its research largesse much more widely, as it pursued the development of high-performance transistors. In its quest for new high-performance products, the military was particularly inclined to take a chance on new firms: in 1959, for example, new firms (those with no background in the older vacuum tube business) accounted for 69 percent of military sales, compared with a 63 percent share of all semiconductor sales.[39]

The military influence was pervasive. In 1958 and 1959, for example, the federal government was directly funding about 25 and 23 percent, respectively, of the R&D (excluding engineering development contracts) undertaken within the U.S. semiconductor industry.[40] If university and federal laboratory work were factored in, along with engineering development funds, and indirect R&D funding embedded in contracts to procure new devices at premium prices, that figure would have been much higher: a congressional committee report estimated that the federal

38. See Misa, "Military Needs," p. 281; and Tilton, *International Diffusion of Technology*, p. 76. The early point-contact transistors were manufactured by attaching two metal wires to the surface of a piece of germanium semiconductor. These devices were very sensitive to environmental factors, had high noise levels, worked only at low frequencies, and could be used only in low-power applications. Point-contact transistors were superseded by so-called junction transistors in the early 1950s, which worked by creating junctions between regions of semiconductor materials treated with impurities. The two principal methods of creating junction transistors involved adding the impurity while growing a single silicon crystal (grown junction transistors) or melting silicon materials already treated with impurities into one another (alloy junction transistors). Although they improved the controllability of semiconductor characteristics, junction transistors still operated only at low frequencies. The diffusion process worked by exposing a semiconductor substrate to a carefully controlled atmosphere of heated, gaseous impurities, which diffused into the surface of the substrate. Very precise control of the composition and thickness of the impurities in the silicon could thus be achieved, and the operating frequencies of the new transistors so produced jumped by two orders of magnitude.

39. Tilton, *International Diffusion of Technology*, p. 91.

40. See Tilton, *International Diffusion of Technology*, pp. 93–94. Tilton stresses that formal R&D contracts were concentrated in older, established electronics firms, but it is clear that substantial government funding flowed into research at newer companies. New entrant Texas Instruments, for example, had about half of its $30 million in R&D spending for 1959—which compares with $70 million shown for the entire U.S. semiconductor industry in 1959 by Tilton's sources—contributed by the government. "Business Week Reports on: Semiconductors," *Business Week*, March 26, 1960, pp. 92–108.

government paid for 85 percent of U.S. electronics R&D in 1959.[41] Two of the largest recipients of U.S. government support in the early days of the industry—Texas Instruments and Motorola—were later to become the biggest producers in the U.S. semiconductor industry.[42]

The U.S. military played a particularly overwhelming role in driving the development and production of high-performance transistors in the late 1950s. At Western Electric, for example, all the diffusion-process transistors manufactured through 1958 (roughly 13 percent of the company's cumulative output of all transistor types) went to military applications, as did 54 percent of alloy junction, 30 percent of grown junction, and 53 percent of point-contact transistors produced through that year.[43]

In addition to R&D funding and procurement, the military provided funds to individual firms to build semiconductor capacity far in excess of what was then required. By 1953, pilot transistor production lines at Western Electric, General Electric, Raytheon, RCA, and Sylvania were funded.[44] Between 1952 and 1959, roughly $36 million was provided to individual firms to build transistor factories that operated at only a fraction of their rated capacity.[45] To use that excess capacity, firms were eager to find new markets for their products.

Foremost among the products in which the new, high-performance transistors were being applied by both military and commercial users were electronic computers. Computers were among the earliest defense systems into which transistors were inserted, and transistor developers such as Bell Labs and later Philco received military contracts to build computers from the high-performance components they had developed.[46]

41. *Coordination of Information on Current Federal Research and Development Projects in the Field of Electronics*, prepared for the Subcommittee on Reorganization and International Organizations of the Senate Committee on Government Operations, 87 Cong. 1 sess. (GPO, 1961), p. 130.

42. John G. Linvill and C. Lester Hogan, "Intellectual and Economic Fuel for the Electronics Revolution," *Science*, March 18, 1977, pp. 1107–14.

43. Misa, "Military Needs," table 3.

44. Tilton, *International Diffusion of Technology*, p. 92; and Linvill and Hogan, "Intellectual and Economic Fuel," pp. 1107–08.

45. Braun and Macdonald, *Revolution in Miniature*, p. 81, report that the industry in 1955 had the capacity to produce 15 million transistors a year, whereas actual output was 3.6 million. Of the $36 million in funding for production capacity, $11 million was provided for alloy junction transistors in 1952, another $15 million for diffused base transistors in 1957, and $10 million for integrated circuits in 1959. Linvill and Hogan, "Intellectual and Economic Fuel," pp. 1108–09.

46. On the intimate link between semiconductor and computer development see

Commercial computers soon followed military computers as the primary market for American semiconductors. Table 1-6 shows how computer manufacturers have continued to be the single largest category of commercial users of semiconductors in the United States from 1960 to the present.[47]

The U.S. military was thus a critically important early market for semiconductors, dominating sales through the 1950s and well into the 1960s before finally stabilizing around 10 percent of the overall U.S. market in the late 1970s and 1980s. Yet the next major development in the semiconductor industry, the integrated circuit, was even more dependent on military customers for its rapid growth in sales than the transistor had been.

Before the announcement of the IC in 1959, the military had funded efforts aimed at developing miniaturized electronics with functional characteristics similar to those of ICs, but the companies that developed the IC avoided military funding in order to ensure that the technologies developed remained privately owned. Nonetheless, the initial stimulus to creating the IC was the announced intention of the armed services to provide a significant market for components with the appropriate characteristics.[48]

Even more than it had in transistors, the military dominated the market for ICs. Whereas the federal government's share of the market for discrete semiconductors peaked briefly at almost 50 percent in 1960, its share of the IC market remained above half of total sales until 1967. For close to a decade, the market for ICs was dominated by military systems.[49]

The first major production application of the IC was in a general purpose, stored-program computer running the Minuteman II guided

---

Flamm, *Creating the Computer*, pp. 15–19, 92–94, 119–20, 122–23, and *Targeting the Computer*, pp. 49–51. See also Misa, "Military Needs," p. 282, note 59.

47. Note that much of the output of "communications equipment" historically went to military customers. Also, the rapid increase of semiconductor consumption within the business services sector in the 1980s probably reflected the explosive growth in demand for standardized microprocessor memory chips integrated into computer systems by value-add resellers and computer retailers and probably should be considered a subset of demand for chips used in computers.

48. See Norman J. Asher and Leland D. Strom, "The Role of the Department of Defense in the Development of Integrated Circuits," IDA paper P-1271, Arlington, Va.: Institute for Defense Analyses, pp. 1–7.

49. In 1962, for example, all ICs produced were used in defense systems. That share fell to 85 percent in 1964, 53 percent by 1966, and 37 percent by 1968. See Tilton, *International Diffusion of Technology,* table 4-8.

Table 1-6. *U.S. Industrial Consumption of Semiconductors,*
by Consuming Sector, Selected Years, 1963–87

Percent of total

Sector	1963	1967	1972	1977	1982	1987
Military ordnance	3.0	4.5	1.2	1.6	2.1	1.1
Yarn mills	0.0	0.0	0.4	0.2	0.2	0.0
Misc. plastic products	0.2	0.0	0.0	0.1	0.2	0.2
Power hand tools	0.0	0.0	0.0	0.1	0.1	0.1
Computers and office machinery	23.3	14.4	17.7	17.6	15.8	12.4
Electrical industry equipment	6.0	3.5	2.4	1.9	2.2	0.1
Household appliances	0.1	0.0	0.0	0.0	0.1	0.3
Electric lighting	2.8	0.0	0.2	2.4	0.5	0.5
Radio, TV, audio	15.2	15.9	14.1	12.5	6.4	4.5
Communications equipment	25.7	26.9	20.0	17.5	11.3	3.4
Other electric components	7.2	17.1	7.3	6.8	3.0	5.3
Misc. electric machinery	0.3	0.2	0.9	2.1	2.6	1.7
Motor vehicles	3.3	1.8	2.3	3.4	2.7	6.1
Aircraft	4.1	0.2	8.5	3.6	7.2	1.4
Other transport equipment	0.0	0.2	0.1	0.0	0.1	0.0
Scientific and measuring instruments	2.1	1.8	1.5	2.4	2.1	10.3
Optical and photo	0.3	0.5	4.9	8.3	13.7	8.8
Misc. manufacturing	0.0	0.4	0.5	0.2	0.3	1.0
Transportation and warehousing	0.9	0.8	0.7	0.3	0.3	0.4
Communications services	0.0	1.4	1.9	4.0	7.5	9.1
Personal services	4.3	8.0	7.5	7.4	9.1	10.1
Business services	0.0	0.9	8.0	7.6	12.8	23.1
Health, education, nonprofit	0.0	0.4	0.0	0.0	0.0	0.0
State and local enterprises	0.0	0.0	0.0	0.0	0.0	0.0
Other	1.1	1.2	0.0	0.0	0.0	0.1

Sources: Author's calculations using data from Bureau of Economic Analysis, *Input-Output Structure of the U.S. Economy: 1963*, vol. 1 (Department of Commerce, 1969) and *1967* (1974); *The Detailed Input-Output Structure of the U.S. Economy, 1972* (1979) and *1977* (1984); *The 1982 Benchmark Input-Output Accounts of the United States* (1991); and *Benchmark Input-Output Accounts of the United States, 1987* (1994).

missile.[50] Awarded in 1960, the contract for this D37 computer—the first built using ICs—alone accounted for about one-fifth of industry sales in 1965.[51] ICs were not shipped in commercial computers until 1965, four years after the first D37 rolled off the production line.

From 1959 through the late 1960s, military and space contracts powered a surge in American IC production. The guidance computer for the

50. Texas Instruments engineers put 94 percent of the computer's electronics on a set of thirteen custom-designed integrated circuits. See Richard C. Platzek and Jack S. Kilby, "Custom-Integrated Circuits—A Military Computer Application," *Proceedings of the IFIP Congress 65*, vol. 2 (Washington: Spartan Books, 1965), pp. 425–26.

51. Asher and Strom, "Role of the Department of Defense," p. 21.

Table 1-7. *Sources of R&D Funds in the U.S. Semiconductor Industry,*
Selected Periods, 1958–76

Millions of dollars unless otherwise specified

Period	Federal funding	Company funding	Total	Federal funding as percent of total
1958–69	495	569	1,064	46.5
1970–76	207	1,144	1,351	15.3
1958–76	702	1,713	2,415	29.1
1958–74	930	1,200	2,130	43.7

Sources: For 1958–69, 1970–76, and 1958–76 see Industry and Trade Administration, *A Report on the U.S. Semiconductor Industry* (Department of Commerce, 1979), p. 8. For 1958–74 see John G. Linvill and C. Lester Hogan, "Intellectual and Economic Fuel for the Electronics Revolution," *Science*, March 18, 1977, pp. 1107–08.

National Aeronautics and Space Administration's Apollo spacecraft gave a potent kick to Fairchild's IC profits, much as the D37 had helped boost Texas Instrument's rising star. Other important contracts went to Motorola, Westinghouse, Signetics, and RCA.[52] Premium prices paid on these military contracts fueled the rapid application of the technology to commercial products.

Industry participants have estimated that between 1958 and early 1970s the federal government directly or indirectly funded between 40 and 45 percent of all industrial semiconductor R&D (table 1-7). This share was to fall sharply in the 1970s as the technology matured, as technically advanced components moved into a booming commercial mainstream, and as the government's share of the overall market declined. Nonetheless, in key areas of semiconductor technology—for example, devices using exotic substrate materials, computer-aided chip design and manufacturing process modeling, rapid turnaround production technology, and innovative reduced instruction set computer (RISC) chips—military support continued to prime the pump of semiconductor electronics in the 1970s and 1980s.[53]

By the late 1980s, government funding of semiconductor-related research held at a stable level of $450 million to a little over $500 million.[54] The government share of the overall chip market, after dipping in the

52. See Asher and Strom, "Role of the Department of Defense."

53. For references to these programs, see Flamm, *Targeting the Computer*, pp. 70–72; and Flamm, *Creating the Computer*, pp. 18–19.

54. See Federal Interagency Staff Working Group, *The Semiconductor Industry* (Washington: National Science Foundation, November 1987), p. 31; and Congressional Budget Office, *The Benefits and Risks of Federal Funding for SEMATECH* (September 1987), p. 60.

Table 1-8. *Federal Contracts as Share of Total U.S. Semiconductor Shipments, Selected Years, 1963–87*

Year	Percent	Year	Percent
1963	35.5	1973	5.8
1967	27.6	1983	4.1
1971	12.7	1987	10.7

Source: Bureau of the Census, *Shipments to Federal Government Agencies, 1963*, MA-175 (Department of Commerce, 1965), *1971* (1973), *1973* (1975), *1983* (1985), *1987* (1989); 1987 is the most recent available report.

early 1980s, climbed back into the 10 percent range in the late 1980s (table 1-8). However, this number—based on direct sales of chips to government and to primary and subcontractors building government systems—significantly understates government's total (direct and indirect purchases) share of the semiconductor market. A calculation using an input-output table to estimate the semiconductor content of commercial equipment procured by the government suggests that in the mid-1980s, purchases by defense agencies alone—direct and indirect—accounted for over one-quarter of U.S. semiconductor shipments.[55]

A significant change, however, had occurred in the relationship between the military and the semiconductor industry. Whereas in the 1960s defense systems had ridden at the leading edge of semiconductor technology, by the late 1970s they trailed far behind the standard set in a booming commercial market.[56] Perceiving a threat, the military proposed a program designed to push the technological frontier in semiconductors and close the gap between military and commercial electronics. The huge and costly effort, known as VHSIC, the Very High Speed Integrated Circuit program, accomplished neither objective, but in the process it set the stage for an important shift in the articulation of objectives for government investments in semiconductor technology.

55. Computer runs from the Defense Economic Impact Modeling System (DEIMS) input-output model estimated that defense spending accounted for $274 million in direct and $3,791 million in indirect consumption of semiconductors in 1986, about 26 percent of U.S. shipments in that year. See Directorate for Information, Operations, and Reports, Department of Defense, *Projected Defense Purchases, Detail by Industry and State, Calendar Year 1986 through 1991* (1987), p. 51, and *Calendar Year 1991–1997*, p. 58. The DEIMS model shows $66 million in direct and $3,447 million in indirect sales in 1985 (about 21 percent of U.S. shipments) and $206 million and $3,906 million in 1991 (just under 20 percent of U.S. shipments).

56. See John A. Alic and others, *Beyond Spinoff: Military and Commercial Technologies in a Changing World* (Harvard Business School Press, 1992), p. 269, for a perceptive analysis of this shift.

For the first time, the potential impact of an R&D program in improving the competitiveness of the American industry, although not a primary goal of Defense Department planners, played an important role in building political support for the program. The strategic objectives of U.S. policies affecting semiconductor technology had begun to swing away from the purely military considerations of the cold war and toward economic conceptions of a strategic industry or policy. This swing was to pick up momentum rapidly throughout the 1980s, breaking through an important political and ideological barrier in 1987 with the formation of the Sematech (semiconductor manufacturing technology) consortium, a large-scale R&D effort with explicitly commercial objectives.

Conceived in a moment of crisis for the semiconductor industry, during the initial months of the Semiconductor Trade Arrangement, Sematech marked an extraordinary shift in policy for an American administration for which government intervention was ideological anathema. Although framed in terms of national security, the Sematech initiative was clearly intended primarily to improve the health of a broad base of U.S. commercial semiconductor manufacturers (many of whom also supplied the Department of Defense). Together with the market share targets accompanying the STA, Sematech constituted a truly radical change in American policy toward its high-technology industries.

What had brought all this about was the extraordinary development of the Japanese semiconductor industry in the previous decade. For the first time, an American postwar high-technology success story—the semiconductor industry—stood at the brink of decisive competitive defeat at the hands of a foreign competitor. How this came to be (for the most part, within the space of a single decade), and how policy responses articulated in terms of strategic arguments were to become a staple of American economic debate, is—it shall be argued—one of the most important episodes in the economic history of the postwar international system.

CHAPTER TWO

New Competition: The Japanese Ascent in Semiconductors

JAPANESE COMPANIES have worked for over forty years to master leading edge semiconductor technology. Through the mid-1970s these enterprises lagged several steps behind the pioneering American firms that had founded the industry. But in less than a decade, by the mid-1980s, the mantle of overall leadership in semiconductor *manufacturing* (although not in design) had clearly passed to Japanese producers, and this development was accompanied by profound shifts in the marketplace.

The pattern of Japanese investments in semiconductor technology in many respects resembled that observed in computers.[1] As they did in computers, government laboratories took the first steps in semiconductor R&D in the early 1950s (although unlike in computers it was small Japanese firms who rushed the new components to market, embedded in innovative consumer electronics products). During the 1960s high trade barriers were put in place, shielding the domestic market. As the private sector took on more of the burden of technology development, the government labs shifted toward more long-term projects. As in computers, a large amount of government aid in the mid-1970s (following the infusion

---

1. For more details see Kenneth Flamm, *Creating the Computer: Government, Industry, and High Technology* (Brookings, 1988), chap. 6; and *Targeting the Computer: Government Support and International Competition* (Brookings, 1987), chap. 5.

39

of significant resources into the computer industry) cushioned the shock
of imminent liberalization of the domestic market for Japanese producers. By the early 1980s Japanese companies had arrived at the technological frontier in key products and were making their presence felt in global
markets.

Changes in Japanese policy in semiconductors often seem to have
trailed closely behind those in the computer industry. Careful scrutiny of
the historical record suggests a compelling explanation: the computer
industry has historically been the primary focus of government policy
toward the electronics industry in Japan, and assistance to domestic semiconductor producers was primarily designed to further long-range objectives in computers.

First Steps

The earliest development of semiconductor technology in Japan took
place within two Japanese research institutes, one affiliated with Nippon
Telegraph and Telephone Corporation (NTT, the quasi-public telecommunications monopoly that had exclusive control over Japan's national
telephone network through the early 1980s), and the other with the Ministry of International Trade and Industry (MITI). The Electrical Communications Department had been spun off from the Ministry of Posts'
Electrotechnical Laboratory in 1948, attached to what was to become
NTT, and renamed the Electrical Communications Laboratory (ECL);
the Electrotechnical Laboratory (ETL) proper was reorganized within
MITI in its current form in 1952.[2] These two public research institutes
(in ECL's case, quasi-public) initially led, and continue to exert a major
role in, the development of semiconductor technology in Japan.[3] The first
transistors built in Japan were fabricated within ECL and ETL.[4] The exit
of key researchers from these laboratories and their move to the private

2. See Electrotechnical Laboratory, *Guide to ETL, 1983–1984* (Ibaraki, Japan, 1983).
3. Until 1985 NTT was a public corporation. In that year it was reorganized as a private
corporation in which the Japanese government owned one-third of the shares; the remaining
two-thirds of the company's stock was sold off to the Japanese public over a ten-year period.
4. There is some dispute over who was first. ETL conducted a point-contact transistor
experiment in 1951, but ECL apparently produced the first properly functioning device.
See Makoto Watanabe, "Electrical Communications Laboratories: Recent LSI Activities,"
Japan Telecommunications Review (January 1979), pp. 3–8; and Yasuzo Nakagawa,
Semiconductor Development in Japan (Tokyo: Diamond Publishing, 1985), pp. 22–31 (in
Japanese).

sector, where they helped organize company development programs in semiconductors, played an important role in the early development of the Japanese industry.[5]

Over the years 1949–51, both government research institutes began research into transistors, and they soon constructed working devices duplicating the Bell Telephone Laboratories' pioneering 1947 discovery. By 1953 three large Japanese electrical conglomerates—Hitachi, Toshiba, and Nippon Electric Corporation (NEC)—had begun research on transistors.[6] Ironically, these companies—the largest established players in the Japanese electrical machinery sector—did not rush to commercialize the new technology. Despite the very generous terms on which the Bell Laboratories made know-how about the new technology available to all comers, no Japanese companies were among the first licensees of the transistor.[7]

5. For example, two prominent semiconductor technologists left NTT to join Sanyo and Nippon Electric Corporation in 1957; a senior semiconductor specialist left MITI's ETL to join Sony Corporation in 1961. Nearly thirty semiconductor researchers left the NTT labs in 1962. See Nakagawa, *Semiconductor Development in Japan*, pp. 282–84.

A historically significant research program in semiconductors was also launched at Tohoku University about this time; Tohoku to this day remains a major academic resource for the Japanese semiconductor industry.

6. Nakagawa, *Semiconductor Development in Japan*, p. 286.

7. Under the auspices of a 1949 Joint Services contract to develop transistors for defense applications, a five-day symposium was convened in September 1951 by Bell Labs and Western Electric, the manufacturing arm of American Telephone and Telegraph Company (AT&T), to disseminate information to the military and their contractors, including industrial and university representatives. Representatives of twenty-two American universities and eighty-six other American companies attended. Two individuals affiliated with foreign firms, serving in their governments' technical liaison offices to U.S. government agencies, attended this meeting: one came from International Standard Electric, in England, and the other from the French subsidiary of ITT Corporation.

A second symposium, explicitly designed to teach "the art and science of transistor technology, theory, and practice as it was known" to commercial licensees was held in April 1952. Representatives of a grand total of thirty-seven firms are known to have either attended the 1952 Bell Labs technology transfer symposium or signed licensing agreements with the Bell Labs by the end of 1952. Eleven of these firms were foreign (twelve if one counts ITT as a foreign-based firm—ITT was U.S.-owned and headquartered but sold its products in foreign markets). Five were British (Automatic Telephone and Electric, British Thomson-Houston, General Electric, Pye, and English Electric), four were German (Siemens and Halske, Telefunken Gesellschaft, Felton and Guilleaume Carlfswerk, and Brush, an American affiliate of Intermetall), one was Dutch (N.V. Philips), and one Swedish (L. M. Ericsson). In addition, representatives of three European subsidiaries of ITT attended the eight-day April 1952 symposium: Laboratoire Centrale de Télécommunications (France), Standard Telephones and Cables (Britain), and Süddeutsche Apparatfabrik (Germany). See F. M. Smits, ed., *A History of Engineering and Science in the Bell System*, vol.

Instead, it was small Japanese companies, not the giants, who took the lead with the new technology. The first company to produce a working transistor was Kobe Kogyo, a small machinery producer,[8] whose technology chief had come to the United States in 1951 and been introduced to the new device during visits to RCA and the Bell Labs. A bootleg research project was started on his return. By the spring of 1952 an experimental transistor had been built and the needed technology acquired and licensed from RCA. Hitachi and Toshiba signed licensing agreements with RCA in that same year. Because of cross-licensing arrangements between Bell and RCA then in effect, a separate license from Bell was not required in order to use the basic Bell patents.[9] Indeed, use of patented transistor concepts was initially cheaper via the RCA channel: the Bell Labs required a $25,000 advance toward future royalties in exchange for patent rights, while RCA did not.[10]

RCA's licensing of its electronic componentry in 1952 was followed by the transfer of its monochrome television system to its Japanese partners in 1953.[11] RCA was willing to not only sell patent rights, but also actively transfer know-how. To service its Japanese licensees RCA set up a Tokyo engineering facility in 1954.[12] In 1960 RCA expanded its presence to include a Japanese research laboratory.[13] In 1962 RCA sold rights to its

---

6: *Electronics Technology (1925–75)* (Murray Hill: AT&T Bell Laboratories, 1985) pp. 28–29); Thomas J. Misa, "Military Needs, Commercial Realities, and the Development of the Transistor, 1948–58," in Merritt Roe Smith, ed., *Military Enterprise and Technological Change: Perspectives on the American Experience* (MIT Press, 1987), pp. 267–68. Foreign licensing and seminar participation have been established from unpublished lists contained in communications from D. S. Hochheiser, AT&T Archives, June 1990 and February 1991; also John E. Tilton, *International Diffusion of Technology: The Case of Semiconductors* (Brookings, 1971), pp. 102, 104, 106. Attendance records for the 1952 symposium may not be absolutely complete; the only surviving historical record of participation is a series of group pictures of attendees.

8. Its name had been changed from the Kawanishi Machine Manufacturing Company. See Nakagawa, *Semiconductor Development in Japan*, p. 75.

9. See Tilton, *International Diffusion of Technology*, p. 77.

10. See William F. E. Long, "Price and Nonprice Practices Under the Uncertain Conditions of Rapidly Improving Technologies—A Case Study," Ph.D. dissertation, Department of Economics, George Washington University, June 1967, pp. 24–25. However, payment of the $25,000 to Bell entitled early licensees to attend the important April 1952 symposium in which detailed transistor know-how was transferred to them. See also "Business Week Reports On: Semiconductors," *Business Week*, March 26, 1960, pp. 93–94.

11. See U.S. Tariff Commission, *Television Receivers and Certain Parts Thereof*, TC publication No. 436 (1971), p. 13.

12. See Jack Baranson, *The Japanese Challenge to U.S. Industry* (Lexington Books, 1981), p. 40.

13. See "RCA Prepares to Open Far East Research Lab," *Electronics,* July 24, 1960,

color television technology to Japanese consumer electronics makers.[14] Japanese electrical engineers later characterized access to RCA electronics technology as a "treasure mountain."[15]

The year 1952 also marked the visit to the United States by Masaru Ibuka, who with Akio Morita had in 1946 started a small Japanese electronics company, Tokyo Telecommunications Engineering, later renamed the Sony Corporation. Ibuka and Morita had been occupied with the development of electronic circuitry for the Japanese military during the war, and after the war's end their government connections were to prove useful in getting their fledgling company off the ground.[16] During his American visit, Ibuka learned of developments in transistor technology and the terms of the Bell Labs licensing offer. In 1953 Morita was sent to America to secure a license and needed technology from Bell, and he returned with a provisional signed agreement and a mountain of technical information.

In 1950, however, at a time of great scarcity of foreign exchange, MITI had been given authority to approve or reject all contracts for technology with foreign companies.[17] MITI repeatedly used this authority over the next twenty years, together with its control over approval of direct investment in Japan by foreign companies and a licensing system for imports, to improve the terms on which Japanese electronics companies could obtain access to foreign technology, and to force foreign companies to transfer or license their technology to Japanese enterprises in exchange for access to the Japanese market. In this case MITI did not approve the licensing arrangement signed by Morita until 1954; it approved similar agreements with Bell Labs for Toshiba, Hitachi, and Kobe Kogyo later

---

p. 11. The initial focus of its Japanese research effort was on basic studies of solid state phenomena.

14. See U.S. Tariff Commission, *Telephone Receivers and Certain Parts Thereof*, p. 13. Heavy and Chemical Industries News Agency, *Foreign Investments in Japan*, 2d ed. (Tokyo, 1967), chap. 5, shows virtually every major Japanese manufacturer of television receivers and parts signing licensing and technical assistance contracts with RCA in the color television area in 1962.

15. Nakagawa, *Semiconductor Development in Japan*, p. 97.

16. Ibuka's contacts were instrumental in initial sales of voltmeters and broadcast equipment to Japan Broadcasting (NHK), Japan National Railways, and other government agencies in the early years of the company. See Nick Lyons, *The Sony Vision* (Crown, 1976), p. 14.

17. See Merton J. Peck and Shuji Tamura, "Technology," in Hugh Patrick and Henry Rosovsky, eds., *Asia's New Giant: How the Japanese Economy Works* (Brookings, 1976), p. 535.

in that year.[18] MITI was later to use its control over technological tie-ups to obstruct entry into the market by other Japanese firms.[19]

Many early concerns over patent rights were to become moot in 1956. In that year AT&T signed a consent decree with the U.S. Department of Justice that required all existing AT&T, Bell Labs, and Western Electric transistor patents to be licensed free of charge to all interested domestic companies. Since Bell Labs had essentially dominated technology development up to that point, this had the effect of placing the mainstream transistor technology of the day—junction transistors—into the public domain.[20]

In other respects as well, U.S. semiconductor technologists were relatively open about permitting Japanese engineers unencumbered access to their technology in the 1950s and 1960s. One individual posted at a Japanese trading company's New York office, for example, estimates that he made arrangements for visits by approximately 3,000 Japanese electronics engineers during a three-year assignment.[21] Japanese researchers also were given free access to American semiconductor research conferences, and came in large numbers.[22] The methods some of these Japanese visitors used to gather information may have contributed to a certain degree of resentment visible in later years on the American side.[23]

18. See Nakagawa, *Semiconductor Development in Japan*, p. 285.

19. Mitsubishi, Sanyo, Fuji Electric, and Oki were reportedly hindered from entering the transistor market by MITI's controls over technology transfer. See NHK, *Electronics-Based Nation: Japan's Autobiography*, vol. 2 (Tokyo, 1991), p. 41 (in Japanese).

20. As Tilton, *International Diffusion of Technology*, pp. 76–77, points out, however, continuing Bell Labs R&D on semiconductor technology, and the resulting patents, made it imperative for semiconductor firms to work out licensing agreements covering new advances with AT&T.

21. The favorite sites for visits were the Bell Labs, RCA, Western Electric, and General Electric plants. NHK, *Electronics-Based Nation*, vol. 1.

22. A recent Japanese television documentary series featured the recollections of an American engineer present at some of these meetings. At one large International Electron Device Conference meeting in the 1960s, for example, the American researcher recalls that about 20 percent of the attendees were Japanese. Often, he notes, all of the authors on Japanese papers with multiple authors attended, apparently for the purpose of monitoring and thoroughly documenting the presentations of other attendees. See NHK, *Electronics-Based Nation*, vol. 2, pp. 347–53.

23. Some of the anecdotes publicly recalled on the Japanese documentary series would be regarded by an American audience as reflecting poorly on their narrator. For example, a top technologist at NEC described how, when he visited General Electric in 1960, GE executives refused to disclose know-how for new products under development, on the ground that GE's existing technology contract with NEC only covered items currently in production. Angered, the NEC technologist cultivated direct contacts with young GE

Concentrating on acquiring existing know-how rather than creating new technology, Japanese chip makers also faced a business environment significantly different from that of American producers. Unlike firms in the United States, where a massive military market sparked the growth of the U.S. electronics industry, Japanese companies could not look to government sales as their engine of growth. Commercial applications instead had to spur development. NTT's labs pointed the way when they publicly exhibited a prototype transistor radio and a transistorized hearing aid in 1953. Although the Regency Corporation was first to actually market a transistor radio, in the United States in 1954, Sony and Kobe Kogyo quickly followed this lead. In Japan, Kobe Kogyo was the first to develop a commercial transistor radio, which it announced in January 1954, but the reception of the crude set was poor, and the radio was not a success.[24] The transistor radio that Sony brought to market in late 1955 was of a markedly higher quality and set Sony on a successful growth path that ultimately transformed the company into today's multinational giant.

The key to Sony's success was its early mastery of the technology needed to mass-produce transistors. By 1956 Sony transistor production had topped 300,000 units, and a year later it had more than doubled, to 800,000 units. Sony's closest competitor was Toshiba, which began mass production of transistors in 1956. Kobe Kogyo did not begin mass production until 1957, and Hitachi and NEC did not build large-scale transistor plants until 1958. When the large companies did enter the game, though, they came in force, and their presence was quickly established. An avalanche of transistors roared down (figure 2-1). In 1958 Toshiba completed a single plant that was to produce 1.2 million transistors per month.

By 1959 Toshiba was the largest Japanese semiconductor producer (with 26 percent of the market), followed by Matsushita (16 percent), Hitachi (15 percent), NEC (15 percent), Sony (11 percent), and Kobe

---

engineers involved in the project and so obtained the needed know-how. In another instance, a former Toshiba researcher described how, forbidden to take notes while visiting a New York office, he would feign a need to go to the bathroom, where he would write important points down. He did this so often that those with him thought he had a stomach ailment. See NHK, *Electronics-Based Nation,* vol. 1, pp. 264–76, which contains material from this series of NHK documentaries.

24. One of Kobe Kogyo's engineers, Leo Esaki, moved from Kobe Kogyo to Sony in 1955, where he was to discover the tunnel diode, which earned him a Nobel Prize in 1973 and became Japan's first significant technical contribution to semiconductor technology.

Figure 2-1. *Output of Transistors and Diodes by Japanese Manufacturers, September 1954–March 1956*[a]

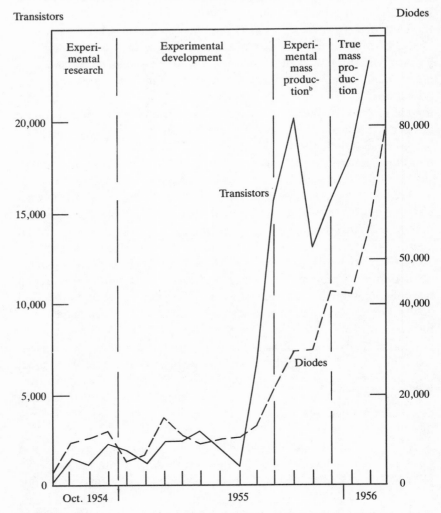

Source: *Twenty-Year History of the Electronics Industry* (Tokyo: Dempa Publications, 1968), p. 110.
a. Manufacturers are Toshiba, Sony, Kobe Kogyo, Hitachi, and NEC. Mitsubishi, Fuji Tsushin, and Matsushita were engaged in experimental research only.
b. Begins with sale of Sony transistor radios.

Table 2-1. *Transistor Output of U.S. and Japanese Companies during Transition to Overseas Production, 1957–68*

Millions of units unless otherwise specified

	Output		Percent of Japanese transistors used in radios	Percent of Japanese transistor radios exported
Year	United States	Japan		
1957	29	6	67	n.a.
1958	47	27	67	n.a.
1959	82	87	55	77
1960	128	140	48	70
1961	191	180	41	67
1962	240	232	34	76
1963	300	268	35	81
1964	407	416	33	69
1965	608	454	30	75
1966	856	617	26	86
1967	760	766	23	83
1968	883	939	20	90

Sources: For the United States, Electronic Industries Association, *Electronic Market Data Book, 1977* (Washington, D.C., 1977), table 77, p. 115; for Japan, John E. Tilton, *International Diffusion of Technology: The Case of Semiconductors* (Brookings, 1971), pp. 156–57.

n.a. Not available.

Kogyo (5 percent). Despite Japan's lag in the technology needed to produce the most advanced products, imports into its highly protected domestic market came to only 2 percent of consumption.[25]

The late 1950s marked a period of explosive growth in Japanese transistor production. In 1957 Japanese unit output of transistors amounted to about one-fifth of American production; by 1959 Japanese output had passed U.S. levels. Until 1960 one-half to two-thirds of Japanese output was incorporated into transistor radios, and through 1961 better than two-thirds of these were exported (table 2-1). In those days, transistor fabrication was a highly labor-intensive activity, and relatively low Japanese wage rates made it possible for Japanese companies to compete successfully on price, particularly in relatively low-quality transistors for

25. These figures are reported in Tilton, *International Diffusion of Technology*, p. 144. Japan's lag in high-performance transistor technology for use in industrial applications (such as communications equipment) was apparent in an internal battle that took place within NEC, when it made a decision to support the construction of production facilities for transistors. Research staff favored the use of indigenously developed transistor technology, while manufacturing engineers advocated the use of more advanced American transistor technology. A bitter conflict was ended in 1957, when NEC's chairman chose to go with the American technology. The episode is detailed in Nakagawa, *Semiconductor Development in Japan*, pp. 91–93.

use in consumer electronics, where cost, not performance, was critical. In the late 1950s and early 1960s, for example, the consumer electronics market was supplied almost exclusively with low-frequency germanium transistors produced with relatively cheap alloy junction and grown junction technologies, while the computer, defense, and industrial markets were the destination for more advanced high-frequency transistors made with sophisticated diffused base technologies.[26] The most advanced high-performance transistors were sold almost exclusively to the computer industry.

The emphasis on mass production of low-end consumer products, where cost, not quality, was paramount, was a successful strategy in the 1960s. Some in the United States dismissed any notion that Japanese producers might pose a threat in the higher performance applications (in computers and military systems) that drove the U.S. industry's technological advance. It would be incorrect, however, to conclude that the early Japanese effort in semiconductors was exclusively oriented toward consumer electronics. There was some attempt to use Sony point-contact transistors when MITI's Electrotechnical Laboratory constructed its first transistorized computer in 1956, but there were problems with these components, and imported products ultimately seem to have been used.[27] A successor to this machine was built in 1957, which successfully made use of more advanced junction transistors made in Japan by Hitachi, as well as germanium diodes produced by NEC.[28] The computer design was transferred to Japanese companies and was the basis for the first commercial computers shipped by them.[29] But the performance of the components used in these machines lagged well behind those used in the United States, where newly developed surface barrier transistors were setting new standards of speed and reliability.

---

26. See Long, "Price and Nonprice Practices," pp. 30–36. For a description of the different techniques, see chapter 1, note 39 of this volume.

27. Takahashi writes that an improved version of a Sony transistor was selected for the computer (the ETL Mark III) design, but the official history of computer development at the ETL notes that point-contact transistors imported from the United States were actually used. See Osamu Ishii, "Research and Development on Information Processing Technology at Electrotechnical Laboratory: A Historical Review," vol. 45 (1981), p. 315 [contract translation by IBRD]; and Shigeru Takahashi, "Early Transistor Computers in Japan," *Annals of the History of Computing,* vol. 8 (April 1986), p. 146.

28. Takahashi, "Early Transistor Computers in Japan," p. 149; Ishii, "Research and Development on Information Processing Technology," p. 315.

29. See Flamm, *Creating the Computer,* pp. 175–76.

Early Exports in Electronics

The mid- to late 1950s are an early breakpoint in the history of the semiconductor industry. On the technical side, Bell Labs made public new technologies for making transistors, while other techniques developed outside the Bell Labs emerged in the commercial market.[30] These new methods soon displaced older technologies in higher performance applications. Transistors based on a silicon semiconductor material, which offered higher performance (although initially at a higher cost), quickly began to replace germanium transistors in industrial applications—such as computers—where better performance justified a premium price. Most important, prototypes of the first integrated circuits (ICs) were built in 1958–59, a development which was to transform the electronics industry over the decade of the 1960s.

In the consumer market, Japanese firms made their first great push into exports, triggering a cycle of reaction and response that in many respects was repeated over the next three decades. The initial vehicle for Japan's export push was the transistor radio (table 2-1). In the early 1960s the push into transistor radios was augmented by exports of transistorized versions of other consumer electronics products, particularly televisions, tape recorders, audio equipment, and, in the late 1960s, transistorized calculators.[31]

The rapid Japanese expansion in consumer electronics and the sharply increasing production of relatively inexpensive, lower performance transistors used in those applications led to sweeping changes in the U.S. electronics industry. Figures 2-2 and 2-3 show rapid changes in the composition of American transistor output over the decade spanning Japanese entry into inexpensive transistors and the consumer electronics in

30. As part of its 1956 consent decree, AT&T had agreed to license all existing patents royalty-free to any domestic company, and not to sell semiconductors in commercial markets (that is, excluding defense and space applications).

31. On Japanese consumer electronics exports see "Japan Boosts TV Set Output," *Electronics*, February 26, 1960, p. 48; "Japanese TV Sets Arriving This Week," *Electronics*, April 29, 1960, p. 32; "Sales and Imports of Radio-TV Rise," *Electronics*, May 19, 1961, p. 12; U.S. Department of Commerce, Bureau of Domestic Commerce, *The U.S. Consumer Electronics Industry* (1975), pp. 8–9, 11. On calculators see U.S. Department of Commerce, Bureau of Domestic Commerce, *The Impact of Electronics on the U.S. Calculator Industry, 1965 to 1974* (November 1975), pp. 13–15, 44; and Badiul Alam Majumdar, *Innovations, Product Developments, and Technology Transfers: An Empirical Study of Dynamic Competitive Advantage, The Case of Electronic Calculators* (University Press of America, 1982), chap. 5.

Figure 2-2. *Volume of U.S. Transistor Production, by Type, 1955–65*

Percent

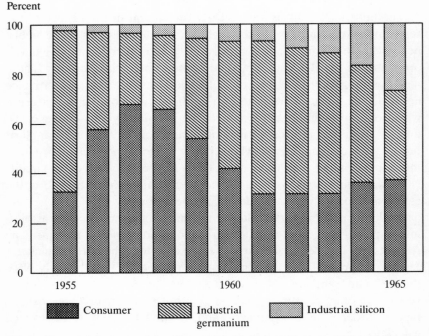

Sources: Electronic Industries Association, *Electronic Industries Yearbook* (Washington, 1959), p. 38; Electronic Industries Association, "Factory Sales of Transistors," monthly series, and "Sales of Transistors by Market," series of reports. Cited in William F. E. Long, "Price and Nonprice Practices under the Uncertain Conditions of Rapidly Improving Technologies: A Case Study," Ph.D. dissertation, George Washington University, 1967, tables 1, 6, 10.

which they were incorporated. There was a sharp decline in the share of transistors for consumer applications (mainly germanium for most of this period) in U.S. output, in terms of both value and quantity; this decline coincided with the Japanese push into consumer electronics exports that began after 1957.

Relatively more of U.S. output shifted into higher performance transistor types used in industrial applications, especially the recently introduced silicon transistors.[32] Rapidly declining costs made the new silicon

32. Over this period the principal performance difference between transistors for the consumer market and those for the industrial market was the ability to operate at high frequencies. Transistors made using grown or alloy junction technology generally operated at low frequencies, while transistors manufactured using newer diffused base techniques operated at much higher frequencies. Virtually all consumer applications required only

Figure 2-3. *Value of U.S. Transistor Production, by Type, 1955–65*

Percent

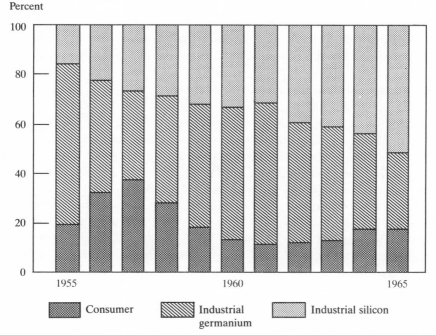

Consumer   Industrial   Industrial silicon
                  germanium

Sources: See figure 2-2.

transistors more attractive for low-end applications over time, and some resurgence by American producers in the market for consumer-use transistors was evident by the mid-1960s. In a pattern that was to recur later, the United States passed Japan in transistor output in the early 1960s as demand shifted to more technically advanced silicon transistors. Japan then raced to catch up in the new technology, ultimately surpassing the United States again in total transistors produced in the late 1960s. This phenomenon—of new technology providing the basis for improved competitive performance by American producers vis-à-vis foreign rivals man-

---

low-frequency operation, while industrial applications typically required high-frequency operation. Higher performance transistors could typically be used in consumer applications with less exacting operational requirements but would not normally be an economic choice because of their greater cost. However, some rejected, "not-up-to-spec" high-performance transistors were adequate for consumer applications, and such rejects were sold into the consumer market. See Long, "Price and Nonprice Practices," pp. 30–31.

ufacturing older products at lower cost—was (as will be seen below) repeated again in the 1960s and in the 1970s.

For the most part, the Japanese government's role in the early development of the Japanese semiconductor industry was as a mediator of contacts with markets outside Japan and as a gatekeeper to the Japanese market. Although the government exercised significant regulatory powers in managing technical and trade contacts with the outside world, and government labs played an important part in Japanese technology development, it was largely the private sector that took the first initiatives in the newly emerging solid state electronics technology, with an eye on markets for consumer electronics products.

The Japanese government was to greatly increase its role in shaping Japan's electronics industry in the 1960s. It was computers, however, and not consumer electronics, that led the government to take significant steps to stimulate technological advance in the Japanese semiconductor industry. MITI's attempts in the 1960s to help Japanese computer manufacturers catch up with developments abroad were to be the backdrop of a successful drive in the 1970s to the technological frontier in semiconductors.

The Seal of Approval

A landmark event for the Japanese electronics industry was the passage of the Law on Temporary Measures for Promoting the Electronics Industry (the Electronics Industry Promotion Law) in 1957.[33] (The basic

33. The Electronics Industry Promotion Law (Denshinho) was actually preceded by a Law on Temporary Measures for the Promotion of Specified Manufacturing Industries in 1956, which, like the 1957 electronics law, provided for low-interest loans from the Japan Development Bank (JDB) for investments in modernization and technology within selected industrial sectors. Machine tools, auto parts, and other basic machinery were among the nonelectronic sectors enjoying favored access to JDB capital. From 1956 through 1960 slightly less than 10.6 billion yen in loans was extended to these sectors. This compares with 2.2 billion yen in loans to the consumer electronics sector granted over the seven years from 1957 through 1963.

The Specified Manufacturing Industries Law was renewed in 1961 and again in 1966, expiring in 1971. The Small Business Finance Corporation (SBFC) joined the JDB in providing subsidized credit to the machinery industry, with loans extended by these two institutions amounting to 53.8 billion yen over 1961–65, and 48.9 billion yen over 1966–70. In 1964 the electronics law was extended for another seven years. Loans over this period amounted to 12.1 billion yen, with the bulk of the funds going into electronic components.

In 1971 the electronics and machinery industry laws were combined into a single mea-

structure was extended by successor laws: the Machinery and Electronics Industry Law, 1971; the Machinery and Information Processing Industries Law, 1978.[34]) Ironically, the 1957 law was primarily intended to promote the consumer electrical and electronics products industries, and only after its renewal in 1964 did the emphasis shift to computers and semiconductors.[35] Perhaps the law's most important provision was its establishment of an Electronics Industry Division within MITI, and an Electronics Industry Council to coordinate policymaking with the Japanese electronics industry. This established MITI as the bureaucratic patron—in Japanese parlance, *genkyoku*—under whose jurisdiction the electronics industry fell. As Komiya writes, a *genkyoku* typically strives for "its industry to be orderly and organized and for there to be no disruptions of any kind."[36] The contrast between this bureaucratic perspective and the mercurial ups and downs characterizing the history of the highly entrepreneurial U.S. electronics sector is noteworthy.

The Electronics Industry Promotion Law, in a pattern that was to be repeated in later legislative frameworks for that sector's promotion, established three classes of support efforts. For important technologies, R&D subsidies were to be given. The second type of program provided low-interest loans from the government-run Japan Development Bank

---

sure, the Law on Temporary Measures for the Promotion of Specified Electronics Industries and Specified Machinery Industries (Kidenho). Over the seven years through 1977, the electronics industry received approximately 20 percent of some 70.6 billion yen in loans under the new law. In 1978 this measure was revised and extended as the Law on Temporary Measures for the Promotion of Specified Machinery and Information Industries (Kijoho). Loans from the SBFC and the JDB amounted to approximately 11 billion yen per year, with approximately 80 percent going into electronics.

See Seiritsu Ogura and Naoyuki Yoshino, "The Tax System and the Fiscal Investment and Loan Program," in Ryutaro Komiya, Masahiro Okuno, and Kotaro Suzumura, eds., *Industrial Policy of Japan* (Academic Press, 1988), pp. 147–48.

34. The 1957 Denshinho law was extended in 1964, then followed by the 1971 Kidenho law, which in turn was followed by the 1978 Kijoho (see the previous note.) Although the Kijoho expired in 1985, another law, the Josokuho (established in 1960, amended to include many new provisions for investment subsidies, and reenacted in 1986) provides legal authority for many of the sorts of subsidies to the computer and semiconductor industries granted under the Kijoho. All recent JDB investments in information technology and related industries, for example, have been made under the framework of the Josokuho. See Japan Electronic Computer Corporation (JECC), *Computer Notes* (Tokyo, 1989), pp. 62–63 (in Japanese).

35. See Koji Shinjo, "The Computer Industry," p. 342; and Ogura and Yoshino, "The Tax System," in Komiya and others, eds., *Industrial Policy of Japan*, p. 147.

36. See Komiya, "Introduction," in Komiya and others, eds., *Industrial Policy of Japan*, p. 11.

(JDB), and later from its sister institutions, for initial production and subsequent capacity expansion in specified products.[37] The third type of support provided additional JDB loans and special depreciation breaks where "industrial rationalization"—that is, improvements in quality or production technology—was desirable.[38] The law also authorized selective exemption from the antimonopoly law, allowing MITI to establish research and production cartels.[39]

Although semiconductor-related R&D and production rationalization were thus supported virtually from the start, the resources applied to these ends were clearly quite small.[40] All electronics-related R&D subsidies averaged a little over 230 million yen (just under $650,000) per year during the seven-year period ending in fiscal 1963. Similarly, JDB loans to the electronics industry accounted for an annual average of 324 million yen (about $900,000) over the same period.[41] As previously remarked, funding mainly went into consumer electronics—only a small portion of these funds was applied to semiconductors.

In contrast, subsidized JDB loans to other machinery industries authorized by similar legislation amounted to over 2 billion yen per year over 1956–60, and almost 11 billion yen annually over 1961–65. Thus, the "temporary measures" of 1957 initially led to only minimal new subsidies for semiconductors. More important, perhaps, was the rigorous policy of

37. It is frequently argued that, in addition to the funds themselves, and the signal to invest in the promoted sector provided to highly regulated private financial institutions (which was quite significant in the capital-short Japanese economy of the 1960s, but less so in the late 1970s and 1980s) or the direct subsidy element in loans at below-market rates, these funds provided an implicit government guarantee for private sector loans to these sectors.

38. This discussion follows Shinjo, "The Computer Industry," p. 342.

39. MITI is reported to have used this law to block entry into semiconductors by some Japanese latecomers, including Sanyo Electric. See Nakagawa, *Semiconductor Development in Japan*, pp. 101–03.

The authority to sanction legal cartels in electronics was not often invoked. Over the years 1957–70 some six legal "rationalization" cartels were established under the auspices of the Electronics Industry Promotion Law: three over the period 1964–66 and three over the period 1969–70. This compares with some 114 established in machinery under the machinery promotion law over the same period. See Yutaka Kosai, "The Reconstruction Period," in Komiya and others, eds., *Industrial Policy of Japan*, p. 47.

40. Categories of products that were promoted, and the years over which the support extended, are shown in Badiul A. Majumdar, "Industrial Policy in Action: The Case of the Electronics Industry in Japan," *Columbia Journal of World Business*, vol. 23 (Fall 1988), pp. 27–29.

41. Majumdar, "Industrial Policy in Action," p. 30–31.

trade protection for the electronics industry formalized and justified by passage of the Electronics Industry Promotion Law.[42]

Protecting the Japanese Market

Notwithstanding Japan's early successes in exporting transistors and consumer electronics to overseas markets, virtually all segments of the internal Japanese electronics market were tightly walled off from foreign competition by this time. The implements used to construct steep walls around the Japanese market were forged during the immediate postwar years, with the approval of the U.S. occupation authorities.

The two principal measures used to protect the domestic market, the 1949 Foreign Exchange Control Law (FECL) and the Foreign Investment Law (FIL) of 1950, were parts of a complex of laws affecting foreign exchange remittances and export-import trade.[43] The FIL essentially required case-by-case approval by MITI of all foreign investments in Japan and was used to effectively block all foreign investment not deemed to be in the national interest.[44] The law also provided the legal rationale for an extensive system requiring prior approval of technological agreements between Japanese and foreign companies that generated foreign exchange remittances. MITI exercised tight controls over licensing, royalty, and technology transfer agreements, in order to influence the pricing and composition of technology imports and the structure of high-technology industries in Japan.[45]

The FECL and related measures had the effect of giving MITI considerable discretionary power over foreign trade transactions. By 1960

42. See Motoshige Itoh and Kazuharu Kiyono, "Foreign Trade and Direct Investment," in Komiya and others, eds., *Industrial Policy of Japan*, pp. 147, 160.

43. Other elements included the Export Trade Control Law (ETCL), the Import Trade Control Law (ITCL), and the rules of the "Standard Method of Payments" governing imports. On the operation of these measures, see Lawrence Krause and Sueo Sekiguchi, "Japan and the World Economy," in Patrick and Rosovsky, eds., *Asia's New Giant*, pp. 411–27, 451–52; Mark Mason, *American Multinationals and Japan: The Political Economy of Japanese Capital Controls, 1899–1980* (Harvard University Press, 1992), pp. 159–61; and Dennis J. Encarnation, *Rivals Beyond Trade: America versus Japan in Global Competition* (Cornell University Press, 1992), pp. 46–50.

44. On the mechanics of how this system was used, see Mason, *American Multinationals and Japan*, pp. 154–61; Encarnation, *Rivals Beyond Trade*, pp. 47–50.

45. See Terutomo Ozawa, *Japan's Technological Challenge to the West, 1950–1974: Motivation and Accomplishment* (MIT Press, 1974), pp. 16–24, 52–66; and Peck and Tamura, "Technology," 544–58.

the foreign exchange control system had evolved into a broad system of MITI-administered import quotas. Mounting foreign criticism led to Japan's first steps toward trade liberalization at the end of the 1950s, when 257 commodities, covering 44 percent of all Japanese imports, were first freed from licensing requirements.[46] The 1960 program of Trade and Exchange Liberalization Guidelines initially considered selected electronics items for liberalization, but MITI later withdrew electronics goods from the list of items to be liberalized.[47]

Tight control over foreign imports of and investment in semiconductors was maintained throughout the 1950s and early 1960s. The limitations on investment were particularly important in ICs, products that often required considerable technical support from manufacturers. Without a local subsidiary to provide that support to customers or distributors, an effective sales effort was difficult, if not impossible.

In 1962, for example, Robert N. Noyce, then one of the key technical figures at Fairchild Semiconductor and co-inventor (with Jack S. Kilby of Texas Instruments) of the integrated circuit, visited Japan to arrange an investment in a local production facility and discuss the sale of licenses to use Fairchild's key patents to Japanese firms.[48] (Together, the TI and Fairchild patents were the key intellectual properties at the center of the emerging IC market.)

Noyce, and Fairchild, were completely outmaneuvered in their initial foray into the Japanese market. MITI flatly refused to permit Fairchild to invest in its own Japanese factory. Frustrated, Noyce then contacted an NEC executive he had met in the United States at a defense semiconductor conference,[49] complained about MITI's reaction, and set about negotiating with NEC over patent rights. Informed of the need for NEC to secure approval of the terms from MITI, Fairchild was pressured to reduce its royalties and fees; only later did the U.S. company discover

46. Krause and Sekiguchi, "Japan and the World Economy," p. 414.
47. See "Japanese Put Off Freeing Electronics Imports," *Electronics*, July 8, 1960, p. 11. The initial, "temporary" delay of three years was to stretch into sixteen years; quantitative restrictions were not completely removed from electronics imports until 1976.
48. This episode is described in Nakagawa, *Semiconductor Development in Japan*, pp. 135–40.
49. Noyce apparently met Mr. Naganuma, the NEC executive, at a conference described in NHK, *Electronics-Based Nation*, vol. 2, pp. 253–56: the U.S. Navy–sponsored Semiconductor Devices Research Conference, held in Boulder, Colorado. Although the conference was closed to the general public, some foreign engineers including Naganuma were allowed to attend.

that NEC's own president also chaired the MITI licensing approval advisory committee responsible for this decision![50]

A deal was finally struck in 1963. Noyce negotiated an exclusive license with NEC for patent rights in the Japanese market. The agreement so completely transferred the Japanese rights to Fairchild's IC technology to NEC that it was later used to block attempts by Fairchild to set up a Japanese sales office. NEC also charged other Japanese companies for their use of covered technologies. It was a poor deal for Fairchild, but it reflected the then-prevailing climate for foreign investors in Japan. That same year, TI began its attempt to enter the Japanese market, and took a much tougher bargaining stance. But TI was not to gain even a small degree of access to the Japanese market until half a decade later, in 1968, after an exhausting battle.

Through 1963 only seven foreign semiconductor producers had been permitted to invest in the Japanese market, in all cases through joint ventures in which they held a minority interest, and in all cases with extensive transfer of technical know-how a part of the deal. Four of these ventures were broad tie-ups in electrical equipment and electronics, which also covered semiconductors. Over the period from 1952 through 1962 the Dutch company Philips NV had acquired a 35 percent interest in Matsushita Electronics, the semiconductor-producing affiliate of Matsushita.[51] From 1953 to 1960 ITT's International Standard Electric subsidiary had increased its holdings in NEC to 17.7 percent. General Electric (GE) had built up a 5.6 percent interest in Toshiba from 1953 to 1959.[52]

Three new ventures were exclusively concerned with semiconductors. From 1957 to 1961 International Rectifier had established a 39 percent stake in Nippon International Rectifier. In 1961 Raytheon Company had invested in New Japan Radio, a semiconductor-producing joint venture with Japan Radio (Raytheon's share had increased to 33 percent by 1963). In 1962 Mitsubishi TRW (with a 20 percent stake held by TRW Incorporated, and another 20 percent held by Pacific Semiconductor) and

50. See Mason, *American Multinationals and Japan*, pp. 196, 325. Mason's account is based on interviews with Noyce and then–executive vice president of Fairchild, Richard Hodgson.

51. This equity was finally sold back to Matsushita in 1993. See Douglas R. Sease, "Abreast of the Market: With Confusing Results for 1st Quarter, It Isn't Surprising the Market Is Stalled," *Wall Street Journal*, May 3, 1993, p. C1.

52. See Heavy and Chemical Industry News Agency, *Foreign Investment in Japan*, pp. 375, 377.

Komatsu Hoffman (46 percent owned by silicon rectifier maker Hoffman Electronics) had been formed to produce specialized semiconductors.[53]

A critical moment in the Japanese semiconductor industry's relations with the outside world came in 1963, when Texas Instruments—co-owner with Fairchild of the intellectual birthrights to the IC—decided to apply for permission under the FIL to set up a wholly owned Japanese manufacturing subsidiary. TI had earlier approached Sony with a proposal to set up a manufacturing joint venture, but negotiations had collapsed in 1959 when Sony insisted on exclusive rights to sell the output.[54] TI had also reportedly contacted NEC in 1960, when top executives were visiting Japan, about a joint venture in semiconductors. The proposal had been rejected outright.[55]

At roughly this same time, International Business Machines Corporation (IBM), using its vast patent portfolio and the promise of technology transfer as a bargaining chip, had managed—after years of negotiations and acceptance of various restrictions on its activities—the highly unusual feat of negotiating with MITI the establishment of a wholly owned computer manufacturing facility in Japan in 1960.[56] TI's interest in establishing a Japanese facility was reportedly motivated at least in part by the prospect of supplying IBM (with which TI had excellent relations elsewhere in the world) and its new Japanese facility with components.[57]

After numerous presentations in Tokyo in 1963, TI filed a formal application in January 1964.[58] Greeted with outright hostility by both

53. See Heavy and Chemical Industry News Agency, *Foreign Investments in Japan*, chap. 5. See also Organization for Economic Cooperation and Development (OECD), *Electronic Components, Gaps in Technology* (Paris, 1968), p. 186.

54. Mason, *American Multinationals in Japan*, p. 176.

55. Nakagawa, *Semiconductor Development in Japan*, pp. 125–27. TI then reportedly joined other U.S. semiconductor producers in petitioning for restrictions on Japanese transistor imports, in what appears to have been the first trade policy initiative taken by the U.S. chip industry (discussed in chapter 3).

56. More details and an analysis of this episode may be found in Flamm, *Creating the Computer*, pp. 180–82; Mason, *American Multinationals in Japan*, pp. 187–91; and Marie Anchordoguy, *Computers Inc.: Japan's Challenge to IBM* (Harvard University Press, 1989), pp. 22–24.

57. Nakagawa, *Semiconductor Development in Japan*, p. 154.

58. The most complete account of this affair from the American side may be found in Mason, *American Multinationals in Japan*, pp. 176–87. Mason was given access to the TI corporate archives in preparing his research.

MITI and the Japanese press, the application was pigeonholed—neither approved nor disapproved—while MITI stalled for time.[59]

The Integrated Circuit Arrives in Japan

MITI's interest in electronics was rapidly increasing in the early 1960s. The market for electronic computers was booming, and it was becoming clear that this was destined to be a major industry.[60] After IBM had been permitted to set up an affiliate manufacturing computers for the Japanese market in late 1960, MITI actively urged three Japanese computer makers to join a cooperative R&D program—the FONTAC project—to develop important computer technologies. This project, begun in 1962, was the first cooperative research project established among competing Japanese electronics manufacturers. Although FONTAC was a relatively small-scale project (it was largely finished by 1964 and received government funding of 338 million yen, or $930,000),[61] it was the prototype for later government-industry collaboration on joint R&D programs.

The introduction by IBM of an innovative new family of computers—the System 360 product line—in 1964 shook up both the mushrooming computer industry and Japanese policy. IBM injected a whole new generation of advanced semiconductor components—integrated circuits—into the commercial mainstream, and this and other innovations rendered

59. It is interesting to note that in Mason's account from the U.S. perspective (*American Multinationals in Japan,* p. 180) the Japanese industry is portrayed as pressuring MITI to resist TI's demands. In Nakagawa's version (*Semiconductor Development in Japan,* pp. 154–66), told from a Japanese perspective, MITI is credited with having a long-range vision of a competitive threat that the domestic industry did not yet see, and organizing an industry all too prone to cutting individual deals on the side with TI. According to Nakagawa, Hitachi, Mitsubishi, and Toshiba all held secret talks with TI. According to Mason (p. 323), Matsushita attempted to cut its own deal through Philips, which had an equity share in (as noted above) and extensive cross-licensing arrangements with Matsushita Electronics.

60. See Flamm, *Creating the Computer,* chap. 6.

61. The project spent a total of 1.126 billion yen, of which 787 million yen was provided by the participating companies (NEC, Fujitsu, and Oki). The Computer Technology Research Association in which these companies joined together as members was not actually dissolved until 1973. *Thirty Year History of Mining and Manufacturing Industry Technology Research Associations* (Tokyo: Japan Industry Technology Promotion Association, 1991), p. 70. See also Shinjo, "The Computer Industry," p. 343.

the Japanese FONTAC effort obsolete virtually overnight.[62] TI's continuing political fight to secure the right to establish a Japanese subsidiary also spotlighted the importance of ICs for the nascent computer industry.

Although the Japanese were already experimenting with ICs by this time, development had proceeded slowly. The first crude, experimental IC built in Japan was constructed inside MITI's Electrotechnical Laboratory in 1960. The private sector lagged well behind: NEC developed its first experimental IC in 1962.[63] There was little demand in Japan. Lacking the huge defense effort in solid state communications and computer components that was driving technology development in the United States, IC production in Japan was not to expand greatly until the solid state calculator came to market in the late 1960s.[64]

The first major push to actually produce ICs came after IBM's 1964 announcement of its System 360 series. Shocked into action, MITI launched several initiatives. Some $80,000 (29 million yen) in grants for R&D on ICs for use in computers was hurriedly made available to Japanese semiconductor makers in 1965.[65] Even more significant, a ten-year, 6-billion-yen program of subsidized loans from the JDB to Japanese chip producers was announced.[66] And in April 1966 an outright ban on all foreign ICs with more than thirty-four circuit elements was put into place.[67]

Even as TI's application for a wholly owned subsidiary sat on hold, MITI signaled that it was willing to permit some degree of foreign participation in the most sophisticated niches of the Japanese semiconductor

62. Actually, the IBM computers did not use true "monolithic" (that is, etched from a single piece of silicon) ICs. Discrete semiconductor components were instead bonded to a ceramic substrate to form so-called hybrid ICs. Although competitors were to soon introduce computers using true monolithic ICs, the IBM System 360 line nonetheless marked a significant step forward in the sophistication of components used in commercial computers.

63. See Martin Fransman, *The Market and Beyond: Cooperation and Competition in Information Technology Development in the Japanese System* (Cambridge University Press, 1992); and Charles Cohen, "Japanese Components Men Stress Integrated Circuits and Diodes," *Electronics*, May 4, 1962, pp. 20–21. Mitsubishi brashly announced the availability of a complete line of ICs in 1961 but was unable to produce a functioning complete circuit until 1964. See Nakagawa, *Semiconductor Development in Japan*, pp. 130–35; NHK, *Electronics-Based Nation*, vol. 3, pp. 83–93.

64. However, Mitsubishi's early IC effort was driven by contracts from the Japan Self Defense Agency to downsize communications equipment. NHK, *Electronics-Based Nation*, vol. 3, p. 92.

65. Yasuo Tarui, "Japan Seeks Its Own Route to Improved IC Techniques," *Electronics*, December 13, 1965, pp. 90–98. The funds were made available to the "big six" of the Japanese computer industry: Fujitsu, Hitachi, Toshiba, NEC, Oki, and Mitsubishi.

66. Nakagawa, *Semiconductor Development in Japan*, p. 158.

67. "Onward and Upward," *Electronics*, October 3, 1966, p. 257.

market, as long as it was effectively controlled by domestic interests. In September 1964 American semiconductor specialist Bernard Jacobs formed Kyodo Electronics Laboratories, in a joint venture with five smaller Japanese electronic component companies.[68] Kyodo quickly became the only significant purely merchant IC producer in the Japanese market—by 1966 it was second only to NEC in sales in the infant Japanese IC market.[69]

Although Japanese chip makers had invested in experimental IC development in the early 1960s, mass production was not to start until 1966, a full half decade behind the American industry.[70] The first ICs coming off these new production lines were for the most part used in the latest computer models manufactured by the computer divisions of the same large companies.[71] But in 1966 imports dominated the emerging market, accounting for almost two-thirds of Japanese IC consumption (figure 2-4).[72]

More powerful means were needed to counteract the mounting competitive threat from rapid technical progress in American computers. In 1966 MITI launched a new program of large-scale national R&D projects: among the first three projects announced was an effort to develop a "Very High Speed Computer System" (VHSCS).[73] The five largest Japanese computer producers—Fujitsu, Hitachi, Nippon Electric, Toshiba, and Oki, which also were among the largest semiconductor makers in Japan—were teamed in an ETL-directed cooperative project to build a large time-sharing computer system to compete with the IBM System 360 Model 67.

68. The five Japanese companies were Toko, Nippon Chemical Condensor, Koden Electronics, Pioneer Electronics, and Alps Electric. See Tarui, "Japan Seeks Its Own Route," p. 91. Jacobs owned 25 percent of the venture's equity and received additional compensation for technical assistance. See Heavy and Chemical Industries News Agency, *Foreign Investments in Japan*, 2d ed., pp. 48, 372.

69. See "Onward and Upward," pp. 257–58.

70. Nakagawa, *Semiconductor Development in Japan*, p. 280. By late 1966 only NEC, Kyodo Electronic Laboratories, and Fujitsu were producing ICs (in small quantities); Hitachi and Toshiba had just completed manufacturing facilities. "Onward and Upward," pp. 257–58.

71. "Onward and Upward," pp. 257–58.

72. In figure 2-4, IC exports from the United States by captive producers such as IBM and the U.S. affiliates of Japanese companies (growing rapidly since the 1980s) are counted as imports, while production by Japanese affiliates of U.S. companies, such as TI, IBM, and Motorola, are counted as domestic output.

73. For more detailed references to MITI projects in computer technology development and their outcomes over this period, see Flamm, *Creating the Computer*, pp. 184–92.

Figure 2-4. *Import Shares of Japanese Semiconductor Consumption, 1966–94*

Percent

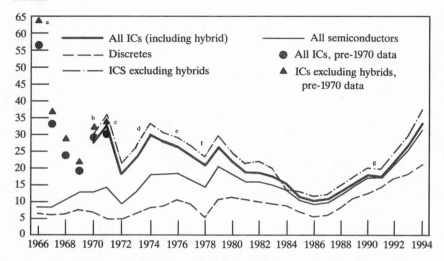

Sources: Pre-1970 measures of IC production from Machinery Promotion Association Economic Research Institute, *Japan-US Semiconductor Industry Research Survey* (Tokyo, 1980), p. 114. Post-1970 production data from Research and Statistics Department, MITI, *Yearbook of Machinery Statistics*, various years. Import data from Japan Tariff Association, *Japan Exports and Imports, Commodity by Country* (Tokyo, various years). Import share of consumption is defined as value of imports divided by apparent consumption (Value of production plus value of imports less value of exports). Hybrids were not a separate category in trade data until 1989. Hybrids are excluded in production data only.
  a. April 1966: ban on foreign ICs with more than 34 circuit elements.
  b. September 1970: Liberalization of ICs with less than 100 elements.
  c. Late 1971: MITI crackdown on LSI IC import licenses.
  d. 1973: removal of licensing of ICs with less than 200 elements.
  e. December 25, 1974: liberalization of IC imports with more than 200 elements.
  f. 1975: liberalization of computer ICs.
  g. U.S.-Japan Semiconductor Trade Arrangement.

As part of this project, Fujitsu, Hitachi, and NEC shared a 400-million-yen ($1.1 million) grant to develop next-generation integrated circuits: so-called LSI (large-scale integration) ICs, containing up to 10,000 transistors on a single chip.[74] This was the largest infusion by MITI of R&D funding into semiconductors yet, and it achieved important results.[75] In addition to fueling progress on LSI within all three

74. Yasuo Tarui, "ICs in Japan—a Closeup," *Electronics*, May 13, 1968, p. 106. Although there is no exact definition, medium-scale integration (MSI) generally is taken to mean ICs with up to 1,000 circuit elements, while very large scale integration (VLSI) devices have up to 100,000 circuit elements.
75. MITI's ETL, which directed the project, developed a metal oxide semiconductor

companies, the funding helped Hitachi develop high-performance logic chips and NEC develop IC memory chips (Japan's first). Chips utilizing semiconductor technology developed in this project soon showed up inside computer systems shipped by Hitachi, NEC, and Fujitsu.[76]

The increased interest in electronics is clearly visible in the fragmentary statistics that are available on MITI subsidies to investment and R&D in the electronics industry in the 1960s. The Electronics Industry Promotion Law had been renewed for another seven years in 1964 and was later combined with the Machinery Industries Promotion Law in 1971 (and again extended in 1978).[77] As before, the three major means of subsidy under these revised measures were R&D subsidies, special depreciation tax breaks, and subsidized loans from the public development banks.

Table 2-2 shows the volume of subsidized loans from public financial institutions extended under these successive laws in the relevant multiyear periods.[78] In years prior to 1964, subsidized loans to the electronics industry averaged one-seventh to one-thirtieth of levels going to the machinery industry annually. Furthermore, little of this funding went to areas other than consumer electronics. After 1964, electronics funding jumped to about one-sixth of machinery industry lending (with the major share going into electronic components), and after 1971 to one-quarter (and going mainly to IC producers). After 1978, annual lending in electronics was quadruple the levels going to machinery companies. Electronics in general, and semiconductors in particular, were clearly being pushed closer and closer to the front of MITI's industrial policy queue in the late 1960s and 1970s.

(MOS) device structure known as the diffusion self-aligned (DSA) circuit, which producers used extensively in following years. The first Japanese IC memory, a 144-bit static random access memory (RAM) chip, was developed by ETL and manufactured by NEC in 1968. An electron beam exposure system was developed in cooperation with JEOL, now the top supplier of electron beam production equipment to the semiconductor industry. This system was the very first such machine developed by JEOL; the second and third systems built were shipped to Sanyo and Toshiba. A step-and-repeat photolithographic system using laser alignment was built by Nikon for ETL as part of this project. This design was later used in the very first commercial wafer stepper system built by Nikon, which was to become a top supplier of such machines to IC makers a decade later. Author's interview with Yasuo Tarui, director of the MITI VLSI project, March 18, 1992; Osamu Ishii, "Research and Development"; and Machinery Promotion Association Economic Research Institute, *Japan-U.S. Semiconductor Research Survey* (Tokyo, 1980), p. 30.

76. See Flamm, *Creating the Computer*, pp. 189–90, notes 43–44.

77. See note 34.

78. These statistics capture only lending to industry under the Electronics Industry Promotion Law and exclude funds borrowed under the auspices of other measures.

Table 2-2. *FILP Loans to Japanese Machinery and Electronics Industries, 1956–84*

Billions of yen unless otherwise specified

Period	Machinery industry		Electronics industry		IC Facility Investment		Electronics loans as percent of IC facilities investment
	Total	Average annual	Total	Average annual	Total	Average annual	
1956–60[a]	10.6	2.1	...	...	...	...	...
1957–63[b]	...	...	2.2	0.3	...	...	...
1961–65	53.8	10.8	...	...	...	...	...
1966–70[c]	48.9	9.8	...	...	...	...	...
1964–70[d]	...	...	12.1	1.7	44.1	6.3	27.4
1964–68[e]	...	...	4.9	1.0	...	...	...
1969–70[e]	...	...	7.2	3.6	...	...	...
1971–77[f]	56.5	8.1	14.1	2.0	123.1	17.6	11.5
1978–84	15.4	2.2	61.6	8.8	1,772.1	253.2	3.5

Sources: Seiritsu Ogura and Naoyuki Yoshino, "The Tax System and the Fiscal Investment and Loan Program," in Ryutaro Komiya, Masahiro Okuno, and Kotaro Suzumura, eds., *Industrial Policy of Japan* (Tokyo: Academic Press, 1988), pp. 147-48; Machinery Promotion Association Economic Research Institute, *Japan-U.S. Semiconductor Industry Research Survey* (Tokyo, 1980), pp. 98, 119 (in Japanese); MITI unpublished twelve-company survey; and *Electronic Industry Almanac 1975* (Tokyo: Dempa Publications, 1974), p. 176.

a. The machine tools industry was the major recipient.

b. Loans were mainly to producers of consumer electronics goods.

c. Period saw a rapid increase in loans to auto parts manufacturers.

d. A major proportion of this lending went to electronic components manufacturers.

e. Japan Development Bank lending specifically budgeted for electronics.

f. Lending was mainly to auto parts manufacturers (within machinery) and integrated circuit producers (within electronics).

Figure 2-5. *Japanese Tax Incentives and Subsidized Lending for Electronics Development, and Investment in Electronics by Major Companies, 1961–71*

Billions of yen

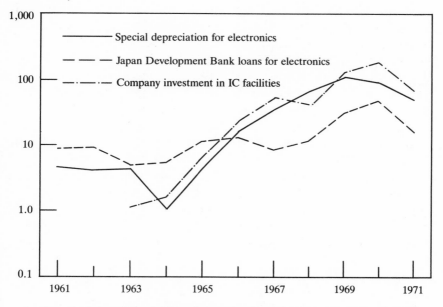

Sources: *Electronic Industry Almanac, 1975* (Tokyo: Dempa Publications, 1974), p. 176; and Machinery Promotion Association Economic Research Institute, *Japan-US Semiconductor Industry Research Survey*, pp. 98, 119.

These subsidized funds, furthermore, were significant in relation to resources the private sector was investing in semiconductors in the 1960s and 1970s. Although not all these funds were going into semiconductors, it is useful to compare the total flows with IC facilities investment expenditures undertaken by major Japanese semiconductor firms. As table 2-2 shows, subsidized electronics loans exceeded a quarter of IC facilities investments in the late 1960s, and stayed above 10 percent of IC plant investments in the 1970s, before dropping under 4 percent in the 1980s.

The importance of subsidized loans and tax breaks can be examined in greater detail for the decade of the 1960s and early 1970s. Figure 2-5 compares the volume of special accelerated depreciation given to the electronics industry from 1961 through 1971 with major companies' investments in IC plant and equipment in those years. Because the marginal corporate tax rate was quite high throughout this period (roughly

55 percent),[79] the tax benefits would also have been quite large. However, this special depreciation was spread over all qualifying electronics-related investments (including the computer industry), not just ICs—unfortunately, further disaggregation is not possible.

Figure 2-5 also compares JDB loans in electronics with investment in IC facilities. Again, subsidized loans are relatively large in relation to investments. Like the special tax breaks, the JDB loans also went to electronics firms making products other than ICs, but we do know that the largest share of the funding after 1964 did go to IC producers. Also, since the JDB was not the only public lender to the industry under the auspices of the Electronics Industry Promotion Law, other relevant subsidized loans are not being captured in this figure.

In contrast to depreciation benefits and loan subsidies, MITI R&D subsidies to electronics under the Electronics Industry Promotion Law were much less important. Figure 2-6 compares all electronics R&D subsidies with IC-related R&D spending by major Japanese producers. If it is recalled that significant R&D subsidies in semiconductors began only after 1965, it is clear that these monies could not have been particularly large in relation to spending levels within the industry. However, these R&D subsidies capture only MITI monies spent under the Electronics Industry Promotion Law—they do not cover large-scale national R&D projects, internal funding of the ETL, or funding provided under other statutes. Nonetheless, it is probably reasonable to suggest that for semiconductors in the 1960s, in the aggregate, tax breaks and subsidized loans, and not R&D subsidies, were the most significant instrument for MITI's promotion of the semiconductor industry.

In fact, R&D subsidies in semiconductors seem to have been focused on relatively narrow themes, and principally on supporting national policy in computers. But while the big ticket semiconductor R&D projects were in digital ICs and were clearly designed to strengthen the Japanese computer industry, now at the center of MITI's electronics strategy, some resources went into ICs oriented toward other applications. Analog circuits for use in communications and consumer electronics were funded at Fujitsu, Toshiba, Hitachi, Shiba Electric, Nippon Columbia, NEC, Matsushita, and Toyo Communication Equipment in the latter part of the 1960s.[80] A MITI-funded consortium of five television manufacturers,

79. See Ogura and Yoshino, "Tax System and the Fiscal Investment and Loan Program," p. 126.

80. Tarui, "ICs in Japan," p. 103.

Figure 2-6. *Japanese R&D Subsidies for Electronics Development, and R&D in Electronics by Major Companies, 1961–71*

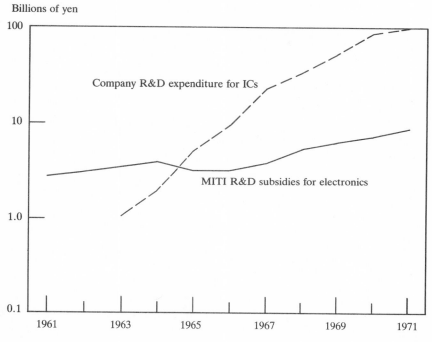

Sources: See figure 2-5.

seven component firms, four universities, and two Osaka-area research institutes began a successful joint R&D effort on ICs for use in television sets in 1966.[81] A substantial ($172,000, or 62-million-yen) project at Hayakawa Electric (later renamed Sharp) pioneered the miniaturization of electronic calculators in a design using LSI ICs around this time.[82] Mitsubishi, the largest supplier of ICs to Japanese calculator manufacturers at the time, used the MITI funding it received for its work on the Sharp

81. The joint research group included Osaka Onkyo, Hayakawa Electric (later renamed Sharp), Matsushita Electric, Mitsubishi Electric, Sanyo Electric, Elna-Fox Electronics, Hoshi Electric, Kobe Industrial, Matsuo Electric, Murata Manufacturing, Nichicon Capacitor, Sodensha, and Osaka University. See Eizi Sugata and Tashihiko Namekawa, "Integrated Circuits for Television Receivers," *IEEE Spectrum,* May 1969, p. 74.
82. "Japan: LSI Calculator," *Electronics,* November 11, 1968, p. 307.

calculator chips for its LSI research.[83] This push into calculators, indeed, was integral to the next major twist in the saga of Japanese semiconductor development.

A Secret Truce in the "Patent War"

Just as the Japanese electronics industry was faced with new and potentially devastating changes in its global competitive environment, Texas Instruments' stalled 1964 application to enter the Japanese market flared up into a hot dispute. Tensions had escalated in mid-1965 when TI's Japanese investment applications were published, some five years after their initial filing.[84] All Japanese semiconductor producers opted to oppose the TI application, and the "patent war" began in earnest. But whereas most Japanese chip makers were apparently willing to make some sort of deal, NEC—armed with its exclusive Japanese rights to the Fairchild IC patents—opted for an aggressive legal counterattack in Japan.

By 1966, however, Japanese firms had begun to eye the United States as a market for a new generation of consumer electronic products containing ICs. In the fall of 1966, after beginning shipments of a new radio containing ICs, Sony was forced to discontinue sales by the threat of patent infringement suits from TI.[85] In that same year, Hayakawa Electric (Sharp) began to mass-produce calculators.[86] The product appeared to have enormous export potential but would be vulnerable to legal attack

83. Mitsubishi decided, however, not to commercialize these experimental chips, and Sharp began to talk to foreign suppliers. The linkups with foreign suppliers of LSI chip sets for calculators are described below. "Japan: LSI Calculator," p. 308; and NHK, *Electronics-Based Nation*, vol. 3, p. 233.

84. Full patent rights were not granted to TI until 1989 (although some limited patent rights were approved in 1977). Andy Zipser, "Texas Instruments Gets Japanese Patent; Analysts See Sizable Addition to Revenue," *Wall Street Journal*, November 22, 1989, p. 1A.

85. Mason, *American Multinationals in Japan*, pp. 183–84.

86. Sharp's entry into calculators was closely linked to MITI policy. In the early 1960s Sharp had actually developed a prototype transistorized computer system but, on advice from MITI, decided not to enter the business. MITI instead had suggested that Sharp make smaller products using digital logic designs that could be mass-produced. By 1964 Sharp had developed its first transistorized calculator. Commercialization of desktop calculators was encouraged with subsidized loans in 1964–65; further investments in production facilities for calculators using ICs were made eligible for subsidized loans and special depreciation benefits from 1966 to 1970. See NHK, *Electronics-Based Nation*, vol. 3, pp. 129–39; and Majumdar, "Industrial Policy in Action," pp. 28–29.

in the United States, where TI's IC patents were recognized. These fears were realized when TI warned Sharp about patent violations in 1967.

MITI, alarmed by the threat to exports, finally responded to TI in August 1966 on its 1964 application for a Japanese affiliate. MITI offered to permit an equal partnership with a Japanese firm, provided that TI also agreed to license its patents to Japanese companies at reasonable rates and accept MITI restrictions on its output in the first three years of production.[87] TI turned down the offer. MITI nonetheless reportedly began to work on persuading NEC to drop its adamant opposition to the TI patents, and the agency secretly began to search for a possible Japanese joint venture partner for TI.[88]

Mitsubishi, supplier of the ICs used in the Sharp calculator (and the largest producer of calculator ICs in Japan), was extremely concerned, and the company sent a team to Texas to seek a solution to the Sharp-TI patent conflict.[89] Mitsubishi was particularly interested in a solution involving some manufacturing tie-up with TI because of its own weakness in leading edge semiconductor technology compared with other Japanese semiconductor producers.[90] Mitsubishi made some new and interesting proposals to TI and offered advice on negotiations with MITI, fully expecting that it would emerge with a new relationship with TI.[91]

At this point, however, MITI intervened. Rather than simply continue the exchange with Mitsubishi, TI's negotiating team in Tokyo was asked to speak with other Japanese companies. Sony soon came forward with a proposal for a temporary 50-50 joint venture under TI control, to be converted into a wholly owned affiliate at a later date, in exchange for giving Japanese firms the license to use the TI patent portfolio. MITI quickly agreed to go along with the Sony proposal. Mitsubishi was shocked to learn it had been unceremoniously dumped; from the TI perspective, it was all rather mysterious.[92]

87. NHK, *Electronics-Based Nation*, vol. 3, p. 229; Mason, *American Multinationals in Japan*, p. 182; and Nakagawa, *Semiconductor Development in Japan*, p. 159.

88. Nakagawa, *Semiconductor Development in Japan*, pp. 164–66.

89. Nakagawa, *Semiconductor Development in Japan*, p. 226, writes that Mitsubishi had 70 percent of the Japanese market for bipolar (the most important pre-MOS LSI technology) ICs used in calculators in Japan. The visit, at the end of September 1967, was "unannounced and unexpected and uninvited," according to one TI executive. Mason, *American Multinationals in Japan*, p. 184.

90. Nakagawa, *Semiconductor Development in Japan*, pp. 161–63.

91. See Nakagawa, *Semiconductor Development in Japan*, p. 163.

92. TI, in fact, never did figure out why Sony stepped forward and took the initiative. Mason, *American Multinationals in Japan*, pp. 184–85; 322–23, note 169.

In fact, MITI had orchestrated the Tokyo negotiations behind the scenes. MITI had approached Akio Morita of Sony and asked him to start a joint venture with TI for the sake of the future of the Japanese electronics industry.[93] In exchange for permission to enter Japan, TI negotiated a 3.5 percent royalty rate for the right to use its patents. This rate would apply to all Japanese semiconductor makers other than NEC: under continuing pressure from MITI to settle its dispute with TI, NEC finally agreed to a considerably reduced payment for use of the TI patents.[94]

This negotiated truce to the "patent war" was concluded in late 1967 and announced in early 1968. On the surface, the public settlement appeared to reflect TI's complete capitulation to MITI, since it had agreed to a 50-50 joint venture with Sony, indicated that it would consult with MITI on its production levels, and licensed its crucial patents to its Japanese rivals—elements all present in the 1966 MITI offer rejected by TI. In reality, however, in what may have been the first use of the tactic in a semiconductor trade negotiation, two "secret side letters" guaranteed TI its principal objective, namely, the right to set up a wholly owned Japanese affiliate. One of the letters, from the Ministry of Finance, stated that the interministerial committee it chaired would give "favorable consideration" to a TI purchase of Sony's equity share after three years; in the other letter Sony for its part agreed to sell its stake at that time.[95]

Battles in the LSI Market

The three-part program adopted by MITI in 1965–66—an outright ban on imports of advanced ICs, a hefty program of subsidized loans and tax breaks for capacity expansion, and a significant increase in semiconductor-related R&D subsidies—gave Japanese producers a substan-

93. Morita is quoted to this effect in Nakagawa, *Semiconductor Development in Japan*, p. 164. Sony was apparently tapped for a variety of reasons: MITI's lack of confidence in Mitsubishi's semiconductor technology capabilities, TI's history of confidence-building contacts with Sony, and a top MITI official's personal relationship with Sony's Morita (both came from the city of Nagoya).

94. See Nakagawa, *Semiconductor Development in Japan*, pp. 165–66; Mason, *American Multinationals in Japan*, pp. 185–86; 323, note 170. Actually, since the patents were not formally granted until much later, these royalties represented "advance payments."

95. Mason, *American Multinationals in Japan*, p. 186.

tial boost in the mid-1960s.[96] By 1969 imports accounted for less than a quarter of apparent Japanese IC consumption (see figure 2-4). But as it had before, rapid technological change was to again threaten these gains in domestic makers' market share.

Calculator sales had expanded rapidly in the late 1960s, and continued development of this product led to the next major crisis for the struggling Japanese semiconductor industry. In the late 1960s the technological focus was on replacing the relatively large number of chips in current models with a much small number of more highly integrated LSI chips. In the United States the production technology for LSI had developed rapidly, and American semiconductor manufacturers were sweeping world markets with low-cost LSI products.

Japanese semiconductor producers initially had a great deal of difficulty mastering LSI technology; they were also strapped financially by the cost of investing in the whole new generation of production facilities required to produce the more advanced chips. An advanced fabrication technology, MOS (metal oxide semiconductor) technology, was also becoming important in the manufacture of LSI chips, but the production process for MOS devices was very demanding. Japanese chip makers lagged well behind American companies in mastering it.[97] The costs of using patents owned by American chip makers put Japanese producers at a further competitive disadvantage, forced to pay about 10 percent of sales as royalties to their American competitors.[98]

96. My account in this section for the most part closely follows those of Nakagawa, *Semiconductor Development in Japan*, pp. 190–205, and NHK, *Electronics-Based Nation*, vol. 3, pp. 264–316, which are only available in Japanese.

A radically different Japanese view of the strength and behavior of Japanese semiconductor companies at the time of the "calculator wars" of the late 1960s and early 1970s has also appeared in the English-language literature. Note, in particular, the view expressed by Takuo Sugano (University of Tokyo) and Michiyuki Uenohara (NEC): "the Japanese coauthors assert that Japanese companies were ready sooner than the Americans to undertake large-scale commercial production of MOS ICs. NEC foresaw a potential mass market for the desk-top calculator, and, in cooperation with Hayakawa (the predecessor of Sharp), developed calculators using MOS ICs. They completed a commercial model in 1966, and the success of this venture helped to establish the practicality of the MOS IC." Michiyuki Uenohara, Takuo Sugano, John G. Linvill, and Franklin B. Weinstein, "Background," in Daniel I. Okimoto and others, eds., *Competitive Edge: The Semiconductor Industry in the U.S. and Japan* (Stanford University Press, 1984), p. 15.

97. See, for example, NHK, *Electronics-Based Nation*, vol. 3, p. 311; U.S. Department of Commerce, *Global Market Survey: Electronic Components* (1974), pp. 5–6, 81–82; Franco Malerba, *The Semiconductor Business: The Economics of Rapid Growth and Decline* (Frances Pinter, 1985), pp. 100–02, 151–52.

98. "IC Makers Shaken at the Dawn of IC Liberalization: Request for 7 Billion Yen

Japanese calculator makers, who appreciated the advantages of highly integrated MOS LSI chips for their products, had embarked on ambitious projects to redesign their calculators using the new technology. Although the domestic chip industry was receiving subsidies from MITI to develop the technology, progress was slow, and local manufacturers were unable to supply the chips needed by local calculator manufacturers. Calculator producers, led by Sharp, then turned to American IC suppliers, who—after initial reluctance—soon developed a thriving business providing MOS LSI calculator chips to the Japanese calculator industry.[99] In 1969 Sharp shipped the first LSI calculator, and a new cycle of rapid technical innovation hit the marketplace.[100]

Booming Japanese calculator production provided American companies with a rapidly growing market where—faced with an absence of domestic competition in MOS LSI—they were finally permitted to sell their products in volume. North American Rockwell (later renamed Rockwell International) had forged a supply relationship with Sharp, Canon struck a deal with Texas Instruments, and Ricoh turned to AMI.[101] In the early 1970s the rapidly growing consumer electronics market exploded, and foreign-made semiconductor sales in Japan soared. Figure 2-4 documents this surge in imports. By 1971 over 35 percent of

---

Subsidy and the Rush to Establish Their Own Technology for Mass Production," *Nihon Keizai Shimbun*, April 18, 1973.

99. A famous story often told in Japan relates how the chief executive of Sharp visited the United States in 1968 to work out an arrangement for an American chip company to supply MOS LSI chips needed in its new calculator designs. He visited eleven manufacturers, including Fairchild, TI, Motorola, AMI, National Semiconductor, RCA, Philco, and Sylvania. All rejected his proposal, because the volumes he required were too high for their current capacity, which was largely tied up with defense production. Finally, just as he was leaving the United States, executives at North American Rockwell had him paged at the Los Angeles airport, to tell him they had reconsidered their initial decision to reject his request and would work with him. The details of this story may be found in Nakagawa, *Semiconductor Development in Japan*, p. 192; NHK, *Electronics-Based Nation*, vol. 3, p. 264. The Rockwell-Sharp collaboration was to continue well into the early 1970s. See "How to Cut a Pocket Calculator in Half," *Electronics*, February 1, 1971, p. 104. Other Japanese calculator makers quickly followed Sharp in striking deals with American chip suppliers. Nakagawa, *Semiconductor Development in Japan*, p. 192.

100. Machinery Promotion Association Economic Research Institute, *Japan-U.S. Semiconductor Industry Research Survey*, p. 30.

101. See Nakagawa, *Semiconductor Development in Japan*, p. 195; NHK, *Electronics-Based Nation*, vol. 3, pp. 264–316.

Japanese IC consumption was imported. American chip makers then held three-quarters of the mushrooming Japanese MOS LSI market.[102]

First Steps toward Liberalization and "Dumping"

Calculator development continued to proceed apace, taking advantage of rapid improvements in chip manufacturing technology. By 1970 one-chip calculators had been introduced. One fruitful outcome of the collaboration between Japanese calculator makers and American chip suppliers was the development of the first single-chip microcomputer, in 1971.[103] But even as competitive pressure on Japanese firms intensified at the high end of the market, in calculator LSI, a new attack emerged at the low end.

Intense foreign pressure on Japan to liberalize its trade regime had mounted in the late 1960s and early 1970s. In September 1970 Japan removed older, less advanced ICs (chips with fewer than 100 circuit elements) from the "negative" list of products requiring prior approval for importation. Despite hefty tariffs (12 percent), imports came pouring in at prices more than 20 percent below previous levels in the Japanese market.[104] The competition was intensified by the severe recession the semiconductor industry experienced in 1970–71.

The "low price offensive"—American chip makers hawking these older logic chips at bargain prices—subjected their Japanese competitors to withering fire, at both the top (in calculator LSI chip sets) and the bottom (standard logic) ends of the market simultaneously.[105] This led to what were perhaps the first charges of "dumping" in U.S.-Japan IC trade.

102. Arthur Erikson and Charles Cohen, "Japanese Electronics Firms Search Out New Markets to Pierce Economic Fog," *Electronics,* November 22, 1971, p. 132.

103. See NHK, *Electronics-Based Nation,* vol. 3, pp. 318–21; and Robert N. Noyce and Marcian E. Hoff Jr., "A History of Microprocessor Development at Intel," *IEEE Micro,* vol. 1 (February 1981), pp. 9–13.

104. "Cheap IC Selling by U.S. Firms Posing Problem; Dumping Mooted," *Japan Economic Journal,* September 29, 1970, as excerpted in Semiconductor Industry Association (SIA), *Japanese Protection and Promotion of the Semiconductor Industry: Japanese Laws, Government and Industry Documents, and Press Reports Relating to Japan's Promotion of Its Semiconductor Industry, 1967–85* (Washington: Dewey, Ballantine, Bushby, Palmer and Wood, 1985).

105. Similar devastation in the market for standard logic occurred in Europe, where this was the period of the so-called logic wars. A price war broke out among American chip makers, inflicting a severe blow to European producers struggling to stay abreast of their American competition. See Malerba, *The Semiconductor Business,* pp. 110–19.

In the latter part of 1970 MITI and the Ministry of Finance launched informal investigations of possible American dumping of ICs in Japan. The intent was apparently to put pressure on American producers to mute unwanted price competition with Japanese manufacturers; officials suggested to reporters that a "dumping charge may rise to the surface if the cheap sale of ICs continues further."[106] The tough talk apparently had little effect on the hardheaded Americans; within months a continuing "low price offensive" by the American producers had forced Japanese producers to cut back production of these chips by 20 percent.[107] By early 1972 MITI was even floating the idea of a rationalization cartel covering logic ICs with fewer than 100 elements, but this apparently went nowhere.[108]

Meanwhile, even at the protected high end of the market—for highly integrated calculator LSI chips—product-starved Japanese calculator manufacturers braved MITI's wrath and red tape and fought for the right to import needed inputs. The issue, quite simply, was survival. U.S. IC manufacturers had crammed all the electronics for a calculator onto a single chip, and these new products had revolutionized the economics of the calculator business. Advances in ICs and reductions in the number of components used, not the wages of workers assembling hundreds of discrete components, became the key to competitive success. New American players, including some U.S. chip makers, even reentered the electronic calculator business, basing their designs on the revolutionary new chips now hitting the market, and substantially reduced the share of the American market captured by calculator imports.[109]

Chip prices dropped precipitously in the U.S. market, under the pressure of aggressive competition. The fierce competition was also exported across the Pacific. From spring 1970 to early 1971, the prices of imported calculator LSI chips dropped by half.[110]

106. "Cheap IC Selling."
107. "Tension Mounts from Planned U.S. Entry into IC Industry," *Japan Economic Journal*, April 27, 1971 (excerpted in SIA, *Japanese Protection*).
108. *Nihon Kogyo Shimbun*, January 8, 1972 (excerpted in SIA, *Japanese Protection*).
109. U.S. Department of Commerce, *The Impact of Electronics*, pp. 13–18.
110. A contemporary account noted that there was little Japanese chip producers could do about it, because many of the Japanese IC producers themselves purchased LSI chips and other specialized products in their capacity as manufacturers of computer and communications equipment, and because Japanese LSI chips were distinctly inferior to American ones. "Thus, stoppage of cheap U.S. imports by any means may invite opposition from Japanese calculator firms." "Prices of LSI Circuits Reportedly Reduced to Half

The "Calculator War"

Under sustained assault by American competitors, Japanese IC makers counterattacked. In the late 1960s and early 1970s the major producers all built new MOS LSI production lines.[111] Still, the new technology was difficult to master—costs were high and yields low. Japanese calculator makers were inclined to maintain their new relationships with American suppliers. Alarmed, the semiconductor producers branded the calculator manufacturers as unpatriotic; Sharp's chief executive came under particularly strong attack as "a traitor who would waste Japan's precious foreign currency."[112] Sharp was further criticized for taking MITI R&D money while buying foreign chips.[113]

MITI at that point made an example of Sharp by refusing to issue it necessary import licenses. Other calculator and office equipment makers joined Sharp in protesting, arguing that "we are buying parts from the U.S. because they cannot be made at home; we cannot do anything if we are not allowed to import."[114] For Sharp the lesson was clear: a necessary minimum amount had to be purchased domestically.

Even this was insufficient to mollify angry domestic semiconductor manufacturers, who faced a continuing crisis. Japanese chip makers continued to lag at least one or two years behind the state of the art in a very demanding new technology, and even as losses mounted, they were pressed for the financial resources to build new generations of production facilities.[115] The ravages of increased competition from imports were reflected in a rising import share (see figure 2-4) and a decline in the

by U.S. Makers," *Japan Economic Journal*, January 5, 1971 (excerpted in SIA, *Japanese Protection*).

111. NHK, *Electronics-Based Nation*, p. 311; U.S. Department of Commerce, *Global Market Survey, Electronic Components*, p. 82. Large investments in upgrading plants to produce CMOS (complementary metal oxide silicon) ICs were undertaken by most Japanese producers in 1972 and 1973. "IC Makers Shaken."

112. NHK, *Electronics-Based Nation*, vol. 3, p. 311 (author's trans.).

113. On the other hand, Mitsubishi, which was working with Sharp to manufacture ICs for a MITI-funded calculator project, actually declined to move to commercial production. See note 83.

114. Nakagawa, *Semiconductor Development in Japan*, pp. 196–97 (author's trans.).

115. For example, note the evaluations of the semiconductor makers' problems found in Nakagawa, *Semiconductor Development in Japan*, pp. 198–200, and U.S. Department of Commerce, *Global Market Survey, Electronic Components*, pp. 81–82. The *Nihon Keizai Shimbun* commented in 1973 that "it is estimated to take at least 2 to 3 years" for Japanese companies to come up with the technology needed to compete with U.S. IC producers. "IC Makers Shaken."

Table 2-3. *Sales by Japanese Semiconductor Companies to Affiliated and Nonaffiliated Companies, 1966–71*

Millions of yen unless otherwise specified

Sales	1966	1967	1968	1969	1970	1971 (estimate)
Inside company	230	984	4,860	14,576	24,780	28,223
Outside company	380	2,120	5,847	11,008	16,172	20,951
Total	610	3,104	10,707	25,584	40,952	49,174
Inside sales (percent)	37.7	31.7	45.4	57.0	60.5	57.4
Outside sales (percent)	62.3	68.3	54.6	43.0	39.5	42.6

Source: Machinery Promotion Association Economic Research Institution, *Japan-U.S. Semiconductor Industry Research Survey*, p. 124.

share of Japanese companies' chip output sold externally (that is, to customers other than in-house equipment-producing divisions). Table 2-3 shows that the external share of Japanese chip producers' sales dropped from 68 percent in 1967 to about 40 percent in 1970.

MITI ultimately came down firmly on the side of the Japanese chip producers (who were parts of the same companies producing computers, the centerpiece of MITI electronics policy) and severely chopped back licenses for imports of American calculator chip sets. In late 1971, even as Japanese calculator production headed toward a new annual peak of 1.3 million units, the head of the MITI electronics and electrical machinery division made clear the ministry's determination to choke off imports: "If we'd approved all the import applications, there would have been enough kits this year to make 4 million calculators."[116]

A typical experience was that of Fairchild, which in 1972 was about to sign a $500 million contract with a Japanese company, only to have MITI pull the plug on the deal.[117] Even successful import contracts faced a daunting gauntlet of bureaucratic obstacles. Top executives at Busicom, the Japanese manufacturer that commissioned Intel Corporation to design and build the first single-chip microprocessor for use in a calculator design, had to make dozens of daily trips to MITI, the Bank of Japan, and the Ministry of Finance in order to secure permission to pay Intel for its development cost, and had to plead with administrators to let the

116. Erikson and Cohen, "Japanese Electronics Firms," pp. 125, 132.
117. NHK, *Electronics-Based Nation*, vol. 3, p. 18.

company import a product deemed vital to its future. When the first prototype of the Intel design arrived in Tokyo, it took over a month for it to clear customs at Haneda airport.[118]

Less typical was the experience of Texas Instruments Japan, which in 1971, consummating its secret 1968 bargain with MITI and Sony, was transformed into the only wholly owned affiliate of an American semiconductor maker inside the Japanese market. In that same year TI Japan began shipping a one-chip calculator IC that sold like hotcakes—at its peak, 300,000 units per month.[119] Although ostensibly perched comfortably behind the protective walls thrown up around the Japanese market, TI continued to fight a silent guerrilla war against the bureaucracy. In 1972, for example, a top Japanese executive quit Mitsubishi and was recruited by TI. Shortly before he was to be hired, he was approached by MITI and warned that, if he joined TI, MITI "would act accordingly."[120] Heeding these warnings, he and others in his group joined another Japanese company.

Further Liberalization and Crisis

The MITI clampdown on calculator LSI imports was successful in providing domestic IC makers some degree of shelter from the competitive storm. The year 1972 marked a strong recovery in the semiconductor business, but by any measure it produced absolutely incredible results in the Japanese semiconductor market. Figure 2-7 presents the internal market share estimates of a Japanese semiconductor company and shows that the Japanese import market share in ICs plunged by almost 40 percent from the previous year, to 22 percent of total consumption. (See also figure 2-4, which portrays identical changes.) There are conflicting explanations for why this occurred. One obvious one is that, as documented above, MITI had openly launched a highly successful administrative offensive against imported ICs by late 1971, applying pressure to both exporters and their potential Japanese customers.

Another explanation, offered by an NEC semiconductor executive in a book that was widely read in Japan, is that the U.S. suppliers simply failed their Japanese customers. According to this account, defective

118. See NHK, *Electronics-Based Nation*, vol. 4, pp. 131–54.
119. NHK, *Electronics-Based Nation*, vol. 3, p. 325.
120. Nakagawa, *Semiconductor Development in Japan*, p. 229.

Figure 2-7. *Japanese Consumption of Integrated Circuits and Import Market Share, 1968–74*

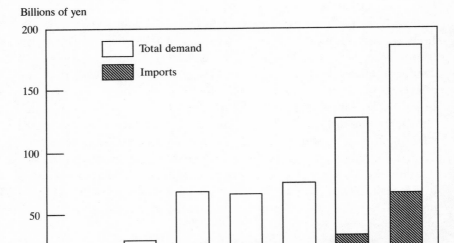

Billions of yen

Source: Contemporary internal estimate by a Japanese semiconductor manufacturer.

calculators returned to Japanese manufacturers around 1971 were found to have defective U.S. chips. Because U.S. IC makers refused to take their defective product back on the grounds that their contract did not require them to do so, the NEC executive alleged, Japanese customers learned that "it was much safer to do business with Japanese makers, who have better warranty and service."[121]

121. Author's trans. The NEC executive went on to blame the quality problem on cheap offshore assembly of these ICs in Southeast Asia: too much emphasis on cost, too little attention to quality control. Interestingly, no Japanese semiconductor *user* is quoted as supporting this version of events. Indeed, one user interviewed by Nakagawa describes the lesson taught over this period as being that a certain minimum amount had to be purchased from domestic chip makers, but notes that even this left Japanese semiconductor producers unhappy. Nakagawa, *Semiconductor Development in Japan,* pp. 197, 203–04.

But NEC had its own quality problems with LSI, according to other accounts in this same volume. After copying Rockwell chips as a last resort, NEC finally managed to develop its own ten-chip LSI calculator kit, but was unable to sell it. When a Japanese calculator producer finally tried to use it, the NEC LSI chips were found to be defective. Ibid., pp. 199–200.

Yet another explanation was offered by a former NEC executive in a 1991 interview with the author.[122] According to this account, U.S. producers simply failed to deliver the volumes Japanese customers demanded, and created a perception that American customers were being given first priority in a tight market. Product-starved Japanese users then "begged" Japanese IC makers to start producing needed chips, and ever since then Japanese calculator makers have relied on domestic producers.[123]

Admirers of the famous Japanese film *Rashomon* might be tempted to suggest, as did director Akira Kurosawa in that work, that there might be an aspect of truth in all three accounts. The only relatively certain facts are the mute percentages depicted in figure 2-7, and those numbers show foreign IC makers regaining a lofty 36 percent share of the Japanese market by 1974; this suggests that U.S. quality or delivery problems could not have been so severe as to permanently depress American companies' sales in Japan. Although it is not logically impossible for U.S. quality and delivery problems to have coincided with a MITI crackdown on calculator LSI chips in late 1971 and 1972, MITI officials, at least, are on contemporary record as opining that imports would have soared absent import restrictions.

But, then, why was import market share allowed to resume its upward course in 1973 and 1974? One important factor may have been that, early in 1973, the U.S. government substantially increased political pressure on Japan to continue liberalizing IC imports. In February of that year the U.S. Special Representative for Trade threatened to bring a case for adjudication by the General Agreement on Tariffs and Trade (GATT), in which it seemed inevitable that Japan would lose.[124] In March 1973 the

122. Author's interview with former NEC executive, March 1991.

123. But a third NEC executive offered a rather different perspective on imported computer ICs in 1974. Speaking at a panel discussion, NEC's executive director remarked that "in the past, Japanese minicomputer makers have relied on American ICs, and MITI gave administrative guidance, putting these things on the negative list . . . if ICs for microcomputers were not placed on the negative list, capital investment would not have been possible by Japanese makers. This meant that since MITI put up the negative list and gave administrative guidance, it was possible for us for the first time to stand on our own feet." "Plunging Into IC Liberalization," *Nikkan Kogyo*, December 12, 1974 (excerpted in SIA, *Japanese Protection and Promotion*).

124. "U.S. Presses Japan with Appeal to GATT for Liberalization of Electronic Computers and IC; Heading toward Majority Vote; Notification Given at Time of Special Representative Eberle's Visit to Japan," *Sankei Shimbun*, March 1, 1973 (excerpted in SIA, *Japanese Protection*). The United States had threatened to file a complaint with the GATT

U.S. government made IC imports a major issue at a meeting of the Organization for Economic Cooperation and Development (OECD) in Paris.[125]

In short, Japanese policies in computers and semiconductors were being subjected to increasingly intense scrutiny from abroad. The outside world continued to put pressure on the Japanese government to further liberalize its economy, and throughout the early 1970s it debated how to cope with these foreign demands.[126] In computers and in its sister sector, integrated circuits, the debate was particularly acrimonious. IBM had announced a new generation of computer systems using advanced LSI components (the System 370 series) in 1970, and once again the Japanese computer industry perceived itself under fierce assault. The continuing pressure to liberalize the domestic market, coupled with accelerating technological competition, combined to create an atmosphere of acute crisis for the Japanese electronics industry.

MITI had taken the initiative in computers in 1971. Warning that foreign machines would sweep through the Japanese market, wipe out domestic computer makers, and leave "knowledge-intensive industries . . . seized by foreign capital," and predicting that the "center of Japan's nerve system will be controlled by foreign capital," MITI successfully argued for delay in full liberalization of foreign investment in computers until 1976, and postponement of removal of restrictions on computer imports until 1975.[127] In exchange for this protection, at MITI's insistence

---

if the liberalization were delayed, and U.S. officials publicly complained before the OECD. "Liberalization of Electronic Computers and ICs Generally Decided; MITI to Hasten Implementation Measures," *Asahi Shimbun*, March 1, 1973; "Industrial Circles' Plan Out of Question—Liberalization of Electronic Computers, Eberle Expresses Strong Dissatisfaction," *Sankei Shimbun*, March 27, 1973; and "Import Liberalization during 1975; Cabinet Decision on Electronic Computers Expected Today," *Nihon Keizai Shimbun*, June 15, 1973 (all excerpted in SIA, *Japanese Protection*).

125. "Industrial Circles' Plan Out of Question—Liberalization of Electronic Computer; Eberle Expresses Strong Dissatisfaction," *Sankei Shimbun*, March 27, 1973 (excerpted in SIA, *Japanese Protection*).

126. Japan had accepted Article XI of the GATT in 1963 and Article VIII of the Articles of Agreement of the International Monetary Fund (IMF) in 1964, and had joined the OECD in 1964. A three-year phased program calling for removal of some quantitative restrictions and partial capital liberalization had been introduced in 1968, and a further phased liberalization of trade and foreign investment was announced in 1971. See Krause and Sekiguchi, "Japan and the World Economy," pp. 425–30; and Itoh and Kiyono, "Foreign Trade and Direct Investment," pp. 164–66.

127. "Contents of MITI Views on Electronic Computer Industry," *Asahi Shimbun*, July 13, 1971 (excerpted in SIA, *Japanese Protection*).

the six major Japanese computer manufacturers were combined into three groups of two companies each, with each group to develop complementary models of advanced computer systems. This cartelization of computer production was to be supported by a generous MITI subsidy to cooperative research efforts in each of the three groups. The R&D subsidy program, dubbed the "3.5 generation" program, targeted the development of commercial products over the 1972–76 period.[128]

The three groupings of firms showed varying degrees of willingness to work together. Fujitsu and Hitachi, organized into the "Ultra Advanced Computer Development Technology Association," received a 27-billion-yen government subsidy over the fiscal 1972–76 period and formed a joint venture to produce IBM-compatible computer peripherals and terminals (Nippon Peripherals). That venture remained active until 1986.[129] Semiconductor R&D was one of the activities funded by this subsidy, but actual research topics were assigned by the association, then carried out (except in their joint venture in peripherals) within each firm separately. The avenues and extent of technical information sharing between companies are little discussed but could not have been great—after a joint product announcement in 1974, the two companies were soon locked in head-to-head competition, with each announcing models that targeted markets ostensibly assigned to the other.

The Mitsubishi-Oki tie-up was a substantially smaller-scale affair. The two companies, formed into the "Ultra Advanced Computer Technology Research Association," received 6.4 billion yen in subsidies over fiscal 1972–76.[130] Both companies had dropped out of the mainframe computer business by the mid-1970s.

Perhaps the tightest of the three groupings was that formed by NEC and Toshiba. Their New Computer Series Technology Research Association was given a government subsidy of 21.6 billion yen from 1972 to fiscal 1976.[131] NEC and Toshiba actually formed a joint venture company,

128. The name alluded to the view that IBM's System 370 was only half a generation more advanced than its so-called third-generation System 360. For further details, see Flamm, *Creating the Computer,* pp. 193–95; *Targeting the Computer,* pp. 135–36.

129. The Fujitsu-Hitachi research association spent a total of 88 billion yen over the 1972–76 period, of which 61 billion yen was contributed by the companies. In addition, another 67 billion yen (entirely contributed by the companies) was spent over the 1975–81 period. See *Thirty-Year History,* p. 77; Flamm, *Creating the Computer,* pp. 194–95.

130. The total budget for the research association was 22.2 billion yen over fiscal 1972–76, plus 3.4 billion yen over an additional period extending through fiscal 1981. A government subsidy was only granted during the first period. *Thirty-Year History,* p. 79.

131. The research association's total budget was 60.5 billion yen from 1972 through

NEC Toshiba Information Systems (NTIS, which became a member of their research association), to develop and market a line of computers. R&D into ICs was one of the areas in which the research association was active. Although the R&D was assigned to research groups established within individual member firms, at least some research topics were in part jointly investigated.[132]

This aid to computer firms was designed as short- and medium-term assistance to help them cope with liberalization, and as a stimulus to greater interfirm cooperation.[133] At the same time, a new round of resources was injected into longer term R&D, under the direction of MITI's Electrotechnical Laboratory. Another large-scale national R&D project in computers, the "Pattern Information Processing System" (PIPS), was launched in 1971. R&D on MOS LSI design and manufacturing technology, including the design and manufacture of a sixteen-bit single-chip microprocessor (demonstrated in 1977),[134] 1K fast and 4K slow static RAM chips, and a 16K DRAM, were carried out as part of the first portion (1971–75) of the PIPS program.[135] About 3.5 billion yen was expended in the area of "devices and materials" by the ETL and its private contractors as part of the program over 1971–75.[136] Almost 90

---

fiscal year 1976; 19.5 billion yen was spent over an "additional development" period from 1977 through fiscal 1981. *Thirty-Year History,* p. 78.

132. *Thirty-Year History,* p. 78.

133. A call by MITI Minister Yasuhiro Nakasone for "promotion of further tieup than the present three groups" in early 1973, for example, was explained to reporters by MITI officials to mean that these firms "cannot stand competition with the huge manufacturers of the United States, unless they deepen mutual tieup in commonization of the software and in taking shares in production and joint development of peripheral apparatus, and unless they either unified into a home-production manufacturer like Britain's ICL in the future or are concentrated at least into a group for specific purposes (Oki-Mitsubishi) and an alignment of electronic computers proper (remaining four companies). Explaining the Minister's statement of the 7th to the press, a MITI leader said, 'For the present, this does not necessarily mean unification of industry circles, but the emphasis of the administrative guidance lies in promoting tieup among the three groups.'" "Cooperation in Concentration Sought—Statement by MITI Minister," *Asahi Shimbun,* March 8, 1973 (excerpted in SIA, *Japanese Protection*).

134. NEC was the first Japanese company to announce a commercial sixteen-bit microprocessor, in 1978. Machinery Promotion Association Economic Research Institute, "Japan-U.S. Semiconductor Industry Research Survey," p. 30.

135. See Electrotechnical Laboratory, *Pattern Information Processing System: National Research and Development Program, 1978* (Tokyo, 1978), pp. 27–30; and Ishii, "Research and Development on Information Processing Technology."

136. These are my estimates derived from graphs found in ETL, *Pattern Information Processing System,* p. 5. This compares with total spending of 17.4 billion yen over the years 1971–78.

percent of the semiconductor funding went into contracts with Japanese companies; NEC, Hitachi, and Toshiba were the contractors on the LSI chips.[137] NEC was later to become the first (in March 1977) and largest-volume Japanese producer of the 16K DRAM; in 1980 Toshiba was to market a commercial version of the sixteen-bit microprocessor it had developed under contract for the PIPS project.[138]

Integrated circuits also benefited from policy changes set in motion in 1971, by being exempted from immediate liberalization, although the delay was less: 1974 for both liberalization of foreign investment (100 percent owned subsidiaries automatically approved) and removal of residual licensing restrictions on imports of ICs with more than 200 elements.[139] Unlike in computers, however, no major plan to subsidize the technological development of ICs apart from activities within the 3.5 generation and PIPS programs was announced, reaffirming that policy in semiconductors was viewed by MITI primarily as a component of computer policy.

As the semiconductor manfacturers' crisis deepened in 1971, however, MITI began to consider more direct measures designed specifically to help the industry cope with the effects of accelerating technological competition and the continuing effects of the partial liberalization of 1970. In the fall of 1971 a plan to support the establishment of a joint IC design venture among major domestic IC manufacturers was floated and then abandoned.[140] In late 1971 an industry group considered the formation of a cartel among Japanese IC makers, to restrict production and allocate products among potential producers. For a period in early 1972 MITI

137. ETL, *Pattern Information Processing System*, pp. 3, 5. The research themes (and contractors) in devices and materials were semiconductor lasers (NEC, Toshiba), reversible photosensitive materials (Sanyo, Fujitsu, Konishiroku), spatial modulation devices (Hitachi, Matsushita, Hoya Glass), magnetic bubble domain devices (Hitachi, Hitachi Metal, NEC, Tohoku Metal), and large-scale integrated circuits (Toshiba, Hitachi, NEC).
138. The chronology given in Machinery Promotion Association Economic Research Institute, "Japan-U.S. Semiconductor Industry Research Survey," p. 31, shows NEC as the first company to sell a 16K DRAM, in March 1977. On Toshiba's commercialization of its PIPS microprocessor technology, see Hajime Iizuka, "Design and Implementation of a Microprocessing Unit with a Flexible Architecture," in T. Kitagawa, ed., *Computer Science and Technologies, 1982* (Tokyo and Amsterdam: Ohm and North-Holland, 1982), pp. 22–38.
139. Import quotas on ICs with 200 elements or fewer were removed in 1973. U.S. Department of Commerce, *Global Market Survey: Export Opportunities for Electronics Industry Production and Test Equipment* (GPO, 1974), p. 69.
140. "MITI Envisages Establishment of Joint Firm of Makers for Blueprinting ICs," *Japan Economic Journal*, August 3, 1971 (excerpted in SIA, *Japanese Protection*).

apparently supported the cartel concept for more mature products.[141] By the spring of 1972, however, the idea was abandoned after encountering bitter dissension within the industry over the fine points of the proposed cartel's organization.[142]

Continuing attempts to delay the impending liberalization were decisively defeated by mid-1973; restrictions on imports of ICs with fewer than 200 elements were also removed in that year. IC imports subsequently jumped sharply, almost back to their 1971 peak of more than a third of Japanese consumption (see figures 2-4 and 2-7). MITI Minister Yasuhiro Nakasone—the same Liberal Democratic Party (LDP) politician who years later was to be chosen prime minister—floated the idea of further consolidation of the three sets of computer groupings into a single national computer firm, along lines followed in Europe (MITI officials actually mentioned Britain's ICL as a model).[143] As before, however, the large electronics companies resisted the idea of a forced merger of their computer divisions into a single national champion.[144]

The political forces calling for consolidation of activities among companies in response to imminent liberalization widened their focus, to ICs, shortly thereafter. Industry had called for a delay in liberalization and a new 7-billion-yen subsidy for semiconductor technology development.[145] Instead, a new MITI proposal for IC manufacturers to reorganize production, with each specializing in particular types of chips, was revealed in mid-March 1973 (just before the next fiscal year was to begin).[146] In June, after the Japanese cabinet refused to further delay liberalization,

141. *Nihon Kogyo Shimbun*, January 8, 1972 (excerpted in SIA, *Japanese Protection*).

142. A news report on the proposal, considered within Japan's Electronics Industry Association, specifically notes that "trying to find out which maker is superior in technology stands to be difficult owing to the swift pace of technology in this field. As such, [industry quarters] felt that reaching a conclusion on such aspects was going to take some time." "Electronics Industry Plans Cartel for Types of Products Using IC," *Japan Economic Journal*, December 14, 1971 (excerpted in SIA, *Japanese Protection*). By March 1972 the idea of a special Kidenho cartel had been abandoned, with firms disagreeing over how to determine which producer would manufacture what types of ICs, and the four top companies resisting the cartelization of more advanced chips. "Cartel Controlling IC Production Types Will be Difficult to Implement for FY1972," *Nikkan Kogyo Shimbun*, March 23, 1972.

143. "'Cooperation in Concentration Sought,' Strengthening of Tieup among Three Groups—MITI Policy," *Mainichi Shimbun*, March 8, 1973 (excerpted in SIA, *Japanese Protection*).

144. "Will Not Carry Out Another Reorganization—Electronic Computer Industry Circles," *Nihon Keizai Shimbun*, March 8, 1973 (excerpted in SIA, *Japanese Protection*).

145. "IC Makers Shaken."

146. "MITI Heading toward Reorganization of IC Enterprises; Adjustment of Produc-

"shocked" IC producers were informed of MITI's decision to begin re-organizing their ranks.[147] The companies instead proposed increased subsidies and the enactment of an antidumping law to deal with anticipated imports of low-priced U.S. chips.[148]

What finally emerged at the end of 1973 out of the blizzard of proposals and counterproposals from the LDP politicians, MITI, and the electronics industry was a compromise—some degree of interfirm cooperation in R&D in ICs, in exchange for sharply increased subsidies. Even earlier, NEC and Toshiba, which had been teamed by MITI in the New Computer Series Technology Research Association, had yielded to the pressure and agreed to develop computer memories jointly as part of their common computer effort; bitter rivals Hitachi and Fujitsu, joined by a shotgun marriage in another development group, had publicly committed themselves to joint development of ICs to be used in their computer products (although they apparently did not follow through).[149] As its contribution to the 1973 compromise, MITI unveiled a two-year, 35-billion-yen ($129 million) package of development subsidies for the IC industry, organized into five areas. Three themes involved specialized manufacturing technologies for LSI ICs used in computers: NMOS (negative metal oxide semiconductor), CMOS (complementary MOS), and silicon gate MOS LSI. The remaining two projects were to develop linear ICs for use in industrial applications. Two of these projects were to be organized as joint development efforts among several companies.[150]

tion Fields Planned for Six Exporter Firms in Preparation of Liberalization," *Nihon Keizai Shimbun*, March 15, 1973 (excerpted in SIA, *Japanese Protection*).

147. "IC Industry Circles Shocked; Caught Between U.S. Offensive and Liberalization," *Nihon Keizai Shimbun*, June 15, 1973; and "IC Industry Reorganization Adjusting Production Areas," *Nihon Keizai Shimbun*, June 15, 1973, p. 6 (both in SIA, *Japanese Protection*).

148. "IC Industry Circles Desire Establishment of Legislative Measures by Government—Prevention of Selling at Low Price," *Nihon Keizai Shimbun*, June 16, 1973 (excerpted in SIA, *Japanese Protection*).

149. "Three Electronic Computer Groups Strengthening Tie-Up, Planning on Joint Development of ICs and Mutual Supply of Equipment," *Nihon Kogyo Shimbun*, June 25, 1973 (excerpted in SIA, *Japanese Protection*). NEC and Toshiba apparently did do joint R&D, whereas Hitachi and Fujitsu apparently did not. An official history of Japanese technology research associations notes that in the NEC-Toshiba grouping, an R&D section was established "in each member firm, which independently and jointly (partially) pursued its assigned topic." In the Ultra Advanced Computer Development Technology Research Association (the Fujitsu-Hitachi grouping), in contrast, a research system was established "within each company and engaged in a topic assigned to each." *Thirty-Year History*, pp. 77–78.

150. "IC Industry Looks to the Government for Aid; Is It Quick Remedy or Only

The structure of the "IC Liberalization Countermeasure" aid program of 1973–74 was intended to stimulate the reorganization of the IC industry, but it had only limited success in encouraging the desired trend.[151] Eight of Japan's twelve domestic IC producers were involved in the program, but only five were finally organized into joint development efforts. Two of these companies, Mitsubishi and Oki, had already been cooperating in the development of ICs for use in computers as part of the computer liberalization subsidy program; the IC program extended this cooperation into NMOS LSI technology development. Joint efforts apparently included creation of a "semiconductor consultative council" to promote joint IC development and exchange technical information, and joint development of an NMOS LSI microprocessor.[152] Similarly, the second group, composed of Fujitsu, Sharp, and Kyodo Electronics Laboratories (Kyodo, still the only pure merchant Japanese producer, was a supplier to both Fujitsu and Sharp) formed a "summit conference" and a joint technology committee to promote joint development and information exchange.[153]

Narrow Escape?" *Nihon Keizai Shimbun,* November 5, 1973; "IC Subsidies to be Granted to 8 Companies; Mitsubishi-Oki and Fujitsu-Sharp-Kyodo Electronic Groups Come to Fore; Hitachi, Toshiba, and Japan Electric to Carry Out Unilateral Development," *Nihon Kogyo Shimbun,* November 29, 1973 (both excerpted in SIA, *Japanese Protection*). One of the joint development efforts was the work on NMOS LSI, carried out by Mitsubishi and Oki; the other was development of multipurpose linear ICs for use in industrial applications, jointly undertaken by Fujitsu, Sharp, and Kyodo Denshi Gijutsu Lab. See *Computer White Paper, 1974 Edition,* p. 22.

151. "MITI expects that groups of companies engaging in joint development will maintain their cooperative structure in other fields henceforth without sticking to those kinds of items which became the objects of a subsidy." "IC Industry Circles to Make Every Possible Effort for Development of Technology; Subsidy Amounting to 1.8 Billion Yen for Fiscal 1974 to be Given Shortly to Eight Companies, Including Hitachi and Toshiba," *Nikkan Kogyo Shimbun,* March 20, 1974. See also "IC Subsidies to be Granted," *Nihon Kogyo Shimbun,* November 29, 1973; and "Joint Development for IC New Model Becomes Active; Mitsubishi-Oki; Fujitsu-Sharp, etc.," *Nihon Keizai Shimbun,* January 6, 1974 (all excerpted in SIA, *Japanese Protection*).

152. Mitsubishi and Oki were already cooperating on development of ICs for use in computers as part of their participation in the 3.5 generation program. Joint work on their IC promotion subsidy project—development of an NMOS microprocessor LSI—was to be divided, with Mitsubishi developing the LSI chip itself and Oki the low-cost package. *Nihon Kogyo Shimbun,* February 19, 1974 (excerpted in SIA, *Japanese Protection*).

153. *Nihon Kogyo Shimbun,* February 19, 1974; and "IC Industry Circles to Make Every Possible Effort for Development of Technology; Subsidy Amounting to 1.8 Billion Yen for Fiscal 1974 to Be Given Shortly to Eight Companies, Including Hitachi and Toshiba," *Nikkan Kogyo Shimbun,* March 20, 1974 (both excerpted in SIA, *Japanese Protection*).

On Christmas Day 1974, MITI finally delivered on its promise to liberalize IC imports by the end of 1974. At almost the last possible moment, ICs with more than 200 circuit elements (other than computer ICs) were finally freed from licensing requirements.[154] The Japanese were far from confident of the ability of their producers to compete freely with imports; it was generally believed that the domestic industry lagged a year or two behind the foreign competition in technological terms.[155] A monthly publication of the Electronics Industries Association of Japan pessimistically observed in March 1975 that, although Japanese producers had caught up with the United States in production of MOS LSI for electronic calculators, "the Japanese industry is thought to be two or three years behind the U.S., even though it has been protected by Government measures."[156]

Japan had also largely liberalized foreign investment in ICs in 1974, and firms were now for the most part free to set up wholly owned subsidiaries. The process of liberalization was finally completed in 1975, when the last import restrictions on ICs used in computers were lifted. Wholly owned foreign investment in computer chips was not permitted until the very end of that year (table 2-4). Liberalization of capital investment was not entirely unencumbered: hostile acquisitions of existing firms, for example, were still barred, and advance notification—and prior approval—of investments were still formally required.[157] This was not to change until the foreign investment law was revised in 1980. But by the mid-1970s the formal doors to the Japanese market had finally swung open for U.S. semiconductor companies. (It should be noted, however, that though imports could now enter without a license, a stiff import duty of 12 percent, as opposed to 6 percent in the United States, was levied on ICs through the end of the decade.)

---

154. "Overall Import Liberalization Tomorrow, MITI Aid to Industry to Continue," *Nihon Keizai Shimbun*, December 24, 1974 (excerpted in SIA, *Japanese Protection*).

155. "Do We Advance toward a Reorganization of the IC Industry?" *Nihon Keizai Shimbun*, December 24, 1974; and "Consideration of an Emergency Tariff," *Nikkan Kogyo Shimbun*, December 21, 1974 (both excerpted in SIA, *Japanese Protection*).

156. "Competition May Short-Circuit Japanese IC Industry," *Journal of the Electronics Industry of Japan*, March 1975 (excerpted in SIA, *Japanese Protection*).

157. Although Japan had made international commitments not to obstruct such investment, it retained the formal internal legal power to block investment. Author's interview with Mr. Nishibe, MITI, and Mr. Sakamoto, Ministry of Finance, Tokyo, March 1992.

Table 2-4. *Japanese Measures to Liberalize Trade and Investment in the Electronics Industry, 1962–76*

Products	Capital liberalization		Import liberalization	Technology transfer liberalization
	50 percent	*100 percent*		
Electronic computers				
Major components	8/4/1974[a]	12/11/1975[a]	12/24/1975	7/1/1974
Peripheral device	7/1/1974	7/1/1974	7/1/1974	7/1/1974
Memory or terminal	8/4/1974	12/11/1975	12/24/1975	7/1/1974
Other parts[b]	8/4/1974	12/11/1975	2/1/1972	7/1/1974
Other components	8/4/1974	12/11/1975	12/24/1975	7/1/1974
Software	12/1/1974	4/1/1976	[c]	7/1/1974
Integrated circuits				
Fewer than 100	8/4/1971[a]	12/1/1974[a]	9/1/1970	6/1/1968
Fewer than 200	8/4/1971	12/1/1974	4/19/1973	6/1/1968
More than 200	8/4/1971	12/1/1974	12/25/1974	6/1/1968

Source: Kenneth Flamm, *Targeting the Computer: Government Support and International Competition* (Brookings, 1987), p. 254; and MITI Machinery and Information Industry Bureau, *Current Condition of the Electronics Industry* (Tokyo, September 1985), p. 8 (in Japanese).
 a. Including integrated circuits for computer.
 b. Input device, output device, communications control, and so forth.
 c. No quotas applied.

Into the Japanese Market

Even before it had become clear that political pressure was finally about to succeed in pushing open the door into the Japanese market, American semiconductor firms had begun to position themselves for the big event. Fairchild (in 1969) and National Semiconductor (before 1972) had attempted an end run around investment restrictions by setting up wholly owned manufacturing affiliates in U.S.-occupied Okinawa, where the investment regime was more liberal than on the main islands of Japan. The companies were gambling that, when Okinawa reverted to Japan in 1972, they would be permitted to retain their wholly owned Okinawa manufacturing affiliates (despite a legal maximum of 50 percent foreign ownership in IC manufacturing affiliates after the initial round of liberalization in 1971) and achieve unhindered access to the Japanese market.[158] The bet failed to pay off. National was forced to close its operation, while Fairchild was compelled to enter into an ill-starred 50 percent joint venture with TDK Electronics, which finally closed in 1977 after years of problems.

158. This episode is recounted in Mason, *American Multinationals in Japan*, pp. 222–23.

Motorola, the second-largest global vendor of semiconductors (behind Texas Instruments) in the early 1970s, also met with frustration in the 1970s and early 1980s. It had established a sales office in Tokyo in 1962 but found itself unable to market its products to Japanese customers without a Japanese manufacturing facility.[159] Reluctant to alter a company policy of avoiding joint ventures in selling its high-technology products in overseas markets, Motorola nonetheless had decided to take the plunge into a 50 percent–owned Japanese joint venture in semiconductors in 1971, when Japanese policy was changed to permit such ventures. After much work, a joint semiconductor manufacturing venture with Japanese audio producer Alps was set up in 1973, but it collapsed in 1975 during an economic downturn, leaving Motorola again without a local factory. Gun-shy, Motorola went back to a pure sales effort until 1980, when it decided to acquire Toko, a small domestic producer of electronic components.[160] Advised by Japanese political and business circles to go slow on outright acquisition, Motorola formed a 50 percent joint venture with an agreement to acquire Toko's equity within three years, in a deal reminiscent of TI's 1968 bargain with MITI and Sony. The wholly owned subsidiary was formed after two years and renamed Nippon Motorola, but sales growth was mostly disappointing. Despite twelve years of investment efforts, Motorola still felt it was stuck outside the core of the Japanese electronics market in the early 1980s.

Despite their frustrating track record in joint ventures, American firms continued to make forays into the Japanese semiconductor market throughout the 1970s. In 1971 AMI set up a marketing office. Advanced Micro Devices set up a Japanese sales office in 1974, followed by Signetics and ITT in 1975, Intel in 1976, and Zilog in 1977.[161] The head-on confrontation with the American IC makers was clearly building toward a climax after 1975, when the last residual restrictions on trade and investment disappeared.

159. This account of Motorola's experiences is drawn from Mason, *American Multinationals in Japan,* pp. 220–31; National Research Council, *U.S.-Japan Strategic Alliances in the Semiconductor Industry: Technology, Transfer, Competition, and Public Policy* (Washington: National Academy Press, 1992), pp. 91–101; David B. Yoffie and John Coleman, "Motorola and Japan (A)," Harvard Business School case 9-388-056 1987, rev. 1989.

160. Both Toko and Alps had lost assets in semiconductor manufacturing when Kyodo Electronics Laboratories, the startup founded as a joint venture by American Bernard Jacobs and seven Japanese companies in 1964, disappeared in the mid-1970s.

161. *Japan Electronics Almanac 1983* (Tokyo: Dempa Publications, 1983), p. 202.

The rising crescendo of concern over imminent liberalization jumped to a higher level of tension in the mid-1970s, when a new "technological shock" was revealed. In 1974 senior officials at NTT became aware of two startling technical developments in the United States. An antitrust case against IBM, launched by Telex in 1973, had resulted in public disclosure of IBM's plans to develop a so-called "Future System" computer in the early 1980s, using a 1-million-bit (1-megabit, or 1M) IC memory chip, at a time when the state of the commercial art was a chip holding only 1,000 bits.[162] Furthermore, on a visit to AT&T's Bell Labs, the president of NTT had observed an experimental system to etch the finely detailed masks used to make very large scale integration (VLSI) chips with small enough elements to pack onto a megabit memory. Shocked into action by how far ahead American semiconductor R&D appeared to be, NTT decided in 1975 to launch a major technological initiative designed to close the technology gap.[163]

NTT Arrives on the VLSI Scene

Although MITI was the formal bureaucratic patron of the Japanese IC industry, NTT had, since its establishment in 1952, played an important (though considerably less visible) role in supporting the domestic industry. Barred legally from manufacturing its own equipment (as AT&T, its American counterpart, was not), the Japanese telephone company—organized as a special, public corporation—had long been engaged in electronics R&D, transferring the resulting technology to and developing products with its Japanese suppliers in the private sector. Fueled by subsidized loans from Japan's public postal savings bank system and the compulsory purchase of its bonds by new telephone subscribers, NTT had enormous financial resources.

162. NHK, *Electronics-Based Nation*, vol. 4, p. 356.
163. Anchordoguy, *Computers Inc.*, p. 139, drawing on published interviews with NTT President Yonezawa; and Yoshio Nishi, "The Japanese Semiconductor Industry," in Björn Wellenius, Arnold Miller, and Carl J. Dahlman, eds., *Developing the Electronics Industry* (Washington: World Bank, 1993), pp. 123–30; and the unpublished slides from the oral talk on which this work is based. See also VLSI Technology Research Association, *VLSI Technology Research Association: Retrospective of the Past Fifteen Years* (Tokyo, 1991), pp. 80–81. Fransman, *The Market and Beyond*, p. 57, and "NTTPC and Computer Makers Plan Super LSI to Curb IBM Advance," *Japan Economic Journal*, April 15, 1975 (excerpted in SIA, *Japanese Protection*), contain variants on this basic story.

Until the early 1980s, when it became the subject of protracted trade negotiations between the United States and Japan, NTT's procurement was effectively closed to non-Japanese suppliers.[164] Indeed, even many Japanese suppliers were frozen out of NTT contracts, which for the most part went to a small group of favored companies, the so-called "Den-Den" family of suppliers. Within this small group, an even smaller group of four principal suppliers—NEC, Fujitsu, Hitachi, and Oki—received the bulk of the NTT contracts. In 1968, for example, these four firms accounted for 70 percent of NTT's procurement.[165] This share was to decline over the next decade, but remained around 50 percent in the early 1980s.[166]

NTT's support for Japanese electronics manufacturers has been organized quite differently from that provided by MITI. NTT's large internal research laboratories conducted basic research and transferred the results of that research to selected suppliers. The transfer of technology typically occurred through joint projects with suppliers to develop equipment using new technology, and through the exchange of technical personnel between supplier companies and the NTT laboratories.[167] Support for R&D by its suppliers has not taken the form of contract R&D performed by these companies for NTT.[168] Instead, funding for supplier R&D has been structured as a part of development and procurement

164. Indeed, the president of NTT's widely quoted response to some of these foreign pressures was to declare that "the only thing we could consider buying overseas would be [telephone] poles and mops." "High Technology Gateway: Foreigners Demand a Piece of NTT's $3 Billion Market," *Business Week*, August 9, 1982, pp. 40–44.

165. Anchordoguy, *Computers Inc.*, p. 42, citing a Japanese study.

166. See U.S. International Trade Commission, *Foreign Industrial Targeting and Its Effects on U.S. Industries, Phase 1: Japan* (Washington, 1983), p. 152; and "High Technology Gateway," pp. 42, 44. Data in these sources show NEC, Fujitsu, Oki, and Hitachi accounting for between 42 and 49 percent of NTT's purchases over the years 1980–82. Companies affiliated with NEC and the Sumitomo trading group (to which NEC belongs) account for another 5 percent of NTT procurement, pushing the big four share above 50 percent. In 1981 NEC was the largest single supplier, accounting for about 20 percent (25 percent with other Sumitomo group affiliates added on) of NTT purchases. Fujitsu accounted for about 13 percent of purchases, Oki 7 to 8 percent, and Hitachi 6 to 7 percent.

167. Kozo Yamamura, "Joint Research and Antitrust: Japanese vs. American Strategies," in Hugh Patrick, ed., *Japan's High Technology Industries: Lessons and Limitations of Industrial Policy* (University of Washington Press, 1986), p. 194; and Jon Sigurdson, *Industry and State Partnership in Japan: The Very Large Scale Integrated Circuits (VLSI) Project* (Lund, Sweden: Research Policy Institute, 1986), p. 100.

168. Published R&D survey statistics, for example, confirm that only relatively small amounts could have been paid out to external organizations for contract R&D in the 1970s and 1980s. See Flamm, *Targeting the Computer*, p. 139, note 19.

contracts for new equipment.[169] Patents resulting from these joint development projects have generally been jointly owned by NTT and the company concerned, so that companies have been free to use technologies developed with NTT assistance.[170]

NTT's early role in Japanese transistor R&D (it boasts of having built the first Japanese germanium transistor, in 1952) as well as the manner in which significant know-how was transferred to the private sector has already been noted. NTT largely focused its semiconductor research through the mid-1960s on the conservative goal of improving transistors used in transmission systems. In the late 1960s, however, NTT began research on electronic switching. Its first studies of the use of ICs were not undertaken until 1965, when a project to design a family of IC logic chips for use in switching systems was begun.[171]

Within the large semiconductor makers, all of which were Den-Den family suppliers, development—and procurement—of ICs used in NTT equipment were of considerable importance. In 1968, for example, NEC had three lines of advanced digital ICs under development: one for "a government-sponsored, large scale integration project" (presumably the LSI component of MITI's Very High Speed Computer System national R&D project), another for use in NTT's electronic telephone exchanges, and a third for use in its next-generation advanced computer systems.[172] Five Japanese manufacturers—Hitachi, Toshiba, NEC, Mitsubishi, and Fujitsu—worked on ICs for use in NTT's "DEX 2" electronic switching system, developed in 1969. Experience on this project is credited by Japanese semiconductor producers with being especially important in establishing rigorous quality and reliability standards for domestic products.[173]

Through the 1960s, though, NTT put little emphasis on support for the Japanese electronics industry beyond its relatively narrow interest in

169. Ira C. Magaziner and Thomas M. Hout, *Japanese Industrial Policy* (Berkeley: Institute of International Studies, University of California, Berkeley, 1980), pp. 107–08; and Sigurdson, *Industry and State Partnership*.
170. Sigurdson, *Industry and State Partnership*; and Yamamura, "Joint Research and Antitrust."
171. Watanabe, "Electrical Communications Laboratories," p. 4.
172. See Tarui, "ICs in Japan," p. 105. The NTT logic design, known as controlled saturation logic (CSL), was used in its D-10 electronic switching system. See Watanabe, "Electrical Communications Laboratories," p. 4.
173. NHK, *Electronics-Based Nation*, vol. 4, p. 14.

creating a domestic supplier base for its specialized telecommunications equipment designs. For example, NTT resisted MITI pressure to use domestic computers; in 1971, 109 of its 172 computer systems were supplied by foreign companies.[174] In the late 1960s, however, NTT began to take a wider view of its domain and started to fund data communications R&D. Its first big jump into the field was the decision to design and build a large time-sharing computer, the Dendenkosha Information-Processing System (DIPS-1), with a development effort initiated in 1968. A measure of NTT's relatively narrow support for IC development up to that time was the fact that the first DIPS models (built by Fujitsu, Hitachi, and NEC) used ICs and other pieces of technology developed for MITI's VHSCS project, in which these same companies participated.[175]

As its Japanese electronics suppliers faced the trauma of market liberalization in the early 1970s, however, NTT began to adopt a substantially broader view of its responsibilities toward ensuring their financial health. When MITI funded the 3.5 generation computer project in 1971 to fend off the IBM threat, NTT stepped in and agreed to coordinate its plans for a follow-up to the DIPS system with the MITI programs, creating an additional 5-billion-yen ($14 million) subsidy for development of essentially the same computer technologies.[176] NTT's increasing concern over the competitive fortunes of domestic electronics companies, now forced to grapple with technologically advanced foreign rivals on a relatively open playing field, was an essential element of its decision to plunge into VLSI research on a large scale.

The NTT VLSI program begun in 1975 was an unprecedented, large-scale effort, with a budget of 20 billion yen ($67 billion) over three years.[177] NTT's internal laboratories cooperated with those of Fujitsu, Hitachi, and NEC, the suppliers of NTT's DIPS computers, in developing the basic technologies—materials, processes, circuits, and systems—needed to manufacture computer chips. Production of a 64K DRAM

174. Anchordoguy, *Computers Inc.*, p. 31.

175. See Flamm, *Targeting the Computer*, p. 129; and Anchordoguy, *Computers Inc.*, pp. 54–56.

176. Anchordoguy, *Computers Inc.*, pp. 121–22.

177. The 20-billion-yen budget is reported in *U.S.-Japan Study Group*, p. 30, and Koki Inoue, "The VLSI Technology Research Association and the Development of the Semiconductor Industry," *Kikai Keizai Kenkyu* (*Machinery Economics Research*), no. 23 (Tokyo: Machinery Promotion Association, 1992), p. 63. See also Anchordoguy, *Computers Inc.*, p. 138.

was chosen as the vehicle for creating and testing the new process technologies.[178] The program met its goals: by 1977 a 64K DRAM had been produced and significant progress made in photolithographic, X-ray, and electron beam systems used to manufacture integrated circuits.[179] By that time, however, the NTT effort was no longer the only such program, or even the largest one, on the scene.

The VLSI Project

MITI was equally disturbed by the reports from the United States that suggested that, once again, the Japanese computer industry was about to find itself set back by another round of rapid technical advance overseas. The challenge was worsened by the fact that, with liberalization almost complete, there would no longer exist a sheltered Japanese market in which domestic producers could at least temporarily escape the rigors of competition at the leading edge.

As discussed earlier, a complex political dance was already under way over the future of the Japanese computer industry. Some forces within the LDP, the ruling political party, wished to force the computer makers into a single national champion firm, along the lines already being followed by British and French government policies. The six largest Japanese computer makers—all of them (except Oki) actually the computer divisions of large and powerful Japanese conglomerates—resisted this idea vigorously. In 1971, however, with Kakuei Tanaka as MITI minister, they had been forced to reconfigure themselves as three loosely allied pairs of firms in exchange for financial assistance from MITI to cope with liberalization. But, as we have seen, these alliances were imposed from outside, and the strongest companies did everything they could to preserve their independence. In 1973 MITI responded by using a program of subsidies to IC R&D to encourage closer relations among the firms. Only the Mitsubishi-Oki pairing (between the two weakest companies in the Japanese computer industry) had responded to this carrot by structuring a cooperative research project; the others had simply gone their

178. Watanabe, "Electrical Communications Laboratories," p. 7.
179. Watanabe, "Electrical Communications Laboratories," pp. 7–8; and "VLSI Memory Chip of Extremely High Concentration Is Developed," *Japan Economic Journal*, May 2, 1978, and "Responding to Criticism of the Closed Nature, Complete Opening of Super LSI Patents of NTT," *Keizai Sangyo Shimbun*, February 8, 1980 (both excerpted in SIA, *Japanese Protection*).

own separate ways with their shares of the IC research subsidies.[180] Complicating matters further from MITI's perspective was NTT's increasing interest in computers and computer-related ICs, which historically had been MITI's responsibility.

When news of the double-barreled IBM-AT&T technological threat from America hit Japan in 1974–75, important voices within the Japanese industrial system clamored for action. NTT advanced further into MITI territory with the announcement of its 1975 VLSI program. The industry begged for huge new subsidies but was opposed by fiscal conservatives within the government. Powerful groups within the LDP repeated their call for a single Japanese national computer champion.[181]

Whether MITI was serving as a buffer between the conflicting interests of LDP politicians and the private sector, or the bureaucrats were "mere puppets for important political interests," or "the politicians were really only 'cheering squads' (*oendan*) for the bureaucratic armies" will probably forever remain a topic for vigorous debate among students of Japanese political economy.[182] Whatever the reality, a three-sided negotiation among MITI, the electronics companies, and LDP politicians took place. A compromise, brokered by the former head of the Dietmen's League for Promotion of the Information Industry, was struck, giving the industry a large subsidy for VLSI research.[183] In exchange, the firms would recon-

180. Fujitsu, however, did join a cooperative effort with two electronics companies not involved in computer production, Sharp and Kyodo Electronics Laboratories. Also, although not a formal part of the IC subsidy program, a certain amount of cooperation in R&D appears to have gone on between Hitachi and Fujitsu through their Nippon Peripherals joint venture, and between NEC and Toshiba through their NTIS joint venture company.

181. The account given below draws heavily upon Masato Hashizume, "Tracing the VLSI Research," in VLSI Technology Research Association, *VLSI Technology Research Association: Retrospective*, pp. 80–87. Hashizume, a former section chief from the electronics industry division of MITI, was appointed head councilor of the VLSI Research Association. See also NHK, *Electronics-Based Nation*, vol. 4, pp. 358–61.

182. These different views of the Japanese political system are marvelously exposited in Chalmers Johnson, "MITI, MPT, and the Telecom Wars: How Japan Makes Policy for High Technology," in Chalmers Johnson, Laura D'Andrea Tyson, and John Zysman, eds., *Politics and Productivity: The Real Story of Why Japan Works* (Ballinger, 1989), pp. 197–200.

183. The little-known Dietmen's League for Promotion of the Information Industry had earlier consulted with MITI when the original 1973–74 IC R&D subsidies, and additional assistance for the computer industry, had been negotiated. See "MITI Informally Decides on Liberalization Countermeasures Expenses—Electronic Computers: 43 Billion Yen for Three Years: Aid to Be Extended for Development of New Models: Total Amount

figure themselves into just two private industry laboratories for the purposes of the subsidy and do a significant amount of R&D jointly in a
single common facility.

The research effort was organized around five companies: Fujitsu,
Hitachi, Toshiba, NEC, and Mitsubishi. (Both Toshiba and Oki dropped
out of mainframe computer manufacture in the mid-1970s, but only Oki,
technically and financially the weakest of these companies, was excluded
from the project.[184]) Technical personnel from MITI's ETL were put in
charge of the common facility, dubbed the VLSI Joint Laboratory, which
gathered personnel from all five private sector participants and the ETL.
NTIS, the NEC-Toshiba joint venture formed in 1971, which already had
established a joint research laboratory system, became one of the two
participating private laboratories. A new entity known as Computer Development Laboratories (CDL), with Fujitsu, Hitachi, and Mitsubishi as
participants, was created as the second industry laboratory.[185] The entire
project was budgeted at 74 billion yen ($236 million), to be spent from
1976 to 1979; of that total MITI was to contribute 40 percent. The overall
objective of the program was to develop the technologies needed to
manufacture a 1M DRAM, or the equivalent, by 1985.[186]

Quite simply, this was by far the largest infusion of R&D subsidies
ever received by the Japanese semiconductor industry, in both absolute
and relative terms. The VLSI project accounted for almost 40 percent
(over 50 percent if the NTT projects are counted as well) of Japan's

of Subsidies to Reach 77.1 Billion Yen," *Sankei Shimbun*, August 14, 1973 (excerpted in
SIA, *Japanese Protection*).

184. Various explanations have been advanced as to why Oki got dropped and not
Mitsubishi. The most common explanation is that Oki was simply too technologically
inferior. See "Responding to Criticism of the Closed Nature, Complete Opening of Super
LSI Patents of NTT," *Keizai Sangyo*, February 8, 1980 (excerpted in SIA, *Japanese Protection*); and Anchordoguy, *Computers Inc.*, pp. 140–41, quoting MITI sources. Other reports
suggest that OKI's close technical ties with American computer maker Sperry Rand through
their Oki-Univac joint venture were a factor. See "Computer Industry Seen Revamped;
Joint VLSI Development Agreed," *Japan Economic Review*, August 15, 1975 (excerpted in
SIA, *Japanese Protection*). Oki was reportedly given 64K DRAM technology by NTT to
compensate for its exclusion. Other firms outside the NTT VLSI project, including Toshiba,
Mitsubishi, and Matsushita, also reportedly lobbied for access to the NTT work. See
"Responding to Criticism."

185. Mitsubishi was the weakest company in terms of its level of IC technology. Author's
interview with Yasuo Tarui, March 1992. Its presence in CDL probably posed little threat,
then, to either Fujitsu or Hitachi. Also, Anchordoguy, *Computers Inc.*, p. 140, cites a
popular Japanese history of the period as suggesting that Fujitsu saw this as a chance to
penetrate the Mitsubishi *keiretsu* grouping's large computer market.

186. Nishi, "The Japanese Semiconductor Industry," p. 5.

Table 2-5. *Japanese Expenditures on Integrated Circuit R&D, by Funding Source, 1963–79*

Hundred millions of yen unless otherwise specified

	Total industry IC R&D expenditure	Govt. IC R&D subsidies		Total VLSI program expenditures by MITI and companies[a]	NTT IC R&D program expenditures[b]	Percent of total IC R&D expenditure accounted for by:			
		Total	MITI VLSI program			Govt. alone	Govt. + NTT	MITI VLSI program	NTT + MITI VLSI
1963	1.1	...	...	...	...	...	...	...	...
1964	2.1	...	...	...	...	...	...	...	...
1965	5.6	4.0	...	...	...	41.7	41.7	...	...
1966	10.0	4.0	...	...	...	28.6	28.6	...	...
1967	24.0	4.0	...	...	...	14.3	14.3	...	...
1968	35.0	4.0	...	...	...	10.3	10.3	...	...
1969	55.0	4.0	...	...	...	6.8	6.8	...	...
1970	87.0	2.0	...	...	...	2.2	2.2	...	...
1971	99.0	1.0	...	...	...	1.0	1.0	...	...
1972	120.0	3.0	...	...	...	2.4	2.4	...	...
1973	170.0	35.0	...	...	...	17.1	17.1	...	...
1974	185.0	35.0	...	...	...	15.9	15.9	...	...
1975	215.0	20.0	...	...	66.7	6.6	28.7	...	...
1976	243.0	50.0	34.1	84.8	66.7	13.9	32.4	23.6	42.1
1977	245.0	90.0	84.4	202.7	66.7	22.4	39.0	50.5	67.1
1978	380.0	105.0	99.2	251.1	66.7	19.0	31.1	45.5	57.6
1979	548.0	75.0	68.2	198.0	66.7	10.9	20.5	28.7	38.4
1976–79	1,416	320	285.9[c]	736.6	266.7	16.0	29.3	36.8	50.1

Sources: Machinery Promotion Association Economic Research Institute, *Japan-U.S. Semiconductor Industry Research Survey*, p. 119, and *U.S. Japan International Comparison of the Semiconductor Industry* (Tokyo, 1981), p. 125; MITI unpublished twelve-company survey; VLSI Technology Research Association, *VLSI Technology Research Association: Retrospective of the Past Fifteen Years* (Tokyo, 1990), p. 66; Koki Inoue, "The VLSI Research Association and the Progress of the Semiconductor Industry," in *Economic Machinery Survey*, no. 23 (Tokyo: Machinery Promotion Association Economic Research Institute, August 1992), p. 63 (in Japanese); Organization for Economic Cooperation and Development, *Electronic Components* (Paris, 1968), p. 178; and author's calculations.

Total IC R&D expenditure sums industry, MITI, and NTT funds.

a. Program began in 1976.

b. Program began in 1975.

c. From 1983 to 1987, some 85 hundred million yen was paid back to the government.

national IC R&D effort in the late 1970s (table 2-5).[187] In addition to the subsidies shown in table 2-5, in-kind contributions of personnel and equipment from MITI's own lab worth 500 million to 1 billion yen, exemptions from tariffs on machinery and equipment worth about 550 million yen, and reductions in taxes on fixed capital further augmented government support for the VLSI project.[188]

There is a virtual consensus in Japan that the effort was an exceedingly important boost to Japanese semiconductor technology and played a key role in making Japanese firms leaders in semiconductor manufacturing in the 1980s.[189] Certainly, tangible progress in DRAM manufacturing, in particular, followed in the wake of the VLSI projects. The first 16K

187. My figures suggest that the VLSI project was a smaller fraction of overall IC R&D than does a widely cited calculation by Ryuhei Wakasugi, because I am using revised MITI statistics that increased estimates of private IC R&D in 1979. See "Research and Development and Innovations in High Technology Industry: The Case of the Semiconductor Industry," *Japanese Economic Studies,* vol. 17 (Fall 1988), p. 15.

188. Inoue, "VLSI Technology Research Association," p. 63.

189. Typical appreciations are the following: "the most successful among [technology research cooperatives] has been the VLSI Technological Research Cooperative. The reason that it has been assessed as a success is, first, that there have been achievements in regard to technologies that are common throughout the semiconductor industry as a whole. . . . Second, there has been an efficient interfirm transfer of the common technologies." Wakasugi, "Research and Development," pp. 18–19.

"Consequently, the VLSI Program contributed greatly to the Japanese semiconductor industry both directly, in terms of its research and development achievement, and indirectly by building a new culture for R&D planning and collaboration." Nishi, "The Japanese Semiconductor Industry," p. 126.

"Today there exists a consensus that the VLSI Project in Japan, established in 1976 as an engineering research association, had an exceptional success in promoting technological development. . . . The project resulted in firmly raising the level of VLSI manufacturing technology of the five participating companies. . . . This is a contributing factor for the increasing shares taken by these companies in the world market for memory circuits. However, the VLSI project also had a profound effect on companies outside the group of 5 companies; today Matsushita has become a major actor in the field.

Furthermore, the project similarly raised the technical level of two major groups of supporting companies . . . not only have Japanese VLSI companies established themselves among the world leaders but its crystal and VLSI equipment manufacturers have also established themselves in the top league." Sigurdson, *Industry and State Partnership,* pp. 62–63.

"In the middle of the 1980s, Japanese companies caught up with their U.S. counterparts, and had stayed in a dominant position in the memory chip market represented by 4M DRAM and 16M DRAM until the beginning of the 1990s. It is well known that the 'VLSI Research Association' assisted by MITI contributed to the success of the Japanese semiconductor industry during this period to a great extent." Ryo Sakamoto, "Will the Japanese Version of 'Sematech' Save the Japanese Semiconductor Industry?" *Foresight* (November 1995), pp. 94–95.

DRAM, introduced by the U.S. producer Intel in 1976, kicked off four generations of DRAMs, through the 1M chip introduced in the mid-1980s, in which the focus for innovation was on perfecting manufacturing technology with which to squeeze a well-understood physical design into ever smaller spaces.[190] Improvements in manufacturing technology were precisely the focus of the VLSI projects, and they provided an entry point for Japanese manufacturers into world semiconductor markets.

Beginning in 1978, after the completion of the first NTT VLSI project, the first significant innovations in DRAM design and manufacturing from Japanese companies (from both NTT and project participant NEC) were publicly unveiled, in the form of the first prototype 64K DRAMs. The success of the first NTT project paved the way for a second 20-billion-yen ($95 million) three-year project, ending in 1980.[191] In the next generation of DRAM, the 256K chip (appearing in 1980), Japanese companies (again NTT and NEC) were even to take the lead as the first worldwide to unveil the new product. These new 256K DRAM designs, in fact, were directly linked to the results of the VLSI projects.[192] From that generation on, Japanese producers clearly dominated the introduction of new product and process technologies in DRAM manufacture.[193]

As the first large-scale experiment in which rival companies' engineers actually worked together in the same laboratories, the MITI VLSI program's apparent success also made it the model for even more intensive experiments with joint research, in Japan and abroad. However, recent revisionist assessments of the VLSI project in the West have argued for two qualifications. First, in assessing the project's technological accomplishments, it has been suggested that the "Japanese did not appear to have made any major breakthroughs. In most areas, [an American VLSI

190. See Betty Prince, *Semiconductor Memories: A Handbook of Design, Manufacture and Application,* 2d ed. (John Wiley, 1991), pp. 211–12. The earlier development of the 1K and 4K DRAMs has been characterized as a period of innovation in the basic physical design of memory cells used in a DRAM, rather than in the process technology used to implement this design.

191. Inoue, "The VLSI Technology Research Association," p. 63.

192. NEC's design used technology developed jointly with Toshiba as part of the MITI VLSI project. See Steve Galante, "Japanese Semiconductor Firms Hustle to Short-Circuit Image as Imitators," *Asian Wall Street Journal,* June 16, 1980, p. 4. Toshiba, on the strength of its burgeoning VLSI skills, was to join the other three firms in the "club" on yet a third NTT VLSI project, started in 1981, which targeted the 1M DRAM. "NTT Has Accepted Toshiba to Join VLSI Project," *Nikkei Sangyo Shimbun,* October 4, 1982, p. 1.

193. See Prince, *Semiconductor Memories,* pp. 218–73, for a historical retrospective on DRAM innovations that supports this assessment.

expert interviewed] felt that the Japanese had simply extended their technology in ways comparable to developments that had already occurred in the United States."[194]

Second, it has been argued that in the VLSI project, as in previous cooperative research programs sponsored by the Japanese government, there was "very little *research* cooperation between the participating Japanese companies, *at least insofar as this refers to the joint generation and sharing of technological knowledge.*"[195] Both of these points merit some brief consideration.

The Technical Impact of the VLSI Project

It is important to remember that the VLSI project was conceived of as a "catch-up" program, designed to bring Japanese producers up to a standard of competence already assumed to be well under development in the West. It was *not* intended to produce "major breakthroughs" in the fundamental technologies of the integrated circuit; it did, however, do a remarkable job of extending and improving on technological concepts already developed or under development overseas.

Figure 2-8 shows twenty-two "major results" of the MITI-subsidized VLSI project, as identified by the VLSI Technology Research Association's recent history of the project, and where they fit into the overall process of IC manufacturing. Twelve of the advances were in equipment used in processing the silicon wafers on which ICs are etched, two in testing the processed wafer, two in manufacture of the large silicon crystals from which the wafers going into the manufacturing process are sliced, two in technologies used to design chips, and six in demonstrating prototypes of specific types of ICs. Thus, sixteen of the twenty-two were demonstrations or improvements in equipment, or design tools, used in the manufacture of any VLSI chip, and only six the development of working prototypes of specific chips.

194. Uenohara and others, "Background," in Okimoto and others, *Competitive Edge,* pp. 38–39, also cited prominently in Fransman, *The Market and Beyond,* p. 84. It should be noted that this assessment was made *before* the VLSI effort had actually ended. As will be documented below, although the government subsidy ended in 1979, the companies involved continued in a "private sector edition" of the project, using private funds only, from 1980 to 1986, to further develop and commercialize the results of the initial, subsidized project. Neither Okimoto and others nor Fransman seem even to be aware that the cooperative effort extended beyond 1979.

195. Fransman, *The Market and Beyond,* p. 58.

Figure 2-8. *Major Results of the Japanese VLSI Project, 1975–90*

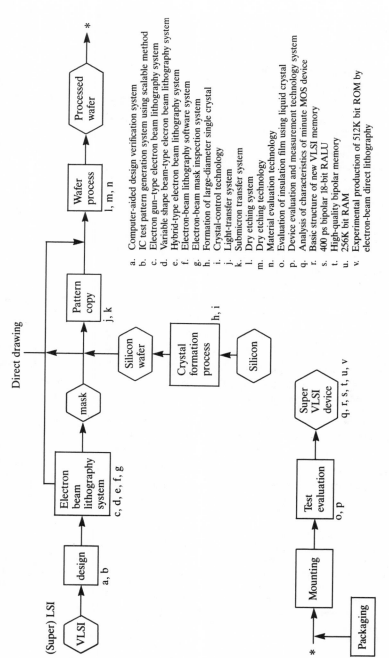

a. Computer-aided design verification system
b. IC test pattern generation system using scalable method
c. Electron gun–type electron beam lithography system
d. Variable shape beam–type elecron beam lithography system
e. Hybrid-type electron beam lithography system
f. Electron-beam lithography software system
g. Electron-beam mask inspection system
h. Formation of large-diameter single crystal
i. Crystal-control technology
j. Light-transfer system
k. Submicron transfer system
l. Dry etching system
m. Dry etching technology
n. Material evaluation technology
o. Evaluation of insulation film using liquid crystal
p. Device evaluation and measurement technology system
q. Analysis of characteristics of minute MOS device
r. Basic structure of new VLSI memory
s. 400 ps bipolar 18-bit RALU
t. High-quality bipolar memory
u. 256K bit RAM
v. Experimental production of 512K bit ROM by electron-beam direct lithography

Source: VLSI Technology Research Association, *VLSI Technology Research Association: Retrospective of the Past Fifteen Years* (Tokyo, 1990).

In the view of the former director of the VLSI Project's Joint Research Laboratory, the four most important achievements of the program were advances in electron beam machines (used to indirectly etch patterns for, and directly write, circuit features on silicon chips); optical "steppers" (photolithographic machines used to imprint patterns on chips); the technology used to grow single silicon crystals, then slice and polish the silicon wafers used in chip making; and techniques used to characterize and measure silicon crystals.[196] Another project participant's list of major research areas shows six topics in semiconductor manufacturing processes, three in materials, three in design and testing, two in packaging, and four in particular device structures.[197]

Consistent with all these lists is the central thesis of the most detailed published study of the VLSI project: that the most visible, concrete effects of the program were in creating technical information flows between Japanese equipment and materials suppliers and IC producers.[198] It was in the manufacturing infrastructure for the Japanese semiconductor industry that the most important advances appear to have occurred.

This is evident to some extent in the way the VLSI project's budget was spent. Figure 2-9 breaks out expenditure by the VLSI Technology Research Association over the four years of the subsidized MITI project, as well as seven additional years of wholly privately funded cooperative research that continued the project after 1979 (this point is elaborated below). Expenditure on facilities, materials, and procurement contracts averaged about 70 percent of the project's cost over the years 1976–79. Yasuo Tarui, the project's leader, estimates that 30 to 40 percent and perhaps even more of the project's budget went into the development of new equipment.[199] Another study estimates that a quarter to a third of the project's funding was spent on the purchase of the most advanced semiconductor manufacturing equipment available in the United States.[200]

196. Author's interview with Yasuo Tarui, March 1992.

197. Nishi, "The Japanese Semiconductor Industry," table 9.2.

198. Sigurdson's excellent *Industry and State Partnership* monograph appears to be the only published English-language study focusing on the actual technical content and results of the VLSI project. Fransman's *The Market and Beyond* is almost entirely devoted to the formal organizational aspects of Japanese cooperative R&D programs.

199. Author's interview with Yasuo Tarui, March 1992.

200. Kiyonori Sakakibara, "Organization and Innovation: A Case Study in Japanese Semiconductor Project," MIT Sloan School of Management, December 1982, p. 14.

Figure 2-9. *Expenditures on VLSI Project and Subsequent Cooperative Research, by Category, 1976–86*

Percent

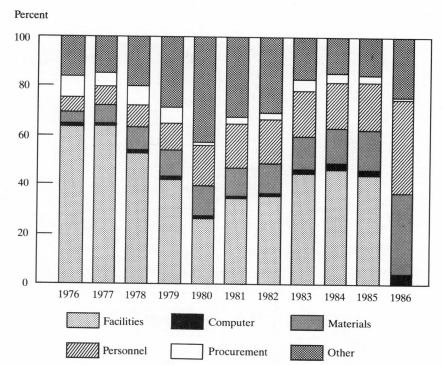

Facilities	Computer	Materials
Personnel	Procurement	Other

Source: VLSI Technology Research Association, *VLSI Technology Research Association: Retrospective*, p. 66.

Ironically, the most direct evidence for the link between the VLSI project and Japanese semiconductor technology shows up in the commercial activities of firms that were not even members of the VLSI project. These equipment and materials firms cooperated and subcontracted with the formal project participants to produce advanced materials and equipment that had not previously been available to the Japanese industry (figure 2-10).[201] The importance of these linkages is visible in the subsequent success of Japanese equipment and materials suppliers in global markets with products utilizing their VLSI project experience.

201. This figure, from Sigurdson, *Industry and State Partnership,* p. 120, is also partly reproduced in Fumio Kodama, *Analyzing Japanese High Technologies: the Techno-Paradigm Shift* (London: Pinter Publishers, 1991), p. 89.

Figure 2-10. *Linkages within VLSI Industrial System*

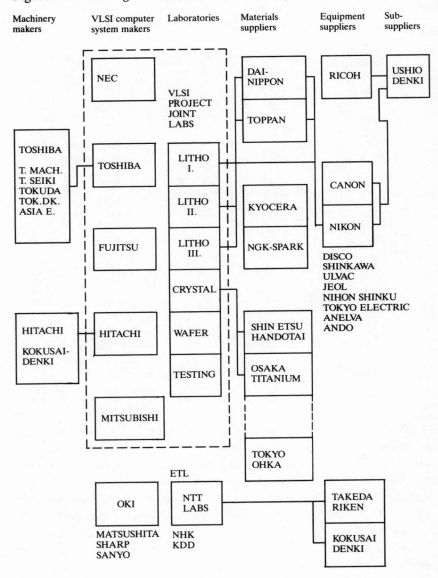

Source: Jon Sigurdson, *Industry and State Partnership in Japan: The Very Large Scale Integrated Circuits (VLSI) Project* (Lund, Sweden: Research Policy Institute, 1986), p. 120.

In wafer processing equipment, two key examples are those of Nikon and Canon, which, marketing wafer processing systems initially developed in collaboration with the VLSI project, went from minuscule market shares in 1981 to over half of the global market for optical lithography equipment used in semiconductor manufacture by 1985.[202] Another equipment maker with a major presence in the global market, gained at least in part through technology developed for the VLSI project, is Toshiba, which in 1978 began marketing an electron beam pattern generator developed for the project.[203] Dainippon Printing and Toppan Printing (producers of masks used in etching features on chips), Kyocera (today the world's largest manufacturer of ceramic packages for ICs), NGK Spark Plug (the second-largest manufacturer of ceramic IC packages), and Ushio Denki (today the world's largest supplier of high-intensity lamps used in photo-etching machines for ICs) are other examples of firms gaining a commanding presence on the world market after emerging from the VLSI project experience. Other firms—such as JEOL in electron beam equipment, Ulvac in ion beam implanters, Advantest (formerly Takeda Riken) in IC testers, and Kokusai Electric in deposition and etching equipment—underwent a similar transformation after their experience with NTT's VLSI project.[204]

In the late 1970s in Japan, chip manufacturing equipment, previously purchased from U.S. vendors, was to an increasing extent displaced by domestic equipment. In 1979 Japanese equipment manufacturers began a push into foreign markets; within four years their global market share had increased by two-thirds in wafer processing equipment (to 25 percent) and had tripled in test equipment (to 21 percent).[205] This rapid

202. Presentation of John Poate prepared for a National Research Council seminar, "Advanced Processing of Electronic Materials in the United States and Japan," Washington, June 4, 1986. In 1985 Nikon had 35 percent of the global optical lithography equipment market, and Canon 17 percent. The leading American producer, GCA (Geophysics Corporation of America), had 30 percent of the market. In 1981 GCA had sold 73 percent of wafer steppers shipped globally. See also Sigurdson, *Industry and State Partnership*, pp. 86–93; Jay S. Stowsky, "Weak Links, Strong Bonds: U.S.-Japanese Competition in Semiconductor Production Equipment," in Johnson and others, *Politics and Productivity*, p. 263; and "VLSI Projection Aligner by Canon," *Electronic News*, November 8, 1978, p. 28.

203. See John Hataye, "Toshiba Machine to Market E-Beam Pattern Generator," *Electronic News*, July 24, 1978, p. 42.

204. Sigurdson, *Industry and State Partnership*, pp. 93–105, 120; and Stowsky, "Weak Links," pp. 265, 268–69.

205. See John Hataye, "Japan Equip. Firms Eye Exports to U.S.," *Electronic News*, December 3, 1979, p. 20; and U.S. Department of Commerce, *A Competitive Assessment of the U.S. Semiconductor Manufacturing Equipment Industry* (1985), p. 4.

growth was to continue: by 1989 Japanese producers accounted for 44 percent of global wafer processing equipment sales (with shares in individual market segments as high as 74 percent for optical steppers, and 60 percent in diffusion furnaces).[206]

Similar histories are evident in semiconductor materials. All five of the large Japanese silicon wafer manufacturers worked closely with the VLSI project; advances in silicon wafer manufacturing technologies have already been identified as one of the key accomplishments of the project.[207] Between 1977 and 1984, Japanese manufacturers doubled their share of world silicon wafer output, from 17 to 34 percent.[208]

In short, the circumstantial historical record suggests that the VLSI projects of both MITI and NTT had direct links to significant improvements in semiconductor manufacturing technology in Japan, and that much of this progress was reflected in a surge of Japanese equipment and materials suppliers into the world marketplace. The IC manufacturers who were the principals in these programs were to score equally impressive gains in global sales in the 1980s, largely on the basis of standardized commodity products with cost structures dominated by manufacturing (such as memory chips) rather than complex, design-intensive products (such as microprocessors) where design and performance details were as important as—if not more important than—production cost.

206. Department of Commerce, *National Security Assessment of the U.S. Semiconductor Wafer Processing Equipment Industry* (April 1991), p. 26.

207. Author's interview with Yasuo Tarui, March 1992. Sigurdson, *Industry and State Partnership*, pp. 76–81, credits links forged between silicon wafer manufacturers and IC producers during the VLSI project with the former's rapid advance on the world market. The information apparently flowed two ways. Sigurdson quotes an employee of Shin Etsu Handotai (SEH), the largest wafer maker: "5–6 years back the IC makers in Japan were behind (U.S.). (SEH) had good contacts with customers in the U.S. We could know the secrets. Japanese IC makers were eager to know and could (provide) some secrets. Japanese (companies) very happy with our information" (p. 79).

208. See Remo Pellin, "Semiconductor Market," in Department of Energy, *Proceedings of the Flat-Plate Solar Array Project Workshop on Low-Cost Polysilicon for Terrestial Photovoltaic Solar-Cell Applications*, DOE/JPL 1012-122 (Pasadena: Jet Propulsion Laboratory, 1985), p. 407. Wafer production is measured in millions of square inches of silicon. If only semiconductor grade wafers are counted (and lower quality wafers used to make solar cells excluded), the global market share of Japanese wafer producers stood at more than 40 percent in 1984. See Strategies Unlimited, *Silicon Wafer Industry Assessment* (Mountain View, Calif., 1986), p. 124.

Technical Cooperation among Japanese Manufacturers

The VLSI project was the first case in which a MITI-supported research association set up its own, independent joint laboratory to undertake research, and the first to mix within a common space researchers from the different participating companies. Its success provided the impetus for later initiatives to establish joint industry R&D facilities.

However, it has rightly been pointed out that perhaps 15 percent of the project's total budget was spent within the joint R&D laboratory, with the remainder expended within the two company-run joint research programs and the individual companies, under the direction of the VLSI Technology Research Association.[209] Actually, the share of the work carried out by the joint lab may have fallen below even this low number; as table 2-6 suggests, personnel in the joint lab as a share of the project's total personnel dropped rapidly from 18 to 11 percent over the four-year life of the project. Furthermore, the topics pursued within the joint lab were specifically chosen to be "common" and "basic" technologies far removed from actual commercial device development.[210]

Even within these constraints, organizational walls between different research groups in the joint lab were erected to protect perceived proprietary interests: Hitachi, Fujitsu, and Toshiba, for example, were reluctant to have their researchers work with other companies in the area of lithography equipment because each of these companies had active commercial efforts under way for these products. As a result, each of these companies' researchers were segregated within their own competing lithography laboratory within the overall joint lab, joined by researchers from NEC and Mitsubishi (who were not in direct competition) but not the other companies.[211] One breakdown of a group of patent applications resulting from the project shows only 16 percent of these patents involving researchers from more than one company.[212] Given these facts, one revisionist critique has argued that there was limited joint generation and sharing of knowledge within the VLSI project.[213]

209. Fransman, *The Market and Beyond*, p. 80.
210. For an interesting analysis, see Soichiro Seki, "Intra-Industry R&D Consortium—Lessons from Japanese Cases," Brookings Institution, May 1992; on the choice of themes for the joint laboratories, see Yasuo Tarui, *IC no Hanashi* (Tokyo: Nihan Hoso Shuppa Kyokai, 1982), p. 144.
211. Fransman, *The Market and Beyond*, pp. 65–67.
212. Tarui, *IC no Hanashi*, p. 149.
213. Fransman, *The Market and Beyond*, pp. 76–79.

Table 2-6. *Shares of Total VLSI Research Association Project Personnel, by Lab Group, 1976–86*

Percent

	1976	1977	1978	1979	1980	1981	1982	1983	1984	1985	1986
Joint Lab	17.7	14.0	12.1	11.3	0	0	0	0	0	0	0
CDL[a]	21.3	38.7	45.9	45.8	30.5	25.0	24.6	18.3	18.8	19.2	0
Fujitsu	0	0	0	0	10.7	11.5	12.9	16.1	13.8	14.1	19.2
Hitachi	0	0	0	0	13.0	13.9	13.6	17.0	17.5	17.9	26.1
Mitsubishi	0	0	0	0	14.6	15.6	16.1	20.1	19.7	20.1	26.1
CDL and associated companies	21.3	38.7	45.9	45.8	68.8	66.0	67.1	71.6	69.7	71.2	71.3
NTIS[a]	61.0	47.3	42.0	42.9	13.8	13.7	10.7	5.1	5.9	5.8	0
NEC	0	0	0	0	10.4	11.8	13.2	10.3	10.7	8.9	8.8
Toshiba	0	0	0	0	7.0	8.5	8.9	13.0	13.8	14.1	19.9
NTIS and associated companies	61.0	47.3	42.0	42.9	31.2	34.0	32.9	28.4	30.3	28.8	28.7
Companies associated with:											
CDL	0	0	0	0	38.3	41.0	42.5	53.2	50.9	52.0	71.3
NTIS	0	0	0	0	17.4	20.3	22.1	23.3	24.5	23.0	28.7

Source: VLSI Technology Research Association, *History of the VLSI Research Association* (Tokyo 1991), pp. 39, 56, 62.

a. CDL is Computer Development Laboratories; NTIS is NEC Toshiba Information Systems.

But a broader view of matters belies this simplistic conclusion. Joint patents may be useful in tracing those formal joint R&D efforts that produce patents, but there is no reason to believe they are a particularly useful indicator of information sharing. Indeed, the brief sketch earlier in this chapter of the early history of Japanese semiconductor development stressed how useful open conferences, conversations, and observations on visits to labs were in obtaining initial access to technical information on American semiconductor technology.

Furthermore, sharing of technical information between participants in MITI-sponsored projects has always been Japanese policy and practice, even before the VLSI project's experiment with a large joint R&D facility. The form of organization before the VLSI program was more of a hub-and-spoke system, with either the ETL or a research association contracting pieces of the program out to individual companies, then taking pains to disseminate the results through meetings and conferences among the participants. A characteristic feature of the system was that research association members were required to share research results with other members.[214]

The results of ETL's internal research through the early 1960s were generally disseminated through technical workshops in which companies participated. With the advent of the large-scale project system in the mid-1960s, however, and coordinated contract research undertaken by private companies, the ETL began to take a more direct role in planning, coordinating, and disseminating contracted private research for projects in which it was involved.[215] For example, Tarui, describing MITI's first grants for IC development in 1965, notes that "the research accomplished in each project will be available to all companies."

> Producers are also being encouraged by the Microcircuit Technical Committee of the Japan Electronic Industry Development Association. Chairman of the committee is Noboru Takagi, a professor at the University of Tokyo; Tsuneo Momota of MITI's Electrotechnical Laboratory is the vice chairman. Because of their affiliations, they can disseminate technical information widely.[216]

Efforts to disseminate technical information among firms were also clearly on the agenda of the VLSI project, even when joint R&D was

---

214. See Akira Goto and Ryuhei Wakasugi, "Technology Policy," in Komiya and others, *Industrial Policy of Japan*, pp. 198–99.
215. Ishii, "Research and Development on Information Processing Technology."
216. Tarui, "Japan Seeks Its Own Route," pp. 90–91.

not involved. Martin Fransman, principal exponent of the revisionist thesis, himself inadvertently makes this point when he quotes the reply of one VLSI participant to his query about the flow of information between the two private labs (the Fujitsu-Hitachi-Mitsubishi and NEC-Toshiba groupings): "they had open conferences and produced thick reports. But of course the reports did not contain all the relevant information about know-how."[217] It is a long leap indeed to conclude that not sharing *all* technical information means that such sharing was not important for the companies involved.

Moreover, the formal walls constructed between research groups within the joint VLSI laboratory were ultimately, participants have suggested, broken down. Parties, social occasions, and evening drinking sessions were used to break down the formal barriers between areas, and considerable information and ideas did ultimately flow back and forth.[218]

The work carried out in the joint laboratory seems to have been productive. One study identifies major results of R&D carried out in this lab in the areas of electron-beam (E-beam) writing systems, E-beam resists, silicon wafer production, silicon crystal micro defect control, LSI testing equipment, plasma etchers, E-beam mask testers, and lithography equipment. The companies associated with developing and testing this equipment are a veritable who's who of today's semiconductor equipment and materials industry in Japan: in E-beam writing systems, Fujitsu, NEC, Fuji Electronic Chemicals, Hitachi, Toshiba, Toshiba Engineering, Akashi Seisaku, Asia Seisaku, and Nihon Business Automation; in E-beam resists, Toray and Tokyo Oka Kogyo; in wafer production and micro defect control, a consortium made up of Shin Etsu, Komatsu, and Osaka Titanium; in LSI testers, Ando Electronics; in plasma etchers, NEC Anelva; in E-beam mask testers, Hitachi and Hitachi Software Engineering; in lithography equipment, steppers from Canon and Nikon, an ultraviolet projection aligner from Canon, and X-ray components from Rigaku Electronics.[219]

The Canon and Nikon equipment projects were particularly significant activities. In addition to receiving R&D subsidies for the next-generation

217. Fransman, *The Market and Beyond*, p. 81. Fransman also describes some research actually conducted jointly by the two laboratory groupings, but he minimizes its importance.

218. NHK, *Electronics-Based Nation*, vol. 4, p. 358; Sakakibara, "Organization and Innovation," pp. 21–22; Sigurdson, *Industry and State Partnership*, pp. 50–51; and author's interview with Yasuo Tarui, March 1992.

219. Inoue, "The VLSI Research Association," p. 68.

projection aligner, 20 to 25 percent of the R&D funding for a prototype stepper—and informal assurances that the system would be purchased by VLSI project members if it met published specifications—were reportedly received from the VLSI joint lab. Nikon reportedly received subsidies from the VLSI joint lab to cover its effort to reverse-engineer, and improve upon, American company GCA's leading-edge stepper, along with informal assurances that project members would buy the new system.[220]

Although the VLSI project was the first *large* effort to bring companies together to conduct joint research, it had some precedents. MITI's efforts to cartelize the computer industry in the early 1970s did bring company researchers together in joint R&D projects, often under the auspices of the research associations. Bitter rivals Fujitsu and Hitachi did join together on R&D through their Nippon Peripherals joint venture, which was a member of and received funding from the research association administering that piece of the 3.5 generation subsidy.[221] NEC Toshiba Information Systems also set up a joint research system that predated the VLSI project; one of its purposes was R&D into ICs. Joint research by NEC and Toshiba through NTIS on the VLSI project is actually known to have produced at least one important piece of technology, an electron beam machine.[222] Finally, the IC subsidies of 1973–74 were specifically structured to create incentives for joint R&D among the participating companies, although the results were at best mixed.

Perhaps the ultimate test of the utility of the coordinated structure of the VLSI project for the companies involved, however, was the market. By 1979 objections raised to the large scale of the subsidy by U.S. industry and government had made it politically difficult to continue the government-subsidized project, and the joint research lab was shut down.[223] The industry, however, continued a "private sector" edition of

220. Ross A. Young, *Silicon Sumo: U.S.-Japan Competition and Industrial Policy in the Semiconductor Equipment Industry* (Austin: University of Texas IC2 Institute, 1994), pp. 185–88. Propelled by these new systems, Canon and Nikon turned the tables on industry leader GCA. From a 10 percent share of the world stepper market in 1980, they jumped to 80 percent by the end of the decade. U.S. stepper makers' share dropped to 10 percent, and GCA, which invented the stepper, went out of business in 1992. Young, p. 99.

221. In fact, Fransman, *The Market and Beyond*, p. 52, notes how important it was for "cross-fertilization" of ideas for engineers from the two rivals to sit down and work together on research in the joint venture.

222. Anchordoguy, *Computers Inc.*, p. 144.

223. See "VLSI Research Group Is Going to Halt Activities at End of March," *Japan*

Figure 2-11. *Public and Private Funding of VLSI R&D Association,*
1976–86

Billions of yen

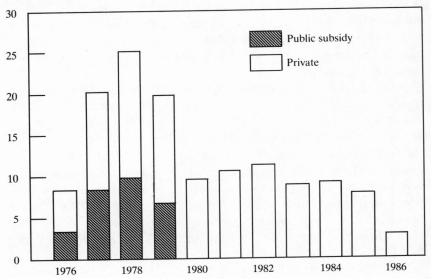

Source: VLSI Technology Research Association, *History of the VLSI Research Association,* p. 66.

the VLSI project for another seven years, funded entirely out of its own
monies, to continue development and commercialization of semiconductor manufacturing technologies first explored under the subsidized program.[224] Figure 2-11 shows the scale of the original and the additional
funding (the latter amounted to about 60 billion yen, or 50 percent more
than the companies' contribution to the original program).[225] Companies
clearly must have found their joint research through CDL and NTIS, and

---

Economic Journal (March 1980, excerpted in SIA, *Japanese Protection*); and Sakakibara,
"Organization and Innovation," pp. 12–14.

224. See "Super LSI Volume Production 3-Year Plan, 10 Billion Yen Invested in First
Year," *Nihon Keizai Shimbun,* July 31, 1980 (excerpted in SIA, *Japanese Protection*); and
unpublished slides for presentation of Yoshio Nishi, "VLSI Technology Perspective," World
Bank, China Electronics Seminar, October 22, 1987.

225. The VLSI project's government funding was actually given in the form of a "con-
ditioned loan" with an obligation to pay back the subsidy if profits were earned on the
technologies and products developed. From 1983 to 1987 the companies actually paid back
some 8.5 billion yen to the government (out of a MITI subsidy of 28.6 billion yen). Inoue,
"The VLSI Research Association," p. 63.

the coordination and information sharing structured through the VLSI Technology Research Association, to have been productive, to continue an effort on this scale. The CDL and NTIS labs continued to be active in joint research in other areas as well.[226]

The utility of VLSI-style joint research clearly seems to have been proven to both government and industry. Immediately after the VLSI joint lab was disbanded, two similar collaborative R&D labs were established as part of new government-industry R&D initiatives.[227] Joint research labs have since become a familiar element of technology policy in Japan.

The Continuing Role of Government

Even if no particular merit is attributed to the joint and coordinated structure of the VLSI projects, it is clear that the resources associated with the MITI and NTT VLSI programs alone must have had an enormous impact on the Japanese semiconductor industry. Table 2-5 shows how resources allocated by the Japanese government and NTT to R&D into ICs went from minuscule amounts to well over 30 percent of total Japanese investments by the late 1970s.

In the 1980s political pressures from abroad forced MITI to cut back the scale of its subsidies to R&D investments in ICs. Table 2-7 shows the considerably smaller scale of the major MITI technology subsidies of the 1980s, particularly in comparison with private sector investments in R&D. As the Japanese industry grew larger and stronger, however, making its presence felt in global markets, this assistance became less critical. The crisis of the 1970s had passed, and the Japanese semiconductor producers, using their newly honed manufacturing skills, were to become the largest global suppliers of commodity memory chips, a product where manufacturing cost was virtually the only thing that mattered.

Instead, the biggest subsidies to the Japanese semiconductor industry in the 1980s came from NTT, the Japanese telephone monopoly. NTT's first VLSI project in 1975 had differed from that of MITI in emphasis,

226. CDL and NTIS were participants in the MITI-sponsored Next Generation Basic Computer Technology program, which carried out research in the areas of software, computer processing of language, and computer peripherals (but not ICs).
227. See Izuo Hayashi, Masahiro Hiano, and Yoshifumi Katayama, "Collaborative Semiconductor Research in Japan," *Proceedings of the IEEE*, vol. 77 (September 1989), pp. 1431–32.

Table 2-7. *MITI Expenditures on R&D Programs Related to Microelectronics, Fiscal Years 1983–92*

Billions of yen unless otherwise specified

Programs	1983	1984	1985	1986	1987	1988	1989	1990	1991	1992
National R&D projects										
Optical measurement and control systems	3.39	2.33	3.44	n.a.	n.a.	n.a.	n.a.	0.47	0.42	n.a.
High-speed computer for scientific and technological uses	1.57	2.25	2.77	2.89	2.95	2.78	2.43	n.a.	n.a.	n.a.
New function elements	1.45	1.48	1.59	1.54	1.40	1.21	1.31	0.77	n.a.	n.a.
Superconducting materials and devices	n.a.	n.a.	n.a.	n.a	n.a	1.06	1.87	2.35	2.78	2.89
Control system for observing objects at molecular level	n.a.	n.a.	n.a.	n.a.	n.a.	n.a.	n.a.	n.a.	n.a.	0.03
Bioelectronics devices	n.a.	n.a.	n.a.	n.a.	n.a.	n.a.	n.a.	n.a.	0.28	0.28
Quantum functional devices	n.a.	n.a.	n.a.	n.a.	n.a.	n.a.	n.a.	n.a.	0.04	0.51
Industrial research support										
Light-activated materials	n.a.	n.a.	n.a.	n.a.	0.18	0.23	n.a.	n.a.	n.a.	n.a.
Micromachine technology	n.a.	n.a.	n.a.	n.a.	n.a.	n.a.	n.a.	n.a.	0.03	0.86
Very advanced processing systems	n.a.	n.a.	n.a.	n.a.	n.a.	n.a.	n.a.	n.a.	2.62	3.16
Nonlinear optoelectronic materials	n.a.	n.a.	n.a.	n.a.	n.a.	n.a.	n.a.	n.a.	0.52	0.53

Extreme environment-resistant materials	n.a.	n.a.	n.a.	n.a.	n.a.	n.a.	n.a.	n.a.	1.70	1.67
Photoreactive materials	n.a.	n.a.	n.a.	n.a.	n.a.	n.a.	n.a.	n.a.	0.42	0.20
Survey of trends in foreign-made semiconductors	n.a.	n.a.	n.a.	n.a.	n.a.	n.a.	n.a.	n.a.	n.a.	0.02
Basic Technology Research Promotion Center										
New Media Community[a]										
Sortech	n.a.	n.a.	n.a.	1.43	1.43	1.43	1.43	1.43	1.43	1.43
OEIC[b]	n.a.	n.a.	n.a.	1.00	1.00	1.00	1.00	1.00	1.00	1.00
Coherent optical communication	n.a.	n.a.	n.a.	0.72	0.72	0.72	0.72	0.72	0.72	n.a.
Amorphous magnetic materials and electronic devices	n.a.	n.a.	n.a.	n.a.	n.a.	0.38	0.38	0.38	0.38	0.38
Total	6.41	6.06	7.80	7.58	7.68	8.81	9.14	7.12	12.34	12.96
Memoranda:										
Japanese industry IC R&D funding	162.10	195.90	231.00	260.10	271.40	320.70	357.50	413.50	415.90	430.50
MITI funding as share of industry IC R&D funding (percent)	3.95	3.09	3.37	2.91	2.83	2.75	2.56	1.72	2.97	3.01

Sources: Japan Electronic Computer Corporation, *Computer Notes* (Tokyo, annual) 1983, pp. 70, 72, 106; 1985, p. 75; 1986, pp. 90, 147; 1988, pp. 78-79; 1989, pp. 76-77; 1991, p. 79; 1992, p. 76 (in Japanese). MITI unpublished twelve-company survey; *Tokyo Electronic Industry Monthly Newsletter* (May 1990), pp. 2-11 (in Japanese); *Japan Electronics Almanac 1989* (Tokyo: Dempa Publications, 1989), pp. 68, 76, 77; Japan Academic Promotion Association, *Academic Monthly Newsletter* (July 1993), pp. 66-74 (in Japanese); and "Omron Electronics Europe Chief Discusses Market Strategy: MITI Machinery and Information Industries Bureau's Budget Detailed," *Tokyo Kikai Shinko*, April 1, 1992.
a. Average annual funding levels for multiyear projects.
b. OEIC is optoelectronic integrated circuit project.

focusing more on devices and device structures, and less on the manufacturing process (although, as described earlier, some important pieces
of manufacturing equipment—testers and etching equipment—came out
of the program).[228] The scale of the overall program was substantially
smaller than MITI's, however, since firms were not required to match
NTT's investment in R&D.

In 1978 NTT began a second three-year VLSI program, this time
focusing on development of a 256K DRAM, and like the first program
budgeted at 20 billion yen.[229] By early 1980 NTT had announced a 256K
DRAM prototype.[230] Yet a third NTT VLSI program was begun in the
fall of 1981, this time targeting 1M DRAMs and computer-aided design
systems for use in designing future generations of ICs.[231] In this third
project, membership was extended beyond the three core members—
Fujitsu, Hitachi, and NEC—to Toshiba. Technology transfer from and
procurement of 256K DRAMs by NTT gave the Japanese semiconductor
makers an early boost in 256K DRAM production in the mid-1980s.[232]

NTT spending on R&D involves a substantial component of joint
research with other companies. Table 2-8, based on data reported by
Kodama, shows that about 28 percent of patent applications jointly filed
by NTT and its suppliers over the 1980–86 period involved two or more
of NTT's suppliers.

By the mid-1980s NTT, through its relationships with its suppliers, had
quietly become the largest source of quasi-public subsidies to the Japanese semiconductor industry. An AT&T semiconductor technologist has
estimated, on the basis of his visits to NTT in the mid-1980s, that about
22 percent of all NTT R&D staff were working on semiconductor technology in 1985, suggesting an internal 1988 R&D expenditure of about

228. Author's interview with Yasuo Tarui, March 1992; Sigurdson, *Industry and State
Partnership*, p. 58; and unpublished slide from Nishi presentation, 1987.
229. "Responding to Criticism of the Closed Nature, Complete Opening of Super LSI
Patents of NTT," *Nihon Keizai Sangyo*, February 8, 1980 (excerpted in SIA, *Japanese
Protection*).
230. See note 190.
231. "NTT Has Accepted Toshiba to Join VLSI Project," *Nikkei Sangyo Shimbun*,
October 4, 1982, p. 1.
232. In 1982 NTT began procuring 256K DRAMs in production quantities (hundreds
of thousands) from Hitachi, NEC, and Fujitsu, using design and manufacturing technology
transferred to these firms at no cost. See Jack Robertson, "Japan Cos. Get NTT's 256K
Skills," *Electronic News*, October 11, 1982, p. 8; and *Nikkei Shimbun*, August 26, 1982,
p. 8.

Table 2-8. *Joint Patent Applications with NTT, by Number of Participating Companies, 1980–86*
Percent unless otherwise specified

Year	Joint patent applications	Number of manufacturers involved (not including NTT itself)			
		One	Two	Three	Four
1980	627	72.9	8.0	1.1	17.8
1981	669	63.7	8.8	3.0	24.5
1982	652	73.5	5.1	1.2	20.2
1983	655	74.4	7.5	3.8	14.4
1984	540	69.3	5.2	0.9	24.6
1985	375	72.0	4.8	4.0	19.2
1986	655	81.2	8.1	3.4	7.3

Source: Fumio Kodama, *Analyzing Japanese High Technologies: The Techno-paradigm Shift* (London: Pinter Publishers, 1991), pp. 98–99. An error in the published data for 1980 has been corrected.

36 billion yen.[233] Table 2-9 estimates Japanese subsidies to semiconductor R&D in 1988 and shows that they amounted to about 18 percent of all public and private spending on semiconductor-related R&D. This ratio is remarkably similar to that resulting from an analogous calculation for the United States in the late 1980s and early 1990s.[234] Although MITI spending was no longer dominant, it was still significant (about one-eighth of all nonprivate spending).

NTT's relationship with private industry was to undergo a significant restructuring in the mid-1980s as a result of the enterprise's privatization

233. David L. Carter, "Estimates of Japanese Government Semiconductor R&D Spending, Analysis," unpublished background material furnished to the National Advisory Committee on Semiconductors, 1990. These numbers do not appear to include external semiconductor research by suppliers funded through development contracts.

234. In 1987 U.S. government spending on semiconductor-related R&D amounted to about 20 percent of the national total. Government spending was $454 million, whereas private industry R&D spending was about $1.8 billion. The latter is estimated to have been 12.9 percent of U.S. company sales of $13.6 billion. The R&D-to-sales ratio is an SIA estimate; U.S. company sales are an estimate from Integrated Circuit Engneering (ICE). Government spending is estimated in Philip Webre, *The Benefits and Risks of Federal Funding for Sematech* (Congressional Budget Office, 1987), p. 60.

In 1992 U.S. Defense Department semiconductor-related spending is estimated to have been about 18 percent of the national total: DOD spending was about $700 million, and private spending about $3.2 billion. The departments of Energy and Defense are estimated to have appropriated $686 million for semiconductor-related R&D in fiscal 1992; see Department of Defense, *DOD Key Technologies Plan* (July 1992), p. 5-28.

See generally Semiconductor Industry Association, *1995 Semiconductor Industry Association Databook: Review of Global and U.S. Semiconductor Competitive Trends, 1978–1994* (San Jose, Calif.: 1995), p. 40.

Table 2-9. *Government and NTT Semiconductor-Related R&D Funding in Relation to Private Japanese Semiconductor R&D Funding, 1988*

Agency or program	Billions of yen
MITI R&D programs related to microelectronic materials and devices	
National R&D projects	
High-speed computer for scientific and technological uses	2.78
New function elements	1.21
Superconducting materials and devices	1.06
Industrial research support	
Light-activated materials	0.23
Basic Technology Research Promotion Center	
Sortech	1.43
OEIC[a]	1.00
Coherent optical communication	0.72
Amorphous magnetic materials and electronic devices	0.38
All MITI programs	8.81
NTT R&D on semiconductors	36.00
Ministry of Education funding	9.00
Science and Technology Agency funding	16.90
Total government and NTT semiconductor-related R&D funding	70.71
Japanese industry IC R&D funding	320.70
Government and NTT funding as a share of all semiconductor-related R&D (percent)	18.07

Sources: *Japan Electronics Almanac 1989*, pp. 68, 76, 77; *Electronic Industry Almanac 1989* (Tokyo: Dempa Publications, 1989), p. 185; AT&T Bell Laboratories–Electronics Technology Planning Center, memorandum, February 1990, pp. 3-5; and MITI unpublished twelve-company survey.
a. OEIC is optoelectronic integrated circuit project.

in 1984. From the privatized NTT emerged a major new, publicly funded institution investing in information technology. A public battle between MITI and the Ministry of Posts and Telecommunications in 1984 over control of Japanese investments in information technology research was resolved with the establishment of a jointly administered Basic Technology Research Promotion Center in 1985.[235] The huge annual dividends from the government's remaining one-third share of the equity of the privatized NTT are funneled into this center, which in turn funds up to 70 percent of joint R&D ventures with private industry. From 1986 to 1990 close to a billion dollars (100 billion yen) of NTT-derived funds were

235. See Johnson, "MITI, MPT, and the Telecom Wars," pp. 227–30; Teruyuki Inoue, *NTT: A Giant Facing Competition and Division in the Information Age*, 4th ed. (Tokyo: Otsuki Shoten, 1992), pp. 108–14 (in Japanese).

plowed into this center. The single largest recipient of these subsidies is the Advanced Telecommunications Research Institute, with a core research staff mostly made up of personnel sent from NTT. The large private corporations that participate in the Institute's projects and fund the balance of its budget also send researchers to work on these research efforts, who then serve as the conduit for technology transfer back into their home organizations.[236]

MITI's stance toward the Japanese semiconductor industry also changed dramatically in the 1980s. The agency lifted formal controls on semiconductor imports and foreign investment for the most part by 1975, but some signs that administrative guidance and other barely visible measures were used to protect domestic producers lingered on through the remainder of that decade. As late as 1979, for example, American chip producers still publicly described alleged incidents in which MITI had forced Japanese customers to cancel orders with American companies and switch to Japanese suppliers.[237] And one could still see the old attitudes at work on occasion, even in the 1980s. Perhaps the clearest example was a MITI initiative, in the early and mid-1980s, to reduce dependency on foreign supplies of the high-purity silicon used to produce semiconductors.

Dependence in Silicon

In 1978, working through its so-called Sunshine Project, MITI set up a special study group within the Japan Electrical Manufacturers Association to examine the low-cost manufacture of polycrystalline silicon (poly), the raw material used to make the silicon wafers on which ICs are

236. Inoue, *NTT: A Giant Facing Competition and Division*, pp. 108–14.
237. "Dr. Hogan tells of a $7 million order Fairchild got recently from an unnamed Japanese customer. 'Our people in Japan gave a party; they were delighted. But one week later the Japanese customer was on the phone literally crying. MITI, he told us, had forced the order to go to Hitachi, even though his people felt we were providing a better product.'" John Reason, "Japan's Electronics Markets: A Pair of Views," *IEEE Spectrum*, vol. 16 (June 1979), p. 52.

"Some reported experience implied Japanese government involvement. It was reported that sales have been cancelled after application to MITI for an import license, and that this was due to telephone calls received by prospective purchasers asking why they were importing when essentially comparable domestic products were available." U.S. International Trade Commission, *Competitive Factors Influencing World Trade in Integrated Circuits* (November 1979), p. 59.

etched.[238] The study group examined technologies for producing the higher purity poly used in manufacturing wafers used in semiconductor production, as well as the lower grade poly used to make solar cells. The basic problem for all Japanese producers was that silicon production was highly electricity-intensive, and electricity in Japan was very expensive.

In October 1980 the New Energy Development Organization (NEDO) was formed as part of MITI's Sunshine Project, to promote technological development. One of its first projects was to develop energy-efficient technology for low-cost polycrystalline silicon for solar cells. The R&D work was divided between the two largest Japanese poly producers, which also happened to be the largest Japanese makers of silicon wafers used in IC fabrication: Shin Etsu Handotai (SEH) and Osaka Titanium Company (OTC). (These same two firms worked closely with the chip producers involved in MITI's VLSI project; see figure 2-10.) OTC was to work on low-cost production of a precursor chemical, and SEH to develop the process for making this material into polycrystalline silicon.[239] Although the OTC-SEH work on low-cost poly production was explicitly aimed at solar grade material, there was a clear intention to apply the work to semiconductor grade material if possible.[240] The very same year that the NEDO silicon project was set up, another Japanese chemical producer, Tokuyama Soda, also began research into poly production.[241]

The extraordinarily rapid rate of growth in the Japanese semiconductor market was apparent to all by the early 1980s, and foreign silicon producers responded by increasing their Japanese presence. Japanese semiconductor production was growing rapidly, and with it imports of poly. By mid-1982 it was clear that, for the first time in the history of the Japanese semiconductor industry, poly imports were going to exceed domestic production. Some in Japan lamented the apparent inability of Japanese firms to be competitive, given the high cost of electricity, and

238. See Toshio Noda, "Processes and Process Developments in Japan," in *Proceedings of the Flat-Plate Solar Array Project Workshop on Low-Cost Polysilicon for Terrestrial Photovoltaic Solar-Cell Applications,* JPL publication 86-11 (Pasadena, Calif.: Jet Propulsion Laboratory, February 1986), p. 213.

239. Noda, "Processes and Process Developments," p. 214.

240. For example, to the question, "Do you think it's possible to use this process to produce electronic-grade silicon?" Noda of OTC replied, "My understanding is that everyone here has a strong interest to see our process applied to produce electronic-grade material. We would like to see that happen, too." Noda, "Processes and Process Developments," p. 231.

241. See "Tokuyama Soda's Polysilicon Business Dept.: Offense with Polycrystalline Silicon," *Nikkei Sangyo Shimbun,* December 18, 1984, p. 6.

went so far as to accuse foreign poly makers of dumping.[242] By the end of 1982 it was clear that imports of poly (roughly 80 percent of which were supplied by the German producer Wacker) had doubled, and Japanese silicon industry leaders had begun to perceive "import dependence" as a serious problem.[243]

Various manifestations of that concern soon appeared. At the urging of the Japanese government, OTC decided to more than double its poly production capacity.[244] Tokuyama Soda was reported in late 1982 to have made a decision to begin producing poly by 1984.[245] By early 1983 it had been reported in the Japanese trade press that financial support from the Japanese government, in the form of low-interest loans from the Japan Development Bank, were being made available to support increased poly production.[246] In the spring of 1983 MITI had also formed an industry study group to examine the silicon industry's "problem" and recommend action; in discussion in the Japanese trade press of the issues to be considered by the study group, the problem of import dependence is given top billing, as "problem number one" of the silicon industry.[247]

The overriding theme of the study group was the rising share of imports in Japanese poly consumption. Japan's semiconductor devices and materials industries were clearly concerned about dependence, as imports climbed past half of consumption.[248] By early 1984 the major recommen-

242. *Raru Metaru Nyuzu* [*Rare Metal News*], July 1, 1982.

243. "Polycrystal Silicon—Imports Will Surpass Domestic Production Next Year," *Rare Metal News,* December 1, 1982.

244. See *Nihon Keizai Shimbun,* July 19, 1982; "Polycrystal Silicon—Imports Will Surpass Domestic Production Next Year," *Rare Metal News,* December 1, 1982; E. Costogue and R. Pellin, "Polycrystalline Silicon Material Availability and Market Pricing Outlook Study for 1980 to 88: January 1983 Update," DOE/JPL-1012-79A (Pasadena: Jet Propulsion Laboratory, February 1983), p. 14. Pellin recalls that in late 1982 or early 1983, when he was working as a consultant to the company, officials at OTC told him that "an understanding between MITI and the Osaka Titanium Manufacturing Company has been reached under which Osaka will increase its plant capacity to 800 annual metric tons in 1985 and to 1800 metric tons by 1990" (p. 13); and Remo Pellin, personal communication to the author, 1989.

245. "Polycrystal Silicon—Imports Will Surpass Domestic Production Next Year." This was also reported in Costogue and Pellin, "Polycrystalline Silicon Material Availability," p. 14.

246. On the provision of JDB loans for new silicon capacity see *Rare Metal News,* March 8, 1983; *Nikkei Sangyo Shimbun,* December 12, 1983; *Rare Metal News,* October 24, 1984.

247. See *Rare Metal News,* March 8, 1983.

248. *Kinzoku Jihyo* [*Metal Industry Review*] no. 1180 (April 25, 1984), pp. 9–10, stresses the fact that imports had reached the magic "50 percent" ratio as a cause for alarm.

dations of the study group report were being implemented, before the report was even finished, and a full fifteen months before the final report was published.[249] These recommendations included expansion of domestic production, encouragement of entry by new Japanese producers, negotiation of long-term contracts with foreign suppliers, direct investment in foreign producers, and construction of foreign plants to take advantage of lower electricity costs. Most of these recommendations were soon translated into action.

In August 1984 it was announced that poly producer Hi-Silicon (a joint venture between OTC and Japan Silicon) would double its annual poly capacity to 1,080 metric tons, and that Tokuyama would also expand its capacity to 1,000 metric tons. Shortly thereafter it became known in the Japanese trade press that both Tokuyama and Hi-Silicon were receiving subsidized industrial technology promotion loans from the JDB in connection with these expansions.[250] Other Japanese silicon producers also scheduled capacity expansions in 1984. These expansions continued even after the silicon market began to turn down in early 1985.

Aside from Tokuyama Soda's pre-study group entry there was a substantial wave of new investment in the silicon business on the part of Japanese steel companies after the release of this report. Nippon Kokkan announced its intention to enter the poly business in July 1985.[251] The next month Kawasaki Steel and LSI Logic announced a joint venture to manufacture wafers.[252] Nittetsu Electronics (owned by Nippon Steel) was willing to talk about its entry into the wafer business, and a technical link to Hitachi, in the second half of 1985.[253]

In addition, there was a substantial increase in foreign acquisitions and investment over this period. SEH and Mitsubishi Metals purchased substantial equity shares in the U.S. silicon producer Hemlock in 1984, while the silicon study group was still meeting. Nippon Kokkan purchased Great Western Silicon from GE in 1985, shortly after the release

249. MITI's study group was named the High Purity Silicon Issues Study Group (Kojundo Shirikon Mondai Kenkyukai). Its *Research Report* [*Chosa Hokokusho*] was published by the Japan Society of Newer Metals Association, a Tokyo industry association, in March 1985.

250. *Rare Metal News*, October 24, 1984.

251. See *Nihon Keizai Shimbun*, July 17, 1985.

252. Avra Wing, "See Japan Challenge to U.S. Wafer Makers," *Electronic News*, July 8, 1985, p. 14.

253. See "International Report: Japan," *Solid State Technology*, vol. 28 (September 1985), p. 20; and *Rare Metal News*, January 8, 1986.

of the group's report. Kawasaki Steel purchased wafer manufacturer Siltec in 1985, in part to supply wafers to its joint venture with LSI Logic. Mitsubishi Metals purchased wafer manufacturer NBK in 1986. OTC acquired American epitaxial wafer manufacturer U.S. Semiconductor in its 1986 fiscal year. SEH began production at a U.S. wafer factory in April 1984, and it finished a plant in England in 1985. NKK announced construction of a plant in Oregon in 1986, but that plan was later suspended when the market turned down.

These responses had a considerable effect: the import share of domestic silicon consumption abruptly dropped to well under 50 percent after 1985, then leveled off. Since the import ratio had been rising in the early 1980s, when the yen had been weak against other currencies (which made foreign products *less* price competitive than Japanese products) and then dropped decisively precisely when the yen strengthened (making foreign products *more* price competitive), this was all the more remarkable.[254]

Considerable activity also continued on the R&D front. By 1984 prototype processing equipment was in operation, and NEDO announced that the construction of a plant with annual capacity of 100 metric tons and using its energy-efficient technology was to begin in 1990.[255] Komatsu Denshi was also reported to have received a contract from NEDO to improve production processes for materials used in silicon production.[256]

NEDO continued through the end of the decade to sponsor work aimed at reducing electricity requirements in poly production. Although the details were not spelled out, NEDO announced in May 1989 the development of a new production technology that reduced the cost of solar grade poly by 50 percent. Interestingly, the attitude toward imports in at least that part of MITI has not changed much. Prominent play was given to the technology's possible role in eliminating dependence on imports of foreign solar grade poly: "Japanese companies have, until now, been totally dependent on imports for procurement of high-grade silicon materials for solar cell production. NEDO's new technology is

254. Indeed, a private antitrust suit (ultimately settled out of court) was brought by American silicon producer Union Carbide in 1988. It alleged that Japanese silicon producers had colluded within the MITI-sanctioned study group to exclude U.S. producers from the Japanese market. See Louise Kehoe, "Japanese Silicon Suppliers Named in Antitrust Suit," *Financial Times*, October 5, 1988; and "Carbide's Poly Charges Echo DRAM Flap," *Electronic News*, October 17, 1988.
255. See Noda, "Processes and Process Developments," p. 231.
256. See *Denryoku Jiji Tsushin*, January 30, 1985.

expected to pave the way for realizing 100% domestic production of solar batteries, according to industry experts."[257]

Aside from occasional episodes like these, however, the most obvious and visible trappings of protection had clearly been lifted in semiconductors. Even complaints about some of the less transparent forms of government intervention, such as administrative guidance, were to drop sharply as Japan turned its attentions fully toward global markets. As a new decade of exports and international competition began in the 1980s, it was a healthy and self-confident Japanese industry that emerged on the world scene. On the other side of the Pacific the American industry faced crisis and self-doubt, as for the first time it confronted a determined, well-funded, and highly capable foreign competitor.

Summary

The portrait painted in this chapter of the development of the Japanese semiconductor industry differs in some important respects from the simplest "Japan, Inc." stereotypes of Japanese industrial policy. A great deal of latent, and sometimes intense, competition within the Japanese semiconductor industry was an important element in its development. Continuing economic battles among Japanese chip companies were moderated from time to time by the Japanese government, which intervened in the name of the collective industry interest, typically in response to an external (foreign) threat.

The first such threat was competition from imports and investment by foreign companies, both of which were stoutly resisted from behind steep protective barriers around the Japanese market. The government blocked all such inroads and used all the weapons at its disposal to force foreign firms to transfer their technology to Japanese partners on the best possible terms for the latter. In the early 1960s, and again in the late 1960s and early 1970s, initially successful forays by foreign chip producers into the Japanese market were blocked and parried by the government at the behest of the domestic industry. This theme—of government providing the means to organize a sometimes intensely competitive and quarrelsome domestic industry into a cohesive united front in response to foreign

257. "NEDO Develops Technology to Produce Silicon Domestically," *Nikkei Sangyo Shimbun*, May 16, 1989, p. 13.

pressures or threats—is pursued along other dimensions in the next chapter.

The basically hostile environment toward foreign investment and imports visible in the 1950s and 1960s was softened considerably, in response to foreign pressure, in the 1970s. By the early 1980s most formal protectionist barriers had disappeared. However, a long tradition of informal "guidance" supplied to industry by the Japanese bureaucracy, coupled with significant regulatory and administrative controls in the economy, continues to provide the means for considerable, informal government influence over private industrial behavior. Such influence was sometimes visible in policies toward semiconductor materials infrastructure in the early to mid-1980s, when reduction of dependence on foreign imports became the objective of an informal, unannounced policy.

Japan's chip producers are mainly internal divisions of larger Japanese industrial conglomerates, and in the 1950s consumer electronics was the major interest of industrial policy in electronics. In the early 1960s the spotlight shifted to computers, and for the next thirty years industrial policy in electronics was mainly organized around a campaign to build a world-class computer industry in Japan. In the mid-1960s it became clear that a competitive semiconductor industry might be needed as a basic element of that plan, and in the 1970s, amid the VLSI revolution in the United States, developing such capabilities for Japan became an essential element of any plan to build an indigenous, self-sufficient computer industry. At that point the policies that later pushed Japan into the forefront of leading edge semiconductor manufacturing technology were put into place. By the mid-1980s Japanese firms had clearly succeeded in semiconductor manufacture (although an equivalent lead in chip design and computer systems continued to elude them).

The first major policy initiatives in semiconductors in the 1960s focused on subsidies to investments in plant and equipment by manufacturers—little money was spent on R&D. In the 1970s the emphasis shifted away from capital investment into R&D, particularly into investments in narrowly focused applied R&D on semiconductor manufacturing equipment and infrastructure. In the mid-1970s a new model of cooperative industrial R&D was introduced and refined in the VLSI projects, and further honed in later initiatives of the 1980s. Bowing to foreign complaints, MITI reduced its relative levels of support for semiconductor R&D, but a considerable part of the slack was picked up by others, particularly NTT, which continues to invest heavily in joint R&D with the major Japanese chip producers. Today, the overall contribution of public and quasi-public (that is, NTT) resources to national investment in semiconductor technology is roughly equivalent to the relative burden borne by government in the United States.

CHAPTER THREE

The Genesis of an American Trade Policy in Semiconductors, 1959–84

THE DOMESTIC semiconductor industry's appeals to the U.S. government for assistance began with the first probes by Japanese transistor producers into the international market in the late 1950s. These pleas have generally been phrased in terms of a defensive policy against some foreign threat. Over the years, however, the description of the threat and of the behavior to be neutralized or counteracted by policy interventions has evolved. The 1986 Semiconductor Trade Arrangement, in a sense, reflected the growing force and sophistication of these strategic arguments for semiconductor trade policy. This chapter reviews just how accounts of the strategic challenge posed by foreign producers have evolved over time, how new institutional mechanisms for dealing with these questions were created in the early 1980s, and how a sharp increase in trade frictions set the stage for the radical experiment in American trade policy set in motion in 1986. One important theme that emerges from a detailed historical examination of U.S. trade policy in semiconductors is that policy has often appeared to have been the cause and not just the effect of changes in the competitive conduct of Japanese semiconductor producers.

A Threat to National Security, 1959

Charges of sales of imports at unfairly low prices date back virtually to the beginnings of the modern semiconductor industry. As discussed in the last chapter, Japanese producers in the late 1950s made large investments in transistor production facilities. In those days there was a fairly sharp dividing line between relatively inexpensive, low-quality, low-frequency transistors used in consumer electronics products, such as transistor radios, which these same Japanese manufacturers were then beginning to manufacture in large volumes, and the relatively expensive, high-quality, high-frequency transistors used in high-performance defense and computer applications.

Early methods of transistor production generally made use of relatively primitive (by today's standards) labor-intensive manufacturing techniques, and open access to the fundamentals of recently developed transistor technology, coupled with relatively low wage rates, made Japanese producers very competitive in the production of low-end semiconductors for use in consumer electronics. Much of the development effort in the American industry, funded by the Department of Defense, was oriented toward increasing performance to better serve high-end military and computer systems markets; markets for lower performance consumer electronic components received relatively little attention.

The first great Japanese export blitz in transistors and consumer electronics at the end of the 1950s was to lead to the first serious clash between the Japanese and American electronics industries. In 1959 a surge of low-priced Japanese transistor imports first hit American markets. Whereas in 1958 Japan had shipped 11,000 units worth a total of $7,000 to the United States, in the first nine months of 1959 it was 1.8 million units worth $1.1 million.[1] In September 1959, citing national security concerns, an American trade association—the Electronics Industries Association (EIA)—petitioned the Office of Civil and Defense Mobilization (OCDM, the executive branch predecessor to the Office of Emergency Preparedness) to impose quotas on Japanese transistor imports.[2] The episode was notable not only as the first shot in a continuing political

1. See "Business Week Reports on: Semiconductors," *Business Week,* March 26, 1960, p. 113.
   2. See "Import Study Nears Showdown," *Electronics,* November 6, 1959, pp. 32–33; "Electronics in Japan," *Electronics,* May 27, 1960, pp. 99–100; and "Washington Rejects Transistor Import Quota," *Electronics,* June 8, 1962, p. 7.

battle over trade policy waged in Washington between American and Japanese electronics companies, but also because it revealed patterns of behavior that were to be repeated continuously over the following three decades.

On the American side, it was a divided industry that petitioned the government for relief. Numerous American companies were by then marketing products manufactured by Japanese partners: for example, in transistor radios, Motorola and General Electric (GE) served as marketers for Toshiba outside Japan, Emerson for Japan's Standard Radio, and RCA Victor and Channel Master for Sanyo. Other American electronics companies, such as IBM, Remington Rand, and Ampex, controlled Japanese subsidiaries and were concerned about jeopardizing their sales in the Japanese market. RCA, in addition to collecting royalty checks from its Japanese licensees, negotiated a licensing arrangement with Hitachi in computers in 1961, in order to sell within the protected local market. A similar arrangement was negotiated between NEC and Honeywell in 1962. Still other American electronics companies had significant ownership stakes in Japanese electronics producers: GE in Toshiba, Westinghouse in Mitsubishi, ITT in NEC, and North American Philips' Dutch parent in Matsushita. As a trade publication noted in mid-1960, "the 'haves' effectively muzzled EIA (both havenots and haves being members) from making a firm presentment to OCDM—and so the subject [of import quotas] quietly languishes."[3]

Within a few years these tensions had worsened, and the American electronics industry had openly split. On one side were U.S. electronic components firms, who favored protection against consumer electronics imports from Japan—in which the bulk of inexpensive Japanese components entering the U.S. market were embedded. On the other side of the chasm sat American consumer electronics companies, who were concerned about protectionist precedents jeopardizing their future access to foreign electronics markets, as well as their supply lines to Japanese manufacturers. By 1968, for example, the EIA was testifying on both sides of trade policy issues, with its components division in favor of protection against electronics imports, and its consumer electronics division opposed (with the exception of Sylvania, which supported protection, and RCA, which took no position).[4] This conflict between compo-

3. "Electronics in Japan," p. 99.
4. Motorola, which produced both equipment and components, belonged to the con-

nents users and producers was to become a persistent theme in later episodes of trade friction.

The appeal to national security concerns was another call to be echoed over the years. In 1960, however, as was already noted, there was a relatively clear dividing line between the types of transistors used in consumer electronics and those used in defense (industrial and computer applications). The Defense Department was at that time pouring substantial resources into the development of advanced semiconductor technology for use in military applications, and even directly subsidizing the industry's investment in new capacity, so it was not surprising that the department's Electronics Production Resource Agency submitted a study to OCDM concluding that adequate capacity existed for both present and future military transistor demand.[5] Narrowly interpreted, defense needs did not provide a particularly compelling rationale for action.

Geopolitical considerations beyond a narrow view of military production requirements also weighed in against a policy of protection. In the early 1960s the United States was concerned to limit Japanese trade with the Soviet Union and cement military cooperation with Japan. Building a strong, friendly bulwark against Soviet and Chinese expansion in Asia was a primary objective in U.S. policy toward Japan, and political and security links were given conscious priority over economic relations.[6]

Even then, however, there was considerable concern over Japan's overtly protectionist trade and investment policies, and the United States and the other industrialized countries exerted steady political pressure on Japan to open up its markets. Pressure from the International Monetary Fund led in 1960 to promulgation of a program of Trade and Exchange Liberalization Guidelines, but the Ministry of International Trade and Industry (MITI) quickly withdrew electronics goods from the list of items to be liberalized.

One concern voiced on the American side at this time—which was to persist to the present day—was that use of protectionist threats against Japanese exports would set a bad precedent, undermining progress toward more open national markets that had been achieved by the inter-

---

sumer electronics division and opposed protection. *Foreign Trade and Tariff Proposals,* Pt. 8, Hearings before the House Committee on Ways and Means, 90 Cong. 2 sess. (Government Printing Office, 1968), p. 3486.

5. "Import Study Nears Showdown," p. 32.

6. See Dennis J. Encarnation and Mark Mason, "Neither MITI nor America: The Political Economy of Capital Liberalization in Japan," *International Organization,* vol. 44 (Winter 1990), p. 37, for references on this point.

national economic system since 1945. The delicate balance between the threats of protection and the open trading system they were intended to secure (a system placed in jeopardy if threats are transformed into actions on too wide a scale) fueled policy debates in 1959, as it does today.

On the Japanese side, the response to the campaign for protection in Washington was to blaze some trails that would become equally well trod over time. A Japanese delegation traveled to Washington in 1959, where it stressed that activities in consumer electronics created no threat to U.S. defense, and that Japanese companies had "no immediate plans to go after the markets for highly specialized transistors."[7] The delegation also made clear its willingness to impose voluntary quotas or other negotiated arrangements. In the late spring of 1960, in response to continuing frictions, MITI proceeded to impose quotas and floor prices on transistor radios exported to the United States.[8] The system was continued in later years.[9]

The issue of third-country exports (evasion of quotas by shipping goods to other countries for reexport to the United States) arose early on and was dealt with administratively by MITI. By 1961 large quantities of Japanese transistors were being exported to Hong Kong (clearly by the same firms whose own exports of transistor radios to the United States were being limited), where they were then assembled into transistor radios and exported to the United States. With sales of inexpensive Japanese transistor radios in the United States undercut by large-scale imports of even less expensive radios made in Hong Kong, MITI suspended transistor exports to Hong Kong in May 1962. A system of quantitative limits on transistor exports to Hong Kong was set up a couple of months later.[10]

7. "Import Study Nears Showdown," p. 33.

8. Quotas for 1960 were set to equal a 20 percent increase over actual exports in 1958 and 1959. Penalties, including cancellation of the quota, were set for firms violating the floor prices or evading controls by exporting to the United States and Canada through third countries. See "Japanese Put Off Freeing Electronics Imports," *Electronics*, July 8, 1960, p. 11.

9. See "Japan Extends Quotas for Transistor Radios," *Electronics*, January 6, 1961, p. 9; and "Japan Eases License Rules, May Cut Transistor Prices," *Electronics*, May 26, 1961, p. 9. By 1971 Japan had voluntary export controls on transistor radio shipments to twenty-nine different countries. See U.S. Tariff Commission, *Trade Barriers*, vol. 5: *Part II: Nontariff Barriers* (1974), p. 255.

10. MITI suspended transistor exports to Hong Kong and Okinawa on the grounds that they were exporting transistor radios to the United States at a price $6 lower than that for equivalent Japanese exports. This episode almost set off a trade war between Hong Kong

Such export cartels had by then been used for some time in postwar Japan, initially in textiles, where they dated back to the early 1950s. By the early 1960s a complex system of legal and administrative measures permitted—indeed, often encouraged (since moderation of "excessive competition" was an avowed objective of industrial policy)—the formation of both domestic and export cartels, regulating pricing, capacity investments, or production levels. By the early 1960s, when similar measures covering transistor radios and transistors were put into place, the Japanese government estimated that roughly 30 percent by value of its exports to the United States were affected by some sort of quantitative control, on domestic or export price or quantity.[11]

Ultimately, the U.S. producers' petition to restrict Japanese transistor imports was rejected in 1962, after a good two and a half years of discussion, on relatively narrow national security grounds. The American semiconductor industry, after an alarming slowdown in 1960–61, was growing rapidly again, American producers were getting the vast bulk of the defense business, and capacity seemed adequate to meet any future surges in demand.[12] Although some U.S. producers continued to complain to Washington that the closed Japanese market was being used as the base for an export push into the American market, the government's attitude was perceived by the Japanese to be that Japan's concentration on the consumer market permitted U.S. firms to concentrate their resources on defense needs, and therefore contributed to national security. As one Japanese observer later put it, the Japanese were pleased that the United States, as a technologically advanced nation, could afford such a fair decision.[13]

Private versus Public Policy: Television Exports in the 1960s

Considerable ambiguity over the contours of official Japanese policy was to develop in the early 1960s, because of an often blurry borderline

and Japan. The U.K. government threatened to suspend Japanese cotton cloth imports by the Crown Colony unless the Japanese lifted their export restrictions on transistors. See "Japanese Transistors Sought by Hong Kong," *Electronics,* June 8, 1962, p. 8; and David Rose, "Hong Kong's Transistor Radio Exports Soar to 100,000 a Month," *Electronics,* September 28, 1962, p. 24.

11. See Eleanor Hadley, *Antitrust in Japan* (Princeton University Press, 1970), p. 387.
12. "Washington Rejects Transistor Import Quota," p. 7.
13. Yasuzo Nakagawa, *Semiconductor Development in Japan* (Tokyo: Diamond Publishing, 1985), pp. 125–27 (in Japanese).

between overt Japanese government mandates and ostensibly voluntary, private actions undertaken by groups of firms responsive to government "suggestions." This phenomenon is well illustrated by the system of price floors devised in 1963 for television receiver exports, the next major irritation to afflict U.S.-Japanese trade relations in electronics.

Japanese production of black-and-white television sets had grown quickly in the late 1950s and by the early 1960s accounted for roughly 60 percent of the value of Japanese consumer electronics output (radios by that time had dropped to about 30 percent of Japan's consumer electronics shipments).[14] As the Japanese market approached saturation in 1960, new strategies were required. The Japanese government attempted to expand the domestic market by setting new standards for color TV and financing new television broadcasting channels.[15] Producers were also encouraged to turn their attention to foreign markets: the U.S. color standard was adopted in order to facilitate exports of transistorized sets to the United States.[16] In mid-1960 MITI announced publicly that Japanese electronics producers were rapidly expanding production facilities for color TV sets primarily intended for export (the sets were thought to be too expensive for domestic consumers, and domestic sales were to be merely a sideline).[17] Japanese TV receivers first began to arrive in the United States in significant quantities in 1960.[18] By 1966 color TV production had risen to a half million sets, with half of that output going to the export market.[19]

Facing mounting trade frictions, seven television-producing members of the Japan Machinery Exporters Association agreed to a private "orderly marketing" agreement, setting minimum export "check" prices for TV sets shipped to the United States in the fall of 1963. In enforcing this system, Japanese television manufacturers also attempted to limit competition with each other in the U.S. market by agreeing to the so-called five-company rule, which required every exporter to limit its U.S. sales to five exclusive customers.[20] Under the Export-Import Trade Law of

14. See Electronic Industries Association of Japan, *Facts and Figures on the Japanese Electronics Industry* (Tokyo, 1986), p. 34.

15. "Japan Moves to Cut Television Set Surplus," *Electronics,* October 14, 1960, p. 11.

16. "Japan Launches Color TV," *Electronics,* January 22, 1960, p. 22.

17. See "Japanese Push Color TV Production Plans," *Electronics,* July 24, 1960, p. 11.

18. See "Japan Boosts TV Set Output," *Electronics,* February 26, 1960, p. 48; and "Japanese TV Sets Arriving This Week," *Electronics,* April 29, 1960, p. 32.

19. Electronics Industries Association of Japan, *Facts and Figures,* p. 35.

20. See Kozo Yamamura and Jan Vandenberg, "Japan's Rapid-Growth Policy on Trial:

1952, private price- or quantity-fixing agreements among firms were legal as long as they were eventually registered with MITI, but a MITI tradition of informal, unwritten "administrative guidance" to firms made even this boundary indistinct. Some details of this agreement were publicly disclosed prior to its implementation, and floor prices were in fact initially set in consultation with MITI.[21]

As this case illustrates, behind-the-scenes efforts to "privatize" actions taken to deter or defuse trade conflict have often been preferred over more public, visible (and perhaps precedent-setting) official interventions by the Japanese government. It is probably safe to argue that, from the early 1960s on, the construction of formal or informal export cartels was the standard response to trade frictions involving Japanese electronics.[22] The formal cartel agreements regulating television exports continued until 1973.[23]

Interestingly, absent continuing direct involvement by MITI in the operation of the cartel, discipline among the Japanese exporters ultimately began to break down. The minimum check price system began to disintegrate by the late 1960s (exporters undercut the minimum prices with a variety of under-the-table kickback schemes to U.S. buyers). The five-company rule, although strictly enforced, could be and was circumvented.[24]

The Television Case," in Kozo Yamamura and Gary R. Saxonhouse, eds., *Law and Trade Issues of the Japanese Economy: American and Japanese Perspectives* (University of Washington Press, 1986), pp. 259–63; and *Brief of Appellants, Zenith Radio Corporation and National Union Electric Corporation,* case nos. 81-2331, 81-2332, 81-2333, United States Court of Appeals, Third Circuit (Philadelphia: International Printing Company, 1983), pp. 14–19.

21. Accounts at the time made much of the distinction between price floors and actual pricing, since "the actual export prices cannot be set because of Japan's antitrust laws." See "Japan Firms Setting Minimum TV Prices," *Electronics,* August 2, 1963, p. 7. A fairly complete discussion of the history of price-setting schemes applied to Japanese TV exports may be found in Yamamura and Vandenberg, "Japan's Rapid Growth Policy on Trial," pp. 238–70.

22. It is not clear, however, that these cartels always worked well. For example, it is known that cheating on check prices on television export prices, in the form of under-the-table rebates to importers, sometimes occurred.

23. In 1974 Zenith Radio Corporation launched a widely publicized antitrust suit against Japanese television exporters.

24. *Brief of Appellants,* pp. 20–28; and Yamamura and Vandenberg, "Japan's Rapid Growth Policy on Trial," pp. 262–63. Despite the rule, some U.S. customers managed to purchase TVs from more than one of the Japanese exporters. A Japanese exporter could also circumvent the rule by naming as one of the five American purchasers its own U.S. subsidiary, which would then be free to sell to any U.S. customer. Kenneth G. Elzinga,

Roughly coinciding with the formation of the TV export cartel was the development of a domestic cartel to fix prices in the Japanese domestic color television market.[25] In 1964, after the continuing depressed circumstances of the TV industry had triggered a round of severe price cuts in the domestic market, a web of discussion groups was formed to set retail prices, rebates, retail and wholesale profit margins, and to discuss such matters as demand forecasts and bottom prices. Production, inventory, and shipment data were exchanged and market shares and output quotas assigned.[26] In 1970 six of the companies were found by the Japan Fair Trade Commission to have broken the antimonopoly law, although they were also found to have ceased the violations. But various of the groups associated with this domestic cartel continued to meet after the export cartel disbanded in 1973. One midlevel group continued to hold monthly meetings until 1977, and the highest-level group (the Okura Group, named after the hotel where the presidents of seven top consumer electronic producers met to discuss these issues for a variety of electric appliances) continued to meet through at least 1974.[27]

The link between the domestic TV cartel and the export cartel has always been controversial. MITI was clearly involved in the design and establishment of what was nominally a privately administered export cartel arrangement and sent representatives to meetings at which actions were taken to set export prices and foreign market shares. If only because

---

"The New International Economics Applied: Japanese Televisions and U.S. Consumers," *Chicago-Kent Law Review*, vol. 64, no. 941 (1988), p. 964; and Franklin M. Fisher, "Matsushita: Myth v. Analysis in the Economics of Predation," *Chicago-Kent Law Review*, vol. 64, no. 941 (1988), p. 973.

25. In the 1950s Japanese producers had set up a cartel to regulate prices in black and white television sets. As David Schwartzman has commented, Japanese TV manufacturers and distributors created a cartel, the Home Electric Appliances Market Stabilization Council, under the aegis of the Electronic Industries Association of Japan in 1956. This group set up a series of retail price agreements and enforcement procedures and attempted to organize production cuts (not fully successfully). In 1957 the Japan Fair Trade Commission decided that the enforcement procedures violated Japan's antimonopoly law and ordered the Market Stabilization Council to halt punitive actions against discounters. (The JFTC did not, however, rule against price fixing per se or force the council to disband.) See Schwartzman, *The Japanese Television Cartel: A Study Based on* Matsushita v. Zenith (University of Michigan Press, 1993), pp. 77–80.

The considerable evidence unearthed in the 1970 and earlier investigations is also cited in Yamamura and Vandenberg, "Japan's Rapid Growth Policy on Trial," pp. 253–57; *Brief of Appellants,* pp. 35–38; and Elzinga, "The New International Economics Applied," p. 964.

26. Schwartzman, *Japanese Television Cartel*, pp. 81-89.

27. Schwartzman, *Japanese Television Cartel*, pp. 87-89.

of the 1970 JFTC decision touching on selected elements of the domestic TV cartel, MITI could hardly have been unaware that there was a parallel cartel structure in place in the shadows for the domestic market. The issue is significant because U.S. television producers always asserted that the relatively higher domestic prices created by the actions of an illegal— but government-tolerated—domestic cartel made it economically feasible for the export cartel to set the much lower foreign prices that did them such serious damage. Allegations of a similar two-tiered pricing structure—low export prices and high domestic prices in a formally or informally protected domestic market—were to be an important factor in U.S.-Japan semiconductor trade friction in the late 1970s.

Competition in the 1970s

Although the immediate competitive threat to the American chip industry was dissipated by its breakneck technical advance, American firms remained most unhappy about being shut out of the Japanese market. As Japanese electronic exports making use of ICs began to enter the U.S. market in the late 1960s, Texas Instruments (TI)—which, barred from establishing a presence in Japan, had refused to license its patents to Japanese companies—successfully jimmied open the Japanese market a notch by threatening to petition to exclude from the U.S. market Japanese exports of electronic equipment using semiconductors that infringed on its patents (a detailed account of this episode may be found in chapter 2). TI was finally permitted to establish a joint venture in Japan, subject to output restrictions, which was then converted into a wholly owned subsidiary in 1971. TI's early presence was the exception that proved the rule of barriers to entering the Japanese market, however.

By the early 1970s rapid innovation had made American chip producers the unchallenged leaders worldwide. In Japan, as was noted in the last chapter, Japanese chip makers were even complaining about U.S. companies "dumping" low-priced chips in the Japanese market, and MITI took active steps to slow the inflow of U.S. imports of ICs.

This was not to say that American chip production did not come under competitive pressure from foreign factories. In fact, the late 1960s and early 1970s marked a rapid transfer of large segments of the American chip industry to offshore locations, as U.S. semiconductor companies moved their relatively labor-intensive assembly and testing operations to foreign subsidiaries or contractors in low-wage countries.[28] A substantial impact from foreign trade was being felt within the U.S. industry: in 1971, for example, former workers at Sprague Electric semiconductor plants petitioned for trade adjustment assistance as U.S. imports of semiconductors assembled overseas surged.[29] In 1973 and 1974 workers at GE transistor and diode factories filed for similar assistance in the face of continuing increases in imports.[30] By the late 1970s approximately 70 to 80 percent of U.S. companies' semiconductor shipments were assembled offshore.[31] But these actions mainly reflected internal restructuring *within* American multinational companies and created little in the way of genuine trade friction likely to stir up international political confrontations.

Under continuing foreign pressure, Japanese quantitative import restrictions on semiconductors had been gradually phased out by 1976. Foreign investment was liberalized in 1974, and most formal trade barriers had vanished by 1980. Restrictive practices in procurement by Nippon Telegraph and Telephone (NTT), the state-owned telecommunications monopoly, as well as in standards, certification and quality requirements, and membership in R&D associations continued to generate complaints by foreign chip makers, however. With the market opening of the 1970s, Japanese chip producers began to face serious competition and were frequently forced to rely on direct intervention by MITI with their customers to fend off American IC imports. As the inevitability of liberalization of the Japanese market became clear, government subsidies to semiconductor R&D were greatly increased in order to help Japanese producers adjust to the oncoming new realities. NTT and MITI launched highly successful cooperative industrial research programs in 1975 and 1976, respectively. The programs were focused on technologies to im-

28. See Kenneth Flamm, "Internationalization in the Semiconductor Industry," in Joseph Grunwald and Kenneth Flamm, *The Global Factory: Foreign Assembly in International Trade* (Brookings, 1985), pp. 68–85.

29. See U.S. Tariff Commission, *Capacitors and Semiconductors: Former Workers of the Sprague Electric Company Plants at—North Adams, Mass., Worcester, Mass., Hillsville, Va., Lansing, N.C., Concord, N.H., Barre, Vt., Grafton, Wis.*, TC publication 395 (May 1971).

30. See U.S. Tariff Commission, *Transistors and Diodes: Workers of the Buffalo, N.Y., Plant of General Electric Co.*, TC publication 588 (June 1973); and U.S. International Trade Commission, *Transistors and Diodes: Workers and Former Workers of the Syracuse, N.Y., and Auburn, N.Y., Plants of General Electric Co.*, ITC Publication 715 (February 1975).

31. Flamm, "Internationalization in the Semiconductor Industry," pp. 83–84.

prove mass production of high-volume chips used in the computer indus-try, particularly DRAMs.

By 1977 it was becoming clear in the United States that the ongoing technology push in Japan was achieving important results and was being accompanied by increased investments in capacity. An initial trickle of imported Japanese DRAMs had begun to surface in the U.S. market. A surge in exports of Japanese chips to the U.S. market, similar to those already seen in consumer electronics and automobiles, seemed likely. Determined to maintain a more effective voice in Washington, American chip producers formed the Semiconductor Industry Association (SIA) as their lobbying arm.[32]

In October 1978 Robert N. Noyce, chairman of Intel, a U.S. DRAM producer, launched the first public salvo across the Japanese bow, charg-ing that Japan's protection of its home market, together with government subsidies, made it possible for Japanese producers to engage in "two-tier pricing." Citing the example of the television market, Noyce argued that firms were free to charge high prices in the sheltered Japanese market, then price exports "as low as they want, since they need only to cover incremental variable costs."[33] Without actually claiming that this was currently occurring in semiconductors, Noyce noted that "if this pattern is repeated in the semiconductor market, the U.S. market would be flooded with underpriced Japanese integrated circuits and LSI products," and U.S. producers would "have no choice but to cut production or go bankrupt."[34]

Noyce's fears were not without factual support. There is some evi-dence, largely anecdotal, that chip prices for leading edge digital ICs tended to be higher in Japan than in the United States through much of the 1970s. As one prominent Japanese manufacturing expert put it: "When production of IC's began in Japan around 1970, IC's made in America always sold for 20% less than the Japanese product. This low price was made possible by the scale merit produced by the huge com-puter industry and the defense industry. When the Japanese companies managed to cut their costs through rationalization of their industry, the

32. After a transcontinental pilgrimage to Washington intended to inform U.S. Trade Representative Robert Strauss that "the Japanese are coming," semiconductor executives were reportedly dismayed to get a "so what?" reaction from Strauss. Author's discussion with industry executives and former government officials, September 1992.

33. See Peter Moylan, "Noyce Rips Gov't as Peril to U.S. Semicon Industry," *Electronic News,* October 9, 1978, pp. 51–52.

34. Moylan, "Noyce Rips Gov't," p. 52.

Table 3-1. *Factory Prices of Finished Integrated Circuits in Europe, Japan, and the United States, 1976*

Dollars

Type of device	West Europe	United States	Japan
Calculator LSI	2.75	2.45	1.86
MOS	3.20	3.00	3.20
Bipolar digital	0.85	0.75	1.36
Bipolar linear	1.10	1.00	0.70
Microprocessor (including memory and support circuits)	150.00	95.00	150.00
Calculator and watch displays (4 mm)			
LED	1.80	1.70	1.50
LCD	5.00	4.50	4.50
Clock displays (15 mm)			
LED	5.50	5.00	5.25
LCD	8.50	8.00	8.00

Source: Mackintosh Consultants, *Market Survey of Semiconductors*, vol. 4: *Applications and Markets* (Study undertaken on behalf of the Ministry of Research and Technology of the Federal Republic of Germany, December 1976), pp. 125–27. Quoted average costs reflect differences in product specifications and volume requirements.

American companies always cut their price even further."[35] Table 3-1, although it does not control for volume and product mix, shows that unit costs for a variety of advanced ICs tended to be significantly higher in Japan in the mid-1970s, while products typically used in consumer electronics (linear ICs, display chips) actually tended to be priced below U.S. levels.

The SIA's activity was successful in prodding a Senate subcommittee, in December 1978, to order the U.S. International Trade Commission (ITC) to launch an informational investigation into the competitive position of the U.S. semiconductor industry.[36] By the time the commission's report was delivered, at the end of 1979, the forecast threat had materialized. Amid a boom in U.S. electronics production, a major shortage of 16K DRAMs had developed over the summer of 1979.[37] Viewing these shortages as an opportunity, three major Japanese producers—Hitachi, Fujitsu, and NEC—had plunged into the U.S. market in force. By the

35. Hajime Karatsu, "Quality Control—The Japanese Approach," paper presented at a seminar, "Quality Control: Japan's Key to High Productivity," Washington, March 25, 1980, p. 9.

36. See Richard Wightman, "ITC Launches Probe of U.S. Semicon Position in Japan, Europe," *Electronic News*, December 18, 1978, p. 44.

37. See Steven Hershberger, "Users Book 16K RAMs Through 1980," *Electronic News*, July 16, 1979, p. 49; and "Schottky, RAM Bind Hits User Revenues," *Electronic News*, September 3, 1979, p. 1.

end of the year they had collectively achieved about a 40 percent share of the U.S. market for 16K DRAMs.[38]

During the nine-month ITC investigation, the SIA's evolving theory of Japanese industrial practices was further elaborated. At an ITC hearing in San Francisco in May 1979, the SIA for the first time suggested that the two-tier pricing scenario was actually occurring, notably in sales of 16K DRAMs.[39] Also, apparently for the first time, it was suggested that, in addition to being an example of what would now be called a strategic trade policy implemented by the Japanese state, Japanese policies contained an explicitly predatory element. Noyce, arguing on behalf of the SIA at the May hearing, articulated a strategic conception of the dangers of dependency on foreign suppliers as follows: "Now one might argue that U.S. consumers benefit from these bargain prices. But we must realistically ask how long such bargain prices last. Middle Eastern oil was a bargain until we became dependent upon it. Similarly, sooner or later the Japanese losses on high density memories will be recouped and I submit that it is foolish to assume any long run benefit to consumers."[40]

What is vague in these statements is whether these strategic calculations are viewed as being undertaken by the Japanese state, with Japanese companies passively responding to changes in state policy, or whether the companies are being accused as active parties to the strategic plan. One might, for example, conceive of "state predation," where state subsidies induce firms to cut prices in order to stimulate exit by foreign rivals.[41] At the time, though, the prevailing conception was one of "Japan, Inc.,"

38. See Lloyd Schwartz, "Mostek Chief: Japan Threatens Industry," *Electronic News,* October 15, 1979, p. 72; and Henry Scott Stokes, "Japan Goal: Lead in Computers," *New York Times,* December 12, 1979, p. D1. Buoyed by surging demand for computers, even IBM was reportedly contacting Japanese chip suppliers in search of 16K DRAM supplies. See John Hataye, "IBM Shopping in Japan for 16K Dynamic RAMs," *Electronic News,* December 24, 1979.

39. See Jim Leeke, "'Practices Abroad Unfair,' SIA Says at ITC Hearing," *Electronic News,* June 4, 1979, p. 106.

40. "Statement of Dr. Robert N. Noyce, Vice Chairman of the Board, Intel Corporation, on Behalf of the Semiconductor Industry Association," Hearings before the U.S. International Trade Commission, San Francisco, May 30, 1979, pp. 21–22.

41. Willig, for example, draws a distinction between "strategic dumping," which relies on national policies to protect exporting companies' home market, in order to gain cost advantages and create monopoly power for the exporters in importing markets, and "predatory-pricing dumping," which is a company strategy to obtain monopoly power in an importing country's market. See Robert D. Willig, "The Economic Effects of Antidumping Policy," Organization for Economic Cooperation and Development, Paris, 1992, pp. 7–8.

with firms and state joined together in a collective strategic plan. In the colorful words of one top American executive, it was the "33 companies in the SIA taking on the sovereign nation of Japan."[42]

"Below Cost" Dumping

The era of two-tier pricing was relatively short-lived. The evidence submitted in 1979 was quite scanty—the head of Mostek, a U.S. DRAM producer, asserted a 20 to 30 percent differential between U.S. and Japanese 16K DRAM prices in 1978 and presented first-quarter 1979 data on five selected sales contracts by a U.S. company to Japanese customers, showing prices significantly above the prevailing prices for equivalent Japanese products sold in the United States.[43] In any event, an attempt to investigate the issue in greater depth would have soon run into the complexities of market structure and distribution patterns for semiconductors in the United States and Japan, which make price-to-price comparisons between the two markets rather tricky.

Briefly put, roughly 70 to 80 percent of sales to U.S.-based semiconductor customers are transacted directly with chip manufacturers through long-term contracts, with deals typically struck months ahead of delivery. The balance is sold through distributors to smaller customers, and as spot sales, often through an active "gray" (secondary) market of brokers, distributors, and other arbitragers. Spot prices typically rise above contract pricing in tight markets and fall below contract prices when demand is slack.[44] Thus, to avoid comparing apples with oranges when searching for price differentials, contract prices should properly be compared with

42. W. J. Sanders, president of Advanced Micro Devices at the time, quoted in Leeke, "'Practices Abroad Unfair,'" p. 106.

43. These data were contained in a confidential submission to the ITC dated August 17, 1979, but appear to have been presented later at a hearing of the congressional Joint Economic Committee in October 1979. See ITC, *Competitive Factors Influencing World Trade in Integrated Circuits,* publication 1013 (November 1979), pp. 70–71; and *U.S.-Japanese Trade Relations,* Hearing before the Joint Economic Committee, 96 Cong. 1 sess. (GPO, 1979), pp. 21, 26. Comparisons are further complicated by the fact that DRAMs are sold on both long-term contracts and on a spot basis, for immediate delivery, and prices can diverge substantially. It is unclear whether an appropriate spot-to-spot or contract-to-contract comparison between the two markets was being made.

44. For detailed evidence on the structure of the U.S. semiconductor market, see Kenneth Flamm, "Measurement of DRAM Prices: Technology and Market Structure," in Murray F. Foss, Marilyn E. Manser, and Allan H. Young, eds., *Price Measurements and Their Uses* (University of Chicago Press, 1993).

other contract prices, not with spot prices, and vice versa. Dumping complaints, historically, have not always drawn these distinctions.

Furthermore, the semiconductor market in Japan has a rather different structure from that in the United States. For the most part, large chip manufacturers in Japan sell directly only to sister electronic equipment divisions of the parent corporation. The vast bulk of external sales to large customers go through authorized sales agents, while smaller customers are served through secondary sales agents who order product from the main sales agents. Even smaller quantities are sold on a spot basis through retailers clustered in selected urban areas, such as Tokyo's Akihabara district.[45] Prices quoted in Japanese trade sources typically refer either to prices to large users through main sales agents or to spot prices in Akihabara. The prices large users pay in Japan are roughly comparable with U.S. contract prices, whereas U.S. spot prices are most similar to Akihabara pricing.

The complexities of direct U.S.-Japan price comparisons never became a major issue, however, because charges of two-tier pricing had a relatively short life. In the months after the San Francisco hearing, demand for 16K DRAMs surged, and as prices soared, complaints about low-priced imports faded away. From mid-1979 until the present day, charges that U.S. prices for Japanese chip imports were below Japanese levels ceased to be an important irritant to trade relations.

Instead, complaints that the Japanese were selling below the cost of production in *both* markets began to emerge.[46] Texas Instruments, the

45. See U.S. Department of Commerce, International Trade Administration, "Japan—Semiconductors/Nonvolatile Memory in Japan—ISA9106," derived from Fuji Keizai Co., "The Semiconductors–Nonvolatile Memory Market in Japan," Tokyo, June 1991; "Japan-Semiconductors/Analog Devices in Japan—ISA9106," derived from Fuji Keizai, "The Semiconductors–Analog Devices Market in Japan," Tokyo, June 1991; and "Japan—Semiconductors/Logic Devices in Japan—ISA9106,"derived from Fuji Keizai, "The Semiconductors–Logic Devices Market in Japan," Tokyo, June 1991.

NEC and Mitsubishi were reported to sell 100 percent of their Japanese sales through sales agents, while Hitachi, Fujitsu, and Toshiba sold 80 to 90 percent of their external sales by this route. Smaller manufacturers such as Sharp were reported to sell as much as 20 to 30 percent of their shipments directly to users, and the remainder through sales agents.

Primary sales agents are typically associated with a particular Japanese manufacturer. From the standpoint of market access, Japanese sales agents rarely handle foreign products that are very similar to, or compete with, products from the Japanese manufacturer with which they are affiliated. Since most Japanese producers offer broad product lines, it is unusual for a primary sales agent to sell imported semiconductors at all.

46. Testifying before a Senate committee in early 1980, Intel's Noyce saw a political

only American producer manufacturing in Japan at the time, had attacked the SIA position in early 1980, asserting that prices received by Japanese producers in their home market were actually lower than American prices for the same Japanese chips; the SIA argued that the Japanese were voluntarily restraining their exports in response to American industry complaints and deliberately creating a soft market at home.[47] Press reports imply, however, that prices in a weakening U.S. market still felt pressure from falling prices in the Japanese market. By mid-1980, for example, Japanese DRAM producers were reported to be starting a campaign of surveillance of shipments to their sales agents, to discourage their resale at very low prices to customers in the U.S. market by gray market traders (and further irritating American producers).[48] By the spring of 1981 Japanese producers NEC (which exported 60 percent of its output to the United States) and Fujitsu had even suspended sales into the U.S. spot market, to lessen the threat of an American antidumping action.[49] Four of the major Japanese IC producers (NEC, Hitachi, Toshiba, and Fujitsu) also announced plans to manufacture DRAMs at U.S. plants, in a complementary bid to reduce trade frictions.[50]

---

intent in this development: "Indeed, Intel buys the 16K RAMs from Japan because we have found that cheaper than to make them ourselves. Now, there is some artificial pricing in that market. That's what I'm suggesting." Senator [Adlai] Stevenson: "Well, that's what I'm getting at. Is there? What do you mean by that?" Dr. Noyce: "Up until the San Francisco hearings, we could buy 16K RAMs in the United States, from Japanese companies at lower prices than we were selling the same product for in Japan." Senator Stevenson: "Is that artificial pricing or are they just more productive and efficient?" Dr. Noyce: "We were meeting the market price in Japan. After the ITC hearings in San Francisco, U.S. prices went up and Japanese prices went down. I think the prices had been artificial there, but that is a very difficult thing to determine." *Trade and Technology, Part III,* Hearings before the Subcommittee on International Finance of the Senate Committee on Banking, Housing, and Urban Affairs, 96 Cong. 2 sess. (GPO, 1980), p. 176.

47. "U.S. Semiconductor Firms Disagree on Import Strategy," *Denver Post,* March 23, 1980, p. 40.

48. See "Exports of Japanese Semiconductors to US at Low Prices Conspicuous; Half Price, Too, through Trading Firms; Manufacturers Strengthen Checking of Destinations," *Nihon Keizai Shimbun,* May 24, 1980, p. 6.

49. "NEC Suspends Shipments of RAM Chips to U.S. Market," *Japan Economic Journal,* March 3, 1981; and "Japanese Electronics Firms Delay Plans to Mass-Produce 64-K Chips," *Asian Wall Street Journal Weekly,* April 6, 1981, p. 14.

50. See "Japan Firms Plan U.S. Production of Advanced Circuits," *Asian Wall Street Journal Weekly,* May 18, 1981, p. 17; Thomas J. Lueck, "NEC Plans $100 Million U.S. Plant," *New York Times,* June 27, 1981, p. D1; and "Top Four Japanese IC Makers Expand U.S. Operations," *Asian Wall Street Journal Weekly,* July 13, 1981, p. 15.

Mounting Frictions: "Quality Dumping"

Japanese chip makers had in fact taken their first hesitant steps toward setting up U.S. production facilities, amid intensifying political criticism, several years earlier. Among the major Japanese companies, NEC had been the pioneer in 1978, when it acquired Silicon Valley producer Electronic Arrays.[51] At least partly in explicit reaction to rising protectionist sentiment, Fujitsu announced an investment in a San Diego manufacturing facility in 1979. In that same year Oki Electric opened negotiations to have some of its products produced in the United States, Hitachi unveiled plans to use a Dallas facility for semiconductor assembly, and NEC began shipping U.S.-made chips from its California plant.[52]

By late 1979 it had become clear that U.S. imports of Japanese chips had roughly doubled in value, compared with the previous year.[53] Amid an unforeseen boom in demand for chips used in computers, and manufacturing yield problems on American production lines for 16K DRAMs, a shortage of DRAMs had developed in the United States.[54] Japanese chip makers rushed into that vacuum. In a parallel development, many Japanese semiconductor equipment makers had begun their first serious probes into the U.S. market. Key production tools developed in cooperation with the NTT and MITI R&D programs in very large scale integration—such as the Canon projection alignment system and the Takeda Riken (later renamed Advantest) high-speed logic tester—were first offered for sale in the U.S. market in 1979.[55] The clearly evident Japanese thrust into the U.S. semiconductor market provoked a rising crescendo of political criticism within the United States.[56]

51. There had been at least two instances of Japanese investments in small American semiconductor companies earlier in the 1970s: Toyo Electronics (which later changed its name to Rohm) acquired Exar Integrated Systems in 1972, and Hattori Seiko acquired Micropower Systems in 1977.

52. John Hataye, "Semicon Makers May Increase Production in U.S.," *Electronic News*, June 11, 1979, p. 24; and "Special Report: Japan is Here to Stay," *Business Week*, December 3, 1979, pp. 81–86.

53. See "Japan IC Exports to U.S. in 8 Months Double to $96M," *Electronic News*, November 19, 1979, p. 58.

54. "Special Report: Can Semiconductors Survive Big Business?" *Business Week*, December 3, 1979, p. 85; and Cheryll A. Barron, "Microelectronics Survey: All That Is Electronic Does Not Glitter," *Economist*, March 1, 1980, pp. 3–4.

55. See John Hataye, "Japan Equip. Firms Eye Exports to U.S.," *Electronic News*, December 3, 1979, p. 20. For a discussion of these products' roots in Japan's VLSI projects see chapter 2.

56. For example, see Lloyd Schwartz, "Mostek Chief: Japan Threatens Industry," *Elec-*

One of the oddest elements in the rising chorus of American complaints was the charge of "quality dumping" that emerged in late 1979, when U.S. and Japanese DRAM prices were undeniably in approximate parity.[57] Reports from American chip users suggested that the Japanese DRAMs they were buying had defect rates one-half to one-third those experienced with comparable American products.[58] U.S. chip makers for the most part did not deny the bad news coming from their customers. Instead, reasoning that higher levels of quality reflected increased expenditure of resources on quality control,[59] the American industry charged that sales of higher quality Japanese products at the same price as American products reflected a form of "dumping." But U.S. companies also increased their own investments in quality control programs.

The Japanese industry's response to this complaint was to hold a well-publicized seminar, "Quality Control: Japan's Key to High Productivity," in Washington in March 1980. The seminar's presentations made the argument that it was superior technology—in the form of quality control techniques imported from American sources and applied and improved upon in Japan—that was responsible for the Japanese quality edge. The seminar also worked to reinforce customer perceptions of the American industry's quality deficiencies, by spotlighting a speaker from Hewlett-Packard Co. who made public data documenting a large quality gap between American and Japanese DRAMs purchased by that company.[60]

tronic News, October 15, 1979, p. 72; Steven Hershberger, "Mostek Chief: Limit Imports from Japan," *Electronic News,* November 19, 1979, p. 42; and Stokes, "Japan Goal: Lead in Computers," p. D1.

57. See Stokes, "Japan Goal: Lead in Computers," p. D1.

58. "Can Semiconductors Survive Big Business?" p. 85.

59. Testifying before the U.S. International Trade Commission in 1979, SIA members argued that "'double' testing techniques of the Japanese were expensive ('It is an economic issue, not a quality issue') and that they considered this 'better deal' given to buyers of Japanese circuits a type of market penetration technique." U.S. International Trade Commission, *Competitive Factors Influencing World Trade in Integrated Circuits,* USITC publication 1013 (November 1979), p. 24. The SIA notes that "To achieve this quality there is a high dependence on the best available tools and automation. This adds to the capital cost, but it may pay in the long run through higher yield and higher productivity. Japan, because of its easier access and lower cost of capital has an inherent advantage." *The International Microelectronics Challenge: The American Response by the Industry, the Universities, and the Government* (Cupertino, Calif., May 1981), pp. 27–28. See also "Japan Makes Them Better," *Economist,* April 26, 1980, p. 55.

60. See Ray Connally, "Japanese Make Quality-Control Pitch," *Electronics,* April 10, 1980, p. 81; and "Japan Makes Them Better," pp. 55–56.

Over the next several years American chip makers suffered through a continuing series of unfavorable quality comparisons with the Japanese by their customers. Although these assessments showed considerable improvement relative to the Japanese competition, American chips were still perceived to lag in quality through the early years of the 1980s.[61] A federal crackdown on fraudulent quality testing and certification practices in Defense Department chip procurement over 1981–82 did little to help this perception.[62] The American semiconductor industry fought back by touting its aggressive adoption of improved quality management practices over this period, and questioning the existence of a quality gap with Japan to the extent portrayed in customer reports.[63] However, at least some American chip producers, as they point to their current world-class standards, frankly acknowledge the existence of a significant quality gap in the early 1980s.[64]

61. See "U.S. RAMs Improve But Still Lag Japan's," *Electronic Engineering Times,* November 10, 1980, p. 1; "Xerox Data Uphold HP Contention that Japanese RAMs are Better," *Electronic Engineering Times,* March 2, 1981, pp. 2, 16; Ray Connolly, "Copy Japanese, U.S. Managers Urged," *Electronics,* April 21, 1982, pp. 106–08; Jerry Lyman and Alfred Rosenblatt, "The Drive for Quality and Reliability, Part 1," pp. 125–28, and "Part 3: Users Push for Quality," pp. 141–43, *Electronics,* May 19, 1981; "HP, Motorola Continue the Debate over Quality and its Management," *Electronic Engineering Times,* May 25, 1981, pp. 14–15; and Larry Waller, "Perception Lag Nags U.S. Chip Makers," *Electronics,* April 21, 1982, pp. 42–43.

62. See "A Crackdown on Chip Quality," *Business Week,* January 11, 1982, p. 110; and Marilyn Chase and Jim Drinkhall, "Grand Jury Bores in on the Faulty Testing of U.S. Defense Gear," *Wall Street Journal,* October 4, 1982, p. 1.

63. Waller, "Perception Lag Nags U.S. Chip Makers," pp. 43–44.

64. Top Intel executive Craig Barrett, for example, was quoted in a 1991 Intel publication as saying that, in the early 1980s, "we realized that while we had always been a leader in technical solutions and product reliability, our product quality levels did not meet world class standards. After benchmarking against the best manufacturers in the world, we learned that we had a long way to go to match up." Katie Woodruff, "Intel's Journey to Total Quality," *Intel Microcomputer Solutions* (January–February 1991), p. 8.

It is worth contrasting this assessment with the public statements of Intel vice-president Willard Kauffman in April 1982: "Kauffman, in fact, questions whether any such gap actually existed to the extent it was perceived by Richard Anderson of Hewlett-Packard Co.'s Computer Systems division. . . . According to Intel's Kauffman, the Japanese still excel in their 'strategy to market quality. . . . We have to sell our quality in the same way. We have nothing to be ashamed of,' he emphasizes." Waller, "Perception Lag Nags U.S. Chip Makers," p. 44.

The customer critique of American chip producers' quality practices in the early 1980s is implicitly supported by U.S. semiconductor companies' advertising and marketing literature in the late 1980s, which stressed the impact of the adoption of rigorous statistical process control procedures on defect rates. Advertising by National Semiconductor Corporation, for example, depicted a decline in defects from 7,800 parts per million (ppm) in

Organizing a Response

In the spring of 1981 the SIA unveiled its program to combat the growing competitive threat from abroad. It identified three key areas for action. One was international in focus: access to foreign markets. The other two were domestic in nature: greater U.S. stimuli to capital formation, and increased investment in U.S. R&D and engineering education. Specific goals with respect to market access in Japan included a reduction of the Japanese tariff on semiconductors from 10 percent to 4.2 percent by April 1982 (rather than in 1987, as scheduled by the Tokyo Round agreement on multilateral tariff reductions under the General Agreement on Tariffs and Trade), and equivalent treatment for U.S. and domestic companies in Japan (national treatment) in "access to financing at competitive rates, bureaucratic processing of subsidiary filings with the government, and ability to recruit top Japanese engineering talent." For the first time also, the SIA called on the Japanese government to create an "affirmative action program" to remedy the effects of past discriminatory practices against foreign firms.[65]

Action quickly followed on several of these issues. In May 1981 the U.S. government announced that it had negotiated the requested reduction in Japanese tariff levels.[66] Congress passed a 25 percent incremental tax credit for R&D in 1981, after the SIA pressed its case for this measure.[67]

The industry itself also took concerted action on the research and education fronts. By late 1981 the SIA had formed the Semiconductor Research Cooperative, a joint effort to funnel R&D funds into basic research projects in universities.[68] And at Stanford University, with joint funding from both industry sponsors and the Defense Advanced Re-

---

1978 to 37 ppm in 1988. National's vice president for quality and reliability was quoted as saying that "We feel our ability to produce and supply high-quality, reliable products is *now* [author's emphasis] second-to-none in the industry." National Semiconductor Corporation, *National Anthem* (1989), p. 7. Similarly, Intel reported a decline in defects from 10,000 ppm in the early 1980s to 200 in 1989, with levels in the 10- to 50-ppm range for mature products. Woodruff, "Intel's Journey to Total Quality," pp. 8–9.

65. Semiconductor Industry Association, *The International Microelectronic Challenge*, pp. 22–23.

66. Clyde H. Farnsworth, "U.S. and Japan Plan Cuts in Semiconductor Tariffs," *New York Times*, May 12, 1981, p. D1.

67. See Andrew Pollack, "Federal Aid Sought for Semiconductors," *New York Times*, May 19, 1981, p. D1.

68. "Electronics Research Projects," *New York Times*, December 27, 1981, p. D1.

search Project Agency (DARPA), a new Center for Integrated Systems was established to train graduate students in microelectronics and focus research on relevant areas.[69] The announcement by MITI in 1981 of its ten-year project to develop a "Fifth Generation" computer prompted calls within both the U.S. computer industry and the U.S. semiconductor industry for the formation of analogous R&D joint ventures among U.S. companies, but the first such venture—the Microelectronics and Computer Technology Corporation (MCC)—was not to come into existence until January 1983.[70] The antitrust concerns that slowed the formation of MCC were not definitively put to rest until Congress passed the National Cooperative Research Act of 1984, which greatly reduced these worries for research joint ventures.

Competition, Collusion, or Predation? The 64K DRAM Wars

Conflict between the U.S. and Japanese industries was to worsen as the next generation of DRAM, the 64K, was introduced in 1981. Japanese producers had diverted resources toward early introduction of the 64K DRAM in mid-1981 and gained a lead over most of their American competitors.[71] Aggressive Japanese pricing in 64K DRAMs and rapidly falling prices stimulated a new round of industry complaints in the United States.[72] By early 1982, with the Japanese share of the U.S. 64K DRAM

69. See "Integrated Systems Center Begins Work," *Stanford Observer,* February 1982, p. 1. DARPA also funded more traditional academic research programs at universities, including the relatively large effort at the Massachusetts Institute of Technology (MIT). See Paul Schindler, "VLSI: Retrieving the Top Spot from Industry," *Technology Review* (January 1983), p. A3, which notes that DARPA funding accounted for $2.7 million of a $5.1 million budget for the MIT microsystems program.

70. "Control Data's Push for a Cooperative Chip," *Business Week,* April 20, 1981, p. 39; Andrew Pollack, "Singing the Semiconductor Blues," *New York Times,* May 24, 1981, p. F4; Marilyn Chase, "U.S. Electronics Firms Consider Joining in Research Venture to Counter Japanese," *Wall Street Journal,* March 1, 1982, p. 6; Ray Connolly, "Updating Antitrust Law to Permit Joint R&D," p. 66, and "U.S. R&D Consortium Takes Shape," pp. 97–99, *Electronics,* March 10, 1982; and William C. Norris, "Cooperative R&D: A Regional Strategy," *Issues in Science and Technology,* vol. 1 (Winter 1985), p. 94.

71. Howard Wolff, "64-K RAM Battle is Murky," *Electronics,* September 8, 1981, pp. 89–90; Alan Alper, "Buyers Hedging on Long-Term 64K Pacts Until U.S. Firms Ramp Up," *Electronic News,* February 8, 1982, p. 1; Andrew Pollack, "Japan's Big Lead in Memory Chips," *New York Times,* February 28, 1982, p. F1; and "A Chance for U.S. Memories," *Business Week,* March 15, 1982, pp. 126–27.

72. See Sabin Russell, "U.S. Suppliers Outnumbered in 64K RAM Competition—for Now," *Electronic News,* August 24, 1981; and "Prices of 64K RAM Drop to One-tenth of Year Ago," *Japan Economic Journal,* September 15, 1981.

market standing at about 70 percent, some American producers, led by Motorola, were pressing the Commerce Department to investigate charges that the Japanese were selling 64K DRAMs below "fair value" (average cost).[73] The particulars of the issues raised are addressed in chapters 5 and 6.

Officials in the Commerce Department at this point decided to "jaw-bone" the Japanese unofficially to do something about tumbling 64K DRAM prices. MITI officials in Washington were warned that the Commerce Department might begin to "monitor" Japanese import prices.[74] In March, however, just as this investigation was beginning, 64K DRAM prices suddenly doubled, Japanese suppliers began rationing U.S. customers, and it was reported that Japanese companies were cutting back U.S. exports in order to blunt moves toward trade restrictions on DRAM imports.[75] Within the Japanese semiconductor industry these reductions in exports are openly acknowledged to have been spurred by MITI guidance.[76] In early April 1982 Japanese DRAM producers actually con-

73. Clyde H. Farnsworth, "Japanese Chip Sales Studied," *New York Times*, March 5, 1982, p. D1; John Eckhouse, "Are Japanese Chip-makers 'Dumping'?" *San Francisco Examiner*, March 5, 1982; and Bruce Entin, "Motorola Asks Inquiry into Japanese Pricing," *San Jose News*, March 10, 1982. Motorola's interest in government action came after a sharply divided SIA reportedly voted down a Motorola proposal to submit a joint anti-dumping complaint. See I. M. Destler and Hideo Sato, "U.S.-Japan High Technology Trade: Politics and Policy," University of Maryland, School of Public Affairs, August 1986, p. 38. Motorola executives also were reported to have wanted the government to take the lead in order to avoid placing itself in an antagonistic position with Hitachi, the top Japanese producer of 256K DRAMs, with which it had close ties. Richard Wightman, "SIA Split on 64K RAM 'Dump' Action; Expect Members to Petition," *Electronic News*, March 22, 1982, p. 1.

74. Clyde V. Prestowitz Jr., writes of his and Lionel Olmer's efforts as officials in the Commerce Department at that time: "Our only tools were bluff and persuasion. Olmer, who was responsible for administering the antidumping laws, warned MITI officials in Washington that the Commerce Department might begin to monitor Japanese chip prices. Our ability to do so was in fact limited, but we hoped to make the Japanese more cautious by suggesting the possibility. It worked for a while. The prices of 64K RAMs began to stabilize, as MITI warned Japanese companies." Clyde V. Prestowitz Jr., *Trading Places: How We Allowed Japan to Take the Lead* (Basic Books, 1988), p. 49.

75. Alan Alper, "See 64K Levels in Line With Demand," *Electronic News*, March 15, 1982; and "Justice Department Investigating Japanese 64K RAM Mktg.; Seek Price, Shipment Data," *Electronic News*, August 2, 1982, p. 1.

76. The Japanese trade publication *VLSI Report*'s timeline of U.S.-Japan trade frictions, for example, sets February 1982 as the date "MITI instructions on dumping began." The Electronic Industries Association of Japan's official industry handbook, *IC Gaido Bukku*, has a timeline also associating February 1982 with the entry "MITI to Japanese

firmed to reporters in Tokyo that they were reducing U.S. exports to alleviate trade friction.[77]

U.S. Commerce Secretary Malcolm Baldrige immediately informed the U.S. press of his initial "favorable" reaction to the Japanese voluntary export restraints, but in an interview for the *New York Times* he warned:

> "They were building much more capacity than they could stand, and we thought there was possibly some evidence of predatory pricing to take over the market."
>
> He added that he thought there was a danger that the nation's security could be threatened if the American computer and telecommunications industries were to become dependent on Japan for supply of the chips, which have been characterized as the crude oil of the 1980s.
>
> "This is something we would not like to see," Mr. Baldrige said, noting that the aim of the Commerce Department investigation was to determine whether the Japanese were building volume in the 64K RAM market as a result of subsidies to their industry or because of policies or practices that insulate it from competition from American companies.[78]

The very next day, MITI and various Japanese company representatives officially denied reports of export restraints.[79] But U.S. prices were to continue well above Japanese price levels through early 1983.[80]

The episode was to take an even more bizarre turn three months later, in July 1982, when MITI was informed by the U.S. Justice Department that Japanese producers were being investigated to determine whether a cartel had been formed to set volume and price levels in the U.S. market.[81] (Prices in the United States had continued to hover at almost double Japanese levels.)[82] An NEC spokesman, apparently unfamiliar

---

industry, no exports that might cause blame for dumping." See "Japan-U.S. IC Frictions," *VLSI Report* (Tokyo: Press Journal, ca. 1988), p. 62; and Electronic Industries Association of Japan, *IC Gaido Bukku* (Tokyo, 1987), p. 62.

77. A. E. Cullison, "Japan Alters Memory Chip Export Policy," *Journal of Commerce*, April 7, 1982; and "64K RAM Exports Are Being Held Down by Makers," *Japan Economic Journal*, April 13, 1982.

78. Clyde H. Farnsworth, "Japan to Cut Export of Chips to U.S.," *New York Times*, April 8, 1982, p. D1.

79. Steve Lohr, "Japanese Deny Any Cut in Chip Exports to U.S.," *New York Times*, April 9, 1982, p. D1; see also Associated Press, "Computer Chip Reports 'Premature,'" *Japan Times*, April 10, 1982.

80. Jack Robertson, "Japanese RAM Power," *Electronic News*, February 28, 1983.

81. "U.S. Will Probe Japanese Makers of Semiconductors," *Japan Times*, July 27, 1982; Steve Lohr, "6 Japan Concerns Focus of Inquiry," *New York Times*, July 27, 1982; "U.S. Probes Sales of Computer Chips by Six Japan Firms," *Wall Street Journal*, July 27, 1982; "Justice Department Investigating," p. 1.

82. "U.S. Won't Indict Japan Semiconductor Makers," *Japan Times*, July 30, 1982.

with U.S. antitrust concepts, responded that "Japanese interests have set relatively high prices for 64-kilobit RAM chips in the U.S. so as not to raise suspicions of dumping"![83] Needless to say, the whole sequence of events left the Japanese somewhat confused. In March, amid loud industry and Commerce Department complaints about excessively low prices, Japanese DRAM import prices had suddenly jumped, only to have the Justice Department announce an investigation of excessively high prices a short time later.[84]

A lawyer speaking for the SIA explained the apparent contradiction as a real-life example of precisely the predatory scenario Robert Noyce had first warned of back in 1979: "They may have committed violations of the dumping laws early on, to buy market share, and now they're getting the payoff by limiting supplies and raising prices."[85] Noyce himself even weighed in on the issue in early 1983: "It is probably correct that the Japanese are selling RAMs in the United States at higher prices. It is a classic case of competitors using predatory low pricing to take the lion's share of a market, and then increasing prices once they dominate that market."[86]

By mid-1983, however, the market for DRAMs was again picking up, and the industry faced looming shortages. Prices in Japan rose sharply, pushed up to U.S. levels.[87] After depositions were taken from Silicon Valley representatives of the Japanese chip makers in the spring of 1983, the Justice Department's antitrust investigation simply faded away.[88] As

83. "U.S. Will Probe Japanese Makers."

84. " 'In the second half of 1981, the U.S. Commerce Department was told by one of our competitors that we were dumping 64K RAMs on the U.S. market,' one semiconductor company spokesman said in Tokyo last week. 'Now, the U.S. Justice department is asking if we are fixing prices and holding back supplies. Well, cartel or dumping, which is it?' " J. D. Kidd, "Japanese 64K Makers Puzzled by U.S. Probe," *Electronic News,* August 22, 1982. See also Andrew Pollack, "Inquiry Puzzles Chip Makers," *New York Times,* July 28, 1982.

85. Thomas A. Skornia, quoted in "A New Front in the War Over Japanese Chips," *Business Week,* August 9, 1982, pp. 22–23.

86. Robertson, "Japanese RAM Power."

87. Sabin Russell and Stuart Zipper, "Motorola Rivals See 64K Woes Pressuring Deliveries, Prices," *Electronic News,* March 28, 1983, p. 52; "Quotations of LSIs Stop Falling, Start Rebounding," *Japan Economic Journal,* May 31, 1983; Sabin Russell, "64K RAM Revival Ends 3-Mo. Lull," *Electronic News,* June 23, 1983, p. 62.

88. Mark Blackburn, "Execs Testify in Computer Chip Probe," *Oakland Tribune,* March 18, 1983; "64K RAM Makers Face Possible Antitrust Charges," *Japan Times,* March 20, 1983. Apparently, however, no depositions were obtained from Japanese management based in Japan.

the industry entered one of its cyclical boom periods, trade frictions receded as an urgent matter requiring attention in Washington.

What was most notable over this period of renewed friction, as the 64K DRAM entered the market, was a subtle shift in the argument about predatory behavior in semiconductors. By 1980 a large and highly visible Japanese R&D subsidy in semiconductors, directly focused on DRAMs and related manufacturing technology, had been ended.[89] Formal quotas had also been ended, and tariffs were low and dropping rapidly.[90] Although complaints about access to the Japanese market persisted, and even intensified in early 1983, it could no longer be claimed that higher prices in the home market enabled Japanese producers to persistently price below average cost in foreign markets. If Japanese producers were losing money in the United States on sales of dumped memory chips, they had to be losing money at home as well.

Thus, the U.S. industry analysis of Japanese predation in 1982 necessarily had to change from one where the Japanese government's policies of R&D subsidy and home market protection created conditions encouraging firms to price exports low, without necessarily requiring explicitly predatory behavior (since home market profits could offset foreign losses). Instead, the argument now was that, to sustain massive losses around the world for a prolonged period of time, Japanese firms must necessarily have adopted an explicitly predatory strategy, with the expectation that in the long run, with the exit of foreign competitors, rents could be collected to offset the initial costs of predation. The appearance of significantly higher U.S. prices for a period beginning in March 1982, and the charge that the fruits of predation were finally being harvested, added an additional element to the mix, namely, collusion: American producers argued that, having achieved a dominant position in the market for DRAMs, Japanese companies were cooperating, either on their own, or with administrative support from their government, to cut back supply on foreign markets in order to collect monopoly rents. Since 1982 this mix of strategic government industrial and trade policy with strategic—

89. Substantial (but considerably less visible) support from NTT and a privately funded cooperative follow-up to MITI's VLSI project, continued, however. See chapter 2 for details.

90. In February 1985 Japan and the United States agreed to end all tariffs on semiconductors, effective the following month. "Japan and US Agree to Abolish Semi-Conductor Tariff Next Month," *Nihon Keizai Shimbun,* February 9, 1985, p. 3.

and collusive—private firm behavior has consistently been presented as the U.S. industry's analysis of Japanese production and pricing practices.

Sectoral Negotiations

As the decibel level mounted in public exchanges between U.S. and Japanese semiconductor companies, pressure for an increasing level of government involvement in trade issues also rose. The U.S. government had already responded to a considerable degree back in the spring of 1981, when several planks of the American chip industry's public policy platform—notably, accelerated bilateral tariff cuts and adoption of an incremental R&D tax credit—had been implemented quite quickly. By early 1982, however, when Japanese producers had displaced American suppliers in the 64K DRAM market to a degree visible to even the most determined optimists, political pressure for further action was building rapidly. In February 1982 senior Reagan administration officials were considering a Defense Department proposal to impose restrictions on imports of 64K DRAMs on national security grounds.[91] As previously discussed, top Commerce Department officials used threats and bluffing to persuade MITI to intervene with Japanese companies.

Processes set in motion over this period were to result in the first formal bilateral negotiations over semiconductor trade issues, soon to become a staple of U.S.-Japan relations. Exchanges between U.S. and Japanese officials led to the idea of forming an ongoing bilateral "working group" concerned with trade frictions in high-technology industries, such as semiconductors, to deal with such issues as market access, export pricing, national treatment, and elimination of tariffs. The formation of this forum was also seen by the U.S. government, and presented to U.S. semiconductor companies, as an alternative to aggressive pursuit of a dumping case.[92] In the spring of 1982 the U.S. and Japanese governments

91. Art Pine, "Computer Chips from Japan May Be Curbed," *Wall Street Journal*, February 5, 1982, p. 6.
92. Author's discussion with former U.S. government officials, September 1992. High-technology sectors other than semiconductors in which there had been some history of trade friction included telecommunications (particularly NTT procurement), the Japanese satellite market, the Japanese computer market, and cooperation on joint R&D projects. See John Sullivan Wilson, "The United States Government Trade Policy Response to Japanese Competition in Semiconductors: 1982–1987," report to the U.S. Congress (Office

formally agreed to the formation of a High Technology Working Group.

The HTWG held its first meeting in July 1982, and the participants reached an agreement to focus on bilateral issues in three sectors: semiconductors, telecommunications equipment, and computers. The principal issue on the U.S. side was access to Japanese markets in these industries, particularly semiconductors. Subcommittees were to be formed to work on specific recommendations for each of these sectors. Semiconductors were the first such sector to be chosen for discussion, and the only sector in which formal agreements were reached under HTWG auspices.[93]

In November 1982 the HTWG completed its first "Agreement on Principles."[94] This framework, accepted by the Japanese and American governments in February 1983, was a broad agreement on the importance of high-tech industries and of free trade and investment in sustaining them, the need to safeguard rules preventing anticompetitive or predatory practices, and a call for equal national treatment in access to markets, and investment promotional programs and subsidies, by both countries.

The only passages of this agreement that specifically mentioned semiconductors called for the formation of a joint task force on high-tech data collection, and set out an outline for a long-term work program using sectoral analyses to identify and resolve sources of trade friction in high-tech industries. Both measures designated semiconductors as the initial case to be considered.[95] In April 1983 a HTWG subcommittee on semiconductors was formally organized, and in July a joint semiconductor data collection effort was initiated.[96]

---

of Technology Assessment, September 1987), pp. 35–37; Prestowitz, *Trading Places,* pp. 49–51.

93. Wilson, "The United States Government Trade Policy Response," pp. 37–41.

94. See "The U.S.-Japan Work Group on High Technology Industries," *SIA Member News,* vol. 1 (February 1984), pp. 1–2, for what may be the clearest account of the HTWG's history in print.

95. See "Recommendations of the U.S.-Japan Work-Group on High Technology Industries" (Washington, February 1983), p. 3, and "Outline for Long Term Work Program," attachments to U.S. Trade Representative, "U.S., Japan Exchange Letters on High Technology Trade," press release, February 10, 1983.

96. Ministry of International Trade and Industry, Subcommittee on Semiconductors, "Study Process of the U.S.-Japan Work Group on High Technology Industries" (Tokyo, November 1983); and International Trade Administration, "Recommendations of the U.S.-Japan Work Group on High Technology Industries, Semiconductors" (U.S. Department of Commerce, November 1983).

Most significant, a formal bilateral agreement on semiconductors was negotiated in November 1983. The agreement covered three main areas. On trade issues, agreement was reached to take steps toward the elimination of tariffs, to continue the ongoing joint data collection effort, and to pursue a variety of other measures to encourage freer semiconductor trade. Especially notable was the declaration that the Japanese government should "encourage Japanese semiconductor users to enlarge opportunities for U.S.-based suppliers so that long term relationships would evolve with Japanese companies." A confidential "chairman's note" specifically outlined four concrete "measures of encouragement"—including, for example, "special consideration" in access to procurement certification tests—to which Japanese chip users would "become subject" in order "to enlarge opportunities for U.S.-based semiconductor producers." U.S. negotiators took these portions of the agreement to be a euphemistic articulation of an oral commitment by MITI to provide "administrative guidance" to Japanese firms to increase their purchases of U.S. suppliers' chips, and indeed the language of the confidential note would seem to seem to support this position.[97] Another critical point was

97. Other steps listed in the confidential "chairman's note" included an agreement to expand "supply sources . . . for some semiconductor products in order to give more business opportunities to U.S.-based semiconductor producers," "opportunities to make direct contacts with semiconductor user divisions of the companies," and "such measures as dispatching to or stationing in the U.S. of staffs responsible for procurement." These measures were described as applicable to "Japanese semiconductor users who are also semiconductor producers," but a concluding sentence in the note promises "best efforts . . . to increase the number of companies which will become subject to the above-mentioned measures of encouragement."

The confidential chairman's note is reproduced in Semiconductor Industry Association, *Japanese Market Barriers in Microelectronics: Memorandum in Support of a Petition Pursuant to Section 301 of the Trade Act of 1974 as Amended* (Washington: Dewey, Ballantine, Bushby, Palmer and Wood, June 14, 1985), p. 91; and in Wilson, "The United States Government Trade Policy Response," pp. 46–47.

Clyde Prestowitz, cochair of the U.S. HTWG representation at the time, argues that the existence of this "secret side letter," with its agreement to "encourage" Japan users to "enlarge opportunities," was the most significant new aspect of the November 1983 recommendations. However, this specific language is also found in the main public text, not just in the confidential note. See Clyde V. Prestowitz Jr., "The New Chip Agreement: Does It Go Far Enough?" *Electronics*, October 1991, p. 25; and Prestowitz, *Trading Places*, p. 156.

The reasons for keeping this note confidential were most likely related to another issue. The note specifically suggests that private business decisions of Japanese firms are to be "subjected" to "measures of encouragement," and therefore acknowledges an explicit MITI role in influencing these decisions. Also, although industry sources say little significance was read into the language at the time, the note applies to *U.S.-based* semiconductor

the Japanese government's agreement to "encourage" NTT to hold meetings in Japan and the United States to explain its certification processes; this the U.S. negotiators understood to be a commitment by the Japanese government to enforce previous undertakings by NTT to open up its procurement to foreign suppliers.[98]

A second portion of the recommendations dealt with investment. Both governments pledged to remove barriers and to open participation in regional investment promotion programs to foreign-owned subsidiaries. A third section on technology sought to promote technology exchanges, increase U.S.-Japan interaction on quality and reliability issues, and improve protection of intellectual property related to semiconductors in both countries. A final section called for a continuing series of bilateral implementation reviews, with meetings to occur at least twice annually.

The most tangible immediate gains resulting from the HTWG, from the U.S. industry's perspective, were in the areas of tariff liberalization and intellectual property protection. By March 1985 the United States and Japan had followed up with an agreement to completely eliminate tariffs on semiconductors, effective the following month. In October 1984 Congress passed legislation, which went into effect in January 1985, protecting semiconductor chip mask layout designs.[99] A similar law was passed in Japan in May 1985.

The main focus of U.S. interest, namely, a larger share of the Japanese market, proved a more elusive goal. Initial results seemed positive: the largest Japanese companies reported that, after discussions with MITI, they had elected to increase their purchases of U.S. semiconductors.[100] Amid a boom in chip demand during late 1983 and early 1984, sales by American companies increased noticeably. Even then, however, American observers complained that "administrative guidance" from MITI was

---

producers (as opposed to all foreign semiconductor producers) and might therefore be interpreted as violating GATT commitments to equal most-favored-nation treatment for all trading partners.

98. "Recommendations of the U.S.-Japan Work Group on High Technology Industries, Semiconductors"; and author's discussion with former U.S. government officials, September 1992. NTT announced a process for testing and certifying semiconductors based on the HTWG recommendation in March 1984. Electronics Industries Association of Japan, *IC Gaido Bukku*, p. 62.

99. The chip mask protection measures also followed a much publicized dispute between Intel and Japan's NEL over the latter's copying of an Intel microprocessor.

100. See Wilson, "The United States Government Trade Policy Response," pp. 47–50; and SIA, *Japanese Market Barriers*, p. 68.

offered only to top and middle management in the six largest companies, with little or no effort invested in contacting smaller users.[101] Moreover, when overall chip sales veered into sharp decline in late 1984, American chip sales in Japan declined even more rapidly, and U.S. companies' market share sagged. Evidence of an active MITI effort to kickstart foreign company sales, or of success in forging long-term relationships, was notably lacking in the declining market. Many in both U.S. government and industry argued that, despite an almost three-year investment in the HTWG effort, little in the way of tangible results had actually been achieved.

Over the long term, though, the HTWG set several important precedents. This set of negotiations institutionalized continuing high-level bilateral negotiations over trade issues in a single high-technology sector. The initial result of this process, the November 1983 HTWG recommendations, was the very first "semiconductor agreement." It marked the first time MITI explicitly agreed to use its leverage with private Japanese firms to promote imports of semiconductors—such voluntary import expansion (VIE) measures were to occupy an increasingly prominent position in the MITI program for dealing with trade frictions as the 1980s progressed. And it was the first time—but not the last—a "secret side letter" was used as a device to avoid some of the difficult issues raised by such commitments. The vagueness of the language in which the negotiators couched these commitments, and the subsequent debates over the extent to which they had been honored, foreshadowed even sharper exchanges over language and meaning in later semiconductor agreements.

As had been true in the past, and would be again, a decline in the global semiconductor market and in the fortunes of U.S. producers was the catalyst for a renewed offensive on trade policy, as the truce forged by the HTWG fell apart. The last such surge in political activity had occurred in the spring of 1982, spurred by the Japanese manufacturers' capture of the lion's share of the 64K DRAM market, weak chip demand, and heavy price cutting. A major study documenting the arguments that the Japanese had subsidized their chip producers, engaged in predatory pricing, and denied U.S. firms free access to the Japanese market had been commissioned at this time.[102] By the time this study was released,

101. Wilson, "The United States Government Trade Policy Response," pp. 48–49.
102. Semiconductor Industry Association, *The Effect of Government Targeting on World Semiconductor Competition* (Cupertino, Calif., 1983).

however, in early 1983, demand—and prices—had clearly begun to rebound and to blunt the sharp edges of American industry complaints. Similarly, in late 1982 the SIA had drafted a Section 301 complaint, alleging unfair Japanese trade practices, based on the charges found in its 1983 report. But by the spring of 1983 the combination of recovery in the semiconductor industry and reasonable progress by the HTWG had led the SIA to shelve its trade case.[103]

Bust eventually followed boom, however, and the post-HTWG bust of 1985 was one of the worst in the industry's history. Trade frictions in semiconductors escalated to an unprecedented level. The events that followed led directly to the landmark U.S.-Japan Semiconductor Trade Arrangement signed in 1986, and a radical realignment of the rules of the game for global competition.

103. See Wilson, "The United States Government Trade Policy Response," pp. 41–42; SIA, "The U.S.-Japan Work Group," pp. 1–2; and Destler and Sato, "U.S.-Japan High Technology Trade," pp. 38–44.

CHAPTER FOUR

The Semiconductor Trade Arrangement and Its Aftermath

THE GRADUALLY ESCALATING tensions over trade and market access between the United States and Japan exploded into open confrontation during the industry slowdown of 1985, triggering a process that culminated in the conclusion of the Semiconductor Trade Arrangement (STA) in 1986. This chapter attempts to construct a coherent account of the actions taken by the U.S. and Japanese governments over the five-year life of the original STA. The next chapter evaluates what apparent impacts of these policies could be observed in the marketplace.

A Thumbnail Historical Sketch

To describe the evolution of the semiconductor trade regime since 1985 requires reviewing a broad range of evidence and accounts in order to assemble the most accurate possible account of the complex political economy of U.S.-Japan semiconductor trade. Such a detailed accounting may exhaust the patience of the casual reader. Readers with a minimal interest in the hows and whys of the changing rules of the game and the detailed evidence supporting my reconstruction of the salient events may wish to skip ahead to the assessment of impacts in the next chapter after reading the balance of this section.

Briefly, the historical period from 1985 through 1991 may be broken into four more or less distinct periods for the purposes of understanding semiconductor trade policy.

Informal Guidance, 1985 to July 1986

In the first period antidumping petitions for three different types of memory chips (as well as an unfair trade practices complaint under Section 301 of the 1974 trade act and a private antitrust suit) were filed in the United States against Japanese chip producers. Over this period politicians and bureaucrats in Japan applied pressure to their domestic industry to slow exports to the United States. This policy met with some success, and prices for the first product acted against—64K DRAMs—stabilized somewhat and even rose in the U.S. market in late 1985. The 64K DRAM dumping case was formally concluded in April 1986, with the U.S. Customs Bureau applying antidumping duties to Japanese 64K DRAM imports. The pressure on 64K DRAMs, an older vintage product, accelerated the shift by both consumers and producers to the next-generation memory chip, the 256K DRAM, which became embroiled in even more heated disputes. Dumping cases in 256K and higher DRAMs and EPROMs (erasable programmable read only memories) and the Section 301 case were suspended at the conclusion of the STA, which was negotiated by late July and signed on September 2, 1986.

Jab and Feint, August 1986 to November 1987

Disagreement over the interpretation of the STA was virtually immediate. Although the Japanese government set up a variety of monitoring and control mechanisms in the fall of 1986, considerable dispute arose over the extent to which price floors set by the U.S. Commerce Department were to be applied to sales in third-country markets. This issue also brought the European Community into the conflict, with the Europeans submitting a complaint to the General Agreement on Tariffs and Trade (GATT) which argued that such restraints were illegal.

In response to mounting political pressure from the United States over third-country dumping, the Japanese government successfully offered guidance to Japanese producers to reduce DRAM output in the first half of 1987. Export control mechanisms were also used to pressure companies to meet minimum export price guidelines. Evasion was initially widespread, and the U.S. government responded by imposing sanctions on imports of selected Japanese products in April 1987.

The Japanese government responded by continuing pressures on firms to hold production down, in an effort to boost prices by restricting supply. Guidance also covered investment in new capacity by Japanese firms. These measures prompted the U.S. government to partially lift the sanctions in June, and again in November 1987 (the remainder of the sanctions were maintained to express dissatisfaction with the slow pace at which it was felt that Japanese companies had increased their purchases of foreign semiconductors).

By late 1987 demand for chips was tightening with a recovery in the computer industry, the main market for these products. Production controls—formally forsworn by the Japanese government in November—had become irrelevant as the industry approached full capacity utilization. Guidance of investment reportedly continued into 1988, however, and Japanese producers reported that a system of regional allocation guidelines for exports was in place by late 1987. Administrative measures were taken that made it more difficult to export chips without the approval of both manufacturers and the government, reducing the possibilities for brokers and arbitrageurs to export chips into the gray market.

The Privatization of Restraints, December 1987 to Mid-1990

By early 1988 a full-fledged shortage of DRAMs was being widely felt in the United States and Europe. Manufacturers increased their surveillance of chip sales to reduce supplies filtering into the Japanese gray market.

By late 1988 and early 1989 semiconductor demand had begun to weaken. In response to a downward drift in DRAM prices, Japanese manufacturers began to cut back output to maintain high prices. Japanese newspapers talked of "coordination structures" among Japanese companies being used to achieve "high price stability." Producers elsewhere continued to expand output aggressively, however, and as Japanese producers cut back, the market share of companies based in third countries (particularly the Korean producer Samsung) rose sharply. As the industry slid into a slump, measures to ration scarce output among competing consumers became superfluous: regional allocation of exports was ended by mid-1989.

Faced with increasing competition in more mature products from foreign producers unrestrained by price floors, Japanese companies seemed increasingly intent on seeking early entry into next-generation products requiring advanced technology, where price competition from less sophisticated rivals was less intense. This strategy encountered problems in the early 1990s, however, as a severe recession slowed the movement toward next-generation memory chips among chip consumers.

Aftermath: The Second Semiconductor Trade Arrangement

A new semiconductor trade arrangement between the United States and Japan, signed in June 1991, ended the DRAM dumping cases and formal Commerce Department calculation of floor prices (the EPROM dumping case continued in a suspended state). A MITI-supervised, fast-track antidumping procedure was created, however, which required Japanese companies to continue to collect the data previously given to the Commerce Department and keep it available on fourteen-day notice, in the event of an antidumping case. There is reason to believe that this effectively created a system of "shadow" floor prices on exports, observed by companies in their export pricing policies. Furthermore, Korean firms, which had presented Japanese suppliers their most aggressive and sustained competition, increasingly became subject to the same price floor discipline, as the outcomes of new dumping cases in the United States and Europe were announced.

Evolution of the Semiconductor Trade Regime

The year 1984 marked a cyclical peak in the semiconductor business. By the late fall of that year, however, semiconductor demand in the United States was weakening rapidly, and the downturn was mirrored in the Japanese market. A series of rapid declines in price for the predominant memory chip of the day, the 64K DRAM, were triggered by the announcement of a sharp cut in its sales price in October 1984 by Micron Technology, a small American memory chip manufacturer.[1] Other man-

---

1. See Micron Technology, Inc., "In the Matter of 64K Dynamic Random Access Memory Components from Japan, Petition for the Imposition of Antidumping Duty," presented to the U.S. Department of Commerce, June 21, 1985 (public version), p. 22; and U.S. International Trade Commission, *64K Dynamic Random Access Memory Components from Japan*, USITC Publication 1862 (Washington, June 1986), p. A-67.

ufacturers, Japanese and American, quickly followed suit, and DRAM prices plunged further into a sustained decline by early 1985.

As the U.S. chip market continued to weaken in early 1985 and their domestic sales faltered, U.S. semiconductor companies began to press complaints in Washington about limited access to the Japanese chip market, where their sales had fallen off even more sharply. Through the spring of 1985 the Japanese market remained relatively robust, and American firms began to worry that their Japanese competitors would continue the trend of record investments in new capacity logged in fiscal 1984, when Japanese investments in semiconductor facilities had more than doubled over fiscal 1983 levels. In May 1985 U.S. trade negotiators reportedly asked MITI to persuade Japanese companies to restrain their investments in new capacity. The request was turned down.[2]

Market Access: The Council of Nine

In the spring of 1985 the debate over the existence of barriers to U.S. companies' access to the Japanese semiconductor market heated up considerably in the United States. Known only to a small number of insiders in American industry and government, some dramatic reports of collusive practices within the Japanese semiconductor industry had surfaced and after investigation been judged credible by the U.S. side.

The failure of U.S. firms to reclaim a larger share of the semiconductor market in Japan after formal trade barriers had been lifted in the mid-1970s had long been shrouded in conflicting explanations, allegations, and suspicions, but no smoking gun had ever materialized. Indeed, as noted in the chapter 3, there was, in the commodity memory chip market, at least, some evidence that Japanese firms had managed by the late 1970s to turn superior levels of quality into a competitive advantage and a reasonable explanation for the Americans' failure to make headway. But what of the rest of the market?

From the perspective of American industry, the stubborn fixity of its Japanese market share coupled with the history of Japanese trade and

---

2. The U.S. request was reportedly characterized as interventionist and asymmetric by the Japanese, who argued that companies themselves should judge whether or not it was appropriate to scale back investment plans, and that it would be one-sided for the Japanese government to do so when the American government was not imposing similar guidance on American producers. See "U.S. Requests Japan to Hold Down Increasing of Facilities," *Nihon Keizai Shimbun*, May 10, 1985, p. 1.

industrial policy in the 1970s constituted sufficient evidence of continuing market barriers. U.S. suspicions had been reinforced in the late 1970s and early 1980s by occasional reports of behind-the-scenes intervention by the Japanese government to discourage foreign sales.[3] But others were unlikely to accept these as solid evidence of a significant problem.

In 1983 when the SIA had made the case that its members had continued to be denied access to the Japanese market, its most dramatic evidence was the history of sales of certain leading-edge products in the Japanese market in the postliberalization era. Data on several innovative new products initially showed explosive growth in U.S. exports to Japan followed by a downward tumble to virtually zero, which coincided with the production of a version of this same product by Japanese manufacturers.[4] The claim was that the moment a Japanese product came on the market, U.S. sales dried up or at best represented a residual demand for what could not be shipped by the new Japanese suppliers. The SIA commented, "this happens even though U.S. firms have been producing longer, and thus should enjoy lower costs than the Japanese."[5]

But no government guidance, private collusion, or other force with a deliberate policy of organizing behind the scenes to exclude foreign suppliers was necessary to explain these facts. The disturbing sales patterns for U.S. chips in Japan might reflect other, perfectly innocent causes (better local sales and service, better quality, or even an irrational customer preference for Japanese brands), any and all of which might be at work.

In 1985, however, a series of reports came forth that had lacked innocent or benign interpretations by even the most agnostic American standards. Reports gleaned from a variety of sources in Japan spoke of the existence of a shadowy Council of Nine consisting of sales representatives of Japanese producers of discrete semiconductors who met pe-

---

3. See note 237 in chapter 2 for examples of such reports. The Consulting Group, BA Asia Limited, in *The Japanese Semiconductor Industry 1980* (Hong Kong, May 1980), pp. 178–79, cites a case in which Japanese power supply makers refused to buy imported capacitors, claiming that NTT required that only domestic components be used in equipment sold to it. Japanese customs practices—capricious misclassification of items and "uplifts" routinely and arbitrarily added to invoiced values—were also cited as an informal barrier to imports during this period. See U.S. Department of Commerce, *A Report on The U.S. Semiconductor Industry* (1979), p. 96.

4. See Semiconductor Industry Association, *The Effect of Government Targeting on Worldwide Semiconductor Competition* (Washington, 1983), pp. 79–84. The examples given are 8080-type microprocessors and programmable read-only memory chips (PROMs).

5. SIA, *Effect of Government Targeting*, p. 79.

riodically in a social setting to set pricing, agree on market shares, and allocate sales to individual customers. At least some elements of the Japanese government apparently knew of these activities, and MITI rep- resentives were reported to have observed or attended some of the meet- ings. Knowledge of the Council of Nine was closely held within both U.S. government and industry, but senior American officials were well aware of it by 1985, and satisfied that it existed.[6]

Because the group was never openly investigated, the extent to which it actually succeeded in affecting competition in the Japanese market for discrete semiconductors will probably never be known.[7] Never issued publicly, the reports circulating behind the scenes nonetheless had an important impact.[8] The SIA was reinforced in its determination to file an official Section 301 trade action. When the government received this complaint, awareness of the existence of the council played some role in hardening its resolve to support and pursue the 301 case.[9] And one can only speculate that the Japanese government may have felt some vulner- ability on these issues and been more willing to negotiate the agreement eventually signed.[10]

Political Pressures, 1985 to July 1986

In June 1985 the Semiconductor Industry Association (SIA), the U.S. industry trade group, filed a Section 301 complaint with the U.S. Trade

6. My account of this episode draws on confidential letters and drafts from U.S. sources that I was permitted to review and interviews with former U.S. government officials. Sources of information about the activities of the Council of Nine, and its very existence, were considered extremely sensitive by both American government and industry.

7. Some within the government apparently argued that such activities were unlikely to succeed in the long run in benefiting their organizers and thus could safely be ignored. Interview with government official, February 1996. As noted in chapter 3, from the early through mid-1980s, semiconductor prices in the Japanese market generally had been near, and at times below, U.S. prices.

8. The major reason why the existence of the council never surfaced at the time is probably the sensitivity of sources of information about its activities. Also, because some Japanese subsidiaries of U.S. companies as well as major Japanese semiconductor producers were alleged to have been involved in these activities, the implications for trade negotiations between the United States and Japan would not necessarily have been clean and sharp.

9. Although by all accounts the existence of the council was only one of several issues, including concerns about national security, that created a consensus to move forward with the 301 case.

10. Shortly after the 301 case was filed, a senior Japanese trade official reportedly told a senior U.S. official that "this case has shaken MITI to the core" because of the history that might be dug up. Interview with former U.S. government official, January 1996.

Representative, alleging that barriers to Japanese market access constituted an unfair trade practice and asking for retaliatory sanctions. Later
that month Micron Technology filed a dumping complaint with the Commerce Department against seven Japanese producers of 64K DRAMs.
By this time chip sales in the Japanese market, too, had slowed considerably, and Japanese producers reported plans to slow their investments
in new capacity, to just under the record amounts logged in 1984.[11]

Meanwhile conflicts over import competition in EPROMs were also
mounting through most of the spring. The leading edge EPROM at that
time was the 256K, and in early 1985 Japanese manufacturers had begun
selling it in significant volume in the U.S. market. A *cause célèbre* was
stirred up when the so-called 10 percent memo, in which Hitachi America
urged its distributors to undercut all rival price quotes by 10 percent, at
a guaranteed 25 percent profit margin, was unearthed and publicized.
(An often unmentioned but important fact is that this memo specifically
urged Hitachi distributors to target Japanese rival Fujitsu, as well as
American makers Intel and Advanced Micro Devices [AMD].)[12] By July
1985, when the Commerce Department began its DRAM dumping investigation, trade frictions in memory chips had intensified to the point
of open conflagration.

The top levels of the Japanese political establishment got involved on
July 17, when, at a meeting of politicians from Japan's ruling Liberal
Democratic Party (LDP) with business organization leaders, a top LDP
politician reportedly urged Japanese semiconductor (and automobile)
companies to restrict their exports "on a voluntary basis, instead of doing
things in a clumsy manner like the Government's taking the initiative."[13]
Within a week Hitachi, the largest producer of 64K DRAMs and the
author of the infamous 10 percent memo, announced voluntary restrictions on semiconductor exports for fiscal 1985. Hitachi's plan to cut exports by 30 percent from 1984 levels was quickly followed by a declaration

11. "Requests for Postponement of Delivery Coming in Succession to Companies Manufacturing Semiconductor Production Equipment; Reduction of Production Plans, Too, is
under Contemplation," *Nihon Keizai Shimbun*, June 23, 1985, p. 4.
12. See Intel Corporation, Advanced Micro Devices, Incorporated, and National Semiconductor Corporation, "In the Matter of: Erasable Programmable Read Only Memories
(EPROMS) from Japan, Petition for the Imposition of Antidumping Duties Pursuant to
the Tariff Act of 1930, as Amended," September 30, 1985 (nonconfidential version), pp.
19–20, 25, 29, and app. 4.
13. See "LDP Even Likely to Request Export Self-Restraint; Clarification Toward
Business World; Automobiles and Semiconductors as Pillars: Intertwined with Opening of
Market for U.S.," *Asahi Shimbun*, July 18, 1985, p. 9.

by NEC that it "plans to reduce exports to the U.S., while increasing its production in the U.S.," and an announcement by Toshiba that it too planned to cut its U.S. chip exports by 20 percent in fiscal 1985.[14] Despite these measures, LDP Policy Board Chairman Fujio publicly stepped up the pressure at the end of July, declaring it necessary to consider the "possibility of restricting exports of automobiles and semiconductors and imposing an export surcharge."[15]

U.S. government investigations of the Micron 64K DRAM dumping case and the SIA's Section 301 petition continued to move forward rapidly through August 1985, when Intel, AMD, and National Semiconductor launched yet another dumping case against Japanese EPROM producers. In September Micron filed a private antitrust case as well, against Japanese DRAM producers.[16]

The chip market meanwhile remained mired in recession: by October, Intel, Mostek, and National Semiconductor had announced their intention to close down facilities and phase out their production of DRAMs.[17] Three Japanese companies with U.S. manufacturing facilities—Fujitsu, Hitachi, and Toshiba—announced postponements in plans to expand their U.S. manufacturing operations, while NEC announced a complete halt in its fabrication of new 64K DRAMs.[18]

14. See "Hitachi to Reduce Semiconductor Exports to U.S. for This Fiscal Year by 30%," *Nihon Keizai Shimbun,* July 23, 1985, p. 1; "Toshiba to Cut Semiconductor Exports to U.S.," Kyodo News Wire story, July 24, 1985; and "Chip Makers to Cut Exports to U.S.," *Japan Economic Journal,* July 30, 1985, pp. 1, 4.

15. See "Also Restriction on Exports of Automobiles and Semiconductors; Policy Board Chairman Fujio," *Yomiuri Shimbun,* July 27, 1985, p. 2.

16. That $300 million suit was dropped in October 1986, a month after the signing of the STA. See *Wall Street Journal,* October 2, 1986, p. 4.

17. Most US firms dropped out of DRAMs *after* the initial cuts in semiconductor exports by Hitachi, NEC, and Toshiba. This contradicts the assertion by analysts affiliated with the SIA that "the move toward production regulation by the Japanese producers' group began . . . [when] Japanese DRAM producers had few competitors left except each other." Thomas R. Howell, Brent L. Bartlett, and Warren Davis, *Creating Advantage: Semiconductors and Government Industrial Policy in the 1990s* (Washington: Semiconductor Industry Association and Dewey Ballantine, 1992), p. 117.

Also, Japanese chip exporters had previously cut DRAM exports in response to political pressure, in 1982, when Japanese world market share was very much smaller than in 1985. Political pressure, rather than a sudden jump in market power, appears to have stimulated supply cuts in both cases.

18. "Construction of Very Large-Scale Integrated Circuit Plant in U.S.; Toshiba Postpones Plan; By Half a Year, or One Year, Due to Semiconductor Depression," *Nihon Keizai Shimbun,* October 23, 1985, p. 9; and "NEC to Adjust Production of 64K DRAMs; Suspends Pre-process Operations; To Digest Semiconductor Products in Stock," *Nihon Keizai Shimbun,* October 31, 1985, p. 9.

It was in this atmosphere of crisis that U.S. and Japanese negotiators held successive, frustrating rounds of talks on semiconductor trade problems in August, September, and October of 1985. By November it was known in Tokyo that the U.S. government was considering initiating an antidumping investigation against Japanese producers of latest-generation 256K DRAMs; for the first time ever Washington itself would be launching the suit rather than waiting for the industry to come forward.[19] Aware of the rumblings in Washington, Japanese manufacturers increased their U.S. export prices for 256K DRAMs that November.[20]

By the end of 1985, though, the worst of the semiconductor recession finally seemed to have passed, and the Japanese press was reporting that key manufacturers were even discussing an expansion of their production of 64K DRAMs. The "production coordination" (as the round of export and production cuts and price increases were described in the Japanese press) launched by major Japanese producers earlier that summer was increasingly effective in pushing up 64K DRAM prices.[21] By December 64K DRAM prices were reported to have risen to 85 cents, from a low of 30 cents in June 1985, and the major Korean producer Samsung, which had actually suspended production of 64K DRAMs to staunch severe losses, resumed shipments under the price umbrella created by the Japanese actions.[22] But the successful efforts by Japanese manufacturers to collectively stabilize prices for 64K DRAMs were too late to halt the administrative machinery now set into motion by Micron's dumping pe-

19. Of course, the government was considering this action because of industry complaints. See "U.S. Heading Toward Antidumping Investigation on 256 Kilobit DRAM of Japanese Make; Conference of Secretaries Advises President; Hard Fight over 'Strategic Goods' Gives Rise to Crisis Feeling," *Nihon Keizai Shimbun*, November 8, 1985, p. 6.

20. "Semiconductor Maker, Full Power for 256K Price Hike, MPU for Price Cut," *Nihon Keizai Shimbun*, January 14, 1986, p. 18.

21. See "Semiconductor Industry Showing Signs of Recovery from Depression," *Sankei Shimbun*, December 5, 1985, p. 6; see also "Following 'Leather,' Also 'Semiconductors' Have Hard Sailing; Japan-US Consultations; MITI Officials in Charge Impatient Without Good Idea," *Tokyo Shimbun*, December 6, 1985, p. 3; "Semiconductor Companies Remain Calm Toward Preliminary Ruling of 'Guilty' on 64K DRAM; Upper-Grade Item Now Attached with Major Importance; Consultations between Japanese and US Governments Are Watched," *Nihon Keizai Shimbun*, December 6, 1985, p. 8.

22. See "Int'l Price Recovery Sparks Semiconductor Industries; Samsung Planning to Expand 64K D-RAM Plant to Meet Rising Demands," *Korea Herald*, December 26, 1985; and "Chip Manufacturers Slate New Facility Investments," *Korea Herald*, February 2, 1986.

By December 1985 regional price differentials appear to have been created in the 64K DRAM market, with Japanese domestic prices falling even as U.S. export prices were rising. See, for example, "Price Reduction Demand for Semiconductors Intensifies," *Nihon Keizai Shimbun*, December 4, 1985, p. 20.

tition. In early December the Commerce Department announced a preliminary finding of dumping in 64K DRAMs.

Higher 64K DRAM prices, coupled with a continuing increase in supply of and falling prices for 256K DRAMs, accelerated the trend for chip users to switch to next-generation 256K DRAM in their electronic systems production. Consumers, Japanese and American producers, and the American government now shifted their attention toward this product. By mid-December the Commerce Department, as threatened, had "self-initiated" an antidumping investigation of DRAMs with densities of 256K or greater.[23] As tensions continued to mount and NEC and Hitachi had "self-reflected upon their excessive competition for mass production which led to the decline of [256K DRAM] prices," the two firms began to cut back on 256K DRAM production as well in late December 1985.[24] As in the case of 64K DRAMs, however, the dumping machinery, once set in motion, was not to be stopped by anything less than a formal government agreement.

By early 1986 Japanese producers were announcing 256K DRAM price increases for their domestic customers (U.S. export prices had been raised the previous November), in a continuing attempt to reduce trade friction.[25] By late January delivery prices for 256K DRAMs had risen in Tokyo, in both the spot and the large-scale contract markets.[26] Japanese users complained loudly about the unnatural price increases induced by *gaiatsu* (foreign pressure).[27] But the administrative gears set in motion

23. This dumping investigation was unusual in two respects. First, it was one of a handful of antidumping cases ever initiated by the U.S. government itself, rather than an industry petitioning the government. Second, it covered 256K and higher density DRAMs, including new products not yet on the market.

24. "Nippon Electric and Hitachi Begin to Curb 256K DRAM Production; Watching U.S. Dumping Investigations," *Nihon Keizai Shimbun*, December 18, 1985, p. 9. See also "NEC, Hitachi Hold Down 256K DRAM Production; To Avoid U.S. Dumping Charge," *Japan Economic Journal*, December 28, 1985, p. 18.

25. See "Super LSI Domestic Shipment Price Hike a Little Over 10%, Consideration of Friction with U.S.," *Nihon Keizai Shimbun*, January 11, 1986, p. 1; "Semiconductor Maker Full Power," p. 18; and "Semiconductor Demand to Recover This Year; From Demand-Price Survey, Price Crash Period Ends; Bottom Spreads Supports Price," *Nihon Keizai Shimbun*, January 14, 1986, p. 14.

26. "Negotiation for Determining Price, Next Month to Be the Peak; Makers Forceful for 256K Price Hike; Users Demand Price Reduction Due to Yen Appreciation," *Nihon Keizai Shimbun*, January 29, 1986, p. 20; and "Semiconductor: Due to Japan-U.S. Trade Friction, Unprecedented Price Rise; U.S.-Made Import Doubtful; Maker Confident in Profit Maintenance," *Nihon Keizai Shimbun*, January 30, 1986, p. 20.

27. "This price increase by Japanese makers, aimed at calming bilateral trade friction,

by the dumping cases continued to turn: by March 1986 the Commerce Department had added preliminary dumping determinations in both the EPROM and 256K DRAM cases to its December 1985 finding on 64K DRAMs.[28] Following the March EPROM ruling, Intel reportedly raised its EPROM prices by an average of 25 percent; Japanese producers concerned about trade frictions followed Intel's lead by raising domestic sales prices, exercising "self-restraint" on low-priced sales of EPROMs.[29]

U.S. and Japanese semiconductor manufacturers also—for the second time—organized an industry "summit" meeting in March 1986. (The first such industry-to-industry meeting had been held in Hawaii in November 1983.) A speech at this meeting, held in Los Angeles, by MITI Deputy Director General Tanahashi of the Machinery and Information Industries Bureau was believed by many Japanese semiconductor executives to have been the origin of the 20 percent market share target for foreign semiconductors later included in a side letter to the STA.[30]

---

has its roots in political judgments beyond simple market principles." "Semiconductors: Abnormal Price Increase Caused as a Result of the U.S.-Japan Trade Friction," *Nihon Keizai Shimbun*, January 30, 1986, p. 20. The article concludes, "One maker pointed out that the year 1986 marks the first time all the makers share a sense of cooperation in the history of the semiconductor industry, unlike in the past when they were competing with each other for higher production and lower prices. They expect that less price competition will bring them larger profits in [fiscal] 1986."

28. The preliminary dumping margins found by the Commerce Department in these cases were relatively high and undoubtedly created further pressure for a negotiated solution to the conflict. For EPROMs, the rates were as follows: Hitachi, 29.9 percent; Toshiba, 21.7 percent; NEC, 188 percent; Fujitsu, 145.9 percent. See "Political Settlement Between Japan and US Expected by Manufacturers; Tentative Decision of 'Guilty' on EPROM," *Nihon Keizai Shimbun*, March 13, 1986, p. 9.

29. "EPROM Prices in Steady Tone—Semiconductors; Japanese Manufacturers Exercise Self-Restraint on Low-Priced Sales," *Nihon Keizai Shimbun*, April 2, 1986, p. 18.

30. In that speech Tanahashi reportedly stated that "it is expected that the total share of U.S.-made semiconductors used by five large Japanese manufacturers will account for 19.5 to 20 percent in 1990." This was interpreted in the Japanese press as a proposal from the Japanese side. See "How Should We Settle Japan-US Semiconductor Friction?" *Nihon Keizai Shimbun*, April 2, 1986, p. 2. It was precisely at this point that reports of a 20 percent market share target by 1990 began to circulate within the Japanese semiconductor industry. See Mikio Fujiwara, "This Is a Side Letter to the U.S.-Japan Semiconductor Agreement," *Bungei Shunju*, May 1988, pp. 124–37 (the author, an executive in a Japanese semiconductor firm, wrote the article under a pseudonym).

On the U.S. side, this March industry meeting prompted an investigation by the Federal Trade Commission. See "Suspicions about Violation of Antimonopoly Law—Guarantee of Share Rate for Semiconductors Imported from US; Fair Trade Commission's View," *Asahi Shimbun*, April 24, 1986, p. 2. The commission's probe was not unusual; it has often examined the outcomes of antidumping agreements.

Birth of the STA

Monthly talks on semiconductor trade issues between the American and the Japanese government began to pick up speed as they approached their sixth session, in March 1986. In late February MITI had reportedly decided to propose a minimum export price system to the American side at the upcoming meeting.[31] With the pressure of a July deadline on the original Section 301 case hanging over them at the meeting, U.S. and Japanese negotiators were reported to have been near agreement on a global price control and monitoring scheme, but a potential deal was blocked by continuing disagreement over an American chip industry proposal to specify a minimum share target for the Japanese chip market, and over measurement issues associated with that plan.[32] Citing Japanese "intransigence," the United States suspended the negotiations on March 28.

By mid-April the proposed global cost and sales price monitoring system was being discussed openly by the SIA.[33] In Japan MITI also publicly floated the outline of a proposed price monitoring system shortly

31. The legal basis for the proposal being considered was "flexible application of the Export Trade Control Ordinance or formation of export cartels on the basis of the Export and Import Transactions Law." "MITI Heading Toward Introduction of Minimum Price System for Semiconductor Exports to US; Proposal Will be Made to US Next Month, at Earliest," *Sankei Shimbun*, February 20, 1986, p. 7.

32. See Louise Kehoe, "US and Japan Poised for Agreement on Semiconductors," *Financial Times*, March 20, 1986, p. 6. Even in its preliminary form the global price monitoring scheme was an immediate target for its critics: a *Financial Times* editorial of March 18 blasted the plan as a "blatantly protectionist" international cartel. Clayton K. Yeutter, then U.S. Trade Representative, responded that "there is no chance whatsoever of that occurring, and any such suggestion would be vehemently opposed by the U.S. government!" See "US, Japan, and Semiconductors," letter to the editor, *Financial Times*, April 25, 1986, p. 21.

It is interesting to contrast Yeutter's comments with the following assessment of objectives by Clyde V. Prestowitz Jr., who as an adviser to Secretary Malcolm Baldrige helped shape the Commerce Department's position in the semiconductor negotiations: "we would have to persuade the Japanese government to force up prices in its home market as well as in third-country markets (that is, markets in countries other than the United States and Japan). This amounted to getting the Japanese government to force its companies to make a profit and even to impose controls to avoid excess production—in short, a government-led cartel.

For the free-traders of the United States to be asking Japan to cartelize its industry was the supreme irony. Yet it was logical." Clyde V. Prestowitz Jr., *Trading Places: How We Allowed Japan to Take the Lead* (Basic Books, 1988), p. 62.

33. See Yoko Shibata and Louise Kehoe, "NEC Bid to Beat US Dumping Rules," *Financial Times*, April 21, 1986, p. 6.

thereafter. MITI's proposed system had three elements: first, administrative guidance to producers on output (using as an instrument a supply-demand forecast committee similar to one used in the Japanese steel industry); second, a company-specific minimum price floor for U.S. exports; and third, a uniform export price floor for exports to other countries.[34]

Also in April 1986 the Commerce Department issued its final dumping determination for 64K DRAMs, effectively ending the 64K DRAM dumping case. Henceforth Japanese imports would pay (company-specific) duties to U.S. Customs. The dumping margins found were 11.9 percent for Hitachi, 13.4 percent for Mitsubishi, 22.8 percent for NEC, and 35.3 percent for Oki Electric.

The suspended semiconductor trade talks were resumed in mid-May, and by the end of that month the essentials of a framework agreement were reported to have been negotiated by MITI Minister Michio Watanabe and U.S. Trade Representative Clayton K. Yeutter. The agreement's basic outline was reported in the press: the U.S. antidumping and Section 301 cases were to be suspended in exchange for the implementation of a system to monitor Japanese pricing of exports (to U.S. and third-country markets) and a Japanese government commitment to assist in doubling foreign chip producers' share of the Japanese market over five years, to 20 percent.[35]

34. See "Lowest Prices by Enterprises to be Fixed—Semiconductors; Japan's Plan to be Proposed for Negotiations with U.S.; Demand-Supply Outlook to Be Announced Four Times per Year; Goal for Increase in Imports Rejected," *Nihon Keizai Shimbun*, April 24, 1986, p. 1.

The fact that MITI proposed the specifics of this system in April, before the STA negotiations were actually concluded (in late July), constitutes proof that the STA did not "cause" the creation of a de facto DRAM cartel, according to a recent SIA analysis: "[the STA causing the organization of a Japanese cartel] was impossible; the joint producers' activity [the late 1985 'production coordination' described above], as well as the plan for the MITI guidepost system, existed well before the arrangement." See Howell and others, *Creating Advantage*, p. 121.

But, as described above, American government and industry complaints, legal initiatives, and other manifestations of continuing trade friction were clearly the proximate "causes" of the political pressure on the industry to jointly cut exports and raise prices in the fall of 1985. (There had, in fact, been similar actions before, in 1982, under MITI guidance, in response to U.S. government complaints.) MITI's April proposal to establish its three-point price monitoring system was part and parcel of the negotiation process that culminated in the STA several months later. The STA itself was the legal framework that regularized and justified these actions as official, bilateral, Japanese and American government policy.

35. See Peter Waldman, "U.S.-Japan Microchip Pact Would Offer Big Boost to Ailing

As the negotiations neared a climax, new interests were interjected into the proceedings. In late April, reportedly concerned that Japanese memory chip exports would be diverted from U.S. to European markets, European chip makers set up a working party to bring dumping charges against Japanese memory chip exports.[36] In June, sensing an opportunity and concerned that their interests were being forsaken, U.S. makers of application-specific integrated circuits (ASICs), which are customized versions of standard logic chips, demanded that—as victims of the same unfair trade practices affecting American memory chip makers—they too be included in the framework of the STA.[37] As the U.S. and Japanese negotiations approached an end-of-June deadline to suspend the pending dumping suits, European chip producers rushed to inform the European Commission that they intended to file an antidumping complaint against Japanese memory chip imports.[38]

The final details required considerable further wrangling. Major sticking points were the scope of the proposed price and cost monitoring system (more precisely, whether just the United States or Europe and Asia as well were to be covered), and the details of how costs were to be determined.[39] As the June 30 deadline came and went, the Commerce Department instituted a "provisional" June suspension of the EPROM dumping case, and in early July the 256K and higher DRAM dumping case was suspended. An intense negotiating process continued, largely centered on the international scope of the proposed price and cost monitoring system and the issue of whether specific market share targets would be stated explicitly or implicitly in the agreement.[40] By late July, however, the essential details had been worked out: "secret" side letters

American Producers," *Wall Street Journal*, May 30, 1986, p. 40; and Jim Van Nostrand, "U.S. Claims Trade Breakthrough in Semi Talks," *Electronic Engineering Times*, June 2, 1986, pp. 1, 16.

36. Christian Tyler, "Europeans in Chip Dumping Check," *Financial Times*, May 1, 1986, p. 6.

37. See Louise Kehoe, "Japanese Targeting Further Chip Market, US Makers Claim," *Financial Times*, June 20, 1986, p. 7.

38. Tim Dickson, "EEC Chip Makers Accuse Japan," *Financial Times*, July 1, 1986, p. 14.

39. Louise Kehoe, "Compromise Likely in US-Japan Semiconductor Talks," *Financial Times*, June 25, 1986, p. 4.

40. Louise Kehoe, "US, Japan Fail to Meet Chip Row Deadline," *Financial Times*, July 2, 1986, p. 6; and Louise Kehoe and Carla Rapoport, *Financial Times*, "Not Long Left to Turn a Chip Truce Into a Treaty," July 9, 1986, p. 5.

specified an "expectation" of a 20 percent Japanese market share for foreign companies within five years, and an agreement in principle to monitor exports into third-country markets, while the main body of the arrangement set out the basics of a price and cost monitoring system.[41]

As a rescheduled July 30 deadline approached, all sides stepped up their pressure. U.S. chip makers launched a new round of complaints of price cutting in EPROMs, in the aftermath of the provisional suspension of the dumping case for that product.[42] Both the Japanese and the American industries reportedly applied eleventh-hour pressure on their governments' negotiating teams to alter the terms of the provisional agreements.[43] In the end, however, an extensive third-country price monitoring system was included in the agreement, applying to 90 percent of chips sold by Japanese producers to sixteen countries worldwide (including the United States, six European countries, and Asia), as was a "secret" side agreement in which Japan recognized "the U.S. semiconductor industry's expectation that semiconductor sales in Japan of foreign companies will grow to slightly above 20 percent."[44] Ironically, this controversial and allegedly secret side letter was quite public right from the start, with the exact wording of relevant sections reported on at least two occasions, just before and after the agreement was completed, in the *Financial Times* of London.[45]

Thus, much of the content of the U.S.-Japan Semiconductor Trade Arrangement of 1986—both public and "secret"—was widely reported from the moment the negotiations ended on July 31. Much less clear was what the agreement actually meant for the chip market and its participants.

---

41. Louise Kehoe and Carla Rapoport, "Tokyo May Agree to Aid US Chip Sales in Japan," *Financial Times*, July 21, 1986; and "US and Japan Near Accord on Semiconductors," *Financial Times*, July 21, 1986, pp. 1, 18.

42. See Brenton R. Schlender, "Japanese Dumping of Chips in U.S. Said to Continue," *Wall Street Journal*, July 25, 1986, p. 8; and Louise Kehoe and Carla Rapoport, "Japan Stepping Up Chip Dumping, Says US," *Financial Times*, July 26, 1986, p. 20.

43. See Louise Kehoe, "Stalemate in Chips Trade Talks," *Financial Times*, July 28, 1986, p. 1.

44. Carla Rapoport, "Japan Yields to US Demands on Microchip Price Monitoring," *Financial Times*, August 4, 1986, p. 1.

45. See Kehoe, "Stalemate in Chips Trade Talks," p. 1; and Rapoport, "Japan Yields to US Demands," p. 1.

Initial Implementation of the Arrangement
August 1986 to November 1987

From the start it was clear that some Japanese chip executives opposed the concessions made by MITI in the STA.[46] Indeed, Japanese company executives and others immediately raised doubts about the legality of both the third-country price monitoring system and the apparent commitment to increase foreign chip sales in the Japanese market.[47]

But even more stringent and direct controls over the Japanese industry had been built into the STA. An explicit, though little discussed, provision of the agreement required MITI to exert direct control over Japanese 256K DRAM exports, to ensure that shipments to the United States through mid-September took place at the "normal" rate.[48] Although this transitional provision was designed to thwart a last-minute rush of shipments by Japanese companies in July and August to circumvent the looming specter of price floors, the U.S. government had clearly set a precedent for tolerating—indeed, requiring—direct intervention by MITI in setting quantities to be shipped by individual companies. Recall also

46. As early as July 7, the chairman of the Keidanren (a business organization representing Japan's largest corporations) expressed "apprehension about the excessively restrictive nature of the agreement." "Japanese Makers Dissatisfied with Chip Trade Accord," *Nikkei News Bulletin*, July 7, 1986. In August a MITI official was quoted as saying, "We still don't think we have [the Japanese industry's] agreement." See Carla Rapoport, "Tokyo Concerned About Setting a Trade Precedent," *Financial Times*, August 4, 1986, p. 4. See also Tim Dickson and Carla Rapoport, "EEC and Japan Criticise Semiconductor Agreement," *Financial Times*, August 2, 1986, p. 1.

47. Brenton R. Schlender and Stephen Kreider Yoder, "U.S.-Japan Semiconductor Agreement Is Expected to Prove Difficult to Enforce," *Wall Street Journal*, August 4, 1986, p. 19; Rapoport, "Japan Yields to US Demands," p. 1; and "Tokyo Concerned about Setting," p. 4.

48. This agreement was contained in a supplementary "secret" memorandum dated August 1, 1986, initialed by Yeutter and Watanabe. The exact text is as follows:

Memorandum Regarding July 31, 1986
August 1, 1986
1. MITI will take appropriate actions to control the volume of exports of 256K DRAMs from Japan to the U.S. for the period between July 2 and September 15. MITI will ask Japanese semiconductor companies to see that shipments in the U.S. and indirect shipments to the U.S. of 256K DRAMs will take place at the normal shipment rate during the said period.
2. For this purpose, MITI has already started the technical work necessary to find out what is the current normal shipping rate. This rate is to be worked out on an objective and unbiased basis by experts by August 15.
3. MITI is also ready to devise in detail the necessary legal or administrative procedures to achieve the aim in [point] 1 as soon as possible.

that, back in April, MITI had publicly proposed a system of administrative guidance to producers on output levels, using a supply-demand forecast committee as an instrument.

At least some MITI officials clearly had even more drastic controls in mind. After interviews with MITI officials, Carla Rapoport of the *Financial Times* reported in early August 1986 that "Further, MITI realizes that price monitoring will be ineffectual if it does not force the industry to cut back on its capital spending plans for enlarging Japan's already huge chip-building capacity. Even more radically, it is understood that MITI may be aiming to force the industry to mothball some of its existing capacity."[49] These thoughts were shortly to be translated into concrete actions.

The immediate benefit to Japanese producers was suspension of two dumping cases (in EPROMs and in 256K and higher DRAMs) and the Section 301 case, in early August 1986. In exchange, the U.S. Commerce Department was to "advise Japanese companies exporting to the US of appropriate fair prices for their products so that they may avoid dumping." For other markets and products, the "secret" side letter required the Japanese government to "monitor company-specific costs and export prices in order to prevent dumping."[50]

It is not widely recognized quite how extensive in scope the original STA of July 1986 was. In addition to covering DRAMs and EPROMs exported to the United States, which were to be priced above "fair values" established by the U.S. Commerce Department, the STA framework called for prices of DRAMs and EPROMs shipped to other, third-country markets to be "monitored" by MITI. In addition, MITI was to monitor "representative" products from seven other classes of semiconductors—MOS SRAMs, ECL RAMs, microprocessors, microcontrollers, ASICs, ECL logic chips, and EEPROMs—using company-specific cost and export price data submitted to the ministry.[51] Other products could be added to the list, after consultation, by either government. The side letter specified that "based upon monitoring or consultation, the

49. Carla Rapoport, "Tokyo Concerned about Setting," p. 4.
50. This is a direct quotation from point 2 of section II of the side letter, describing "Third Country Market Measures."
51. SRAMs are static random access memory chips; ECL is emitter coupled logic; ASICs are application-specific ICs; and EEPROMS are electronically erasable programmable read-only memory chips.

GOJ [Government of Japan] will take appropriate actions available under laws and regulations in Japan, including ETC [the Export Trade Control ordinance], in order to prevent dumping."[52] Thus the STA explicitly called for the use of Japan's export control regulations, administered by MITI, for the purpose of enforcing price floors.[53]

Virtually immediately, in the first part of August, MITI began to invoke the export trade control ordinance, and the ministry completely suspended the issuance of new export licenses for 256K DRAMs, to deal with the "last minute export contract" problem described above (sales in export contracts had roughly tripled in July).[54] Japanese producers were advised to either export at the fair value set by the Commerce Department—not the price set in earlier contracts—or forgo exports.

Implementation of the STA moved quickly in the fall of 1986. In August the Commerce Department had also issued its initial, company-specific "fair values" (known as foreign market values, or FMVs) for DRAMs and EPROMs, and not unexpectedly, Japanese producers faced large jumps in the prices set for their exports to the U.S.[55] On September 2, 1986, the official signing of the STA took place in Washington. By mid-September MITI had set up its so-called Forecast Committee, whose task it was not only to publish forecasts but also to help in "correction of imbalances" between Japanese supply and demand for products cov-

52. This is point 5 of section II, "Third Country Market Measures," of the letter.

53. During the negotiations, the American side first suggested that the Japanese use their export control act for enforcement purposes. Author's interview with a member of the U.S. negotiation team, 1989.

54. "Certificate of Approval for Exports to US Withheld, Semiconductors; For Part Covering Last Minute Contracts; Carrying Out of Promise at Time of Reaching Agreement at Negotiations, MITI; 256 Kilobit DRAMs as Objects," Asahi Shimbun, August 6, 1986, pp. 12–13; "MITI 'Blocked' Big Chip Exports Before Price Rise," Japan Times, August 7, 1986, p. 1; and "MITI to Control July Semiconductor Exports," Nikkei News Bulletin, August 7, 1986.

55. The range of FMVs set for individual companies apparently became much more compressed over time. Company-specific 256K DRAM FMVs initially set in July for the third quarter of 1986 reportedly ranged from $2.50 to $8.00. The second round of FMVs, for the fourth quarter of 1986, reportedly ranged from $2.50 to $4.00, while the third round, set for the first quarter of 1987, spanned $2.50 to $3.00. The FMVs for the second quarter of 1987 were reportedly identical to the first-quarter floor prices, but those in the third quarter of 1987 dropped to the $1.89–$2.75 range (averaging $2.34). See Nikkei Top Articles, December 2, 1986; "U.S. Announces No Revision—Fair Prices of Japanese-Make Semiconductors for April-June Period," Nihon Keizai Shimbun, March 24, 1987, p. 3; and "U.S. Dept. of Commerce Announces Fair Market Prices for ICs," Nikkei News Bulletin, September 28, 1987.

ered by the price monitoring framework.[56] This twelve-person committee, representing manufacturers, users, and outside experts, was to meet
and approve a forecast for the final quarter of 1986, to be released
publicly by MITI on September 30.[57] The side letter explicitly called for
this committee to issue forecasts, which were to generate some controversy (as discussed below).[58] On September 22 MITI established a new
office exclusively devoted to running the export monitoring system.[59]

Bilateral agreement over what the price monitoring system was sup-
posed to do was absent from the start. The price monitoring mechanism
initially set *lower* standards for prices in third-country markets than for
sales to the United States. MITI argued that the STA required it to end
"below cost" pricing in third-country markets, but required the use of

56. "Quarterly Price Monitoring of Semiconductors; MITI Announces Prospects for
Supply and Demand as Well," *Nihon Keizai Shimbun,* September 23, 1986, p. 1.

57. The Forecast Committee met on a quarterly basis to discuss historical and forecast
data supplied to it by MITI. MITI actually compiled the numbers appearing in the forecast,
based on information requested from—and furnished by—Japanese semiconductor producers and users. After this quarterly meeting the supply-demand forecast for the next
quarter was then formally issued by MITI.

Three of the committee members came from Japanese semiconductor producers: Toshiba (representing the major established producers; Toshiba's K. Kadono was also chair
of the electronics policy board of the Electronic Industries Association of Japan, or EIAJ),
NMB Semiconductor (a recent, relatively small and independent Japanese entrant into
DRAM production, represented by T. Tamura), and Texas Instruments Japan. Six industry
members represented user industries: Hitachi's K. Fujiki, chair of the Japan Electronic
Industry Development Association (a computer industry group); NEC's M. Yamauchi,
chair of the Communication Industry Association of Japan's management board; Canon's
K. Yamaji, a member of the board of the Japan Business Machine Makers Association;
Toshiba's F. Ota, chair of the EIAJ's consumer electronics board; Yokogawa Electric's S.
Yamanaka, representing electronic measuring equipment makers; and IBM Japan's M.
Motobayashi. Two of the three industry "experts" were dispatched from the Nomura Research Institute (O. Hayama) and Dataquest Japan (S. Morishita), both well-established
consulting firms with close links to the industry. The third outside expert, Professor Tadao
Miyakawa, of Hitotsubashi University, a frequent consultant to government and industry
on electronics industry forecasts, chaired the committee. See MITI, "Semiconductor Demand and Supply Forecast Committee," unpublished document, n.d., given to the author
in September 1988. Note that TI Japan's participation on a twelve-member Forecast Committee is not mentioned in this document (which lists only eleven names), but was specifically cited in an interview with MITI officials, September 1988.

58. Point 5 of section I of the side letter, covering market access, reads as follows: "Both
Governments recognize the importance of discouraging marketing activities which serve to
undercut the intent of the Arrangement. The Government of Japan will compile demand
and supply forecasts on the Japanese semiconductor market in compliance with its domestic
laws and regulations."

59. "Quarterly Price Monitoring of Semiconductors," p. 1; and "MITI Creates Chip
Export Price Monitoring Office," *Nikkei News Bulletin,* September 22, 1986.

the Department of Commerce's cost accounting procedures only on export sales to the U.S. market. For third countries, MITI argued, the ministry would use its own costing guidelines, which, for example, would recognize lower marketing and distribution costs in non-U.S. markets. After further U.S. pressure, however, MITI made some concessions and agreed to some unification of the two monitoring systems for exports after November 1986.[60] Late that month it notified producers that the trade control ordinance might be applied to third-country exports. Disagreement about whether or not the STA required identical pricing of exports to U.S. and third-country markets was a point of continuing dispute well into 1987.

Enterprising merchants quickly invented a variety of creative mechanisms to evade the newly minted MITI controls.[61] Separate "sales promotion rebates" to customers were granted to lower the effective cost of overseas chip shipments below the invoiced (and monitored) export prices. Brokers plugged chips into sockets on electronic circuit boards so that they could be classified as circuit board rather than chip exports, and thus circumvent MITI scrutiny of chip export prices; the chips could later be pulled from the sockets and used in other products. Large shipments were divided into transactions worth less than a million yen each (the trigger level requiring explicit MITI approval for export prices); MITI responded by dropping the minimum value of transactions requiring approval to a mere 50,000 yen after December 1986.

Domestic dissatisfaction with MITI's monitoring system and production guidelines had intensified in Japan by early 1987, developing into a major source of continuing semiconductor trade frictions. DRAM output in the last quarter of 1986 had vastly exceeded the MITI forecast, and exports to third countries in Southeast Asia had surged. When, at the end of December 1986, the Forecast Committee put out its forecast for the first quarter of 1987, it called for a sharp reduction in output by Japanese producers.

60. See "To Unify Standards for Calculating Original Costs of Semi-conductors; MITI's Policy; Aimed at Calming Down Dissatisfaction over Price Differentials in Exports to Third Nations," *Asahi Shimbun*, November 23, 1986, p. 9.
61. "'One Product Having Three Prices' Distorts Markets; Five Months Since Japan-US Semiconductor Negotiations; Mutilated Restrictions with Loopholes; Third Nations Single Out Agreement for Criticism," *Asahi Shimbun*, December 10, 1986, p. 9; and "Approval Necessary for Over ¥50,000—Semiconductor Exports; Trade Control Ordinance to be Revised; to Prevent Evasion of Price Monitoring," *Asahi Shimbun*, December 12, 1986, p. 9.

Commerce Department officials apparently had hoped for substantial production cutbacks in Japan, but those hopes were initially dashed.[62] Some companies, particularly TI Japan and NEC (then the largest Japanese producer of 256K DRAMs), were reportedly reluctant to follow MITI's new "guidance" on production and export volumes.[63] Certainly, the official MITI forecasts for the first two quarters of the STA's operations tended to be significantly below actual production, and forecast errors were much larger than in later quarters (see figure 4-1 below). U.S. merchant semiconductor producers complained bitterly that Japanese producers continued to export DRAMs to third-country markets at prices substantially below the newly devised FMVs for U.S. imports of Japanese chips. In late January 1987 Commerce Department and USTR officials ended a Tokyo negotiating session with the Japanese government by announcing at a Japanese press conference that they had presented evidence of such transactions to Japanese officials.

Controlling Production

By mid-February it was becoming clear that current production levels would inevitably force further DRAM price reductions on the domestic market; would add to the increasing volumes of chips sold into the gray market, made up of difficult-to-control independent brokers and intermediaries; and would exacerbate the flow of chips around the MITI monitoring system and into third-country markets. On February 18, 1987, MITI took the unusual step of revising its December 1986 forecast downward by 10 percent. (This remains to date the only time that MITI has

62. See John Sullivan Wilson, "The United States Government Trade Policy Response to Japanese Competition in Semiconductors, 1982–1987," report to the U.S. Congress (Office of Technology Assessment, September 1987), p. 119.

63. See "Nippon Electric Switches to Policy of Cutting 256K DRAM Production; Structure for Monthly Production of Nine Million Will Be Established; MITI's Guidance Swallowed for Easing of Friction," *Nihon Keizai Shimbun*, March 6, 1987, p. 9; and "NEC to Cut 256K DRAM Output by 10%," *Nikkei Top Articles*, March 6, 1987. In "NEC Cuts Domestic Output of 256K 40% in March," the *Japan Economic Journal* reported on April 4, 1987, that "while the Government has asked Japanese microchip makers to curtail production to help alleviate the chip trade dispute with the U.S., NEC did not comply with the call until February, saying that the domestic market had no oversupply of memory chips" (p. 19). By March 1987 TI Japan was apparently the lone holdout against production cutbacks among Japanese DRAM producers. See Robert Ristelhueber, "TI Speeding Up U.S. DRAM Output," *Electronic News*, April 20, 1987, pp. 1, 6.

issued a midterm revision to its supply-demand forecast.) The official reason given for the revision was that "the exportation of those products remained small and domestic demand remained sluggish."[64] But MITI officials made it quite clear in open discussions with the Tokyo press corps that the revised forecast "sends a strong signal of MITI's desire" and that MITI was using the forecasts as "administrative guidance" to Japanese chip producers to cut back production.[65] The MITI spokesman was quoted as saying that "we don't punish people if our expectations aren't met, but I am sure it will play a role as a production guideline."[66] Yukio Honda, director of MITI's industrial electronics division, even told *Electronic News*'s Tokyo reporter that MITI's administrative guidance to producers had been cleared with the Japanese Fair Trade Commission.[67]

At least some U.S. officials involved in the chip negotiations welcomed MITI's actions. Quoting one such anonymous official, the *Wall Street Journal* reported that "the U.S. has resisted telling Japan how many chips it thinks Japanese producers should be making because of the issue of sovereignty, but the official conceded that MITI's production guidelines 'could potentially be helpful' in driving up Japanese prices around the world."[68] Some in the U.S. chip industry were also receptive to MITI-ordered production cutbacks. Micron Technology Chairman Joseph Parkinson was quick to praise a later round of March cutbacks to a Japanese reporter as "a correct approach," and he added, "The problem is how quickly the Japanese side will achieve results by carrying out the Japan-

64. MITI, "Revision of the Semiconductor Supply-Demand Forecast," Tokyo, February 18, 1987.

65. See "Actual Production Quotas to be Allocated for Semiconductors," *Asahi Shimbun*, February 17, 1987, p. 9; "Japanese Chipmakers Asked to Cut Production," Kyodo News wire report, February 18, 1987; "MITI to Instruct Microchip Makers to Cut Production," *Japan Times*, February 19, 1987; Peter Waldman, "Japanese Chip Firms Told to Cut Output 10% as U.S. Deadline on Accord Nears," *Wall Street Journal*, February 19, 1987, p. 6; A. E. Cullison and Rose A. Horowitz, "Japan Presses Chip Makers to Cut Back on Production," *Journal of Commerce*, February 19, 1987, p. 1; and "Japan Asks 10% Cut in Chip Output," *New York Times*, February 19, 1987, p. D1. The *Wall Street Journal* article quoted Japanese chip executives as suggesting that MITI was acting less to save the STA than to reduce the impact of the ongoing recession on Japanese firms.

66. Waldman, "Japanese Chip Firms Told to Cut Output," p. 6.

67. Minoru Inaba, "MITI Sets DRAM/EpROM Cuts," *Electronic News*, February 23, 1987; see also "MITI to Urge Semiconductor Manufacturers to Cut Production, for Prevention of Sale of Low-Priced Products to Third Countries," *Nihon Keizai Shimbun*, February 19, 1987, p. 14.

68. Waldman, "Japanese Chip Firms Told to Cut Output," p. 6.

US Semiconductor Agreement, which includes the reduction of production by Japan."[69]

In March, complaints about third-country "dumping" of Japanese DRAMs continued to roll in from American chip makers, and MITI, worried about American retaliation, stepped up its pressure on Japanese producers. Approvals of export licenses for memory chips were delayed, and a new system requiring "export shipment certificates," issued by the chip manufacturer directly to the exporter, was introduced in order to seal off chips sold into the domestic market from resale as exports by gray market sources.[70] (In mid-March new export approvals were reported to have virtually stopped; by mid-April the wait for an export license was running one to two months.[71]) The new measures provoked complaints from both Japanese semiconductor makers and their customers but did not silence continuing complaints from American chip makers and their government. Pressure mounted for further actions to reduce the flow of low-priced chips out of the domestic Japanese market and into foreign markets. Since tighter export controls were not entirely successful (or not working rapidly enough), MITI turned its attention toward further restrictions on supply, in order to raise domestic prices.

In mid-March MITI decided to follow up on its downward February forecast revision by incorporating even greater production cuts into its next, regular second-quarter forecast. MITI also reportedly decided to tighten restrictions on the production of 1M DRAMs, which were just entering the market in significant quantities, by issuing company-specific production guidelines for the first time (such targets were already in effect in other types of DRAMs and EPROMs). MITI officials were quoted in press reports as believing that "continuation of production curtailment should give rise to the feeling that there is a shortage of supply in the market in or about June."[72]

---

69. "Guidance for Reduction of Production Appreciated; U.S. Semi-conductor Manufacturer Board Chairman Obtains Oki Electric Industry's Invoice; 'Quick, Concrete Results Urged,'" *Nihon Keizai Shimbun*, March 22, 1987, p. 3.

70. See "Semiconductor Agreement in Pinch, Complaints Pouring in against Curbing of Exports; US Side Checking into Concrete Retaliatory Measures," *Asahi Shimbun*, March 19, 1987, p. 9.

71. "Semiconductor Agreement in Pinch," p. 9; and "Production Cut Pushes Up 256K DRAM Prices," Nikkei Top Articles, April 21, 1987.

72. See "Curtailment of Semiconductor Production by 20 Percent Will Continue Even After April; MITI Holds 'Still Oversupply,'" *Asahi Shimbun*, March 20, 1987, p. 9.

The building atmosphere of crisis deepened on March 19, when Micron Technology made public the results of a "sting" operation in Hong Kong, in which Japanese producer Oki Electric's sales agents had been lured into documenting sales to a Hong Kong buyer at less than FMV prices.[73] (The following week, Micron was to announce another such case, this one involving a Hong Kong distributor associated with Hitachi.[74]) On March 20 the presidents and vice presidents of the largest Japanese chip makers were summoned to MITI and notified of the forthcoming "guidance" from MITI.[75] Shortly thereafter NEC (previously identified in the Japanese press as resisting MITI production guidance), announced a 40 percent cut in 256K DRAM output.[76] MITI also reportedly followed up on the production cuts by ordering large Japanese manufacturers to raise domestic 256K DRAM prices in early April, and export prices later that month.[77]

It was too little, too late, however, to stave off U.S. retaliation on the third-country exports issue. By early March 1987 a subcabinet group within the U.S. government had agreed to consider sanctions on Japanese imports as a punitive measure, and by April these had been announced.

The involvement of MITI in cutting back Japanese DRAM production was widely reported in the U.S. trade press, since MITI had put pressure

73. See "Evidence of Dumping of Semiconductors; US Company Officially Releases Copy of Invoice of Oki Electric Industry's Corporation in Hong Kong," *Asahi Shimbun*, March 20, 1987, p. 2; "Suspicion of 'Trap' Becomes Stronger—Semiconductor Dumping Case Involving Oki Electric Industry," *Asahi Shimbun*, March 21, 1987, p. 9; and "Oki Electronics' Bargain Sale of Semiconductors; Ignites US Distrust," *Nihon Keizai Shimbun*, March 21, 1987, p. 8.

74. See "Hitachi Is Also Engaging in Dumping; Hong Kong Agent Exports at $1.89," *Asahi Shimbun*, March 26, 1987, p. 1; and "Indicts Hitachi on Semi-conductors, Following Oki Electric; Interviews with US Micron Corporation and Hitachi Ltd.," *Nihon Keizai Shimbun*, March 27, 1987, p. 9.

75. "MITI to Make Utmost Efforts to Avoid Retaliation; US Senate Resolution over Semi-Conductor Friction," *Nihon Keizai Shimbun*, March 21, 1987, p. 1.

76. See "NEC Cuts 256K DRAM Output by 40% in March," Nikkei News Bulletin, March 26, 1987. Toshiba also announced large reductions. "Production Cut Pushes Up 256K DRAM Prices," Nikkei Top Articles, April 21, 1987.

77. Initially a price of 330 yen was discussed, but a level of 280 yen ultimately appears to have been chosen. The 280-yen figure compares with an average of 240–250 yen in mid-March. See "Production Cut Pushes 256K DRAM's Domestic Price Higher," Nikkei Top Articles, April 7, 1987; "MITI Requests Makers to Export 256K DRAM at Over 350 Yen," Nikkei News Bulletin, April 25, 1987; and "Semiconductor Market Wobbling Due to Foreign Pressure (Part 1); Erroneous MITI Scenario," *Nihon Keizai Shimbun*, March 31, 1987, p. 20.

on TI Japan to follow its guidance on production.[78] The company, which produced much of its DRAM output at its Miho, Japan, fabrication line, responded by declaring to the press its willingness to comply with MITI's demands. On April 6, 1987, *Electronic News* reported that MITI had on two occasions thus far in 1987 requested Japanese firms to cut DRAM output, and that TI Japan would slash its output of 256K DRAMs by 13 percent to comply with MITI's wishes. Asked to respond to MITI's contention that TI was resisting its requests, Ramesh Gidwani, a TI group vice president, answered, "We have been asked to reduce production, and we are complying. Does that sound like we are resisting?"[79] TI President Jerry Junkins declared to a stockholders meeting, "Although we are responding to this Japanese directive, we do not believe that cutting production, with the attendant risk of creating an artificial shortage, is the correct approach."[80]

By early April supplies of chips to the Japanese gray market were declining.[81] Prices for 256K DRAMs had begun to rise in Japan despite complaints from Japanese users; a trading firm representative said, "we have to accept the new prices because of the direct guidance of MITI."[82] By mid-April DRAM prices had jumped in Singapore and Hong Kong, and trade officials in those countries added their voices to the chorus of complaints about the STA to those of increasingly vocal European officials.[83] As protests from U.S. and Japanese users over higher domestic and export chip prices began to mount, MITI became considerably more reluctant to spell out the precise nature of its actions in public.

Although MITI was to later argue that the details of its guidance were merely "suggestions," since they were not legally binding, the export

78. It was reported in the Japanese press that TI Japan had been asked by MITI on March 25 to cooperate by reducing 256K DRAM output. See "MITI to Guide Japan TI Into Production Reduction—256K DRAM," *Nihon Keizai Shimbun*, March 26, 1987, p. 9; "Semiconductor Manufacturing Coordination Guidance to Be Applied to Foreign-Capital Manufacturers, Too; MITI," *Asahi Shimbun*, March 24, 1987, p. 9.

79. See "TI Japan to Cut Output of 256K DRAMs by 13%," *Electronic News*, April 6, 1987, p. 4.

80. Ristelhueber, "TI Speeding Up U.S. DRAM Output," p. 6.

81. "Semi-conductor Market Wobbling Due to Foreign Pressure (Part 2—Conclusion); Spot Transactions Continue to Decline," *Nihon Keizai Shimbun*, April 1, 1987, p. 20.

82. "MITI-ordered Production Cutbacks Raise Local Prices of 256K DRAMs," *Japan Economic Journal*, April 18, 1987.

83. "We Ask EC Representative Denman; Semiconductors," *Nihon Keizai Shimbun*, April 11, 1987, p. 7; and "Japanese Semiconductors, Southeast Asia Crying Out Over Shortage; Will Not Tolerate Sparks from Japan-US Friction," *Yomiuri Shimbun*, April 12, 1987, p. 7.

control apparatus was clearly wielded in a fashion designed to guarantee compliance.[84] A franker interpretation of these activities was offered by MITI in April 1987, in a memorandum circulated in Washington in response to the sanctions imposed by the U.S. government. Defending MITI's attempts to raise prices in third-country markets, this memorandum was later to prove a crucial piece of evidence in the GATT decision that the third-country pricing provisions of the STA were illegal.[85] It contained the following description by MITI of its own activities over this period:

> In November 1986, MITI invoked the Export Trade Control ordinance in order to prevent below-cost exports. Thereafter, in January 1987, Japan lowered the minimum level for export licenses from ¥1 million to ¥50,000. In February 1987, Japan increased scrutiny of export license applications for third country exports in order to prevent gray market sales. In March 1987, the MITI minister convened an emergency meeting of the Chairman or President of each of the ten semiconductor companies to impress upon them the importance of avoiding dumping in third country markets.
>
> Other actions have been taken aimed at reducing supplies and squeezing out gray market transactions. In February, MITI exercised administrative guidance to the companies to reduce production during the first quarter of 1987 by 23 percent below fourth quarter 1986 levels. Last month, MITI again exercised administrative guidance to the companies to reduce production still further in the second quarter to 32 percent below fourth quarter 1986 levels.[86]

Reducing "Excessive Competition"

Thus by April 1987 both the production and the export of DRAMs by Japanese companies had been placed under fairly tight MITI controls. The vice chairman of NEC publicly acknowledged MITI's role but called upon Japanese companies "to extricate themselves from the inclination toward excessive competition, as can be seen from the rivalry among ten

---

84. TI Japan, for example, was reported to be experiencing long delays in receiving MITI export approvals. A spokesperson was quoted as saying that TI "didn't know if the delay was intentional or not." See Jack Robertson, "Japan Export Delays Draw Fire From U.S. Makers," *Electronic News,* April 6, 1987, p. 4. A frank discussion of the use of export licensing procedures to control export prices may be found in Fujiwara, "This Is a Side Letter," pp. 124–37.

85. See General Agreement on Tariffs and Trade (GATT), *Japan—Trade in Semiconductors: Report of the Panel* (Geneva, March 24, 1988).

86. MITI, "Japanese Position Paper," Tokyo, April 10, 1987, p. 8.

companies in one market."[87] Although "semi-compulsory measures, such
as reduction of production under MITI's guidance and establishment of
virtual export restrictions" were the principal reason why "industry circles have begun to show conspicuous moves for orderly marketing, such
as correction of the inclination toward excessive competition," talk of
manufacturers making their own, private efforts to curb competitive ex-
cesses increasingly surfaced within the Japanese industry.[88] In early May
one "leader of a big semiconductor company" was quoted as saying that
"it is indispensable for manufacturers to make their own efforts hereafter,
in such ways as to establish prices in accordance with the balance between
demand and supply." Sadao Inoue, a high-ranking Fujitsu executive,
declared, "MITI will probably continue its guidance for some time to
come. However, industrial circles have learned various lessons from recent events. If a clear interpretation of the Japan-US Semi-conductor
Agreement, including the definition of dumping, is established . . . it
may become possible to establish an order in industrial circles in accordance with this interpretation."[89]

The idea of "industry circles" working together in private to avoid
unnecessary trade friction received a blessing of sorts from MITI within
the next several months. The Discussion Council on Future Prospects of
the Machinery and Information Industries, an advisory group reporting
to the Director General of MITI's Machinery and Information Industries
Bureau, produced a draft report in June 1987 which attracted considerable attention in Japan, by recognizing the need for government intervention in private sector decisions on production and investment in order
to prevent trade friction.[90] The report also called for a considerable

87. The NEC's Atsuyoshi Ouchi stated in an interview: "Under MITI's administrative
guidance, the 256K DRAM of Japanese make is subjected to thoroughgoing measures for
the reduction of production and the restriction of exports. Also, the 'gray market' for
transactions through brokers, which market is said to be the cause for sales at low prices,
has almost disappeared." "US Semiconductor Retaliation; Industrial Circles Will Review
International Strategy; Inclination toward Excessive Competition Will Be Improved; Japan's Experiential Rule Shaken," *Nihon Keizai Shimbun*, April 19, 1987, p. 4.
88. "Semi-conductor Retaliation; Manufacturers Showing Signs of Self-Reflection(?!),"
Asahi Shimbun, May 3, 1987, p. 8.
89. "Semi-conductor Retaliation," p. 8.
90. See the description of this report in "Contents of Facilities Investments Also Will
Be 'Supervised'; MITI Will Change Survey Method for Prevention of Friction; Reduction
of Excessive Investments Will Be Urged," *Asahi Shimbun*, January 12, 1988, p. 11.

amount of coordination among rival firms in the semiconductor industry.[91] Issued in its final form in late August 1987, this document specifically called on semiconductor producers to cooperate in planning investments and in matching production to forecast levels of supply and demand.[92] The report was endorsed by MITI officials and later became the rationale for a new survey and informal MITI guidelines on semiconductor investment and production levels, issued in 1988.[93]

These calls for "privatization" of the implementation of the STA came as the new MITI enforcement measures were beginning to have a significant impact on market conditions. By mid-May 1987 MITI's more stringent control regime had helped lift 256K DRAM prices 15 percent on the domestic market.[94] In June 1987, after a meeting with Japanese Prime Minister Yasuhiro Nakasone, U.S. President Ronald Reagan suspended a portion of the retaliatory sanctions on Japanese imports.

91. See "MITI's Council on Machinery and Information Industries Draws up Report Calling for Promotion of 'International Cooperation' by Semiconductor Manufacturers; Establishment of 'Prospects for World Demand' Urged," *Mainichi Shimbun,* June 12, 1987, p. 9.

92. Council on Future Prospects of the Machinery and Information Industries, *Prospects for International Cooperation in the Machinery and Information Industries* (Tokyo: MITI, August 24, 1987) (in Japanese). Concrete proposals for the semiconductor industry included these two points (p. 41):

1. Coordinating Information Gathering and Providing Information on Demand and Supply: . . . Also, in light of past failures, we expect to make an accurate forecast of international demand and supply of semiconductors. It is desirable to promote the idea of establishing an international practice among related industries so that they can exchange ideas and information without breaking any laws.

At present, MITI is making a short-term forecast of demand and supply of semiconductors through discussions at the investigation committee on semiconductor demand and supply forecasts. It is necessary to listen to the opinions of makers, users, and experts on relevant items (of semiconductors), make a short and long term forecast of demand and supply in major markets, and examine the possibility of providing adequate information that will benefit companies' planning for manufacturing and investment.

2. Preventing Dumping: On preventing dumping, we should monitor export prices and costs according to the US-Japan Semiconductor Agreement. Considering the fact that prices below production cost are due to production in excess of actual demand, we should seek cooperation of related companies and maintain a production level that matches with actual demand based on the demand and supply forecast mentioned above. Moreover, we need to investigate whether we need to improve the current antidumping system.

93. "Contents of Facilities Investment," p. 11; and author's interview with Japanese semiconductor industry analyst, November 1989.

94. "Share of Imported 256K DRAM Chips May be Rising," Nikkei News Bulletin, May 13, 1987.

European Reaction

European governments, originally worried about excessive supplies being diverted to their market as a consequence of U.S. import restrictions, instead shifted to requesting that Japan now increase supplies of chips to the European market. By this time both prices and quantities for export to third-country markets, as well as the United States, had been placed under increasingly strict MITI guidelines.[95] At midyear computer industry sales—and chip demand—began to recover, and it was soon clear that a worldwide chip shortage was developing. Prices continued to climb worldwide, and in early June the Reagan administration announced a partial lifting of the sanctions imposed in March, to reward the reduction in "dumping."[96]

As prices climbed in European markets, Europe's already negative reaction to the STA turned even more sour. The perceived injury to European interests was procedural, economic, and psychological. The procedural damage involved the secret, bilateral negotiation of a framework intended to affect international market conditions for semiconductors, without consultation with the Community. Fears of economic damage were provoked by two issues: rising memory chip prices, as MITI tightened its export price "monitoring" mechanisms for semiconductors sold in third-country markets to levels prevailing on exports to the United States; and the implementation of measures designed to increase sales of American firms in Japanese semiconductor markets. (The latter fear was, strictly speaking, unfounded: the infamous 20 percent market share was to be supplied by "foreign-based" firms, not necessarily American companies.)[97] The psychological damage was the implicit message that Europe had ceased to be a player that counted in the international semiconductor industry and could safely be ignored. All three irritations moved Europe to complain to the GATT in September 1986, and when consultations with the United States and Japan proved unsatisfactory, to formally request review of its grievances by a GATT panel in February 1987.

95. "American-Make Semi-conductors Cannot Be Purchased, Even If One Desires to Buy; Supply Falls Short Due to Recovery of Market," *Yomiuri Shimbun*, May 29, 1987, p. 6.

96. Third-country prices for Japanese DRAMs had reportedly risen to within 85 percent of the U.S. FMVs. "U.S. to Partially Remove Tariffs against Japan," *Nikkei Top Articles*, June 9, 1987.

97. Of course, since the side letter containing the 20 percent provision was nominally secret, the Europeans had no way of knowing this for sure.

Why was Europe not consulted, in order to head off these complaints? Europeans certainly realized that the game was afoot in the first part of 1986, and they shot off messages to the Americans and Japanese asking that strictly bilateral talks affecting third-country markets be broken off.[98] The European Community and the United States were at the time embroiled in disputes over agricultural trade issues, and that may have made cooperation on other issues more difficult. And since European producers accounted for only 10 to 12 percent of global chip production, they may indeed have simply been viewed as a marginal player, an unnecessary complication in what were already difficult bilateral talks.

An irony in all this was that a movement was already under way in Europe to lobby for an STA-like outcome for the European market. By May 1986, European chip producers, in envy of the extraordinary political success of the American SIA, had followed its example by forming the European Electronic Component Manufacturers Association (EECA) in order to bring dumping cases against Japanese chip imports.[99] In December 1986 the EECA filed a dumping complaint with the European Community, and by April 1987 the Community had widened the investigation to include both DRAMs and EPROMs.[100] The EECA argued that supplies diverted from the U.S. market by the STA were being shifted to the European market, further depressing prices.

After the STA was signed in September 1986, prices continued to drop in the European market (as in Japan) through the spring of 1987.[101] After sanctions were imposed and more-restrictive production guidelines imposed by MITI in the second quarter of 1987, prices quickly rose. In the fall of 1987 Japanese vendors were informing their European customers that an allocation system had been installed, and by late 1987 prices in the European market had risen substantially above U.S. levels. Through 1988 and into 1989, Japanese vendors continued to tell European cus-

98. Author's interviews with EC officials, summer 1989.

99. Christian Tyler, "Europeans in Chip 'Dumping' Check," *Financial Times,* May 1, 1986, p. 6.

100. "EEC Receives Complaint from SC Manufacturers," *International Herald Tribune,* December 6–7, 1986, p. 13; and "EC to Investigate Claims of Japanese Dumping of EPROMs," *Financial Times,* April 10, 1987, p. 5.

101. A more extensive analysis of movements in European DRAM prices may be found in Kenneth Flamm, "Semiconductors," in Gary Clyde Hufbauer, ed., *Europe 1992: An American Perspective* (Brookings, 1989), pp. 225–92; see also Kenneth Flamm, "Measurement of DRAM Prices: Technology and Market Structure," in Murray F. Foss, Marilyn E. Manser, and Allan H. Young, eds., *Price Measurements and Their Uses* (University of Chicago Press, 1993), pp. 174–81, 191–95.

tomers that MITI restrictions limited their European shipments.[102] Thus in the fall and winter of 1986 prices fell below American levels, then caught up by the fall of 1987 (as European purchasers heard about allocations), and finally rose above U.S. levels by mid-1988.

As a consequence, the STA provoked the extreme ire of both European chip producers and consumers. Producers protested first, as European chip prices dropped below American levels. Then, after American pressure resulted in reduced Japanese shipments to Europe and Asia, and chip prices in Europe rose well above American levels in 1988, European consumers began to shout and scream, as they saw themselves disadvantaged in highly competitive global electronic systems industries. It is not evident that there was a deeper strategy behind the roller coaster path of American-European memory chip price differentials: regional price differentials would inevitably arise under any regional supply rationing scheme.

Strategy or not, Europe was angry. The legal vessel in which the European outrage was deposited was an EC complaint before the GATT about the third-country export monitoring system. In March 1988 a GATT panel ruled that the third-country pricing provisions of the STA were illegal, and the Community stood vindicated. However, a formal resolution of the GATT complaint was not to be concluded until over a full year later, in June 1989. At that point, Japan agreed to monitor *retrospectively* exports to non-U.S. markets, to abolish the Forecast Committee, and to "refrain from interfering with the level of production."[103]

The GATT ruling initially seemed to pose a thorny dilemma for the Japanese. On the one hand, the United States was adamant that it would regard lower prices in third-country markets than in the American market as a subversion of the letter and intent of the STA. On the other hand, the GATT ruling clearly required the Japanese government to cease

102. One cannot, of course, neglect the possibility that this was merely a convenient excuse. However, according to a number of knowledgeable Japanese involved in that country's electronics industry (who spoke to me in late 1989), MITI continued administrative guidance on investment through the spring of 1988, continued to guide the regional allocation of DRAM shipments until mid-1989, and continued to offer opinions on companies' investment plans through at least late 1989.

103. Commission of the European Communities, "Japan to Implement GATT Recommendations on Trade in Semiconductors," press release, Brussels, June 22, 1989. Japan had actually issued a proposed response to the GATT panel report on March 7, but it was apparently not accepted by the Community until June.

foreign market value

taking actions—setting export prices or volumes, or fixing production—that had the effect of setting prices in third-country markets. How could Japan escape this quandary?

The dilemma was solved by events: chip prices had by mid-1988 risen far above FMV levels. Even with no "guidance" whatsoever, prices in Europe (by the measures just offered, inflated above U.S. levels, in turn well above FMVs) would certainly not fall below FMV levels any time soon. Thus, ending official monitoring on European pricing and exports would not lead to clashes with the Americans as long as prices outside Japan remained above FMV levels.

By mid-1989 when the GATT case was finally settled, however, all but the final details had been ironed out of a price undertaking ending the DRAM dumping case brought by the European Community against Japanese firms. Ironically, the Community was to ask MITI to implement the Community's very own set of standards for the monitoring of European prices, thus replacing the earlier GATT-illegal system with an EC-blessed floor price structure.

Japan and the European Community agreed in principle on a settlement of the DRAM dumping case in August 1989. The settlement proposed that a quarterly reference price on DRAMs be calculated on the basis of past historical cost data plus a generous 9.5 percent profit margin.[104] Japanese manufacturers signing the undertaking would be required not to sell below this reference price. By early October 1989 the undertaking had been signed by eleven Japanese manufacturers; it was then passed on to the Council of Ministers for final approval. At the United Kingdom's insistence it was also agreed that the five-year agreement be reviewed in 1991, when the STA was scheduled to expire.[105] Price undertakings by Japanese exporters were accepted, and the DRAM dumping investigation was terminated by the European Commission on January 23, 1990.[106]

---

104. The U.S. Commerce Department added an 8 percent profit margin to a constructed cost of production in setting the FMV.

105. Lucy Kellaway, "Japan-EC Microchip Agreement Hits Snag," *Financial Times*, October 5, 1989, p. 2.

106. An antidumping duty was also set to apply to gray market imports. See Commission Regulation no. 165/90, January 23, 1990, published in *Official Journal of the European Communities*, January 25, 1990, pp. L 20/5–30.

From Shortage to Crisis

As prices continued to edge up in the spring of 1987, chip users' complaints became louder. In its June forecast MITI eased up on 256K DRAM production, allowing producers to increase output by a little over 10 percent.[107] For example, after the new forecast was issued, Fujitsu immediately announced an increase in its 256K DRAM production for the July-September quarter.[108] Some in the Japanese industry at this time believed that tight controls on 256K DRAM production were accelerating the trend for producers to switch over to production of the next-generation product, the 1M DRAM.[109]

With increased 1M DRAM output and consequent price declines, MITI had also opted to decree limits on production increases in that product, in order to stabilize prices.[110] In early September both Toshiba and Hitachi publicly confirmed that they had been scaling back production of 1M parts in response to guidance from MITI. A MITI spokesman in Washington responded defensively that it set "no limitation or guideline to the electronic industry"; a Toshiba representative in Tokyo "clarified" the company's earlier statement by adding that "Toshiba-Tokyo took serious consideration of the MITI demand forecast and judged it should change its own manufacturing schedule and plan."[111]

Even as prices were edging up, MITI's control framework was extended into new areas. After complaints from U.S. producers of application-specific integrated circuits, MITI reportedly decided to extend the price monitoring framework to certain types of customized chips in June.[112] By the fall of 1987 the largest-volume type of ASIC chip, gate

107. See "Japan to Raise Output of 256 Kilobit DRAM by 10%," Kyodo News wire dispatch, June 24, 1987, which reports that MITI officials "instructed" Japanese chip makers to increase production of 256K DRAMs, and quotes them as saying that the ministry would "allow" the growth in response to an increase in overseas demand; see also "MITI to Ease 'Guidance' for Reduced Chip Production," Nikkei News Bulletin, June 10, 1987; and "MITI to Partially Lift Restrictions on Chip Production," Nikkei News Bulletin, June 11, 1987.

108. "MITI Reports Microchip Output Estimate (1)," Nikkei News Bulletin, June 25, 1987.

109. "IC Manufacturers Prepared to Expand Production in Kyushu; No Moderation," *Asahi Shimbun,* June 16, 1987, p. 11.

110. "1M DRAM Price to be Unchanged in June," Nikkei News Bulletin, June 10, 1987.

111. Rufus Baker, "Toshiba Cuts 1-Mb Chip Ships," *Electronic Buyers' News,* September 14, 1987, pp. 1, 80.

112. See "MITI to Keep Closer Watch on ASIC Prices," Nikkei News Bulletin, June 23, 1987.

arrays, had joined memory chips on the list of products for which MITI guidance was holding down production.[113]

As a recovery in the computer industry continued briskly through the fall of 1987, prices for chips continued to be pushed upward. In the late-September forecast for the final quarter of 1987, MITI again raised production quotas for DRAMs: by roughly 5 percent for 256K chips and 80 percent for 1M memories.[114] By October a serious shortage appeared to be on the horizon, and manufacturers were reported to be turning away large orders. Domestic Japanese prices even began to approach the FMV export price floors.[115]

Controls on Investment

The inclination of Japanese chip producers was to expand capacity to meet the looming shortage, but MITI officials were reported as feeling that "The removal of the measures which have been taken to curb investments, at this time, may lead to an oversupply of products."[116] As recovery in the industry continued, MITI shifted much of the focus of its guidance to investment, as companies argued that they should be allowed to revise their fiscal 1987 investment plans upward. Reported *Nihon Keizai Shimbun,* "MITI, which is concerned about the possibility of a flaring up of the Japan-US semi-conductor friction again, maintains rigid supervision over the facilities investments of the major companies. As a result, these companies have drawn up moderate investment plans."[117]

The increasing signs of shortage satisfied the American government that MITI had acted forcefully to increase chip prices in worldwide markets. Indeed, by the fall of 1987 the agonized screams of chip users were being heard loud and clear, and the U.S. government had switched its public posture to one of encouraging MITI not to restrict chip production. On November 3, 1987, in a delicate ballet of communiqués, the U.S.

113. See "Anonymous Round Table Discussion on Silicon," *Kinzoku Jihyo,* no. 1319, November 15, 1987.
114. "MITI Lifts Limit on Chip Production," *Nihon Keizai Shimbun,* September 25, 1987, p. 3.
115. "Semiconductors; Sudden Increase in Demand Leads to Growth of Strong Concern Over Shortage of Supply," *Nihon Keizai Shimbun,* October 2, 1987, p. 20.
116. "Nine Semiconductor Companies Keep Facilities Investments Moderated in Revised Plans as Well, In Accordance with MITI's Guidance," *Nihon Keizai Shimbun,* October 10, 1987, p. 8.
117. "Nine Semiconductor Companies Keep Facilities Investments Moderated," p. 8.

government partially removed more of the spring sanctions and announced that it was satisfied that third-country dumping had ended. MITI announced that it was "imposing no quantitative or other restrictions on the production, shipment, or supply of semiconductors, except MITI continues to exercise export control from the view point of COCOM."[118] This last qualification, it turned out, was crucial, since an extremely rigorous export control system continued in place during 1988, effectively making it impossible for foreign buyers to purchase Japanese chips at prices not directly controlled and administered by the chip manufacturers.[119] The wording is also notable in that it does not directly address the issue of whether extralegal MITI "guidance" or "suggestions," in contrast to legally binding "restrictions," might be influencing Japanese manufacturers' production decisions.

Behind the scenes, the U.S. government was less than unequivocal in its view of the situation. Officials were not eager to see Japanese firms increasing their chip capacity to meet the looming shortage, preferring to see American companies "reenter" the DRAM market.[120] While pub-

118. MITI, "Statement of Ministry of International Trade and Industry Concerning Trade in Semiconductors," Tokyo, November 3, 1987. COCOM, the Coordinating Committee on Multilateral Export Controls, was the body through which the Western alliance coordinated its controls of exports for national security purposes.

119. In 1988 and 1989 foreigners wishing to purchase DRAMs from Japanese vendors were required to register with MITI; MITI would not consider issuing an export license to a Japanese vendor until the application was approved (author's interviews with semiconductor purchasers, 1989). The policy resulted in tight control over the ability of foreigners to gain access to the Japanese DRAM market. From personal experience in attempting to purchase DRAMs on the retail market in Japan, in March and September 1988 in the Akihabara in Tokyo and Den-Den Town in Osaka, I can report that no retail outlet appeared willing to sell DRAMs to an obvious foreigner walking in off the street. When questioned on this point, a MITI official suggested that the vendors may have been concerned about export licensing requirements.

120. This attitude was aided and abetted by the widespread dissemination of a study carried out by the consulting firm Quick and Finan, which forecast that "The tight supply conditions seen in late-1987 [are] only temporary. Several firms have just begun full-scale production and others will soon follow in the first half of 1988. Thus, throughout most of 1988, potential supply capacity will ramp faster than demand under any plausible market scenario." Quoted in Semiconductor Industry Association, *One and One-half Years of Experience Under the U.S.-Japan Semiconductor Agreement* (Cupertino, Calif., March 1, 1988), p. 26.

This view conflicted sharply with the consensus view in Japan at that moment, where semiconductor manufacturers predicted a shortage lasting at least until the spring of 1989. See Mitsuhiro Takahashi, "Producers Slow to React to Chip Shortage," *Japan Economic Journal*, May 7, 1988, p. 1. Needless to say, the view in Japan—where more than 90 percent of merchant DRAMs were produced—was the correct one.

licly proclaiming that they were acting to remove limits on chip production, American trade negotiators continued to press MITI to limit investments by Japanese firms in new capacity well into 1988.[121]

The Control Regime after November

What was the real content of the changes in the export control regime made in November 1987? Organizationally, responsibility for approval of export license applications was moved from MITI's semiconductor monitoring office (charged with surveillance of export prices) to a separate office, so that "there was no feedback from the Monitoring Office to the office dealing with COCOM screening." However, "in cases where export prices were 'extremely below cost,' MITI would express its concern to the companies concerned. . . . Under the old system . . . some misunderstanding seemed to have been created among exporters that delays had been caused by inappropriate pricing. The new system would eliminate such misunderstandings."[122]

What of the infamous production "forecasts"? After November 1987 the production forecast system was henceforth to be described by MITI as

a reference for manufacturers in their production schedules. MITI explained its objective to manufacturers and impressed upon them the need to reflect real demand in their production. Individual companies were expected voluntarily to bring their production almost in line with the forecasts, taking into account the appropriate total production. The forecasts were not legally binding and the Government did not allocate production volume to individual companies. For manufacturers to conspire on production volume was against the anti-trust laws in Japan.[123]

Unfortunately, this description of the operation of the forecast system contrasted rather markedly with the assessment circulated by MITI itself in Washington in April 1987 (and diverges as well from the public utterances of various MITI officials through mid-1987, Japanese press accounts, and the testimonials of others involved in semiconductor trade). Indeed the GATT panel surveying the MITI monitoring system rejected the MITI characterization in its March 1988 report, specifically citing the April 1987 MITI position paper, and concluded that

121. Author's interviews with former U.S. government officials, 1989.
122. The quotations are from a GATT panel's summary of the official Japanese government position. See GATT, *Japan*, p. 9.
123. GATT, *Japan*, p. 10.

196 THE SEMICONDUCTOR TRADE ARRANGEMENT

an administrative structure had been created by the Government of Japan which operated to exert maximum possible pressure on the private sector to cease exporting at prices below company-specific costs. This was exercised through such measures as repeated direct requests by MITI, combined with the statutory requirement for exporters to submit information on export prices, the systematic monitoring of company and product-specific costs and export prices and the institution of the supply and demand forecasts mechanism and its utilization in a manner to directly influence the behavior of private companies. . . . The Panel considered that the complex of measures exhibited the rationale as well as the essential elements of a formal system of export control. The only distinction in this case was the absence of formal legally binding obligations in respect of exportation or sale for export of semi-conductors. However, the Panel concluded that this amounted to a difference in form rather than substance because the measures were operated in a manner equivalent to mandatory requirements.[124]

Indeed, the only immediate, concrete change visible in the operation of the forecast system after 1987 was that from March 30, 1988, on (after the GATT semiconductor panel had issued its report) MITI forecasts were accompanied by text asserting that the forecasts were issued for "reference" purposes only, and not for the purpose of restricting production.[125] Since this assertion had also been made to—and rejected by— the GATT panel with regard to the 1987 forecasts, it was less than compelling.

The Zen of Accurate Forecasting

Certainly the statistical record does not indicate a sea change in the relative accuracy of the "forecasts" of the most politically sensitive products—DRAMs and EPROMs—before and after November 1987. During the final quarter of 1986, before MITI clamped down hard on Japanese producers, the DRAM production forecasts were wildly inaccurate, with production of both 64K and 256K parts exceeding one-quarter-ahead forecasts by margins in the range of 30 to 50 percent of actual production.

124. GATT, *Japan*, p. 42.
125. "Since there seems to be a misunderstanding that this supply-demand forecast is compiled for the purpose of restricting production, we make it clear that it is not compiled for that purpose, but for the reference of related parties and that manufacturers are free to produce more than the production in the forecast. Decision on production fully depends on producers." MITI, Machinery and Information Industries Bureau, "Semiconductor Supply-Demand Forecast for 2nd and 3rd Quarter 1988," Tokyo, March 30, 1988, p. 1.

Since MITI reacted by "revising" the first-quarter 1987 DRAM forecast downward by 10 percent in midquarter, it is perhaps not surprising to find that production fell short of the original first-quarter forecast for DRAM chips by margins clustering rather tightly around 10 percent and by 15 to 20 percent for EPROMs (figures 4-1a and 4-2a). And until MITI began issuing detailed guidance on company-specific production levels for the 1M DRAM, in the second quarter of 1987, the one-quarter-ahead forecast error for the first quarter of 1987 was spectacularly large for that product, with the forecast falling short by about 35 percent of actual realized output (figure 4-1a). From that point through November 1987, a period of relatively public MITI controls on production, the forecasts issued for 265K and 1M DRAM production levels three and six months out typically fell within 10 percent of actual output, as shown in figure 4-1a. (For 1M DRAMs, with production dominated through most of this period by just one producer, Toshiba, the error was more typically on the order of 5 percent.)

The truly noteworthy feature of the post–November 1987 DRAM production forecasts is that they continued to be extraordinarily accurate, *after* the invisible hand, rather than government's, ostensibly ruled the market. Indeed, at least one explanation offered in early 1988 for the unprecedented jump in memory chip prices clearly flew in the face of these forecasts' accuracy. Some analysts argued that unexpected yield problems were a major factor in the shortage then developing.[126] It was widely believed in the industry that some companies, particularly NEC

126. "At present, manufacturers are hurrying the expansion of [1M DRAM] production. Because of such technical problems as the delay in the improvement of the yield rate, however, production in the July–September period also is likely to increase by only about 2,700,000." "Semiconductor Prices Continue Rising; Shortage of Supply Will Reach Peak in July-September Period," *Nihon Keizai Shimbun,* July 16, 1988, p. 18.

" 'It took chip makers longer than they expected to get good yields of one-megabit chips,' says Jim Feldhan, vice-president of In-Stat Inc., a Scottsdale Ariz. market research firm that tracks the semiconductor industry. Mr. Feldhan says as many as 90% of all one-megabit chips were defective and had to be discarded as recently as early this year, but that lately yields of good chips have improved to around 50%." Brenton R. Schlender, "Chip Prices Fall: Easing of Shortage Seen," *Wall Street Journal,* July 18, 1988.

"Currently, major makers are shipping an estimated 12 million units of 1-M DRAM chips monthly.

The industry had planned to reach that level four to six months earlier. Apart from Toshiba, however, producers were unexpectedly slow to boost production. Then, companies' average yield rate, a production yardstick that measures the percentage of useable chips, stood at between 40% and 50%." Tadashi Tamaki, "Heavy Investment in Chip Factories Spurs Fear of Glut," *Japan Economic Journal,* October 1, 1988, pp. 1, 5.

Figure 4-1. *Errors in MITI Forecasts of Japanese DRAM Production, Quarterly, 1986–88, and Semiannually, 1987–94*[a]

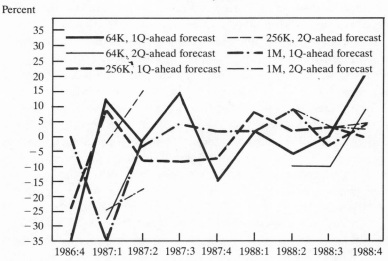

A. Quarterly forecasts

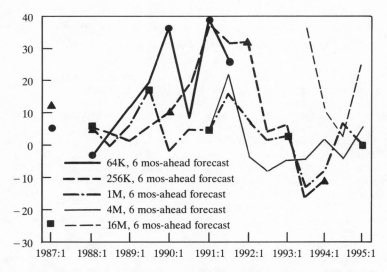

B. Semiannual forecasts

Source: MITI, "Semiconductor Supply-Demand Forecast" (quarterly); after 1988, semiannually.
a. Forecast error is defined as forecast less actual as a percent of actual. A positive error represents an excess of actual over estimated production.

Figure 4-2. *Errors in MITI Forecasts of Japanese EPROM Production, Quarterly, 1986–88, and Semiannually, 1987–94*[a]

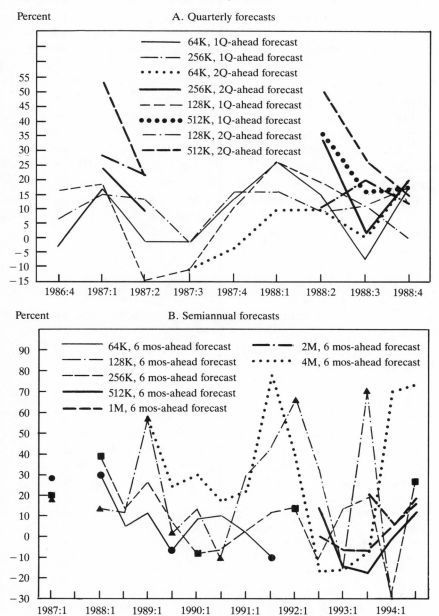

Percent   A. Quarterly forecasts

- 64K, 1Q-ahead forecast
- 256K, 1Q-ahead forecast
- 64K, 2Q-ahead forecast
- 256K, 2Q-ahead forecast
- 128K, 1Q-ahead forecast
- 512K, 1Q-ahead forecast
- 128K, 2Q-ahead forecast
- 512K, 2Q-ahead forecast

Percent   B. Semiannual forecasts

- 64K, 6 mos-ahead forecast
- 128K, 6 mos-ahead forecast
- 256K, 6 mos-ahead forecast
- 512K, 6 mos-ahead forecast
- 1M, 6 mos-ahead forecast
- 2M, 6 mos-ahead forecast
- 4M, 6 mos-ahead forecast

Source: MITI, "Semiconductor Supply-Demand Forecast" (quarterly); after 1988, semiannually.
a. Forecast error is defined as forecast less actual as a percent of actual. A positive error represents an excess of actual over estimated production.

and Hitachi, had had some difficulties raising their manufacturing yields in their early production of the 1M DRAM.

But if such yield problems, in the aggregate, played a significant role in creating shortages, of 1M DRAMs in 1988, it is hard to see how they could have been unexpected, given the accuracy of the MITI forecasts. Since manufacturing time from starting a silicon wafer on the line to a finished memory chip is typically 60 to 75 days, to adjust production scheduling in midquarter to compensate for unexpected yield problems during that same quarter and still hit the three-month forecast would be most difficult.[127] To accurately hit both three-month and six-month forecast targets, as happened with both the 256K and the 1M DRAM production forecasts for 1988, would seem to require quite an accurate estimate of yields over a roughly six-month period. Thus, a continuous history of coming within 10 percent of three- and six-month production forecasts, in a product requiring over two months of processing on the production line, suggests that unanticipated yield problems could not have been a major issue for the industry as a whole in 1988.

Although yields were undoubtedly low in the initial stages of 1M DRAM production (as has been true for every generation of chip), it is hard to reconcile big surprises in yields with the pinpoint accuracy of the MITI production forecasts. Thus both one- and two-quarter-ahead DRAM production forecasts exceeded actual output by about 9 percent in the second quarter of 1988 (figure 4-1a). At other times during 1988, forecast output for 1M DRAMS stayed within 2 to 4 percent of actual production. At their worst, unanticipated yield problems could only have been a marginal factor.

Japanese chip manufacturers quoted in the Japanese trade press, on the other hand, gave producers' restraint in production and investment the lion's share of the credit for the 1988 DRAM crisis. Reacting to reports of a shortage of semiconductors and American user complaints,

127. The "unexpected yield problems" thesis attained its most extreme form in the assertion that the earthquake that Japan suffered on December 17, 1987, by shaking up chip plants and yields, was a proximate cause of the price surge of early 1988. This story was given wide circulation by George Gilder in "How the Computer Companies Lost Their Memories," *Forbes*, June 13, 1988, p. 81. MITI's Forecast Committee actually met on December 15, two days *before* the big earthquake, to discuss DRAM production during the first quarter of 1988. That forecast was extremely accurate: actual production fell short of the forecast by only -1.6 percent for 64K parts, -7.7 percent for 256K parts, and -1.9 percent for 1M parts.

a top Hitachi semiconductor executive offered the following assessment of the shortage's causes in March 1988:

> It is mainly memory chips which are short. The cause is that, just at the time when Japanese manufacturers, who are the major suppliers, held down production out of fear of escalating the semiconductor friction, demand for office automation equipment, such as personal computers, etc., increased sharply. This situation will continue throughout this year. However, we cannot tell about the next year and after. From this sense of non-transparency, various manufacturers are cautious about facilities investments, and as a result, the tight situation of demand and supply in the market is being spurred.[128]

By historical standards the official MITI DRAM production forecasts issued from 1987 to 1989 were extraordinarily accurate. Before that period and after, divergence between predicted output levels and actual ones for either one quarter (figure 4-1a) or six months (figure 4-1b) ahead showed noticeably greater variation.[129] The same observation may also be made for EPROMs (figure 4-2).

The Privatization of Restraints, December 1987 to Mid-1989

Certainly it was a kinder, gentler, and considerably more discreet control system that continued after MITI's November 1987 disavowal of production controls. By this point, however, production controls had begun to fade as a serious issue, as demand for memories surged. Firms had been permitted to increase production steadily over the second half of 1987, and by the end of that year many Japanese semiconductor producers were approaching capacity limits. Increasingly, limitations on new capacity investments and the allocation of production between domestic and regional export markets became the focus for scrutiny by Japanese government regulation of the industry. Administrative guidance of investments and a regional allocation system for exports, already visible in 1987, were to become important factors during the emerging electronics boom of 1988.

---

128. "One Year Since Retaliation on Semiconductors; Interview with Government and Industry Circles on Prospects," *Mainichi Shimbun*, March 26, 1988, p. 9.

129. The MITI forecasts were available for one quarter ahead only from the first through third quarters of 1987 and were switched to a six-month forecast at the end of 1988.

Regional Allocation of Exports

By the fall of 1987 a regional allocation system for exports had been put in place. (See the section on "European Reaction" above.) In 1988 this system was to create large price differentials between DRAMs offered for sale in domestic Japanese markets and those shipped in foreign markets. The allocation system was not ended until mid-1989.[130]

Guidance on Investment

As previously remarked, it was widely reported in the Japanese trade press that MITI had exercised administrative guidance over manufacturers' revisions to their fiscal 1987 investment plans in the fall of that year. That fall had also marked the publication of a report by a MITI advisory council calling for a greater government role in ensuring that company production and investment plans did not create trade friction. Top MITI officials had endorsed the report. In January 1988 MITI replaced its twice-yearly survey of Japanese industrial investment with a much more detailed and rigorous questionnaire, in order to better monitor investments in trade-sensitive sectors, particularly semiconductors, and advise businesses when investments were judged to be excessive.[131]

130. Author's interviews with Japanese semiconductor executives and analysts, November 1989.

131. See "Contents of Facilities Investments," p. 11. The *Asahi* reporter wrote, "The surveys until now have been conducted by means of questionnaires with major emphasis on the monetary amounts of investments. In the coming survey, however, MITI will examine the contents of investments as well, and ask about such details as the kinds of products and the amount of production of the respective items. It has decided on revision of its survey method, from the standpoint that the trade friction, which is represented by the Japan-U.S. semiconductor problem, has often been caused by excessive investments on the part of Japanese enterprises. Depending on circumstances, MITI will ask enterprises for reduction of the investments, which it has judged to be clearly excessive as a result of its survey. . . . In August of last year, the 'Discussion Council on Future Prospects of the Machinery and Information Industries,' which is a personal consultative organ of the Director General of the MITI Machinery and Information Industries Bureau, drew up a report which recognized the necessity for the Government to intervene in the decisions to be made by enterprises on production and investments, for the purpose of preventing the occurrence of trade friction."

MITI's reaction to the possibility that enterprises might object was the following: "The survey is to be conducted with the cooperation of enterprises in all cases. The main purpose of the survey is to grasp the reality. Even if MITI judges investments to be excessive, it will only take such steps as to urge caution."

Eight months later, an editorial in the *Japan Economic Journal* explicitly noted the

January 1988 also marked a significant appreciation of the yen. Conscious of continuing scrutiny from the United States, Japanese chip makers reportedly raised their U.S. export prices from 5 to 10 percent on contracts signed in that month.[132] Coupled with a recovery in the computer industry, an increasing demand for memory chips, and the delayed effects of restrictions on new capacity investments in 1987, these price increases touched off an extraordinary spiral of rising chip prices. In early 1988 spot prices for DRAMs in the United States soared to historically unprecedented levels—the price of the best-selling 256K DRAM tripled over a four-month period—and the U.S. computer industry was plunged into crisis; producers scrambled for supplies of critical memory chips.

Despite the emerging DRAM crisis, however, Japanese output increased relatively slowly. In the Japanese business press, sporadic reports suggested that MITI was continuing to offer occasional informal guidance to producers on production plans after November 1987.[133] Through the close of the 1987 fiscal year in March 1988, MITI maintained a relatively tight rein on investment in new capacity. With the approach of summer, and a worsening shortage, restrictions on capacity increases were apparently relaxed.[134] In the late spring, as plans for fiscal 1988 were disclosed,

---

existence of the investment control system: "While the Japanese government was busily setting up an export price monitoring system and instituting a series of production, and plant and equipment investment controls, the demand from Japanese and American semiconductor users rebounded sharply." "Editorial: Chip Pact Obsolete," *Japan Economic Journal*, August 13, 1988, p. 22.

132. "Growing Moves to Raise Export Prices, Centered on Exports to US; Semiconductor Prices Upped by Five to Ten Percent," *Nihon Keizai Shimbun*, February 1, 1988, p. 1.

133. One newspaper ("Prices of Japanese-Make Semiconductors Rise Sharply in East Asia; Almost Twice as High as Level at Start of This Year," *Nihon Keizai Shimbun*, December 23, 1987, p. 18) reported that "manufacturers maintain a cautious attitude toward expansion of production under MITI's guidance." The same paper ("Listen to GATT's Decision," *Nihon Keizai Shimbun*, March 7, 1988, p. 2) later noted that "MITI is carrying out strict administrative guidance, such as to allot production quotas for the respective items to manufacturing companies, etc., in an effort to make the agreement truly effective." See also "Semiconductor Shortage; Prices of Personal Computers and Word Processors Rising Gradually," *Asahi Shimbun*, June 23, 1988, p. 3. This article, published seven months after MITI declared that it would not formally restrict production, explains a current semiconductor shortage in part by saying, "On the other hand, however, it is also strongly viewed that MITI's guidance toward various manufacturers for their reduced production— as a counter-measure for the Japan-US semiconductor friction—took effect to excess."

134. Tight restrictions were reportedly maintained through March 1988 (author's interview with Japanese industry source, 1989). Another Japanese analyst close to the semicon-

204 THE SEMICONDUCTOR TRADE ARRANGEMENT

an average 38 percent increase in semiconductor capital investment was announced by Japanese chip producers.[135] In the face of a mounting shortage, Japan's semiconductor business community viewed these investment levels as quite restrained.[136] The *Japan Economic Journal* even explained the continuing shortage of semiconductors in the following terms: "One reason for this is the makers' restraint in investment. They seem to be acting, or not acting, in concert."[137]

"Profitability Instead of Market Share"

Around this period, in mid-1988, the Japanese trade press began to routinely publish analyses suggesting that manufacturers were acting much more collusively than in the past, and that firms were consciously acting with restraint in order to increase profitability for the industry as a whole.[138] These two themes—restraint in investment, linked in part to

ductor industry whom I interviewed at this time told me of reports then circulating that chip makers were submitting investment plans to MITI for informal review.

135. "Big Semiconductor Companies to Increase Facilities Investments for Fiscal 1988 by 38%; Stress to be Laid on 1M DRAM," *Nihon Keizai Shimbun*, May 27, 1988, p. 8.

136. Commenting on the proposed increases in investment, the *Mainichi Shimbun* wrote, "However, whether or not the shortage of stocks will be dissolved by such increased production is not clear. As for the reason why, semiconductor industry circles carried out facilities investments to excess and increased their production on a large scale, centering on fiscal 1983 and 1984, and they suffered from a big depression in 1985. There was such a bitter experience, and they generally agree on the view that 'there was a tragedy in the fact that various companies were mainly producing memory chips in those days' (NEC Chairman Atsuyoshi Ouchi). In view of facilities investments, too, the amount of NEC's facilities investments is one-third of that in those days (¥140 billion in fiscal 1984), and Toshiba—about one-half (¥148 billion in fiscal 1984). Moreover, it is because all manufacturers remain cautious about production leaning toward memory chips, as is represented by such words as 'We want to maintain 20% for [production of] memory chips.' Accordingly, the shortage of memory chips is likely to be carried over to next year." "Semiconductor Shortage Serious; Going beyond Production Increase Setup," *Mainichi Shimbun*, June 29, 1988, p. 8.

137. Naoshi Tamaki and Akira Kikuzumi, "Shortage of DRAMs Causing Headaches," *Japan Economic Journal*, July 2, 1988, p. 4.

138. For example, a May 7 analysis in the *Japan Economic Journal* gave three reasons for the sluggish investment in chip production. Said the reporter, "one reason is the fact that manufacturers are loath to repeat the debacle of 1984–85 [when demand shrank drastically after a buildup in capacity]. Another reason is the fact that Japanese semiconductor makers are making comfortable profits from the present arrangements based on the Japan-U.S. semiconductor agreement signed in the autumn of 1986 at the urging of the Ministry of International Trade and Industry. Production cutbacks enforced by the agreement have sharply boosted the market prices of semiconductors and Japanese manufacturers *have come to value profits more than market shares* [emphasis added]. A third reason, on

MITI attitudes, and more cooperative behavior ("profitability instead of market share")—were to surface often in Japan over succeeding months.[139]

An article in *Nihon Keizai Shimbun* in June 1988 elaborated on this theme.

> The silicon cycle has completely come undone. The reason for this is the constraint on investment on the part of semiconductor makers, done in the name of "cooperation." Koji Kobayashi, chairman of NEC, asserts that "The semiconductor manufacturers will not let the tragedy of 1984 happen again. . . . According to one of the top men at a leading chip manufacturer, 'I would not want this to be construed as a cartel. Manufacturers have become more open in sharing information, so that there is more coordination. In the old days, we used to send industrial spies to gather information about the activities of our rivals, but recently, that practice has vanished.' For example, Toshiba announced that 'Mass production of the next-generation 4 megabit chips is not planned for this year, and we can not predict when it will start.' Such openness should dispel any lack of trust within the industry, and is related to efforts to keep excessive competition under control. . . . It does not appear that the chip makers' move away from memory chips and their cooperative efforts to avoid excessive competition are temporary phenomena." The shortage of memory chips is thus expected to continue for a long time.[140]

The same themes were echoed in an August 1988 analysis in the *Japan Economic Journal*. After noting that capital investment in fiscal 1988 was expected to reach 430.8 billion yen, compared with 762.8 billion yen during the boom year of fiscal 1984, the reporter quoted Yukio Honda, director of MITI's industrial electronics division, as saying, "This figure mirrors the manufacturers' prudent stance toward capacity expansion for fear of another recession in the future." The article continued:

---

the other hand, is Japanese manufacturers' fear that any large-scale equipment investments will rekindle trade friction with U.S. industry." Takahashi, "Producers Slow to React," p. 10.

139. The *Japan Economic Journal* ran an editorial on May 28, for example, that commented, "It is unnecessary to note that semiconductor manufacturers are aware that the government's supply-demand forecasts mean nothing less than quantitative production controls, though the ministry takes the position that it is simply one of the many forecasts it routinely employs. Similarly, the legality of the government's semiconductor export price monitoring system is also ambiguous. There is no doubt that the monitoring of exports by the overseas subsidiaries of Japanese semiconductor manufacturers has no legal justification." "Editorial: Scrap the Chip Pact," *Japan Economic Journal*, May 28, 1988, p. 22.

140. "Supply-Demand Structure for Semiconductor Industry Shifts," *Nihon Keizai Shimbun*, June 8, 1988, p. 9.

Chip makers have long been vying for larger market shares, as mass production leads to remarkable cost reductions. Manufacturers embarked on capacity expansion in peak years which repeatedly caused overproduction and a resulting recession. Any coordinated action among manufacturers was unthinkable. But after the 1986 Japan-U.S. chip pact, MITI led the manufacturers to reduce IC production by about 30% in early 1987. "The pact helped to virtually create a coordinated production control by chip makers that we have never seen before," a broker said. . . . The Japan-U.S. microchip agreement so far has failed to achieve improved access for U.S. semiconductor makers to the Japanese market. But it seems certain that the pact helped Japanese chip makers strengthen their profitability through production control.[141]

Japanese semiconductor analysts in Tokyo also stressed both the trend toward more oligopolistic behavior in the chip industry and the government's role in encouraging it. Nomura Securities echoed the familiar "profits instead of production volume" theme in a September 1988 analysis.

In addition, equipment investment for research and development in this area involves major risks. This leads to an oligopolistic market, in which a few major firms dominate. The top Japanese firms tend to benefit from this kind of situation. Moreover, the Japanese-U.S. treaty on semiconductors— which is expected to be amended slightly—is seen as bringing about a stabilization of prices and should contribute to ensuring sustained profit earnings. The switchover from policy of expanding production volume to a situation where management places an emphasis on profitability will also contribute to market stability.[142]

The same rhetoric was repeated in a conversation I had with an official in MITI's industrial electronics division on September 26, 1988. After a long and inconclusive discussion of what sort of exchanges between MITI and the Japanese industry might be considered "controls" or "guidance" or even "suggestions," I was told by the official that it was the position of the Japanese government in general, and MITI in particular, to encourage chip firms to stress profitability instead of market share.

Probably the closest thing to an official confirmation of a Japanese government policy of tolerating, if not encouraging, collusive restraint in production and investment among domestic producers came at a Novem-

141. Shigehisa Shibayama, "Chip Shortage Expected to Last Through '89," *Japan Economic Journal*, August 13, 1988, p. 5.
142. Nomura Securities Co., Ltd., "Economic Insight: Heavy Demand for Memory Chips," *Japan Times*, September 17, 1988, p. 9.

ber 1988 Washington meeting between U.S. and Japanese trade negotiators. The American side's summary memorandum for that session described the following exchange:

> Finally, USG [U.S. government] expressed concern about the role of Forecast Committee and its impact on market. USG stated many feel Forecast Committee influences or sets production and investment levels rather than simply forecasting. USG said language used by GOJ [government of Japan] to describe Forecast Committee suggests a broader role, and asked how exactly Forecast Committee can "stabilize" market. USG also asked how it is that Forecast Committee has been as accurate as it has in projecting production levels.
>
> GOJ responded that there is no reason for USG concern about Forecast Committee. GOJ stated forecasts are accurate because they are provided by companies themselves which are then aggregated by Committee. GOJ stated Japanese industrial society is very competitive and each member of Forecast Committee is very interested in how much its competitors produce. GOJ said Forecast Committee had "freed the Japanese semiconductor companies from unnecessary competition" after period two years ago during which companies competed in supply and manufacturing.[143]

As the shortage of DRAMs intensified in the late summer and early fall of 1988, the major Japanese chip producers ultimately announced upward revisions of their capital spending plans.[144] Even after those revisions, however, capital spending was far (about 40 percent) below outlays during the 1984 boom in absolute terms, despite the substantial growth in both the sales base and the cost of equipment. October estimates of fiscal 1988 spending on capital equipment by Japanese chip companies as a fraction of sales were up somewhat over 1987, but well short of pre-STA spending levels. Furthermore, a significantly smaller proportion of this equipment investment went into DRAM capacity than during the 1984 chip boom.[145]

Curiously, these relatively modest increases in revised capital investment plans for 1988 inspired a campaign in the English-language press against a forthcoming chip glut. The *Japan Economic Journal,* in a prominently displayed, October 1, 1988, front page article, decried the "heavy

143. "Subject: US-Japan Semiconductor Consultations, November 17–18, Washington," memorandum, Office of the U.S. Trade Representative, Washington, n.d.

144. Some analysts in Japan claimed that the investment plans publicly announced by semiconductor companies exceeded actual investments by a significant amount. See "Chips in Short Supply," *Tokyo Business Today,* September 1988, p. 9.

145. This point was made to me by a MITI official on September 26, 1988.

investment in chip factories" as spurring fears of a glut—yet the revised investment plans given for seven Japanese semiconductor companies were up only a very modest 6.9 percent over the initial "restrained" plans described by that same newspaper, back in July, as unlikely to allow production to keep pace with demand.[146] TI Chairman Jerry Junkins, in an interview with a Japanese reporter, stepped up the pressure by arguing that "from now on, there will be a slowdown in increases in demand. So manufacturers should be responsible for their facilities investments so that they will not disturb the order of prices and market trends."[147]

MITI and Japanese producers continued to predict a shortage well into 1989, however, and manufacturers announced further increases in revised fiscal 1988 facilities investment.[148] By the end of October revised investment plans for the six largest Japanese chip producers were up 24 percent over initial fiscal 1988 levels.[149] The increase prompted prominent U.S. press coverage of Japanese producers "fighting each other" in a bruising race to invest in new capacity.[150] But through the end of 1988, and into early 1989, demand in Japan continued at robust levels, relatively high prices were sustained, and industry insiders predicted continued shortages.[151]

146. See Tadashi Tamaki, "Heavy Investment in Chip Factories Spurs Fear of Glut," *Japan Economic Journal*, October 1, 1988, pp. 1, 5; and Tamaki and Kikuzumi, "Shortage of DRAMs Causing Headaches," p. 4.

147. See "Concerned About Increase in Japan's Investments; Interview with US TI Chairman," *Nihon Keizai Shimbun*, October 4, 1988, p. 8; and "IC Maker Fears Output Rise Might Kindle Dumping," *Japan Economic Journal*, October 15, 1988, p. 19.

148. With the release of the fourth-quarter 1988 forecast, MITI officials predicted further shortages of 1M DRAMs through at least March 1989. "1M DRAM Microchips to be in Short Supply," *Japan Economic Journal*, October 15, 1988, p. 17. In a widely cited interview, Toshiba's Tsuyoshi Kawanishi told U.S. industry executives in late September that memory chip shortages could be expected to last another three years. "Prolonged Chip Shortage Seen," *Financial Times*, September 28, 1988, p. 26; "Chips Shortage Will Last 3 Years," *Business Times* (Malaysia), September 28, 1988, p. 13.

149. "Chip Makers to Hike Investment," *Japan Economic Journal*, November 5, 1988, p. 12.

150. David E. Sanger, "A New Japanese Push on Chips," *New York Times*, November 9, 1988, p. D1.

151. See "Semiconductor Memory Chips; High Prices Sustained, Due to Tightness of Demand and Supply," *Nihon Keizai Shimbun*, November 9, 1988, p. 20; "1M DRAM Chip Prices to Continue to be High in Dec.," *Nikkei Top Articles*, November 23, 1988; and "Semiconductor Prices on Spot Market; High Prices Continuing for Chips of High-Speed Processing;" *Nihon Keizai Shimbun*, January 24, 1989, p. 20.

Clampdown on the Gray Market

The period from late 1987 through early 1989 saw a continued attempt by Japanese chip producers, with government backing, to dry up unauthorized sales in the gray market. MITI export regulations—which required foreign chip buyers to register with MITI, and obtain a certificate from manufacturers attesting to the origin of exported chips, before the ministry would grant an export license—were potent bureaucratic barricades blocking off access to the Japanese gray market by foreign purchasers.

The major manufacturers also moved on their own in 1988 to reduce supplies filtering into the Japanese gray market. Producers increased their surveillance of chip sales, to watch for unauthorized resale of their products by authorized buyers. Retaliation for such gray market transactions could serve to discourage them. In 1988 and 1989 reports of these efforts to reduce gray market sales surfaced with increasing frequency in the Japanese trade press.[152]

---

152. "The spot market prices of semi-conductor memory chips have been sharply rising from the bottom prices in the first half of last year, as various manufacturers have been tightening their supply to the spot market because the Japan-US semiconductor friction came to arise. Especially, that the market prices having been rising since the beginning of this year reflects the fact that the manufacturers' side has been strengthening their monitoring of memory chip sales in order to prevent the channeling thereof through their affiliated routes. At one time, memory chips flowing back from users to the market and memory chips removed from such products as used personal computers were seen on the market for sale. In this case, too, it is said that 'there is almost nothing because manufacturers are strengthening their watch' (spot market dealer)." "Semiconductor Memories—Spot Market Prices Continuing to Rise; 256K Prices Double from Beginning of This Year," *Nihon Keizai Shimbun,* October 4, 1988, p. 20.

"On the spot market, centering on Akihabara, Tokyo, the shortage of DRAM chips has become still more serious than ever . . . there is a voice saying as follows: 'Manufacturers have been tightening their monitoring of sales, and so the influx [of 256K DRAM chips] into the spot market through their authorized agents has stopped completely. We cannot answer inquiries.' (spot market dealer)." "Semiconductor Memory Priced One Stage Higher; Demand Remains Vigorous," *Nihon Keizai Shimbun,* December 13, 1988, p. 20.

"There is a voice saying that 'High-speed processing [DRAM] chips will not appear on the spot market because domestic manufacturers are strengthening their monitoring of sales' (spot market dealer in Akihabara, Tokyo)." "Semiconductor Prices on Spot Market," p. 20.

"Industry experts said Japanese manufacturers are restricting the outflow of high-speed 256-kilobit DRAMs into the spot market." "Spot Prices of Microchips Up Slightly," Nikkei Top Articles, February 9, 1989.

Figure 4-3. *Year-to-Year Changes in Fisher Ideal Price Indexes for DRAMs and EPROMs, 1972–89*

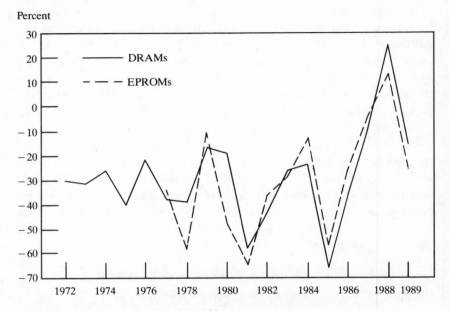

Percent

Source: Based on data in chap. 5, figures 5-2 and 5-3.

Reaction in the United States

By early 1988 a full-fledged shortage of DRAMs was being widely felt in the United States and Europe. As prices soared, substantial differentials between Japanese large-user prices and foreign contract prices appeared. (On the gray market, inherently much more difficult to control, available data suggest that U.S. and Japanese spot prices were roughly equalized.) These differentials persisted throughout 1988 and into 1989, as Japanese manufacturers increased their surveillance of chip transactions by their sales agents, in order to reduce supplies filtering into the Japanese gray market.

Figure 4-3 shows just how extraordinary changes in Fisher Ideal price indexes for DRAMs and EPROMs over this period look when compared with earlier years. These indexes refer to average worldwide sales prices; U.S. contract prices soared by a considerably greater margin, and spot prices jumped even higher, roughly quadrupling in the first months of 1988 (chapter 5 describes the construction of this chained index of year-to-year Fisher Ideal price comparisons and presents detailed price analyses). As American users howled in pain over enormous, unprecedented price increases, criticism of the STA mounted.

American chip producers responded by arguing that reductions in supply and increases in price were the consequence of a successful campaign of predation, rather than an outcome created or facilitated by the STA. One influential defense articulated this view in a particularly concise way:

> Any economist will tell you that we shouldn't complain about foreigners dumping, because consumers benefit. The one exception is if foreign firms can put domestic firms out of business, and then raise prices. If it is costly to re-enter the business (like it is to restart DRAM production), foreign firms can gain monopoly profits at the consumer's expense.
>
> . . . Rather than signaling a bankrupt trade policy, today's shortages in DRAMs should remind us that dumped products in an industry like semiconductors usually lead to higher prices and limited availability if domestic suppliers are allowed to be destroyed.[153]

As in earlier episodes during the 1980s, the industry's analysis of rising U.S. prices was that monopoly power created through earlier expenditures on a successful campaign of predation was finally being exploited. In the spring of 1989, SIA vice chairman Wilf Corrigan pointed out to a House of Representatives subcommittee that, while the sanctions had compelled Japanese producers to stop dumping, "what [the Japanese] did then was to begin to double and triple prices, and to act like you'd expect a cartel to act." Because earlier dumping had forced other producers out of the market, argued Corrigan, "they could afford then to raise prices with impunity and say that they were doing it to comply with the agreement."[154] Such claims of predatory Japanese behavior were further reinforced when it became evident that market prices had stayed well above the Commerce Department's floor prices (the FMVs) in 1988 and 1989.

---

153. David B. Yoffie, "Chip Shortage: Don't Blame the Pact," *Wall Street Journal,* June 21, 1988. Yoffie, a professor at the Harvard Business School, became a member of Intel's board of directors in 1989.

154. Wilfred Corrigan, in testimony before the Subcommittee on Oversight and Investigations, House Committee on Energy and Commerce, March 1, 1989, as reported in "Japan Should Be Targeted by Super 301 for Failure to Honor Chips Pact, SIA Says," *International Trade Reporter,* vol. 6 (March 8, 1989), p. 290. The published transcript of his testimony contains a slightly different, and somewhat garbled, variant of these remarks. See *Unfair Foreign Trade Practices,* Hearings before the Subcommittee on Oversight and Investigations of the House Committee on Energy and Commerce (GPO, 1989), p. 107.

"Coordination Structures" and "High Price Stability," 1989 to 1990

Demand for semiconductor began to weaken in late 1988 and early 1989, and by the late spring of 1989 the semiconductor industry had entered another recessionary period. This weakening in demand, together with the rapid expansion of DRAM production by Korean producer Samsung, was to radically alter the operation of the semiconductor trade regime.

Price Stabilization

With demand weakening, DRAM prices began to drift downward in early 1989. Although sharper declines had been expected, contract prices for 1M DRAMs to be delivered in the second quarter of 1989 remained relatively stable.[155] Demand for memory chips continued to weaken through the late spring of 1989, yet prices remained relatively high. What was happening?

Unpublished MITI statistics of monthly semiconductor shipments by Japanese producers show that 1M DRAM shipments were basically flat from March through May of 1989.[156] The official forecast data show that 1M DRAM production during the first half of 1989 rose to an output rate roughly 32 percent higher than that of the last quarter of 1988; inventories of 1M DRAMs jumped by an unprecedented 3.2 million units over the first half of 1989.[157] Thus, Japanese government data suggest that, after increasing production of 1M DRAMs during the first quarter

155. Use of the expression "high price stability" began appearing in the Japanese press around early 1989. For example, "it has become definite that the 'high price stability' will be sustained for the time being, while being backed by steady-toned demand." "1M DRAM; 'High Price Stability' to be Sustained in April-June Period, Too," *Nihon Keizai Shimbun*, February 1, 1989, p. 9.

156. Monthly statistics on shipments through May 1989 were readily available when I interviewed a MITI official on this subject in December 1989. These MITI statistics were even more extensive and detailed than the quarterly numbers assembled for the publicly released "forecast."

157. Production of 1M DRAMs during the last quarter of 1988 was roughly 69.6 million units, compared with an average quarterly output rate of 92 million units during the first half of 1989. See MITI, Machinery and Information Industries Bureau, "Semiconductor Supply-Demand Forecast," September 30, 1988, December 23, 1988, and June 22, 1989. The largest quarterly increase in producer inventories of 1M DRAMs prior to this, registered in the first quarter of 1987 (when MITI had first provided producers with "guidance" to reduce 1M DRAM shipments), was 0.5 million units.

Figure 4-4. *Output of 1M DRAMs by Japanese and Non-Japanese Manufacturers, 1989*

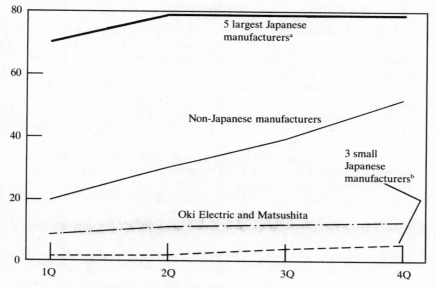

Millions of finished chips

Source: Author's calculations based on unpublished Dataquest data.
a. Toshiba, Hitachi, NEC, Fujitsu, and Mitsubishi.
b. NMB, Sanyo, and Sharp.

of 1989, producers began to cut back production in the late spring, sliced shipments even more, and increased their inventories of parts.

Unofficial production estimates by the semiconductor industry consulting firm Dataquest provide an even clearer picture of what was going on. Figure 4-4 shows Dataquest's quarterly estimates of DRAM production by several groupings of firms.

It is clear that the major Japanese DRAM producers, as a group, cut back production after the second quarter of 1989. (Because of steadily increasing productivity on chip manufacturing lines, constant output actually means cutbacks in production runs.) The additional 1M DRAM chip supply, entering a market increasingly depressed by sagging computer demand, was mostly coming from non-Japanese suppliers, who were rapidly increasing production.

At the time this was happening, 1M DRAM prices were far above the price floors then set for exports (roughly triple the average FMV; see chapter 5). Contemporary press accounts suggested that Japanese chip makers, as a group, were cutting back production to stabilize prices and slow their decline. One analysis of this "high price stabilization" appeared in *Nihon Keizai Shimbun* in June 1989:

> it does not seem that there will be a collapse of prices at one stroke. Manufacturers have established a "co-ordination structure" among themselves in the last several years, and they unanimously say that they "will head toward a soft landing, which will gradually follow a downward curve." Concerning this artificial price policy, American users, which depend on Japanese products for the greater part of their semiconductor procurement, are now beginning to voice criticism. The focus of the Japan-US friction, in which the low prices were a problem, is likely to be placed on the high-price stabilization of Japanese products from now on.
>
> . . . On the other hand, however, users show their dissatisfaction, as follows: "That good showing [high profits of Japanese semiconductor makers] in their settlements of accounts stands on the sacrifice of users." Users demand "a price reduction to the fair market value (FMV)."
>
> . . . In the US too, users came to show obvious moves for taking the price determination right away from Japanese manufacturers, as they are irritated at Japanese manufacturers price determination based on their *dango* [collusion] constitution. The rising of arguments in the US for creating a semi-conductor futures market, is a sign indicating their intention to leave the price-determination right to the market. Also, seven US manufacturing companies' establishment of "US Memories," which is a DRAM joint production company, shows their intention of trying to weaken Japanese manufacturers' price control.[158]

As semiconductor demand continued to weaken over the summer and fall of 1989 and inventories continued to increase, many wondered whether manufacturers could maintain their discipline and refrain from lowering prices to sell their output. But chip makers recognized that it would create a potential threat to the profitability of the industry as a whole if all succumbed to temptation and a round of vicious price cutting, like that seen in 1985, broke out. *Nihon Keizai Shimbun* noted:

> "In order to prevent this cutthroat competition from recurring, we decided not to increase our production," said Goh Kawanishi, special general manager of Toshiba, the leading manufacturer in the semiconductor industry. However, as the general economic environment that surrounds the semi-

---

158. "Users Dissatisfied with 'High Price Stabilization'; Semi-Conductor Manufacturing Companies to Carry Out Price Reduction of 5% from Next Month," *Nihon Keizai Shimbun*, June 28, 1989, p. 2.

conductor industry worsens, many experts wonder how long this "coordination structure" among manufacturers will last.[159]

In the early fall of 1989, as demand continued to weaken, production cuts were widely reported in the Japanese trade press. At the end of July NEC announced that it had scaled back its plans for increasing production of 1M DRAMs.[160] In September Toshiba, Hitachi, and Mitsubishi each reported that they were cutting back their production levels by about 10 percent.[161] In late October NEC announced that it too would actually cut current production levels in the first quarter of 1990.[162]

Although clearly unwilling to reveal the precise details, companies were actually quite open about the trend toward "cooperation." In early 1990, for example, a U.S. government official asked several executives from a major Japanese chip producer why the company was cutting back DRAM production even though the market-clearing price was still several multiples of production cost. One executive replied, "since the Semiconductor Agreement, we [Japanese DRAM manufacturers] have moved from competing for market share to market sharing."[163]

Where the government did continue to play an explicit and overt advisory role to companies was in investments in new capacity. Although rigorous guidance appears to have ended in mid-1988, MITI continued to provide "advice" to producers on capacity plans. In November 1989, for example, MITI was reportedly telling Japanese firms not to increase capacity, because overcapacity in a downturn was likely to lead to firms exporting more, exacerbating trade frictions.[164] And in December 1989

159. "Falling Spot Price: Over-Production and Less Demand for Semiconductors in Japan," *Nihon Keizai Shimbun,* August 3, 1989, p. 24.

160. NEC withdrew its plan to increase monthly production of 1M DRAMs from 6 million to 8 million chips. *Nikkan Kogyo,* July 29, 1989, p. 1; *Nihon Kogyo,* July 29, 1989, p. 1.

161. "Major Semiconductor Manufacturers to Reduce 1M-DRAM Production by 10% from September," *Nihon Keizai Shimbun,* September 14, 1989, p. 10.

162. The announced cuts were from 6 million to 5 million units per month. *Dempa Shimbun,* October 31, 1989, p. 1; *Nikkan Kogyo,* October 31, 1989, p. 1; *Nihon Kogyo,* October 31, 1989, p. 5.

163. Author's interview with a U.S. government official, May 1990.

164. Author's interview at a Japanese semiconductor company, November 1989. Reports of MITI guidance on facilities investments in automobiles and electronics, complete with a threat to invoke the trade control ordinance, were reported in "Temptations of MITI's 'Managed Trade' (Part 1)—Request for Expansion of Imports; Balancing of Exports and Imports Pressed for," *Nihon Keizai Shimbun,* December 14, 1989, p. 7. MITI officials I interviewed during this same period admitted only to talking to producers about the political dimensions of their investments, such as investment locations in the European

several Japanese producers were still getting "opinions" from MITI on how much investment in semiconductor capacity was appropriate.[165] As before, however, U.S. policy appears to have played an important role in encouraging such action. In early November 1989 the U.S. government had urged MITI to restrain Japanese companies' investments in several industries, including automobiles and electronics.[166]

MITI also continued to operate its price monitoring system, although some modifications were made in March 1989, in response to the GATT panel report.[167] Extensive data on exports and unit prices for many different products, including DRAMs, EPROMs, SRAMs, microprocessors, and microcontrollers shipped to nineteen different countries, were collected on a monthly basis from Japanese producers. The semiconductor monitoring office checked export prices against costs (retrospectively) and "advised" companies when there appeared to be problems.[168] By some press accounts, MITI occasionally provided guidance to Japanese companies on pricing of DRAMs and SRAMs in the domestic market, to prevent rampant price cutting and "share the market with overseas makers."[169]

U.S. Memories to the Rescue

In September 1988, as panicked U.S. consumers scoured the globe for DRAMs, the SIA and the American Electronics Association (AEA)

Community. Even that level of intervention was officially denied: "MITI . . . has vigorously denied guiding Japanese companies to spread their investments around European Community countries rather than concentrating them in the UK. 'We have never given any guidance on this issue. It is a matter that is entirely up to companies themselves to decide,' MITI said here yesterday." Ian Rodger, "Tokyo Denies Issuing Anti-UK Guidance," *Financial Times*, December 21, 1988, p. 3.

165. Author's interviews with Japanese semiconductor executives, November and December 1989.

166. See "U.S. Request Contains Contradictions; 'Concerned over Increase in Japan's Facilities Investments,'" *Nihon Keizai Shimbun*, November 5, 1989, p. 3; "Urges Holding Down of Expansion of Facilities Investments; On Automobiles, Iron-Steel, Electronic Equipment, and Ship-Building," *Nihon Keizai Shimbun*, November 7, 1989, p. 1; and "Temptations of MITI's 'Managed Trade' (Part 2)—'Guidance' Also for Facilities Investments," *Nihon Keizai Shimbun*, December 15, 1989, p. 7.

167. See MITI, "Japan's Coordination Policy in Response to the GATT Panel Report on Semiconductors," Tokyo, March 7, 1989.

168. Author's interviews with a Japanese semiconductor executive and with MITI officials, 1991.

169. "MITI Helps Prevent Memory Chip Price Fall," Nikkei Top Articles, May 25, 1990.

created a joint Extraordinary Measures Task Force to formulate a response.[170] In March 1989, while DRAM demand remained strong and prices high, a proposal to form a joint U.S. production consortium made up of both memory chip users and producers was circulated among the membership of the SIA and the AEA. The joint project was conceived as a way to overcome users' DRAM supply problems and producers' capital availability problems.[171] By late June U.S. Memories, a formal joint venture, had been launched with the initial participation of three computer firms (IBM, Digital Equipment Corporation, and Hewlett-Packard) and four semiconductor producers (LSI Logic, National Semiconductor, Advanced Micro Devices, and Intel). Each partner had agreed to take 5 to 10 percent equity shares in the new for-profit venture. In addition, IBM offered its 4M DRAM technology to the consortium, a senior IBM executive (Sanford Kane) took up the reins as CEO of the new venture, LSI Logic's chairman (Wilfred Corrigan) agreed to chair its board of directors, and 1991 was set as the target date for initial production of 4M DRAMs.[172] A broad consensus—including public expressions of support from rival U.S. DRAM producers Motorola, Micron, and Texas Instruments—for the venture seemed to prevail through the summer.[173] In addition to the original seven committed participants, who had all signed letters of intent due to expire in December 1989, additional companies, including AT&T, which actually contributed some seed money but no firm decision, hovered at the margins of the deal.[174]

By summer, however, trouble clouds had begun to collect over U.S. Memories. In late July Cypress Semiconductor Corporation president and industry maverick T. J. Rodgers, complaining that he had not been invited to join, blasted the proposed consortium as a threat to his company before a congressional subcommittee: "As soon as the necessities of the semiconductor market force U.S. Memories to drift outside the dynamic RAM market, they will be a company armed with antitrust immunity

170. "DRAM Co-op: Now's the Time," *Electronic Buyers' News*, July 10, 1989, p. 17.

171. Brian Robinson, "SIA, AEA Urge DRAM Alliance Formations," *Electronic Buyers' News*, March 20, 1989.

172. Jack Robertson, "Confirm Major IBM Role in Creation of U.S. Memories," *Electronic News*, June 26, 1989, pp. 1, 58; and John Thompson and Stacey Peterson, "DRAMs: Made in the USA," *Computer Systems News*, June 26, 1989, pp. 1, 8.

173. Patrick Burnson, "DRAM Joint Venture Draws Praise," *Infoworld*, July 3, 1989, p. 32; and Ann Thryft, "Domestic DRAM Makers Applaud U.S. Memories," *Computer Design*, July 17, 1989, pp. 1, 38.

174. David Roman, "DRAM Co-op Catching On," *Electronic Buyers' News*, September 18, 1989, pp. 20–22.

and government subsidies—competing with my company and the hundreds of other semiconductor companies in the United States that do not enjoy the luxury of immunities and subsidies."[175] Even more important, perhaps, the DRAM market had begun to weaken. In late September, Apple Computer announced it would not join the consortium.[176] By November, Sun Microsystems and Tektronix had announced their disinterest too, even as Rodgers launched another public campaign against the consortium, this time volunteering to license IBM's DRAM technology and deliver products by 1990, well ahead of the consortium.[177]

By December, amid continuing declines in DRAM prices, U.S. Memories managed to conclude its licensing deal with IBM. But as the letters of intent signed by the seven original backers approached their expiration date, and with them commitments to invest capital and purchase the consortium's output, no new firms could be found to take the remaining shares of the consortium. Indeed, additional computer companies such as Unisys, Tandy, Dell Computer, and Everex Systems had gone public with their disinterest.[178]

The final blow to U.S. Memories was delivered at a meeting on January 10.[179] Hewlett-Packard scaled back its initial commitment, even as possible backers AT&T, Compaq, NCR, and Tandem Computers reported that they were still not ready to commit. Unsuccessful in putting together a deal after nine months of intense maneuvering, suffering repeated assaults by critics, and faced with a softening memory market, the consortium was officially terminated and consigned to a historical footnote. The irony of the episode is that in later years smaller groupings of companies announced a variety of joint ventures in the chipmaking business that had many similarities to the U.S. Memories proposal. U.S.

175. David Roman and Brian Robinson, "IC Coops: A Cartel Threat?," *Electronic Buyers' News*, July 31, 1989, pp. 1, 50.

176. Richard March, "Clock Running on Stalled U.S. Chip Consortium," *PC Week*, October 2, 1989, p. 139; and Eric Nee, "Apple Shuns DRAM Consortium," *Computer Systems News*, October 2, 1989, pp. 3, 8.

177. David Roman, "U.S. Memories Suffers Rejections"; and Hugh G. Willett, "Cypress Would Speed IBM Design," *Electronic Buyers' News*, November 20, 1989, pp. 1, 66. The IBM technology actually was licensed to Micron, but Rodgers and Cypress ultimately did not take the plunge into DRAMs.

178. Richard March, "U.S. Memory-chip Group Garners Lukewarm Support," *PC Week*, December 11, 1989, p. 137; and David Roman, "Co-Op Gets IBM License," *Electronic Buyers' News*, December 25, 1989, pp. 1, 4.

179. David Roman, "U.S. Memories Fades Away," *Electronic Buyers' News*, January 22, 1990, pp. 1, 56.

Memories was unique, however, in that its principals consisted exclusively of U.S. companies and that it was responding to a problem conceptualized as a *national* problem of vulnerablility in supply of a key industrial commodity.

Korea Forges Ahead

Continued weakness in demand for semiconductors led Japanese producers to again announce a round of production cuts in early January 1990. The public announcements in Tokyo of cutbacks in DRAM production by the three largest memory chip makers were virtually simultaneous, prompting a prominent report in the U.S. press that they were "acting much like a cartel."[180]

The cuts, publicly justified as an effort to prevent drastic price declines, also roughly coincided with the demise of the U.S. Memories DRAM production consortium. Some in the U.S. industry even charged— unjustifiably, in light of the evidence just presented on 1989 cutbacks— that Japanese companies had flooded the market and forced prices down in an effort to torpedo U.S. Memories.[181]

180. "The contrast was striking: On the day last week when American electronics companies decided to abandon their cooperative venture to make computer chips, their competitors in Tokyo had already moved in lockstep.

"One by one, within hours, Japan's biggest chip makers announced plans to cut their production of one-megabit memory chips. . . . There was far too much supply, each company explained in great detail, and unless production was cut, prices would continue to fall drastically.

"The Japanese companies, despite repeated contentions that they are now each other's fiercest competitors, were acting much like a cartel." David E. Sanger, "Contrasts on Chips," *New York Times,* January 18, 1990, p. D1.

Sanger's account is incorrect on one important point—that Japanese producers chose to reveal their cutback plans just after the proposed U.S. Memories DRAM production consortium failed. The cutbacks described by Sanger were reported in the Japanese press on Sunday, January 7, and in the U.S. press on Tuesday, January 9. See "Drastic Drop of 1M DRAM Price: Semiconductor Makers Reinforce Production Cut (10–15%) Starting This Month," *Nihon Keizai Shimbun,* January 7, 1990, p. 5; and G. Pascal Zachary, "Japan's Biggest Memory-Chip Makers Are Cutting Output in Bid to Ease Glut," *Wall Street Journal,* January 9, 1990, p. B4. The critical meeting at which the U.S. Memories proposal clearly failed was on Wednesday, January 10, and its failure was publicly announced on Monday, January 15. See Stephen Kreider Yoder, "U.S. Memories to Abandon Bid for Chip Venture," *Wall Street Journal,* January 15, 1990, p. B4; "Lessons Linger as U.S. Memories Fails," *Wall Street Journal,* January 16, 1990, p. B1; and Andrew Pollack, "Memory Chip Cooperative Is Officially Declared Dead," *New York Times,* January 16, 1990, p. D1.

181. These charges were widely repeated within the U.S. semiconductor industry. They actually appeared in print in an article by Michael Borrus ("Chips of State," *Issues in*

In early January major Japanese chipmakers simultaneously announced large cutbacks in production of 1M DRAMs.[182] Reducing output did, indeed, stabilize prices, which actually rose in the Japanese market in early 1990.[183] An uptick in demand—widely attributed to increased demand for memory used in newly introduced notebook computers and high-performance workstations—coupled with the production cuts, led to rising 1M DRAM prices and a brief panic over shortages in the U.S. market in March 1990.[184] (Some analysts believe that IBM's new, memory-intensive OS/2 operating system for personal computers, introduced at about this time, suffered in the market as a result of this firming up of memory prices.) Japanese producers responded by reducing second-quarter production of 1M DRAMs less rapidly, but prices remained stable in Japan.[185] In response, Japanese chip makers actually reversed course and increased their output of 1M parts over the summer.[186]

Science and Technology (Fall 1990), pp. 40–48), which claimed that "the Japanese press reported that Japanese firms were creating a glut in the market to lower prices and thus discourage the [U.S. Memories] initiative" (p. 44).

Despite considerable effort, I was unable to turn up any such report in the Japanese press, and in a telephone conversation in November 1990 Borrus, an analyst at the Berkeley Roundtable on the International Economy, acknowledged to me that he could provide no specific reference for this claim. The belief that Japanese makers had engineered a shortage to kill U.S. Memories was, however, fueled by the *New York Times*, which claimed (incorrectly) that production cuts were announced just after the abandonment of U.S. Memories: "No one has suggested that Japan's production cuts and the abandonment of U.S. Memories are directly related. But both were spurred by a growing glut of chips, after two years of huge demand." Sanger, "Contrasts on Chips," p. D20.

As noted above (and in the Japanese press), Japanese producers were actually cutting back on shipments of DRAMs in 1989, in order to shore up prices—the precise opposite of behavior consistent with "creating a glut in the market."

182. See "Drastic Drop of 1M DRAM Price: Semiconductor Makers Reinforce Production Cut (10-15% Starting This Month)," *Nihon Keizai Shimbun*, January 7, 1990, p. 5. "Another Sharp Price Reduction for Semiconductors," *Nihon Keizai Shimbun*, January 18, 1990, p. 24, reported an 11 percent cut in 1M DRAM output at Toshiba, 17 percent at NEC, 22 percent at Mitsubishi, 15 percent at both Fujitsu and Hitachi, and 27.5 percent at Oki.

183. "Price Reduction in Semiconductors Stopped; Positive Result of Production Cuts," *Nihon Keizai Shimbun*, February 27, 1990, p. 24.

184. G. Pascal Zachary, "Computer Firms Face Shortage in DRAM Chips," *Wall Street Journal*, March 9, 1990, p. A4; Corinne Bernstein-Levy, "Cost Reversal Baffles Buyers," *Electronic Buyers' News*, March 5, 1990, pp. 1, 50; and Barbara Darrow, "DRAM Dearth Overblown, Vendors Say," *Infoworld*, March 19, 1990, p. 105.

185. See "1M DRAM Wholesale Market Recovered; Major Manufacturers Began Increasing Production," *Nihon Keizai Shimbun*, June 21, 1990, p. 3.

186. See "Makers Once Again Boosting 1M DRAM Production," Nikkei Top Articles, June 21, 1990; "Wholesale Prices Cut for 1M DRAM; First Reduction in Six Months,"

The resurgence in demand was short-lived in the American market, where prices soon resumed their descent.[187] Because of stronger demand in the Japanese market, where prices were rising, investment plans for fiscal 1990, announced in May, remained at fiscal 1989 levels for Japan's ten largest chip makers.[188] In a relatively tight market, producers' continuing efforts to dry up supplies of chips resold into the Japanese gray market were generally successful.[189]

Continued weakness in overseas markets finally dragged down buoyant Japanese DRAM prices in late 1991.[190] The real news at this point, though, was that significant imports of 1M DRAMs into the Japanese market were forcing Japanese makers to reduce their prices.[191] As Japanese producers had cut back production to maintain prices, other late entrants into the 1M market overseas—particularly Korea's Samsung—had rapidly increased their production and market share. Not subject to the political and legal pressures faced by Japanese chip makers, they were able to sell everything they could make by slightly undercutting the export prices established by their Japanese competitors. By late 1991 this tactic had even begun to put pressure on Japanese vendors' prices in their own home market.

These realities were already evident by the end of 1990. As one respected source commented,

---

Nihon Keizai Shimbun, September 15, 1990, p. 24; and "Major Chip Manufacturers Hesitate in Increasing 1M DRAM Production," *Nihon Keizai Shimbun*, October 4, 1990, p. 3.

187. David Roman, "DRAMs Slide Again," *Electronic Buyers' News*, June 25, 1990, pp. 1, 58.

188. *Dempa Shimbun*, May 28, 1990, p. 1; *Nikkan Kogyo*, May 26, 1990, p. 6; "Future Direction of Sematech," *Nihon Keizai Shimbun*, June 5, 1990, p. 11; and Satoshi Isaka, "Electronics Makers Still Optimistic," *Japan Economic Journal*, June 23, 1990, p. 17.

189. "Spot dealing in semiconductor memory chips has decreased rapidly and some of the buyers are closing their businesses. This is because there is a diminished supply of chips in the spot market due to the manufacturers' instructions to their own authorized distributors. . . .

"The spot market has been said to reflect the trend of demand and supply of semiconductors accurately and a leading index for authorized distributors. While this basic belief has not been destroyed, some point out that 'free price formation has been threatened as manufacturers' influence over distribution has grown, and administrative guidance, backed by the U.S.-Japan Semiconductor Agreement, has strengthened." "Rapid Decrease in Spot Dealing of Semiconductor Memory," *Nihon Keizai Shimbun*, June 7, 1990, p. 24.

190. "Wholesale Prices Cut for 1M DRAM; First Reduction in Six Months," *Nihon Keizai Shimbun*, September 15, 1990, p. 24; and "Sharp Wholesale Price Reduction of 1M," *Nihon Keizai Shimbun*, October 10, 1990, p. 24.

191. See "Spot Chip Prices Continue to Decrease," *Nihon Keizai Shimbun*, October 31, 1990, p. 24.

The reason for the price drop of 1M can be said to be that production capability is larger than actual demand. Although domestic LSI makers have been trying to stabilize the price of 1M through large production cuts, since "the Japanese 1M market share in the U.S. has dropped below 50%" (according to Mitsubishi) production cuts may not be effective. The lower market share in the U.S. is due to the fact that several other makers such as Samsung, Siemens, and Micron Technology have improved their supply capability. Therefore Japanese makers cannot depend on production cuts to stabilize price and must "wait for demand to recover."[192]

By 1991 Samsung was the largest producer of 1M DRAMs in the world; Siemens and Micron had also climbed rapidly up the ranks of producers of this product. Japanese producers reacted not by slashing prices (and profits), but instead cut back on capacity used to produce the older generation product and focused their resources on rapidly shifting to the next-generation memory chip, the 4M DRAM.

Four-megabit DRAMs were first shipped in volume in 1989, and by 1990 they dominated the chip making efforts of Japanese makers. In effect, Japanese producers phased out of the lagging edge product, the 1M DRAM, and shifted into the leading edge product. For 1989 through 1991, Japanese producers' share of 4M DRAM output could only be described as overwhelming (see chapter 5).

However, the industry suffered through continuing depressed demand for computers through most of 1991. Since a large-scale shift toward use of 4M chips would require new computer designs, persistent slowdown in the computer industry also tended to put relatively more pressure on leading edge chip prices, since slower computer innovation meant less demand for advanced chips. No less than 80 percent of some producers' 4M DRAM production, for example, was reportedly sold to American computer companies.[193]

Slowing computer sales in 1991 hit major producers just as they were ramping up to peak production of the 4M DRAM. Where MITI had forecast 20 million units would be demanded domestically, actual domestic sales for the first half of 1991 were only 12.1 million units; MITI's forecast of 32.7 million units sold domestically for the second half of 1991 contrasted with actual sales of 19.4 million. Net exports also fell some-

192. "Japanese Promote Strategic Production Formation and Increased Investment of Plants, Desire an End to US-Japan Semiconductor Agreement," *Nikkei Microdevices*, January 1991, p. 68.

193. "Semiconductor Production Scaled Down, So Is Also Output of Circuit Photocopier," *Japan Economic Review*, August 15, 1991, p. 11.

what short of the forecasts: 42 million units rather than a forecast 39.6 million were shipped in the first half, and 62.9 million instead of a forecast 72.5 million in the second.

Through most of 1991, then, Japanese producers were forced to reduce the rate at which new 4M capacity was to be brought online, and to sell what they did at lower prices more competitive with the continually declining prices of older generation chips.[194] The so-called crossover point, where the price per bit of memory for a new chip reaches rough parity with that of an older generation chip, came in 1991, less than two years after the crossover between the 1M and 256K DRAMs.[195]

Thus by mid-1991, as the Semiconductor Trade Arrangement expired, a new pattern of competition between Japanese memory chip producers and their foreign competitors had emerged. Constrained by export price floors in their ability to compete on price in mature products, and redirected from "competition for market share" into "market sharing" by political pressures, Japanese semiconductor companies were looking to the technological leading edge for their profits. Successful competition increasingly meant being first to achieve significant output of the latest-generation memory chip, and extracting significant profits from that product before less advanced and less politically constrained competitors were able to imitate it, enter the market in volume, and wrest control of pricing from the technological leaders in Japan. The slump in chip demand in 1991, just as the leap into a more advanced product was being made, illustrates that the new strategy has not eliminated the risk that has always been a part of the memory chip business.

Aftermath: The Second Semiconductor Trade Arrangement

In June 1991 a second U.S.-Japan trade arrangement was signed, replacing the first five-year arrangement due to expire in July. The major change in antidumping measures was the termination of the FMV floor

194. "How Far Will 4M DRAM Fall; Demand Slower Than Mass Production," *Nihon Keizai Shimbun,* August 14, 1991, p. 18; and "Further Price Reduction of 4M DRAM," *Nihon Keizai Shimbun,* August 8, 1991, p. 20.

195. Since every generation of DRAMs quadruples the number of bits on a chip, and higher density chips are more desirable because of their lower power consumption, lower interconnection costs, and reduced space requirements, "crossover" is generally defined to occur when newer generation memory chips are five to six times more costly than older generation chips. Because a new generation of chip is introduced roughly every three years, crossover between generations would also be expected to take place at three-year intervals.

price system administered by the Commerce Department and its replacement with a fast-track antidumping procedure, in which Japanese companies were required to collect cost and price data and have it available to the U.S. and Japanese governments on fourteen days' notice in the event of an antidumping case. Data on third-country exports were also expected to be made available to a third-country government should it (possibly at U.S. request) initiate an antidumping case.

The end of the FMV system, however, was less radical a step than it at first appeared. The data on export prices and costs that producers now collected were essentially the same as the data previously requested for the calculation of FMVs, and the format of the FMV calculations was now well established; thus Japanese manufacturers were effectively required to maintain a system of "shadow" FMVs, to be handed over to the Commerce Department on fourteen days' notice should frictions develop. When interviewed in mid-1991, Japanese semiconductor executives told me that their companies were continuing to use their FMV-like calculations as guidelines on export pricing. Through June 1991 MITI was also continuing to monitor export price and quantity data on shipments of sensitive products to major markets, and although smaller producers were dropped from this system after this date, the major producers continue to report these data to MITI.[196] And MITI's revised system of twice-yearly forecasts of semiconductor supply and demand continued unchanged from previous years.[197] Furthermore, the European Community's system of reference prices, supervised in Japan by MITI, continued in effect on DRAMs and EPROMs. Thus a variety of well-established mechanisms designed to constrain pricing in world semiconductor markets continued in place in their original or revised form.

One significant development came to a resolution in early 1993. Dumping cases against Korean exports of DRAMs were filed in the European Community in 1991 and in the United States in 1992. In the United States the International Trade Commission made an affirmative preliminary finding of injury and sent a Micron Technology DRAM dumping petition on to the Commerce Department for further action. In the European

196. Author's interview with a semiconductor executive, Tokyo, July 1992.

197. Although no longer released officially by MITI, the semiconductor forecasts continued to be compiled through 1995 and with persistence (MITI officials at first insisted the forecasts had been discontinued after the signing of the second STA) could be obtained from MITI by the author. EPROMs were dropped from the forecast in mid-1994.

Community a price undertaking in DRAMs was quietly negotiated in consultation with Korean producers and EC member states.

Since Korean producer Samsung's rapid expansion of DRAM production and willingness to undercut Japanese pricing had played an important role in undermining the "high price stabilization" of 1988 and 1989, shackling Korean producers with the same restraints placed on the Japanese had a potentially important impact on global DRAM pricing. If stringent restraints had been placed on the pricing of Korean DRAM imports, it could have had a significant effect on memory chip prices.

Both cases were resolved in the spring of 1993. In the United States, very modest dumping duties were placed on Korean DRAMs: for Samsung, which accounts for the vast bulk of Korean chip production, the duty was only 0.82 percent; Hyundai Electronics was assessed a tariff of 11.45 percent, Goldstar Electron a 4.97 percent levy, and all other South Korean makers a duty of 3.89 percent. These low tariffs had minimal effects. The most significant factor constraining Korean pricing as a result of this finding, in fact, was probably the "monitoring" of Korean prices that followed.[198] In August 1995 the U.S. Court for International Trade ruled that the dumping margins on Korean DRAMs had been incorrectly calculated. The ruling had the effect of removing dumping duties entirely for Samsung and lowering duties for Goldstar and Hyundai to a level of 5 percent or less.[199]

The price undertaking negotiated between the Koreans and the Europeans in early 1993 followed a provisional 10.1 percent antidumping duty on Korean DRAMs in September 1992. The agreement, which required Korean makers to submit detailed quarterly cost data to the European Community, reportedly differed substantially from the much-criticized floor price arrangement negotiated with the Japanese in 1990.[200] Because continuing high prices for memory chips had rendered these

198. See David Roman, "Korean DRAM Case Fizzles," *Electronic Buyers' News*, March 22, 1993, pp. 1, 44; and Bob Davis, "Panel Clears Way for U.S. to Impose Import Duties on Korean Microchips," *Wall Street Journal*, April 23, 1993, p. A11. Indeed, some within the industry charged that the mild penalties on Korean chips were the result of political pressure from computer makers.

199. "Semiconductors Face Lower Antidumping Duties," *Korea Times*, August 26, 1995, p. 8; and "USA: Samsung Electronics Cleared from Anti-dumping Charges," *Korea Economic Weekly*, November 9, 1995.

200. See Linda Bernier, "EC Acts on Dumping," *Electronic Engineering Times*, March 8, 1993, p. 97. The Korean price floors were company specific like the U.S. FMVs, rather than the single common floor applied by the EC to Japanese chips.

floor prices irrelevant, the European Commission instituted the first of a series of suspensions of the requirement for Korean companies to file cost data on European DRAM and EPROM exports as of June 1995.[201]

Even where no official finding results, dumping cases can still have a restraining effect. In early February 1990 Micron Technology took the lead in publicly raising the possibility of a dumping suit against Korean vendors, although no suit was actually filed until 1992.[202] That same month, in apparent response, Samsung announced a 20 percent production cut.[203] Although Samsung was to later resume expanding production of 1M DRAMs as prices firmed up somewhat later in the spring of 1990, a point had clearly been made. That same point had been made earlier, in the 1980s, in successive episodes of Japanese "production coordination" and "self-restraint" triggered by resurgent semiconductor trade friction with the United States. Although perhaps less effective and more transitory than formal trade restraints, political "jawboning" and the threat of official action can have real effects on economic behavior.

201. "EU Suspends Antidumping Rule on Semiconductors," *Korea Times*, August 26, 1995, p. 8. The suspension was extended in January 1996. "EU to Suspend Anti-dumping Regulations on DRAM Chip Imports," *Korea Economic Weekly*, January 3, 1996.

202. See Greg Garry and Hugh G. Willett, "Korean Chip Dumping?" *Electronic Buyers' News,* February 5, 1990, pp. 1, 58.

203. "Samsung Cuts DRAM Production to 20% of Peak," Nikkei Top Articles, February 27, 1990; and "Japan's 1-MB DRAMs Yet to Respond to U.S.'s BB Ratio Upswing," Nikkei Top Articles, May 9, 1990.

CHAPTER FIVE

Effects of the Semiconductor Trade Arrangement on Semiconductor Markets

IT WAS IN SEPTEMBER 1986 that the United States and Japan officially signed the first bilateral Semiconductor Trade Arrangement (STA). Reports in the Japanese press suggested that the formal arrangement was preceded by a period of informal pressure on the Japanese semiconductor industry—what in the United States might be called jawboning—from political and government circles in Japan, reacting in turn to foreign complaints. From July 1985 on, for more than year before the formal signing of the STA, Japanese industry was encouraged to reduce low-priced DRAM exports to the United States and exercise restraint in production. This effort was actually successful in raising 64K DRAM prices, but it came too late to preempt the formal conclusion of the dumping case involving that product, or to preempt the filing of other trade cases (in EPROMs and in 256K and higher DRAMs). The STA, which led to suspension of these later dumping cases, the suspension also of a Section 301 complaint about market access, and the termination of a private antitrust suit, intervened in the market at three distinct levels.

First, there was the public document released to the media, which described a couple of major sets of measures. One section of the agreement described actions to be undertaken by the Japanese and American governments to encourage increased sales in the Japanese market of

227

semiconductors made by foreign-based firms, and both governments agreed that "expected improvement in access should be gradual and steady over the period of this Arrangement."[1] Also as part of the arrangement's market access–related measures, both governments reaffirmed that there should be fair and equitable access to patents resulting from government-sponsored research, and that both governments had "every intention to refrain from policies or programs which stimulate inordinate increases in semiconductor production capacity."[2]

The other broad class of measures in the public version of the STA fell under the rubric of "prevention of dumping." Dumping cases against Japanese DRAMs and EPROMs were to be suspended in separate agreements between the U.S. Commerce Department and Japanese chip companies. These suspension agreements, referred to within the STA itself, created a system of company-specific price floors (foreign market values, or FMVs) on Japanese DRAM and EPROM exports shipped to the United States; the FMVs were to be set and administered by the Commerce Department using cost data provided by Japanese companies.

The text of the STA itself went on to call for creation of a cost and price monitoring system for additional products, supplementing the Commerce Department FMVs on DRAMs and EPROMs. This system called for company- and product-specific cost data to be submitted to MITI, "in accordance with procedures established by MITI." If the U.S. government were to then provide information to Japan supporting a belief that Japanese firms were selling product in the United States at less than a company-specific fair value, after a period of consultation the Japanese government was committed to taking "appropriate actions available under the laws and regulations in Japan to prevent exports at prices less than company-specific fair value." The U.S. government reserved the right to self-initiate new antidumping actions, and Japan agreed to encourage its companies to provide the data given to MITI to the Commerce Department as well in such an event. Initially this monitoring system was to cover a certain number of standard chip products,[3] but it

1. "Arrangement between the Government of Japan and the Government of the United States of America Concerning Trade in Semiconductor Products," September 1986, sec. I, point 2. (Hereafter STA.)

2. STA, sec. I, point 4.

3. The products to be included were MOS (metal oxide semiconductor, a standard lower cost chip technology) static random access memory chips (SRAMs), ECL (a high-

could be extended unilaterally by either government to cover additional items. A final—and most controversial—provision of this section of the STA called for the Japanese government to monitor, "as appropriate," costs and export prices of these same products exported from Japan to so-called third-country markets.

At a second level, beyond this "public" STA, a number of important "secret" side letters established several particularly sensitive measures. In a crucial section of the most famous letter, the Japanese government agreed that the U.S. chip industry's "expectation that semiconductor sales in Japan of foreign capital-affiliated companies will grow to at least slightly above 20% of the Japanese market in five years . . . can be realized and welcomes its realization."[4] On the basis of their oral discussions with Japanese negotiators, U.S. trade negotiators understood this to be a definite commitment by the Japanese government to a foreign market share of at least 20 percent by 1991.[5] Japanese trade negotiators later denied the existence of any such side letter.[6] Other sections of this same secret letter decreed that "the Government of Japan will compile demand and supply forecasts on the Japanese semiconductor market in compliance with its domestic laws and regulations," and specifically noted that Japan's export trade control ordinance was to be used to enforce the third-country market provisions of the agreement.[7] In another secret side letter MITI agreed to deal with a last-minute, preagreement surge in

---

speed, higher cost technology) RAM chips, certain microprocessors, microcontrollers, and ASICs (application-specific integrated circuits). Quotations in text are from STA, sec. II, point 2, paras. 4, 5.

4. The precise wording for the 20 percent market share target was reported in Louise Kehoe, "Stalemate in Chips Trade Talks," *Financial Times*, July 28, 1986, p. 1, and Carla Rapaport, "Japan Yields to US Demands on Microchip Price Monitoring," *Financial Times*, August 4, 1986, p. 1, before the STA was actually signed. Crucial paragraphs from this "secret" side letter were published verbatim in Mikio Fujiwara, "This Is a Side Letter to the U.S.-Japan Semiconductor Agreement," *Bungei Shunju* (May 1988), pp. 124–37, and in *Inside U.S. Trade*, November 18, 1988, pp. 15–16.

5. Author's interviews with U.S. trade negotiators, 1990, 1992.

6. The top Japanese negotiator described the situation as follows: "A rumor circulated that the semiconductor agreement included, aside from the published part, a secret side letter in which Japan promised to raise to 20% the share of American products in the Japanese semiconductor market within five years. I can state categorically that there was no such secret letter in the semiconductor agreement." Makoto Kuroda, "Talking Tough about Trade," *Journal of Japanese Trade and Industry*, no. 6 (1988), p. 31.

7. STA, point 5, sec. 1. The relevant passage (sec. II, point 5) read: "the Government of Japan will take appropriate actions available under laws and regulations in Japan, including ETC [the Export Trade Control ordinance], in order to prevent dumping."

230 EFFECTS OF THE SEMICONDUCTOR TRADE ARRANGEMENT

256K DRAM exports by establishing direct quantitative controls over the volume of these exports.

After the STA went into effect, and for the next five years, there was a virtually constant stream of meetings and contacts between Japanese and American government officials, leading to a number of public follow-up actions and statements. These included the imposition of sanctions by the United States in March 1987, their partial removal in June and November of and the public disavowal of production controls by Japan, and in November the disavowal of ex ante review of export pricing for non-American markets and the dissolution of Japan's supply-demand forecast committee in 1989, in response to a negative ruling by the General Agreement on Tariffs and Trade. These public meetings and actions made the semiconductor agreement's institutional structure a living, breathing creature, whose nature changed substantially over time.

At a third level, beyond any open or secret bilateral agreements between governments, was the complex and evolving array of administrative procedures implemented by both the Japanese government and Japanese companies in order to further the often vague objectives of the STA. As described in chapter 4, a variety of sources suggest that government-led measures included—at various times—controls (nominally, "guidance" to Japanese companies) on production and investment, allocational guidelines on exports to major consuming regions, controls on export pricing enforced through the use of export control procedures to block or delay objectionable exports, less coercive types of "advice" on acceptable pricing levels for monitored products, and export licensing procedures designed to shut off uncontrolled chip exports through the so-called gray market.

Even more controversial is some evidence suggesting that, after 1987, major Japanese producers may have taken private actions—with some degree of coordination taking place outside the framework of administrative measures or guidance provided directly by the Japanese government—in dealing with the objectives set by the STA, particularly as foreign pressure forced the Japanese government to take a less directly interventionist stance. In particular, these actions may have included some element of coordination and information sharing among Japanese companies on plans for production and investment in key products (particularly in coordinated cutbacks in output levels after the spring of 1989), and a crackdown on distributors aimed at preventing unauthorized "leakage" of products into the gray market.

Although both these public and private actions were taken within the political context of the STA, it might reasonably be argued that they were not part of the STA proper. Implementing actions, after all, were not formally defined by the two sides within the agreement. All actions either deliberately implemented (if undertaken or actively encouraged by government) or passively tolerated (if a private response) by either government, in response to the arrangement's often vague language and surrounding politics, will be considered here as part of a constantly mutating STA trade "regime." An ongoing political dialogue between governments and companies shaped all these measures. Rather than forming a static set of policies, the STA was more of a legal and political framework for a complex, dynamic game involving both public and private players. The moves in this ill-defined game were a collection of formal and informal actions—some announced publicly, others undertaken relatively privately—evolving constantly over time.

Impacts on Semiconductor Supply and Demand

The natural place to begin examining the impact of this complex of trade policy measures, and the ensuing maneuvers, on market outcomes is to sketch out the basic facts of supply and demand, before and after the agreement. The structure of demand for EPROMs and DRAMs is considered first, then the basics of supply. In examining demand we must also briefly digress to consider its structure, particularly the segmentation of demand by distinct marketing channels. The bulk of the exposition will deal with DRAMs, whose value (and political visibility) was double that of EPROMs ($2 billion in DRAMs were sold worldwide by merchant producers in 1986, compared with $0.9 billion in EPROMs, according to Dataquest estimates of worldwide merchant revenues). The same basic description, however, also characterizes institutional arrangements linking suppliers with consumers in the EPROM market.

The Structure of Demand

DRAMs, along with other memory chips, have a reputation as the "commodity" product par excellence within the semiconductor industry: they are high-volume, standardized goods, with almost perfect substitution among different manufacturers' offerings the norm. Certainly industrywide standardization has been a feature of competition in DRAMs since the early 1970s.[8] Chips from different manufacturers use the same array of package types and pins and have many common minimal technical specifications. They mainly use the same speed classifications (rated in nanoseconds average access time to a bit). They are generally physically compatible, in the sense that products with appropriate specifications from different manufacturers may be substituted within a given piece of equipment. Although DRAMs are in this sense a commodity product, the actual physical design of the chip's internal structures and many subtle aspects of its performance vary by manufacturer.

Because of these subtle but important variations across producers in DRAM electrical and physical performance parameters, large manufacturers typically put a device through an extensive and expensive qualification process.[9] Some retesting is required every time the manufacturing process for a chip is changed. These costs provide an important economic incentive for systems manufacturers to limit the number of qualified suppliers for a particular application. Quality standards maintained by a manufacturer reduce the need to test components after purchase, and DRAMs are generally shipped to large customers in boxes with quality seals to guarantee factory-set standards. (Physical handling of chips is a major cause of failure or degradation; DRAMs and other high-density circuit designs are very sensitive to static electricity.) Purchasing chips from a new supplier, or outside of manufacturer-controlled sales channels, will generally require expensive additional testing (unless, as sometimes happens, the chips can be purchased in boxes with the original factory quality seals intact).

There are three basic purchasing channels linking the supplier with U.S. users of integrated circuits. First, large electronic equipment manufacturers (original equipment manufacturers, or OEMs) who purchase large volumes of product deal directly with chip manufacturers. Trans-

8. The most prominent exception to this pattern was the 16K DRAM, where two different and incompatible designs coexisted—one with a single power supply voltage, the other with dual voltage requirements. Both configurations were produced by multiple companies, however.

9. Merely qualifying and testing a second source for a part already in use has been estimated by one industry source to cost $120,000. Qualification costs were large enough to prompt at least one group of relatively large computer manufacturers to form a cooperative chip qualification joint venture, in order to pool these costs. And within the electronics purchasing community, talk of the economic pressure to reduce the number of suppliers is a staple of everyday conversation.

actions in this market are generally labeled "contract" pricing, though the nature of these "contracts" is an interesting subject to which I will shortly return. It is not unknown for OEMs to contract for large purchase volumes in order to qualify for volume discounts, then resell the excess over their actual needs to gray market channels.

Second, chip manufacturers maintain a formal distribution network through "authorized distributors" to service lower volume customers. Chip manufacturers warrant the product distributed through this channel, and they often play an important role in the technical support and quality assurance programs offered to the customers. Given the historical fact of continuous yet relatively volatile decline in chip prices, manufacturers have historically offered their authorized U.S. distributors "price protection," assuring them that if they lower their sales prices to meet market competition they will receive a credit reflecting the difference between the distributor's purchase price and the lower sales price to the final consumer. For sales through this channel, the bulk of the risk related to price uncertainty therefore has been assumed by the chip manufacturer. Pricing generally seems to be on a spot basis, although there can be a substantial lead time between orders and deliveries in times of buoyant demand, and distributor prices take on a "contract" aspect.

Finally, there is the so-called gray or spot market. Independent distributors, brokers, and speculators buy and sell lots of chips for immediate delivery. There is also a significant retail market selling directly to computer resellers and users wishing to upgrade computer systems or replace defective parts. Supplies of chips on the American gray market come from chip manufacturers, OEMs, and authorized distributors selling their excess inventories, and from Japanese trading companies and wholesalers purchasing directly from Japanese DRAM manufacturers.[10] Gray market products are not warranted by the manufacturer and have frequently been subjected to unknown handling and quality assurance procedures. It is not unknown for "pulls" (used chips pulled from sockets in finished equipment) and even chip manufacturers' rejects to show up in misrepresented form in gray market channels. On the other hand, products are sometimes available in the manufacturers' original packaging with quality seals intact, in all respects identical to parts purchased on contract.

10. See, for example, U.S. International Trade Commission, *64K Dynamic Random Access Memory Components from Japan*, USITC publication 1862 (June 1986), p. A-12.

In 1985 U.S. industry sources estimated that authorized distributors accounted for about 30 percent of chip manufacturers' DRAM sales.[11] Since gray market sales are often resales of product originally sold through OEM contracts or authorized distributors and double count chips flowing into the gray channel from sources other than chip manufacturers, one must be careful in calculating the share of these different channels in sales to final users. One 1985 estimate held that 20 percent of "the market" (presumably, end users) is accounted for by the gray channel in times of shortage. This is roughly in line with more recent estimates. In early 1989 one industry source estimated that perhaps 15 percent of DRAM sales to final users went through authorized distributors, and an additional 15 percent through the gray market.[12]

An absolutely critical point is that prices typically vary considerably across these different marketing channels. Price comparisons will therefore be affected in an important way by the precise nature of the sales transaction.[13] Spot prices often rise above contract prices when markets are tight; they tend to fall below contract price levels when demand is slack. Figure 5-1, based on data collected by the U.S. International Trade Commission (ITC), portrays contract, distributor, and spot pricing over 1983 to mid-1985, a period that included both tight (1983 to mid-1984)

11. U.S. International Trade Commission, *64K Dynamic Random Access Memory Components from Japan*, USITC publication 1735 (August 1985), pp. A10–A11.

12. The estimate is that of Don Bell, of Bell Microproducts Inc., whom I thank for spending the morning of February 16, 1989, attempting to educate me in the intricacies of the DRAM market.

13. A U.S. International Trade Commission investigation of Korean DRAM sales in the United States, for example, found that *all* Korean sales were made on a spot basis, and found no consistent pattern of Korean products being sold above or below spot sales prices for U.S.-made products. In contrast, an EC investigation of Korean DRAM exports to the European Community over roughly the same period found Korean producers' prices undercutting those of Community producers by 9.3 to 20.7 percent (weighted-average margins). However, the EC report makes no mention of distinctions drawn between contract and spot sales. In principle, and one suspects in practice, the types of transactions used in these price comparisons can make a significant difference in the conclusion drawn.

See "Commission Regulation (EEC) no. 2686/92," *Official Journal of the European Communities*, September 17, 1992, p. L 272/19; and U.S. International Trade Commission, *DRAMs of One Megabit and Above from the Republic of Korea*, USITC publication 2519, (June 1992), pp. A-45, A-50. In 100 OEM spot market price comparisons, Korean prices were below those of the U.S. product in 47 cases, above them in 48, and the two types of product were priced identically in 5 cases. In 67 comparisons of authorized distributor purchases, the Korean product was priced lower in 23 cases, above U.S. prices in 42, and priced identically in 2 cases.

Figure 5-1. *Prices of 64K DRAMs in Contract, Distributor, and Spot Markets, 1983–85*

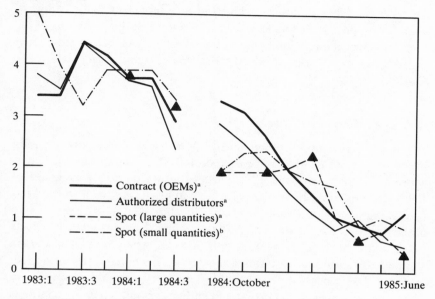

Dollars per chip

Contract (OEMs)[a]
Authorized distributors[a]
Spot (large quantities)[a]
Spot (small quantities)[b]

1983:1   1983:3   1984:1   1984:3   1984:October   1985:June

Source: U.S. International Trade Commission, *64K Dynamic Random Access Memory Components from Japan*, USITC publication 1735 (August 1985), pp. A40–A41. Prices shown are for DRAMs with an access time of 150 nanoseconds.
a. 10,000 to 100,000 units.
b. Fewer than 10,000 units.

and slackening (late 1984 on) market conditions.[14] In four out of six quarters of relatively robust demand (through the second quarter of 1984), spot prices exceeded contract sales prices to OEMs. In seven out of nine months of a period of slackening demand (during and after the last quarter of 1984), spot prices were at or below contract price levels. As will be seen, spot prices soared much farther above contract price levels during the DRAM shortage of 1988 than during the tight market of 1983–84, depicted in this figure.

Further analysis suggests that the quantity commitments that appear to be the economic heart of OEM "contract" pricing should perhaps best

14. See U.S. International Trade Commission, *64K Dynamic Random Access Memory Components From Japan* (1985), pp. A40–A41. The prices shown are for DRAMs with an access time of 150 nanoseconds.

be interpreted as a device to minimize additional and significant qualification and testing costs incurred by chip users, and reduce risks associated with uncertain aggregate demand faced by producers (see appendix 5-A). Thus, long-term chip contracts represent the marriage of
quantity commitments to a forward price in force at the beginning of the
contract. Over the remaining life of the contract, however, contract buyers seem to enjoy something like the "price protection" offered to
distributors.

A Long-Term Perspective on Pricing

For the period up to the mid-1980s there is essentially only one published source of historical price data on DRAMs: Dataquest, an American market research firm. Dataquest publishes quarterly estimates of
DRAM production, by bit capacity, and an aggregate worldwide "average selling price" (ASP) for every capacity chip in large-scale production
at the time of the estimate. The ASP is a "billing" price; that is, it reflects
bills sent out and (one hopes) receipts collected when product is actually
shipped. Because there is often a lag between negotiation of a contract
and actual shipment, current "billing" price is a weighted average of
"booking" prices in both current and earlier periods, when the contracts
were actually negotiated. The Dataquest ASPs are thus probably best
conceptualized as some weighted average (with unknown, variable
weights) of current spot and distributor prices, and current and lagged
contract prices.

There are numerous methodological difficulties in the procedures used
by Dataquest to produce these and other price estimates.[15] Nonetheless,
for all practical purposes, these numbers were until the 1990s the only
consistent, long-term source of detailed information on semiconductor
prices and will be used here to analyze pricing trends.

In looking at long-term trends in memory pricing, one frequent practice in the industry is to calculate a cost per bit of memory, for example
by dividing total revenues by the total bits of memory produced in all
generations of memory chip at any moment in time. This essentially

15. The producer price index series produced by the U.S. Bureau of Labor Statistics
are not useful in this context. A detailed critique of available statistics may be found in
Kenneth Flamm, "Measurement of DRAM Prices: Technology and Market Structure," in
Murray F. Foss, Marilyn E. Manser, and Allan H. Young, eds., *Price Measurements and
Their Uses* (University of Chicago Press, 1993), pp. 157–97.

weights the price of every type of chip by that chip's share of total bits produced. Changes in this price index, unfortunately, confound period-to-period shifts in demand from one type of chip to another with changes in chip prices. Even if all prices remained constant, for example, the aggregate cost per bit might drop significantly if the composition of demand were to shift toward a chip with a lower price per bit.

A better practice is to use a constant set of weights for prices being compared in any two periods. One particularly attractive weighting scheme is the so-called Fisher Ideal price index, which has desirable properties as an approximation to a theoretically ideal price index.[16] Figures 5-2 and 5-3 show annual percentage rates of change for Fisher Ideal price indices for DRAMs and EPROMs (the products subject to dumping investigations), constructed using Dataquest worldwide ASP estimates, along with the widely used (if conceptually less attractive) calculations of aggregate cost per bit. Although there are some differences between the two series, the overall trends are very similar. The same figures also show annual rates of change in two aggregate measures of the quantity of chips produced: aggregate bits of memory and an (again conceptually more attractive) index formed by dividing aggregate expenditure by a Fisher Ideal price index. Again there are some differences, but the overall trends are very similar.

In both DRAMs and EPROMs, annual rates of change in price hit all-time historical highs in 1987 and 1988, after the STA went into effect. Indeed, for the first time in the history of memory chip production, positive annual changes in price were registered in 1988 for both types

16. The Fisher Ideal index I am constructing is the geometric mean of two price indexes calculated using a single weight on the price of every type of chip in two adjacent time periods (the weights are constant in the two time periods); this approach does not confuse the effect of price changes with the shift in usage from one generation of chip to another. The weights used for the first index are the expenditure shares of chips in the starting period, while the second index uses expenditure shares for each type of chip in the ending period: the Fisher Ideal price index giving the price in period 1 relative to period 0 is

$$\sqrt{\sum_i \left(\frac{p_i^0 q_i^0}{\sum_i p_i^0 q_i^0} \right) \cdot \frac{p_i^1}{p_i^0} \bigg/ \sum_i \left(\frac{p_i^1 q_i^1}{\sum_i p_i^1 q_i^1} \right) \cdot \frac{p_i^0}{p_i^1}}.$$

Diewert has shown that this weighting scheme produces a "superlative index"—a second-order approximation to a true, exact price comparison between two periods derived from microeconomic theory. See W. E. Dewert, "Superlative Index Numbers and Consistency in Aggregation," *Econometrica*, vol. 46 (July 1978), pp. 883–900. Indices for adjacent pairs of time periods have been chained together to produce a single, continuous price index.

Figure 5-2. *Year-to-Year Changes in DRAM Prices and Quantities Produced, 1972–89*

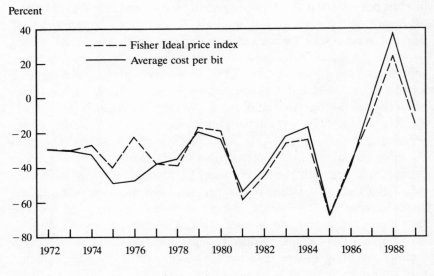

A. Price changes

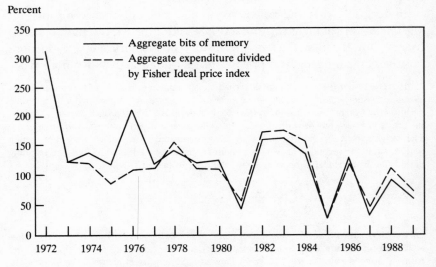

B. Quantity changes

Source: Author's calculations based on unpublished Dataquest data.

Figure 5-3. *Year-to-Year Changes in EPROM Prices and Quantities Produced, 1972–89*

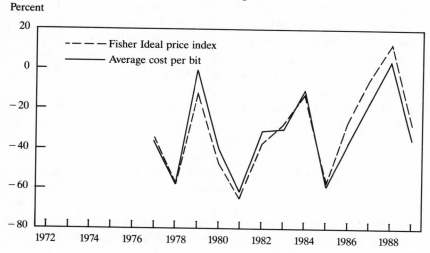

A. Price changes

Percent

- - - - - Fisher Ideal price index
——— Average cost per bit

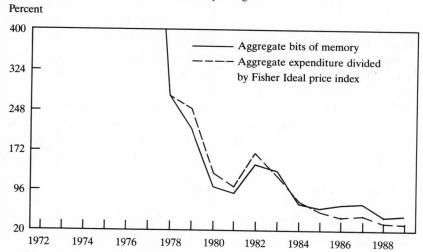

B. Quantity changes

Percent

——— Aggregate bits of memory
- - - - - Aggregate expenditure divided by Fisher Ideal price index

Source: Author's calculations based on unpublished Dataquest data.

of chips. The contrast with the historical record was particularly striking in DRAMs, where previously recorded annual rates of change had stayed within roughly the −20 to −60 percent range. Prices fell by only 10 percent in 1987, then suddenly jumped up by well over 20 percent in 1988.

The deviation from historical trend was so egregious, in fact, that some argued that a secular upward shift in the price of DRAMs was occurring. Within the industry this came to be known as the proposition that future memory pricing would follow the "bi rule" rather than the "π rule."[17] The π rule refers to the fact that, in the past, DRAM prices for each generation of chip had tended to decline asymptotically toward the \$3 level (π = 3.14159 . . .) as mass production of that generation peaked, then to half that level at the end of its life. Since a new generation of chip was introduced on average about every three years, and since each new generation of chip quadrupled the number of bits on a chip, this amounted to a 75 percent reduction in the cost of a bit of memory every three years, or an annual rate of decline of about 36 percent per year. (Remarkably, this is roughly the annual decline in memory bit cost produced by analyses of actual historical data. See figure 5-4 for data on DRAM prices through 1985.) The bi rule ("bi" is also, coincidentally, the pronunciation of the Japanese kanji character meaning "doubling") suggests that in the future every new-generation chip will approximately double in price as mass production peaks. (See figure 5-5 for DRAM pricing after 1985.) Following the previous logic, this means a 50 percent decline in bit cost every three years, for an annual rate of decline of about 20 percent, or about a 50 percent smaller long-run cost decline than under the π rule.

The cyclical sensitivity of semiconductor demand is also apparent in the large swings in price change depicted in figures 5-2 and 5-3.[18] Although 1985—the year when trade friction kicked the political process leading to the STA into motion—had set a historical record for plunges

17. See M. P. Lepselter and S. M. Sze, "DRAM Pricing Trends—The π Rule," *IEEE Circuits and Devices Magazine,* vol. 1 (January 1985), pp. 53–54; Yasuo Tarui, "From the π Rule to the Bi Rule," *Nikkei Microdevices,* no. 27 (July 1987), pp. 165–67; and Yasuo Tarui and Tadaaki Tarui, "New DRAM Pricing Trends: The Bi Rule," *IEEE Circuits and Devices Magazine,* vol. 7 (March 1991), pp. 44–45.

18. Cyclical impacts on semiconductor industry employment are discussed in detail in Kenneth Flamm, "Internationalization in the Semiconductor Industry," in Joseph Grunwald and Kenneth Flamm, *The Global Factory: Foreign Assembly in International Trade* (Brookings, 1985), pp. 101–03.

Figure 5-4. *Prices for Four Generations of DRAMs, Pi Rule, 1974–89*

Dollars per chip

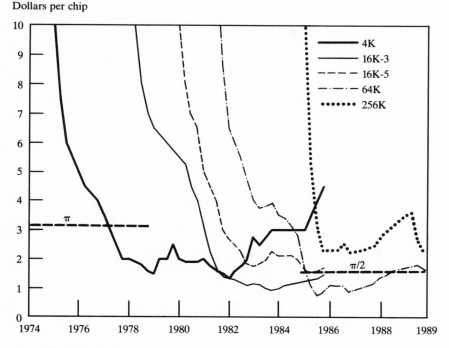

Source: Unpublished Dataquest data.

in DRAM prices, the plunge was only marginally steeper than that seen in 1981. In EPROMs, on the other hand, the price decline of 1985 actually fell short of the drop registered in 1981 and was about the same as that seen in 1978. Rates of increase in chip output also fell close to historical lows in the years after 1986, in both DRAMs and EPROMs.[19]

19. At least one study, commissioned by the Semiconductor Industry Association, the U.S. industry group, argues that the movement of prices in DRAMs after the STA went into effect had returned price levels to their long-term trend by 1989, and therefore that the STA had no apparent deleterious effects in raising prices above long-term trends. See Technecon Analytic Research, Inc., "Impact of the Semiconductor Agreement on DRAM Prices," Washington, December 1990. The logic of this argument, however, suffers from three significant flaws.

First, all forecasts of trend price levels are very sensitive to the choice of base period used to estimate the trend. By choosing a particular base period, one can raise or lower trend growth rates arbitrarily. By excluding the period of sharp recession in 1985 from

Figure 5-5. *Prices for Three Generations of DRAMs, Bi Rule, 1987–95*

Dollars per chip

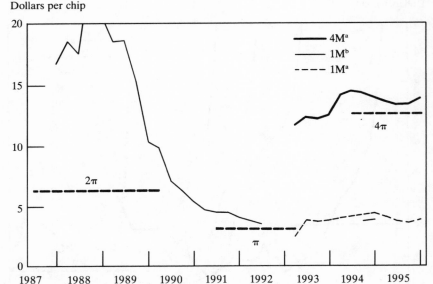

Sources: Author's calculations based on data from Dataquest, "DQ First Monday Report"; Electronic Business Buyer; and Nikkei Telecomm (on-line data service), "Large User Contract Price."
a. Electronic Business Buyer.
b. Dataquest, "DQ First Monday Report."

Regional Semiconductor Price Differentials

A significant factor complicating discussion of DRAM pricing was the appearance of significant regional price differentials in 1987 and 1988, after the signing of the STA. In response to the STA the Japanese gov-

---

the period used to estimate the trend, the rate of decline was reduced in this study.

Second, the "correct" trend in this study is estimated by taking price as a function of cumulative bits of memory produced. Even if this relationship held exactly true to historical trend after the STA went into effect, and price was exactly as predicted by some historical function of the cumulative bits of memory produced, producer behavior in response to the STA would still have raised prices in an "ahistorical" fashion by cutting back on the number of bits produced. That is, looking at a graph of price versus cumulative bits would lead one to conclude that nothing representing a break from historical, structural relationships had occurred—but if producers had indeed cut back substantially on production (and cumulative bits produced) relative to past trends, prices would indeed have risen extraordinarily. In other words, even if some structural relation between price and cumulative bits were unaffected by the events set in motion by the STA, the growth of bits produced over time certainly was, and this latter channel would have been expected to be the primary effect of

ernment began to set floor prices for export sales of DRAMs by Japanese companies. Initially, different standards were set for sales to the U.S. market and other (third-country) export markets; after U.S. protests, however, the systems were later unified. (In response to European protests, the pricing guidelines were separated again in 1989.)

Regulation of Japanese export sales ultimately involved four elements. First, an export licensing system was adopted. This system required de facto government approval of the export price, which was to be set above minimum norms established by Japan's Ministry of International Trade and Industry (MITI). Second, foreign purchasers of Japanese chips were required to register with MITI. Third, all export transactions required a certificate from the original chip manufacturer attesting to the fact that the chips in question were actually manufactured by that producer. Fourth, MITI established informal regional allocation guidelines, to ensure that supplies were not diverted from one export market to favor another.

By most accounts, MITI's guidance was quite effective in setting minimum pricing standards for Japanese DRAM manufacturers' direct export sales. (Because Japanese manufacturers were at this time responsible for between 80 and 90 percent of world DRAM sales, this effectively worked as a floor on price in the global market.) The intent of the second and third elements was clearly to reduce access by foreign purchasers to Japanese gray market channels not under the direct supervision and control of Japanese chip manufacturers. Predictably, prices in the unregulated domestic Japanese market soon dropped below foreign export prices.

The rising differential between U.S. and Japanese DRAM prices was exacerbated by producer actions that, from the standpoint of U.S. consumers, made the true price of their DRAMs even greater than the hefty contract prices reported to market researchers. The practice of tying sales of DRAMs to consumer purchases of other chips that would not otherwise have been bought from Japanese DRAM vendors (particularly ap-

---

restrictions on supply associated with the STA. The Technecon Analytic Research study actually notes that the growth in bits produced (and consumed) in the late 1980s was well below historical trends.

Third, the assumption that some fixed structural relationship exists between price and cumulative bit production is dubious. As will be seen in chapter 6, although yields, and therefore marginal costs, may have a relatively simple relationship to cumulative output, there is no reason to expect prices to follow marginal costs very closely, particularly over periods when production is constrained by available capacity.

plication-specific integrated circuits, or ASICs) appears to have surfaced in 1988 and 1989 and was the source of widespread complaints in the computer industry.[20] Indeed, when the U.S. Memories joint venture was being organized in 1989 (see chapter 4), Wilfred Corrigan, CEO of ASIC producer LSI Logic, made it clear that a major reason he agreed to join the DRAM consortium and chair its board of directors was to acquire some leverage against the widespread use of these tying arrangements and free his ASIC customers from the need to shift to Japanese ASIC suppliers in order to maintain access to needed supplies of DRAMs.[21] Functionally, the cost premium faced by U.S. customers was even larger than the monetary differential alone would suggest.

In apparent response to the appearance of regional price differentials after the STA was negotiated, Dataquest began reporting regional contract pricing for a sample of twenty-five semiconductor components, on a biweekly basis, beginning in early 1987. These data (the Dataquest "First Monday Report") are based on a survey of six to ten respondents, primarily chip manufacturers, in each of six geographic regions.[22] Japanese producers do not report a contract price, and the Japanese price cited by Dataquest apparently refers to "large volume wholesale" prices.

Figure 5-6 compares the Dataquest "First Monday Report" volume contract prices for a 120-nanosecond, 256K DRAM in five regions, averaged over the course of a quarter, with the Dataquest ASPs for all 256K DRAMs through the last quarter of 1988. The increase in regional pricing differentials in 1987 and 1988 is readily apparent. The extraordinarily sharp rise in European 256K DRAM prices relative to other regions is particularly notable, and a little puzzling. (Analysis detailed elsewhere suggests that this sharp rise may be fictitious.)[23]

20. One well-known U.S. computer vendor's head of purchasing described in great detail how they had been forced to order ASICs in order to secure shipments of DRAMs in 1988. Author's interview with U.S. computer systems vendor, 1989.

21. Jack Robertson, "Confirm Major IBM Role in Creation of U.S. Memories," *Electronic News*, June 26, 1989, pp. 1, 58.

22. For DRAMs the survey asks for the current contract price negotiated for three different volumes: 1,000, 10,000, and "volume" (over 100,000). In some published Dataquest reports, however, it is stated that "volume" prices mean greater than 20,000 parts, not 100,000; the higher definition of "volume" for DRAMs is apparently an exception to this rule. If producers have not concluded any contracts for a particular volume, they are asked to estimate the price that would have been negotiated on a contract of that size. I thank Mark Giudice of Dataquest for this description of the "DQ First Monday Report" survey in various conversations during 1989 and 1990.

23. See appendix 5-A and Flamm, "Measurement of DRAM Prices," for evidence on this point.

Figure 5-6. *Quarterly Dataquest Monday Contract Prices and Average Selling Prices for 256K DRAMs, 1987–92*[a]

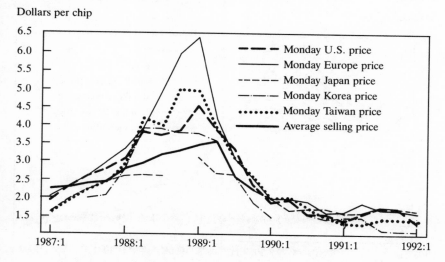

Dollars per chip

Legend:
- – – – Monday U.S. price
- ——— Monday Europe price
- - - - - Monday Japan price
- —·—·— Monday Korea price
- •••••••• Monday Taiwan price
- ——— Average selling price

Sources: Author's calculations based on unpublished Dataquest data; and Dataquest "DQ First Monday Report."
a. Figures are quarterly averages.

Figure 5-7 gives the equivalent time series for 1M DRAMs.[24] All available data seem to show unprecedented, sustained increases in DRAM prices from 1987 on. (Compare this with 64K DRAM pricing during the shortage of 1983–84, for example, in figure 5-1.)

Weekly data on large-user wholesale prices in the Japanese market are also regularly published in *Nihon Keizai Shimbun* (commonly called *Nikkei,* a Japanese business daily roughly equivalent in terms of reputation and stature to the American *Wall Street Journal* or the London *Financial Times*).[25] Various contract (or large-user) pricing data for both U.S. and

24. The "DQ First Monday Report" contract prices refer to 120-nanosecond, 1M parts, in plastic cases, which were the cheapest available 1M part. It is surprising, then, that the Dataquest ASPs—which presumably average these with more expensive parts and with higher priced parts sold through distributors and in the spot market—for at least three quarters, during a period of relatively stable prices, are at or below an average of volume contract prices. This casts some doubt on the reliability of one or the other of these data sources.

25. These data are reprinted on a weekly basis in the *Japan Economic Journal,* the English-language weekly based on *Nikkei.* When possible the former is used here, supplemented by online data available from *Nikkei*'s computerized data base, and selected issues of *Nikkei* itself. Data sampled on a weekly basis were averaged to calculate a monthly time series; monthly series were averaged to create a quarterly series. Monthly (quarterly)

Figure 5-7. *Quarterly Dataquest Monday Contract Prices and Average Selling Prices for 1M DRAMs, 1987–92*[a]

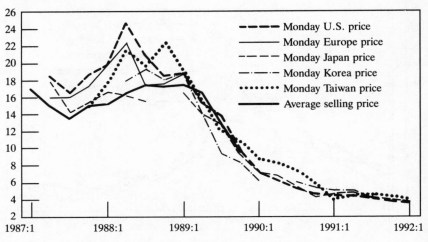

Dollars per chip

Sources: See figure 5-6.

Japanese sales are shown in figures 5-8 and 5-9. Figure 5-8 contrasts the *Nikkei* large-user Japanese wholesale price for 256K DRAMs with "DQ First Monday Report" 256K DRAM contract prices. It is clear that Dataquest's Japanese volume contract price basically follows the *Nikkei* wholesale price reasonably closely (albeit with occasional bursts of noise), and the two are probably different estimates of the same underlying concept. The Dataquest estimates for Japan should probably be regarded as less reliable than the *Nikkei* numbers, because they appear to be subject to frequent retroactive revisions.[26]

estimated yen per chip was converted to dollars using market average monthly (quarterly) exchange rates reported by the International Monetary Fund.

26. Indeed, no two sources for the Dataquest price series appear to give exactly the same numbers, particularly for prices of product sold outside the United States. For this reason, one may speculate that revisions to the exchange rates used to convert prices to dollars (the currency in which Dataquest reports all contract prices) may play an important role in these retrospective changes.

The most dramatic example of retrospective revision of the Dataquest regional contract price estimates involved the U.S. government's use of these data in its submission to the GATT panel investigation of the STA, in order to challenge price data furnished by European companies. The European data, however, were apparently also supplied by Dataquest! The data submitted by the U.S. government, in turn, differed from data furnished to me by Dataquest in February 1989.

Figure 5-8. *Monthly Contract Prices for 256K DRAMs in the United States and Japan, 1985–92*

Dollars per chip

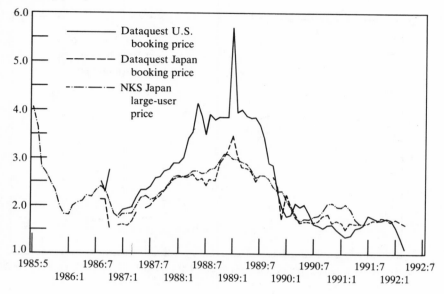

Sources: Dataquest "DQ First Monday Report"; *Nihon Keizei Shimbun* (various issues); and *Nikkei Telecomm* (on-line service).

Figure 5-8 seems to show quite clearly that a large and significant differential between the U.S. and Japanese markets existed in 1988 and most of 1989 in large-volume, direct "contract" sales between Japanese manufacturers and their customers in the two regions. Figure 5-9 compares the *Nikkei* time series on wholesale 1M DRAM prices with Dataquest contract pricing data in the United States and Japan. Figure 5-10 compares a *Nikkei* time series on wholesale 64K DRAM prices with Dataquest quarterly worldwide average sales prices. (U.S. contract price estimates were unavailable for 64K parts.) Both seem to tell basically the same story of significant regional differentials from late 1987 through late 1989. Contemporary accounts in the Japanese business press also report large price differentials between Japanese and foreign markets over this period.[27]

27. For example, on April 9, 1988, the *Japan Economic Journal* reported that Korean chip exports to Japan had dropped steeply, because Korean producers had found prices higher and profitability greater in the U.S. market. Noted the *Journal* in "Korean Chip

Figure 5-9. *Monthly Contract Prices for 1M DRAMs in the United States and Japan, 1987–92*

Dollars per chip

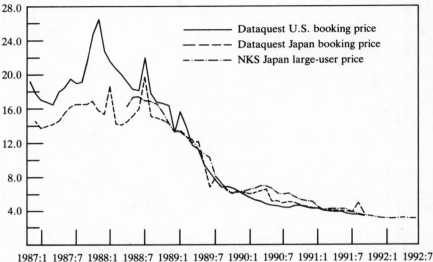

1987:1 1987:7 1988:1 1988:7 1989:1 1989:7 1990:1 1990:7 1991:1 1991:7 1992:1 1992:7

Sources: See figure 5-8.

The data displayed thus far have largely ignored DRAMs sold by distributors and in the spot market. This misses an important dimension of the change in market conditions after the signing of the STA. To

Producers Shun Japan to Harvest Higher Profit in U.S." (April 9, 1988), p. 4, "The reasons for the shift are obvious. With Japanese semiconductor exports to the U.S. down sharply as a result of increased trade friction between the two nations, prices on the underfed U.S. market have begun to skyrocket. . . . At the same time, though supplies in Japan were also dwindling, Japanese companies were reluctant to accept price hikes beyond the typical ¥340 cost for large users. Not surprisingly, Korean makers moved quickly to enter the more lucrative U.S. markets."

On July 18, 1988, *Nihon Keizai Shimbun*'s morning edition carried a front-page story reporting that the price differential between the Japanese and foreign market for a 256K DRAM had widened from 100 yen, as trade friction heated up, to 200 to 300 yen. (An English-language summary of the story was carried in "Foreign Semiconductor Prices Even Higher," *Japan Economic Journal*, July 30, 1988, p. 10. Depreciation of the dollar in the intervening period makes the dollar value of that change in the differential substantially greater.) On August 9, 1988, *Nihon Keizai*'s morning edition carried a story on page 18 reporting that Japanese semiconductor producers had begun stepping up exports, with much higher price tags than these products carried in the domestic market. (An English-

Figure 5-10. *Quarterly Japanese and World Contract Prices for 64K DRAMs, 1985–89*

Dollars per chip

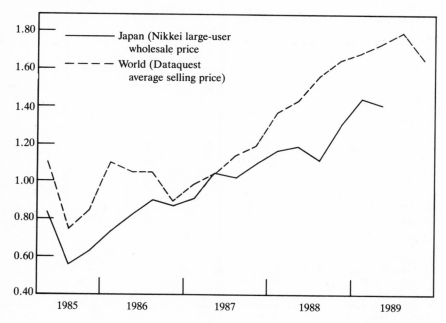

Sources: Author's calculations from unpublished Dataquest data; Dataquest, "DQ First Monday Report"; and *Nihon Keizai Shimbun* (various issues).

remedy this situation I have constructed time series showing retail spot prices for memory chips, beginning in the spring of 1985. To do so, I collected weekly data on sales prices from one of the larger retail vendors of memory chips in the United States during this period.[28] The advertised prices are dated (an important point, since there is typically a substantial

language summary appears in "Japan Chip Makers Step Up Exports Amid Shortages," *Japan Economic Journal*, August 20, 1988, p. 15. The story also notes that observers feared this might exacerbate shortages in the domestic market.)

28. The vendor was Microprocessors Unlimited, of Beggs, Oklahoma; the weekly advertisements were found in the pages of *PC Week* and *Infoworld* magazines. The prices shown refer to the following parts: the 64K DRAM is a 64K × 1, 150-nanosecond, DIP-packaged chip; the 256K DRAM is a 256K × 1, 120-nanosecond DIP package; the 1M DRAM is a 1M × 1, 100-nanosecond DIP package through October 1989, and an 80-nanosecond chip after that month. For comparison purposes, a 100-nanosecond 1M DRAM sold for $10.99 in the first week of November 1989, whereas an 80-nanosecond part sold for $10.99 the following week.

Figure 5-11. *Monthly Spot Prices for 256K DRAMs in the United States and Japan, 1985–91*

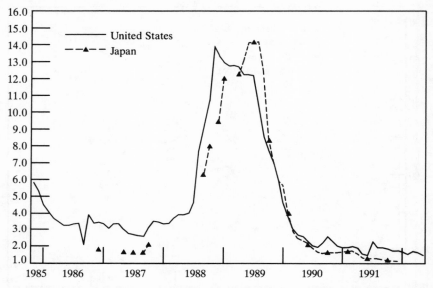

Dollars per chip

Sources: Author's calculations based on published advertisements of Microprocessors Unlimited; *Nihon Keizai Shimbun* (various issues); and Nikkei Telecomm (on-line service).

lag between the submission of advertising copy and its publication); contacts with this vendor have also made it clear to me that these were real prices, that is, product was actually available in stock at these prices. Figure 5-11 shows spot retail prices for 120-nanosecond, 256K DRAMs. The contrast with figure 5-8 is striking: spot retail prices quadrupled between early 1987 and early 1988!

Some data may also be gathered on spot pricing in Japan. Beginning in October 1988, spot price data for 256K DRAMs, of unspecified speed, have been published regularly in *Nihon Keizai Shimbun*. Weekly prices have been averaged over the month;[29] some occasional reports on spot prices have also been collected from news articles in *Nikkei* and are also

29. Prices from the Tuesday or Wednesday morning editions of *Nikkei* were used, beginning in October 1988. Data from before that date are taken from an assortment of occasional *Nikkei* news reports on DRAM market conditions. Monthly exchange rate data published by the International Monetary Fund were used to convert yen prices to U.S. dollars.

Figure 5-12. *Monthly Spot Prices for 1M DRAMs in the United States and Japan, 1986–92*

Dollars per chip

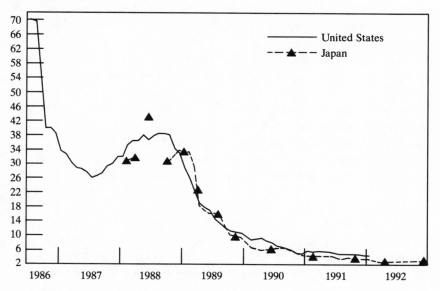

Sources: See figure 5-11.

shown in figure 5-11. The fragmentary data shown here suggest that Japanese spot prices generally followed the upward trajectory of American spot prices, but with a lag, so that some differential persisted through mid-1988. In the fall of 1988 Japanese spot prices seem to have briefly passed American spot prices, only to fall to approximate parity with U.S. prices in 1989. These data suggest that MITI's attempts to wall off the Japanese gray market from the foreign gray market only slowed convergence and did not prevent arbitrage from linking prices in the two markets. This is not completely surprising, since the gray market is virtually by definition a bastion of untamed entrepreneurialism.

Figure 5-12 shows an equivalent retail, spot price series I have constructed for 1M DRAMs (with 100-nanosecond access times), along with data on Japanese 1M DRAM spot prices collected from *Nikkei*. The pattern of regional differentials shown by these more fragmentary data is similar to that for 256K DRAMs. U.S. buyers paid some premium relative to Japanese buyers through mid-1988; Japanese spot prices may have briefly shot past U.S. levels at the end of the year, and 1989 brought with it rough parity. The increase in price from mid-1987 to late 1988 is notable, but far less striking than that for 256K DRAMs.

In DRAMs, then, the overall pattern was one of a significant regional price differential in contract pricing between the domestic Japanese and overseas markets from 1987 through 1988. The differential disappeared in 1989, as semiconductor demand slackened and prices began to fall sharply. In the spot market, in contrast, there was some transitory differential in pricing between the Japanese and American markets, but after some short lag prices essentially tended to converge in the two regions. What this suggests is that government price and export controls were relatively effective in controlling direct transactions between the major Japanese producers and their large customers, but considerably less effective in regulating transactions in the secondary spot market, where arbitrage between the American and Japanese markets approximately equalized prices in the two regions.[30]

Overall regional contract price indexes for DRAMs may be constructed using estimates of contract prices in the United States and Japan for individual products. A Fisher Ideal price index for the United States was constructed by taking quarterly Dataquest contract price estimates in the United States for 256K and 1M DRAMs, and Dataquest ASP estimates for 64K DRAMs, and using expenditure shares constructed using relative volumes of worldwide production of these products at U.S. prices to construct weights for period-to-period comparisons. A chained index built up from quarter-to-quarter Fisher Ideal comparisons was then calculated for the United States.[31] A similar index was constructed for Japan but used *Nikkei* estimates of large-user wholesale prices, and relative global volumes, to produce quarterly Japanese expenditure share weights.[32] Since comparisons of contract prices show U.S. and Japanese

30. Export controls were notoriously porous to small-scale evasions. Chips are compact and light, and a million-yen shipment could fit into a small box or large attaché case. Anecdotes of such irregular traffic abounded in 1987 and 1988.

31. For 256K DRAMs quarterly averages of Dataquest "DQ First Monday Report" price estimates for volume contracts were used throughout this period. For 1M DRAMs Dataquest worldwide ASP estimates were used until the first quarter of 1987, when the Monday contract price series starts; thereafter the latter series was used. For 64K DRAMs Dataquest worldwide ASP estimates were used throughout, since Monday contract price data were unavailable for this product.

32. Quarterly averages of *Nikkei* weekly price quotes for 256K DRAMs were used throughout. For 64K DRAMs average *Nikkei* large-user price quotes were used through the second quarter of 1989, after which Dataquest worldwide ASPs were used. For 1M

Figure 5-13. *Quarterly Fisher Ideal Price Indexes for Large-User Purchases of DRAMs in the United States and Japan, 1986–89*

1986:4 = 100

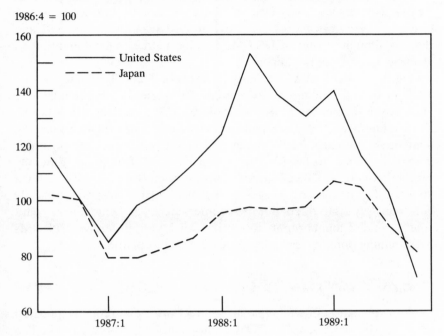

Sources: Author's calculations based on unpublished Dataquest data; *Nihon Keizai Shimbun* (various issues); and Nikkei Telecomm (on-line service).

prices for all sizes of DRAMs at rough parity in the fourth quarter of 1986, the DRAM price indexes for the United States and Japan were calculated taking that quarter equal to 100.

Figure 5-13 shows the relative movement in the two regions of an aggregate price index for large-user purchases of DRAMs from 1986 through 1989. The aggregate differential between the United States and Japan looks very much like the pattern of differentials observed in each of the individual varieties of DRAM, with U.S. prices rising well above Japanese levels in 1987 and 1988, peaking at a level almost 50 percent higher than in Japan in the second quarter of 1988, then falling to rough parity by the end of 1989.

DRAMs Dataquest worldwide ASPs were used through the first quarter of 1987, then replaced with *Nikkei* average large-user wholesale prices beyond that date.

Fewer data are available on the market for EPROMs (in particular, data on Japanese spot pricing are infrequently reported in the business press). Nonetheless, it is clear that a considerably different situation prevailed. Whereas the overall trend toward rising prices in 1987 and 1988 is similar to that in DRAMs, significant regional differentials in contract pricing did not appear.

Figure 5-14 compares quarterly averages of *Nikkei* large-user wholesale prices with Dataquest volume contract prices, where comparative data are available, for 128K and 256K EPROMs. (As in DRAMs, the *Nikkei* numbers probably should be viewed as more reliable than the Dataquest estimates for the Japanese market.) No persistent price differential appears to have emerged; Japanese EPROM prices generally tracked American EPROM prices, though often with a lag of roughly a quarter. The data clearly suggest that Japanese EPROM prices basically followed American prices downward after late 1988. Contemporary Japanese press accounts report that prices for imported foreign EPROMs were pulling domestic price levels down over this period.[33]

The Supply of Memory Chips

An explanation of why regional contract price differentials appeared in DRAMs, but not in EPROMs, must consider supply factors. The industrial organization of the supplier industry for DRAMs and EPROMs is depicted in tables 5-1 and 5-2, which calculate the Hirschman-Herfindahl index (HHI) of concentration, and the share of global supply coming from firms headquartered in different countries, over time.[34] The HHIs calculated here will range in value from one (most

33. "The drop [in EPROM prices] was attributed mainly to price competition among foreign-affiliated makers hoping to expand their market share in Japan." "Prices of EPROM Chips Decline," *Nihon Keizai Shimbun,* April 19, 1989.

"The price of EPROMs was pushed down by competition from foreign manufacturers." "Semiconductor Prices on the Decline," Nikkei Top Articles, April 24, 1989.

"In September . . . 1M chips dipped to 1,050–1,100 [yen]. . . . Some U.S.-produced EPROMs are now selling for around 1,000 yen, according to an official at a semiconductor trading house." "EPROM Prices Drop for the First Time in Six Months," Nikkei Top Articles, October 2, 1990.

"Imported EPROMs are also having an impact on prices . . . excess American output is entering the Japanese market, said a sales agent." "1M EPROM Prices Plummet below 900 Yen," Nikkei Top Articles, November 2, 1990.

34. Because Dataquest identifies chip suppliers by the label on the chip, the reported U.S. share of the DRAM market and the calculated HHI credit chips manufactured by

Figure 5-14. *Quarterly Wholesale Prices for 128K and 256K EPROMs in the United States and Japan, Selected Years, 1985–92*

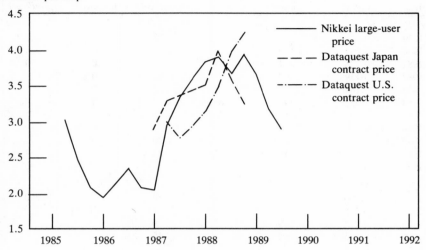

A. 128K EPROMs

Dollars per chip

B. 256K EPROMs

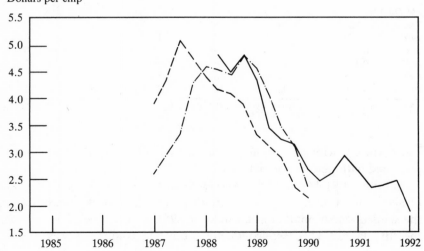

Dollars per chip

Sources: Dataquest, "DQ First Monday Report"; *Nihon Keizei Shimbun* (various issues); and Nikkei Telecomm (on-line service).

Table 5-1. *Market Concentration in DRAMs, 1979–89*

Year	Hirschman-Herfindahl index	Percentage of global production				
		Top five Japanese producers[a]	Other Japanese producers	U.S. producers	European producers	Korean producers
64K DRAMs						
1979	0.525	67	0	33	0	0
1980	0.264	59	0	41	0	0
1981	0.178	67	6	28	0	0
1982	0.129	60	6	33	0	0
1983	0.108	53	7	38	2	0
1984	0.092	48	10	38	4	0
1985	0.091	54	7	31	4	4
1986	0.099	56	11	19	5	10
1987	0.106	44	13	20	6	17
1988	0.170	19	22	21	1	37
1989	0.273	4	17	34	0	45
256K DRAMs						
1982	1.000	100	0	0	0	0
1983	0.265	92	1	7	0	0
1984	0.213	89	3	9	0	0
1985	0.165	82	3	15	0	0
1986	0.135	77	5	15	1	2
1987	0.102	55	11	24	2	9
1988	0.091	46	18	28	2	7
1989	0.078	43	19	25	3	10
1M DRAMs						
1985	0.964	99	0	1	0	0
1986	0.369	87	0	13	0	0
1987	0.347	93	6	1	0	0
1988	0.173	78	10	6	3	2
1989	0.135	54	16	14	5	11
1990	0.110	45	15	21	5	15

Source: Author's calculations based on unpublished Dataquest data.
a. Fujitsu, Hitachi, Mitsubishi Electric, NEC, and Toshiba.

concentrated, with all output produced by a single firm) to zero (approached with perfect competition).

In both EPROMs and DRAMs, levels of concentration (generally at or above the .1 level) were considerably greater than those measured in the semiconductor market as a whole. In 1987, for example, the HHI for the fifty largest U.S. semiconductor-producing companies was .0539,

Korea's Samsung but marketed by Intel under its label to Intel, and Toshiba-made DRAMS sold by Motorola to Motorola.

Table 5-2. *Market Concentration in EPROMs, 1980–89*

Year	Hirschman-Herfindahl index	Percentage of global production				
		Top five Japanese producers[a]	Other Japanese producers	U.S. producers	European producers	Korean producers
64K EPROMs						
1980	0.625	0	0	100	0	0
1981	0.389	11	0	89	0	0
1982	0.273	50	0	50	0	0
1983	0.157	60	1	40	0	0
1984	0.143	65	2	31	1	0
1985	0.136	65	2	30	4	0
1986	0.138	64	0	29	6	0
1987	0.089	44	4	38	13	0
1988	0.114	14	3	54	26	3
1989	0.131	7	3	59	29	2
128K EPROMs						
1982	0.916	0	0	100	0	0
1983	0.405	22	0	78	0	0
1984	0.165	61	0	39	0	0
1985	0.154	77	0	23	0	0
1986	0.155	65	0	34	0	0
1987	0.112	50	1	43	6	0
1988	0.119	29	2	52	17	0
1989	0.153	17	1	58	24	0
256K EPROMs						
1983	0.536	13	0	88	0	0
1984	0.394	18	0	82	0	0
1985	0.202	46	0	54	0	0
1986	0.155	59	0	41	0	0
1987	0.107	47	3	44	5	0
1988	0.097	36	2	50	12	0
1989	0.093	27	1	55	18	0
512K EPROMs						
1984	0.500	0	0	100	0	0
1985	0.479	6	0	94	0	0
1986	0.257	32	0	68	0	0
1987	0.138	57	6	36	1	0
1988	0.105	46	3	42	8	0
1989	0.111	26	2	57	14	0
1M EPROMs						
1987	0.253	80	0	20	0	0
1988	0.161	82	0	18	0	0
1989	0.133	63	0	34	3	0
2M EPROMs						
1988	1.000	100	0	0	0	0
1989	0.882	96	0	4	0	0
4M EPROMs						
1989	0.290	80	0	20	0	0

Source: Author's calculations based on unpublished Dataquest data.
a. Fujitsu, Hitachi, Mitsubishi Electric, NEC, and Toshiba.

Figure 5-15. *Hirschman-Herfindahl Indexes and Cumulative Output for Five Generations of DRAMs*

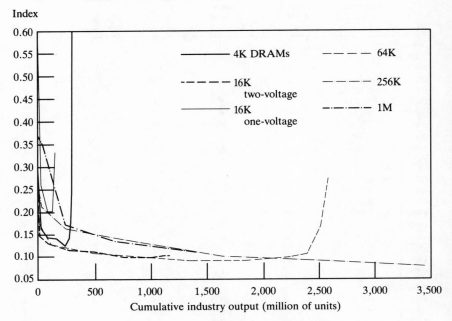

Source: Author's calculations based on unpublished quarterly Dataquest data.

down from .0597 in 1982.[35] What this probably indicates is that semiconductor firms tend to specialize in particular types of products. It is also evident that HHIs typically vary enormously over the product life cycle, with values of one when the very first firm introduces a new-generation chip, dropping over time as others enter the business, then rising again at the end of the life cycle as firms phase out their production of the older product and shift to newer items.

Thus, consideration of whether or not excessive monopoly power exists, or whether the industry is more concentrated than the historical norm, cannot be independent of the stage in the product life cycle. One convenient measure of how advanced in the product cycle a generation of product might be is cumulative output over time. Figures 5-15 and 5-16 show the evolution of HHIs over time for various generations of

35. See U.S. Bureau of the Census, *1987 Census of Manufactures: Concentration Ratios in Manufacturing* (U.S. Commerce Department, 1987), pp. 6–38.

Figure 5-16. *Hirschman-Herfindahl Indexes and Cumulative Output for Six Generations of EPROMs*

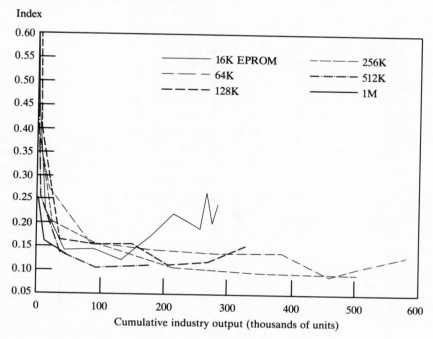

Source: Author's calculations based on unpublished quarterly Dataquest data.

DRAMs and EPROMs, calculated using annual data.[36] Historically, HHIs in both of these products have tended toward the neighborhood of .1 as mass production peaked. Normalizing for position in the product life cycle using cumulative output, across generations of chip, there is little obvious evidence of a secular trend toward significantly increased concentration in EPROMs. One might observe that, during the early years of 256K and 1M DRAM production, the industry was somewhat more concentrated than the historical norm, but by 1989 concentration in both these products looked similar to earlier historical levels.

Although variation in producer concentration may not be much of a potential factor in explaining differences in DRAM and EPROM market conditions over 1986–89, geographic concentration offers greater prom-

36. These calculations use Dataquest estimates of annual production by all merchant companies. Production by exclusively captive producers (with no sales on the open merchant market) such as IBM is excluded.

ise. For the three most recent generations of DRAM, Japanese market shares were 67 percent (in 64K parts), 82 percent (256K), and 99 percent (1M) in 1986, when the STA was signed. In EPROMs, overall, Japanese company shares were considerably lower: 65 percent (128K), 59 percent (256K), and 32 percent (512K). Thus, proportionate reductions in production or exports by Japanese producers would have potentially greater relative impacts on aggregate supply in DRAMs than in EPROMs.

Furthermore, very different adjustments to Japanese production of DRAMs and EPROMs are revealed in estimates of Japanese chip output after the STA was implemented. Some of the best and most detailed semiconductor statistics available cover DRAM and EPROM production in Japan. As a consequence of the STA and the operation of the MITI supply-demand forecast committee, Japanese companies reported a large amount of data to MITI, from which quarterly estimates of Japanese production, consumption, and net exports were compiled through late 1988. After the dissolution of this committee in 1989, MITI continued to compile these estimates on a semester basis.[37]

As in DRAMs, MITI was reportedly issuing guidance to Japanese producers to reduce EPROM production in 1987.[38] As will be seen shortly, moreover, foreign producers rapidly expanded EPROM sales in the Japanese market, and by 1989 Japan's share of world output for many types of EPROMs had fallen sharply and Japan had become a significant net importer of EPROMs. This contrasted with the situation in DRAMs, where Japan remained a large net exporter and continued to dominate world supply.

In looking at production statistics, one should note that output in a fully utilized chip fabrication facility is *not* fixed over time, but instead rises on an S-shaped profile (figure 5-17). This occurs for two basic reasons. First, as fabrication technology gradually improves, smaller feature sizes become feasible. As the chip size is shrunk, more chips can be fit onto the surface of the large silicon wafers on which the processing actually takes place. Three or more such "die shrinks" may typically

37. In addition to including the worldwide production of Japanese companies, these statistics also include the Japanese chip production of U.S.-owned affiliates, including those of IBM Japan, TI Japan, and Tohoku Semiconductor (a Japanese manufacturing operation jointly owned by Toshiba and Motorola).

38. See, for example, "MITI Authorizes Higher Chip Output for 3rd Quarter," *Japan Times*, June 25, 1987; "Semiconductor Price Negotiations Face Rough Going," *Nikkei Top Articles*, August 15, 1987; and "MITI Is to Start Regulating SRAM Prices," *Electronic Times*, September 17, 1987, p. 6.

Figure 5-17. *Yield Plot for IBM's 64K DRAM*

Relative final test yields

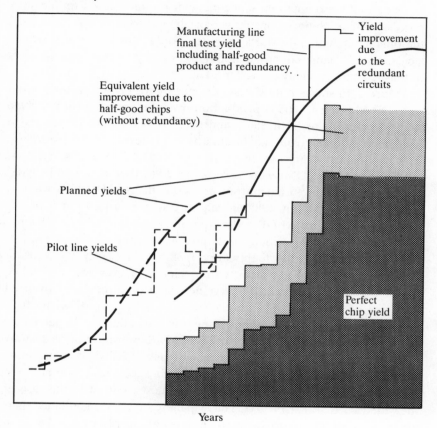

Years

Source: C. H. Stapper and others, "Evolution and Accomplishments of VLSI Yield Management at IBM," *IBM Journal of Research and Development*, vol. 26 (September 1982), pp. 532–45.

occur over the life cycle of the product within a single company.[39] Second, for a given die size, better control over the manufacturing process associated with experience and learning-by-doing creates improved yields— the number of good chips that can be extracted from each batch of processed silicon. Learning economies—predictable declines in unit cost

39. A desirable effect of incrementally smaller chips is gradually improved speed. Thus the access time of the standard 256K DRAM produced by most manufacturers went from 150 to 120 nanoseconds over the 1987–88 period, as the result of die shrinks.

that come with experience—are created as a consequence of the cost declines associated with improved yields.

One should also note that semiconductor production is a very lengthy process, with a relatively long lag—typically two to three months—from the time a batch of chips starts down the production line to shipment to a customer.[40] Changes in production rates are fully reflected in shipments of finished chips roughly a quarter later.

Figure 5-18 shows published estimates by MITI of quarterly Japanese DRAM and EPROM production over the period from late 1986 to 1988. Output of both 256K and 64K DRAMs clearly appears to have been cut back from maximum attainable levels in 1987 and 1988, with particularly sharp cuts in the first half of 1987. Declines in output of all types of EPROMs were reversed in 1987, then resumed in 1988. Figure 5-18 indicates that output of 64K and 256K DRAMs was cut back quite sharply from 1988 on, supporting anecdotal reports of significant cutbacks in 1M DRAM output in the second half of 1989 and first half of 1990, described in chapter 4. Output of all varieties of EPROM declined quite steeply after 1988.

MITI estimates of net exports from Japan are even more revealing (figure 5-19). Japan remained a large net exporter of the most advanced varieties of DRAMs throughout the post-STA period, with 50 to 60 percent of 1M DRAM output going overseas, and over 70 percent of 4M DRAM output exported.[41] In EPROMs, however, imports steadily increased, and by 1991 imports of many types of EPROMs—including the most advanced products—accounted for a substantial portion of the domestic market, and greatly exceeded domestic output in the case of the widely used 512K EPROM. Thus Japanese suppliers maintained their already preeminent position in the global supply of DRAMs; in EPROMs, in contrast, where the Japanese share of world supply was initially considerably smaller than in DRAMs, Japan cut back production sharply and vastly increased imports.

As table 5-3 shows, foreign-based firms' EPROM market share in Japan surged upward in 1990 and 1991. (In these *Nikkei* estimates of

40. In a 1986 submission to the Commerce Department, Texas Instruments gave the following estimates of time elapsed from the beginning of chip fabrication to sale to the final consumer, for a 64K DRAM: wafer fabrication, 20–28 days; die inventory, 7–14 days; assembly and testing, 10–14 days; and finished goods inventory, 20–30 days. See Texas Instruments (1986), pp. 18–19.

41. The share of older vintage 256K DRAM chips exported dropped sharply in 1990 and 1991, however.

Figure 5-18. *Japanese Ouput of DRAMs and EPROMs, 1986–95*

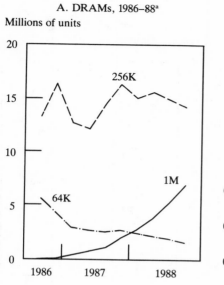

A. DRAMs, 1986–88[a]

Millions of units

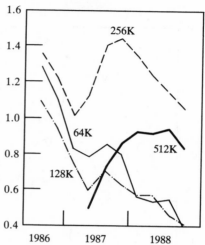

B. EPROMs, 1986–88[a]

Millions of units

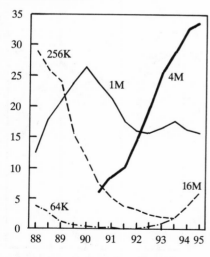

C. DRAMs, 1988–95[b]

Millions of units

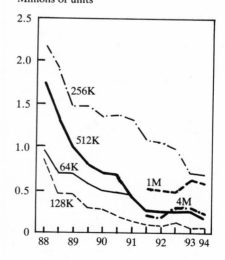

D. EPROMs, 1988–95[b]

Millions of units

Source: MITI, "Supply-Demand Forecasts."
a. Quarterly data.
b. Semiannual data.

Figure 5-19. *Japanese Exports of DRAMs and EPROMs as a Share of Total Output, 1986–95*

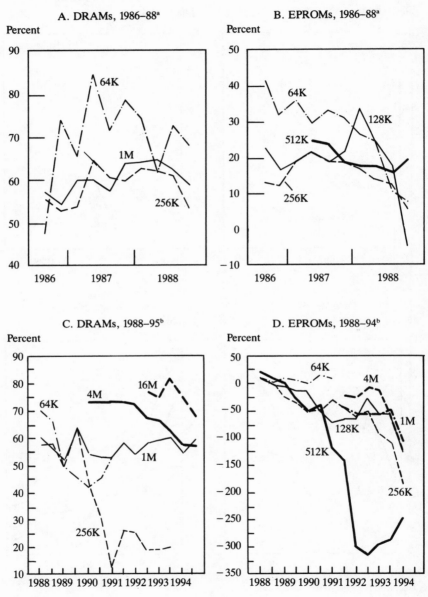

A. DRAMs, 1986–88[a]

B. EPROMs, 1986–88[a]

C. DRAMs, 1988–95[b]

D. EPROMs, 1988–94[b]

Source: MITI, "Supply-Demand Forecasts."
a. Quarterly data.
b. Semiannual data.

Table 5-3. *Shares of U.S. and Japanese EPROM Manufacturers in the Japanese Market, 1988–92*

Percent

Company	1988	1989	1990	1991	1992:1
128K EPROMs					
Intel	10	18	29	n.a.	n.a.
Hitachi	21	16	3	n.a.	n.a.
Mitsubishi Electric	21	16	6	n.a.	n.a.
Fujitsu	20	14	7	n.a.	n.a.
Toshiba	12	11	7	n.a.	n.a.
NEC	10	7	4	n.a.	n.a.
Others	6	18	43	n.a.	n.a.
256K EPROMs					
Advanced Micro Devices[a]	n.a.	n.a.	n.a.	10	10
Intel	7	13	20	7	8
Hitachi	20	20	12	3	3
Mitsubishi Electric	21	15	16	4	4
Fujitsu	22	23	18	5	5
Toshiba	19	20	16	2	2
NEC	10	2	2	0	0
Others	2	7	16	69	68
512K EPROMs					
Advanced Micro Devices[a]	n.a.	n.a.	n.a.	21	22
Intel	7	11	17	18	17
Fujitsu	18	15	27	9	7
Toshiba	16	12	13	4	4
Hitachi	17	11	15	5	5
Mitsubishi Electric	18	13	10	4	4
NEC	24	34	11	2	2
Others	0	3	7	37	39

Source: Unpublished data from Nikkei Telecomm (on-line service). First quarter of 1992 is estimated.
n.a. Not available.
a. Advanced Micro Devices is not broken out separately in 1988–90; in those years it is included in "Others."

EPROM market share, the foreign-supplied EPROMs show up in the sales of Intel, Advanced Micro Devices, and smaller suppliers included in the "other" category.)

All these individual pieces of evidence suggest that, at the margin, Japanese exports of DRAMs effectively determined prices in overseas markets. Since Japan was by far the most important source of supply for this product to world markets, restraint on exports, enforced with significant border controls on most export sales, could create a situation in which foreign prices could rise well above domestic levels.

In EPROMs, in contrast, Japan started with a considerably smaller share of aggregate world supply. The importance of Japanese supplies was to shrink further as Japanese suppliers cut back production of all types of EPROMs. Since imports of EPROMs were freely permitted—indeed, encouraged, in order to increase foreign-based semiconductor sales in the Japanese market, as called for by the STA—and foreign suppliers accounted for a large and increasing share of aggregate global supply, unfettered inflows of these products from abroad effectively pushed Japanese prices to the same levels as in markets outside Japan.

The FMV System

From late 1986 through mid-1991 the U.S. Department of Commerce administered a system of company-specific price floors (FMVs) for U.S. imports of Japanese DRAMs and EPROMs. With the announcement of sanctions against third-country dumping applied to Japan in March 1987, pressure was placed on the Japanese government to ensure that the prices of Japanese products sold in other foreign countries met or exceeded the minimum U.S. price. What effect did the FMVs have on the market price of the affected products?

Although the FMVs set for each individual Japanese company by the Commerce Department were never revealed, it is possible to deduce a range within which these levels varied. Japanese companies were required to file public reports every quarter with the Commerce Department which contained "ranged" estimates of various cost concepts, calculated as the true value plus or minus an error of up to 20 percent (this was done to preserve a degree of confidentiality). The economic interpretation of these cost concepts is subject to considerable debate, but the published estimates nonetheless provide unambiguously useful data. Because these cost estimates were the basis for the procedures by which the Commerce Department set FMVs, some insights about FMVs can be culled from these published reports.

FMVs for a given quarter (call this quarter t) were basically set by the Commerce Department according to the projected cost for the previous quarter $(t - 1)$. This in turn was based on actual cost of production in period $t - 2$. Projected cost in quarter $t - 1$ and actual cost in period $t - 2$ were contained in a report filed with the Commerce Department in quarter $t - 1$.[42]

---

42. See Semiconductor Industry Association, *Antidumping Law Reform and the Semiconductor Industry: A Discussion of the Issues* (Cupertino, Calif., February 1990), for what appears to be the only published description of this methodology.

Let us consider the "average FMV" for some type of chip, defined as the unweighted average across companies and minor product variants of the true FMVs set for Japanese imports. If we treat the "error" added on to the confidential, true-cost estimates to produce public, ranged cost estimates as a random error with zero mean, averaging public ranged cost estimates across minor variants on a specific type of product within a company, and across companies, will produce an unbiased estimate of this average FMV. Similarly, adding 25 percent to the maximum public ranged estimate among the population of ranged average costs for all companies will produce a number that must always be greater than or equal to the greatest true FMV among all Japanese companies, and therefore bounds the true FMV from above. Using these calculations, then, we can construct an upper bound on the maximum company-specific FMV and an estimate of the average FMV across all companies.

The picture is slightly complicated by the fact that—particularly in the initial quarters of the operation of the FMV system—the Commerce Department reviewed and made changes to the cost estimates submitted by the companies in their quarterly reports prior to issuing FMVs. After the system had operated for a while, however, and both the Commerce Department and the companies had iterated onto a set of procedures that produced estimates acceptable to the Commerce Department, FMVs generally were set quite close to the constructed cost projections submitted by Japanese companies.[43]

There is some evidence that actual FMVs had settled down around companies' projected cost submissions by mid- to late 1987. Figure 5-20 shows estimates I have constructed of average FMVs across companies for 256K DRAMs (as well as a bound on maximum company-specific FMV) based on public cost submissions to the Commerce Department (of projected cost in period $t - 1$).[44] These are shown along with ranges for actual FMVs for this product reported in the Japanese trade press in 1986–87. As can be seen, after initial large discrepancies between company and Commerce Department calculations of cost, the midpoint of the range for actual FMV—as reported by the press—settles around the

43. Interview with law firm staff responsible for Japanese company submission to the U.S. Commerce Department, January 1990.
44. Because the type and extent of data reported for every company were different, and often varied over time, the details of how these estimates were constructed are not always obvious. These details are discussed fully in appendix 5-B.

Figure 5-20. *Average Foreign Market Value (FMV) across Companies for 256K DRAMs, 1986–91*

Dollars per chip

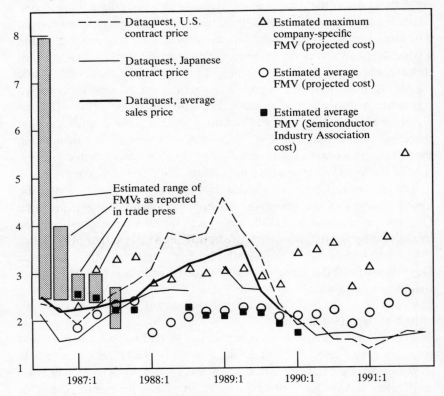

Sources: Author's calculations based on data in table 5B-1; Dataquest, *DQ First Monday Report*; Dataquest average sales price; and Semiconductor Industry Association, *Antidumping Law Reform and the Semiconductor Industry: A Discussion of the Issues* (San Jose, Calif., February 1990), p. 8.

approximate neighborhood of my average FMV around the third quarter of 1987.

Also shown in this diagram are the midpoints of upper and lower bounds on constructed actual cost reported in these quarterly submissions, as compiled by the Semiconductor Industry Association, the U.S. industry's trade association. Actual cost in quarter $t - 2$ may be considered an alternative estimate of FMV in period t.[45] Although the meth-

45. The SIA is not always consistent in how it associates these estimates of actual cost with a time period. In Semiconductor Industry Association, *Anitdumping Law Reform*,

odological details of the SIA's work in producing these estimates have not been published, this figure shows that these estimates are generally quite consistent with my estimates of FMV in period t based on projected cost in period $t - 1$.

This figure shows that by late 1987, and continuing through late 1989, U.S. contract prices for 256K DRAMs had risen substantially above FMVs, and therefore that the FMVs were not constraining U.S. DRAM import prices over this period. From 1990 on, however, at least some Japanese DRAM imports appear to have been priced out of the U.S. market, as 256K DRAM prices fell below average FMVs. Indeed, Japanese 256K DRAM production fell quite sharply from 1990 on. Despite rapid and deep cuts in Japanese production, U.S. prices dropped below the FMV levels.

Figure 5-21 shows a similar pattern in 1M DRAMs. From late 1987 through early 1990, U.S. contract prices stayed above average (and, over most of this period, even the highest-cost Japanese company's individual) FMV. Unlike the case in 256K DRAMs, however, U.S. prices roughly track average FMV over the remainder of the STA's lifetime, and Japanese companies cut neither production nor exports of 1M DRAMs to the degree visible in 256K parts. Cuts in Japanese output were apparently successful in boosting prices to levels at or above the Commerce Department FMVs.

An essentially similar story applies to EPROMs. Figure 5-22 compares Dataquest ASPs with estimated average and upper-bound FMVs for 512K and 1M EPROMs calculated from SIA estimates of upper and lower bounds on FMVs (the mean of the upper and lower bounds is used here as an estimate of average FMV). From late 1987 through mid-1989, market prices exceeded the FMVs, and therefore the FMVs could not have actively constrained pricing of these EPROMs.

Because of this fact, some have argued that the STA itself was not the cause of "abnormally" high prices outside Japan.[46] What this argument ignores, of course, is that production cuts and restraint in investment were clearly associated with the implementation of the STA prior to 1988,

---

actual cost estimates are given for the quarter in which the report was submitted (although the actual cost pertained to the previous quarter), whereas in SIA, *Four Years of Experience under the U.S.-Japan Semiconductor Agreement: "A Deal Is a Deal"* (San Jose, Calif., November 1990), the actual cost estimates are assigned to the quarter in which the cost occurred (that is, the quarter prior to the submission quarter).

46. See SIA, *Four Years of Experience*, p. 65.

Figure 5-21. *Average Foreign Market Value (FMV) across Companies for 1M DRAMs, 1986–91*

Dollars per chip

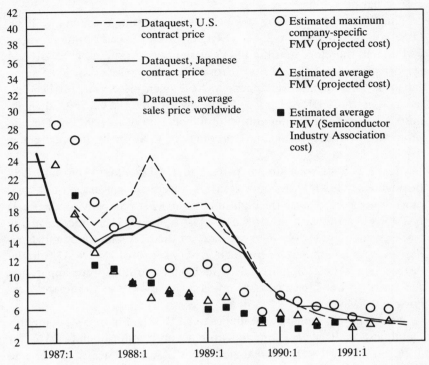

Sources: See figure 5-20; and Semiconductor Industry Association, *Four Years of Experience under the U.S.-Japan Semiconductor Agreement: "A Deal Is a Deal"* (San Jose, Calif., November 1990).

and therefore played some role in the increase in price by reducing aggregate world supply of these chips. It is true, however, that the brakes applied to Japanese supply apparently greatly exceeded the restraint required to raise foreign prices above the Commerce Department FMVs through 1989. The extent to which this reflected mere miscalculation (in particular, underestimation of the recovery in chip demand in 1988, or below-average growth in yields in semiconductor production in 1987 and 1988) rather than a deliberate decision to exploit the new framework, organizing a group of producers accounting for most of global production

Figure 5-22. *Prices and Estimated Foreign Market Value (FMV) for 512K and 1M EPROMs, 1987–90*

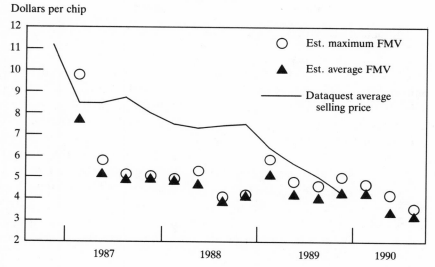

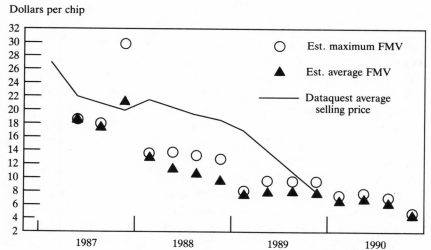

Sources: Dataquest average sales price; and SIA, *Four Years of Experience.*

to seek greater monopoly rents in foreign markets, is probably destined to remain a controversial subject.

Impacts of Regional Price Differentials on Regional Welfare

No matter what the origin of the restraints on production, it is clear that measures related to implementation of the STA had a significant impact in reducing aggregate supply, particularly during and after 1987. Direct restraints on production were in effect through late 1987. Japanese press reports suggest that "guidance" supplied to producers in restraining investment levels continued through at least early 1988, and therefore probably affected supply through at least 1989. Quantifying the impact of these measures on supply, however, would be an exceedingly difficult, if not impossible task, because counterfactual estimates of supply of different types of chips in the absence of these measures would require extensive modeling using data that are not available, in which subtle impacts on the shift of production capacity among types of chips would also need to be included.

The effects of one subset of these measures—the border controls, which created significant regional price differentials—can, however, be assessed in a rough sort of way. For purposes of this analysis I shall assume that Japanese and foreign production of DRAMs can be taken as essentially fixed over the 1987–88 period. More specifically, for foreign producers, output is assumed to be limited by available capacity over 1987–88. For Japanese producers I assume that production controls (in 1987) and capacity (as demand surged in 1988) constrained aggregate output, and that the regional allocation of that output between the Japanese and foreign markets was set administratively. The question I will ask is what the impact on Japanese and foreign welfare would have been had regional price differentials not been created, and had prices for DRAMs been equalized in global markets.

This approach assumes away any effect of regional price differentials in increasing non-Japanese DRAM output over this period. That is, I shall assume that output of other producers in 1987–88 would not have been significantly lower had world DRAM prices been equalized across regions. Given that non-Japanese producers appear to have been close to capacity over this period, and given the long gestation period associated with new investments in semiconductor production capacity (a year or more), this is probably not a terrible approximation. However, if such effects were significant, foreign output would have been lower, and prices higher, had chip prices then been equalized internationally by the removal of Japanese border controls.

Table 5-4 shows my calculations of how welfare inside and outside Japan would have been affected if production and investment restraint had continued, but border controls had been removed. For purposes of this calculation, the world is divided into two regions: Japan and the rest of the world. Unpublished quarterly MITI data on production and export sales by Japanese producers (that is, excluding foreign affiliates producing in Japan) in 1987 and 1988 have been combined with Dataquest estimates of chip production by non-Japanese producers, to produce estimates of production by both Japanese and non-Japanese companies, consumption in Japan and the rest of the world, and the share of Japanese producers in rest-of-the-world DRAM sales.[47]

In making these calculations I have assumed a demand curve with a constant price elasticity of -1.5 (chapter 6 presents evidence on the price elasticity of DRAM demand supporting this assumption). Aggregate DRAM revenues in each of the two regions were estimated by taking quarterly regional consumption, by type of chip (64K, 256K, and 1M), and multiplying by the contract price using the same data used above to construct my regional Fisher Ideal price indexes. Regional DRAM revenues were then divided by the Fisher Ideal price indexes to estimate quantities (measured in dollars of fourth-quarter 1986 DRAM output) consumed. The effect of shifts in aggregate demand unrelated to price is derived on a quarterly basis, as a residual, by combining observed price changes with the assumed price elasticity of demand.[48]

---

47. For 64K DRAMs data excluding foreign affiliates producing in Japan were unavailable. In this case I use published MITI data on shipments and exports, which include the output of U.S. affiliates in Japan, but for this product Japanese production by IBM and TI is believed to have been insignificant. The unpublished MITI data were provided to U.S. government officials and obtained by me. In producing the estimate it is assumed that Japanese producers supply 100 percent of Japanese chip consumption.

48. The methods used to construct table 5-4 are briefly summarized as follows: start with constructed time series on DRAM prices in Japan and the rest of the world (ROW), P_j and P_{row} (the price indexes shown in figure 5-13, with the ROW price the same the as U.S. price), total DRAM production by Japanese producers, S_j, and DRAM sales revenues in Japan (Japanese companies' DRAM production less exports times Japanese DRAM price, summed over all chip types) and ROW (ROW companies' DRAM production plus Japanese companies' exports times ROW DRAM price, summed over all chip types). Dividing regional sales by the regional price indexes produces quantity indexes Q_j and Q_{row}.

Table 5-4. *Estimated Welfare Effects of Removing Japanese Border Controls on DRAMs, 1987–88*

Millions of dollars unless otherwise specified

	1987:1	1987:2	1987:3	1987:4	1988:1	1988:2	1988:3	1988:4
Price elasticity = −1.5								
Price before removal of controls (1986:4 = 100)								
Japan	79.5	79.4	83.2	86.8	95.7	97.7	97.0	97.9
Rest of world	84.9	98.5	104.0	113.4	124.2	153.0	138.4	130.7
Price after removal of controls (1986:4 = 100)	83.3	92.8	97.5	105.7	116.0	137.2	126.7	121.2
Japan								
Change in company revenues	2.4	10.8	14.7	20.6	24.8	67.4	57.5	57.7
Change in domestic sales	52.6	72.0	90.6	176.5	251.7	396.0	393.0	374.3
Change in export sales	−50.2	−61.2	−76.0	−155.8	−226.9	−328.6	−335.5	−316.6
Change in consumer welfare	−5.7	−21.3	−30.0	−43.7	−55.7	−122.4	−110.4	−110.7
Total impact	−3.3	−10.5	−15.3	−23.1	−30.9	−55.0	−52.9	−53.0
As percentage of regional sales without border controls	−1.9	−4.9	−5.3	−5.6	−5.6	−7.0	−6.3	−5.7
Rest of world								
Change in company revenues	−2.3	−9.6	−12.9	−17.4	−20.8	−51.4	−46.4	−48.9
Change in consumer welfare	5.9	23.8	33.5	50.1	63.7	154.5	132.5	128.3
Total impact	3.6	14.2	20.7	32.7	42.8	103.2	86.1	79.3
As percentage of sales without border controls	1.5	4.4	4.9	6.3	6.5	10.4	8.0	6.1

Price elasticity = −2

Price before removal of controls (1986:4 = 100)								
Japan	79.5	79.4	83.2	86.8	95.7	97.7	97.0	97.9
Rest of world	84.9	98.5	104.0	113.4	124.2	153.0	138.4	130.7
Price after removal of controls (1986:4 = 100)	83.3	93.0	97.7	106.0	116.3	138.3	127.3	121.6
Japan								
Change in company revenues	2.4	11.7	15.9	22.9	27.7	78.6	65.2	63.8
Change in domestic sales	52.6	72.5	91.3	177.8	253.5	402.6	397.6	377.8
Change in export sales	−50.2	−60.9	−75.4	−154.9	−225.8	−324.0	−332.4	−314.0
Change in consumer welfare	−5.6	−20.8	−29.2	−42.3	−54.0	−115.2	−105.3	−106.8
Total impact	−3.2	−9.1	−13.3	−19.4	−26.3	−36.6	−40.2	−43.0
As percentage of regional sales without border controls	−1.8	−4.3	−4.6	−4.7	−4.7	−4.6	−4.8	−4.6
Rest of world								
Change in company revenues	−2.3	−9.2	−12.4	−16.6	−19.9	−47.6	−43.7	−46.6
Change in consumer welfare	5.8	23.2	32.8	48.6	61.7	146.1	126.9	124.0
Total impact	3.6	14.0	20.4	31.9	41.8	98.5	83.2	77.4
As percentage of regional sales without border controls	1.5	4.3	4.8	6.1	6.3	9.8	7.7	5.9

Sources: Author's calculations based on sources and price indexes produced in figure 5-13; unpublished Dataquest data on DRAM production; unpublished MITI data on shipments and exports by Japanese producers.

Removal of border controls was then simulated by calculating the quantities of aggregate DRAM consumption required in each region to equalize DRAM prices in Japan and the rest of the world, taking world DRAM output as fixed. Prices would rise in Japan and fall in the rest of the world. Japanese consumers clearly lose by paying higher prices, while Japanese producers gain from increased global profits (despite increased shipments, revenues fall in export markets; these are more than offset by increased revenues on smaller quantities sold in the Japanese market). But when the loss in Japanese consumer welfare is added to the gain in producer revenues (and therefore profits, since total output is constant), Japan as a whole loses from price equalization, by an amount generally ranging from about 5 to 7 percent of Japanese DRAM consumption with a single world price. Thus, the border controls as a whole worked to increase Japanese welfare by lowering the price to Japanese chip consumers at the expense of Japanese producers (who would have received

---

Assume demand for DRAMs in Japan and ROW is given by a constant price elasticity demand function with elasticity -1.5, $Q_j = Z_j P_j^{-1.5}$, where Z_j is a Japan-specific constant that varies over time, reflecting shifts in aggregate demand within Japan. A similar equation describes demand in the rest of the world. Z_j is calculated in every quarter by dividing quantity Q_j by $P_j^{-1.5}$ (and Z_{row} analogously). Given the two Q's and the two P's (corresponding to border controls and different prices inside and outside Japan), the effect of the removal of border controls given the same total production is to produce a single new world price P' and new quantities Q_j' and Q_{row}' consumed in Japan and the rest of the world, which must solve the system of three equations in three unknowns:

$$Q_j' = Z_j P'^{-1.5}; \quad Q_{row}' = Z_{row} P'^{-1.5}; \quad Q_j' + Q_{row} = Q_j + Q_{row}.$$

Solving for P', we have

$$P' = \left(\frac{Q_j + Q_{row}}{Z_j + Z_{row}}\right)^{\frac{-1}{1.5}}.$$

After border controls are lifted, Japanese companies' revenues change by

$$(P' Q_j' - P_j Q_j) + [(S_j - Q_j')P' - (S_j - Q_j) P_{row}],$$

where the first term is the change in domestic revenues, and the second term (in brackets) is the change in export revenues. For ROW producers, change in sales revenues from their production is given as

$$(P' - P_{row})(Q_j + Q_{row} - S_j).$$

With the constant elasticity demand function, the effect on consumer welfare of a change in price from P_0 to P' equals

$$\int_{P_0}^{P'} Zp^{-1.5}dp = Z\left(\frac{2}{\sqrt{P_0}} - \frac{2}{\sqrt{P'}}\right).$$

greater global revenues by diverting chips into higher-priced foreign markets). The benefits to Japanese consumers more than offset profits forgone by producers.

Conversely, the border control system worked to reduce aggregate welfare in the rest of the world. Non-Japanese producers clearly gained, but their gains pale in comparison with the loss inflicted on consumers in the rest of the world market—increased profits for rest-of-the-world producers amount to about one-third of the welfare loss suffered by consumers in the rest of the world market as a consequence of the price differentials so created. The total impact of border controls on the rest of the world's welfare is a loss equivalent to roughly 5 to 10 percent of the value of DRAM consumption with a single world price.

As expected, however, the net gain to Japan is less than the net loss to the rest of the world. These results are relatively robust to changes in the assumed price elasticity of demand. Table 5-4 shows that changing the assumed price elasticity from −1.5 to −2, near the upper bound for empirical estimates of this parameter, changes the outcome of the simulation only marginally.

The relatively small magnitudes of the welfare changes induced by border controls in this exercise provide a dramatic contrast with much larger estimates by others of monopoly rents associated with STA-induced price movements. For 1M DRAMs alone, market analysts' calculations suggested that the value of sales exceeded costs in 1988 by anywhere from $1.2 billion (for Japanese companies only) to close to $2 billion (all suppliers).[49] It is probable that 1M parts alone accounted for one-third of Toshiba's operating profit in 1988, one-fifth for Oki Electric, 17 percent for Mitsubishi Electric, 15 percent for NEC, and 13 percent for Fujitsu (Nomura Research estimates). By 1989 the concept of "bubble money"—supernormal profits due to abnormal scarcities of product—was reportedly being used in Japanese industry circles to describe the profits being made on DRAMs. Estimates of "bubble money" being collected in DRAMs by early 1989 hovered around $3 billion to $4 billion per year.[50]

49. The figure for Japanese companies is calculated by converting a fiscal 1988 operating profit of 150 billion yen to dollars at 130 yen to the dollar. See Barclays de Zoete Wedd Research, *Japan-Electronics, Semiconductors* (Tokyo, December 1988), p. 17. The figure for all suppliers is according to the consulting firm In-Stat. See Richard McCausland, "Semiconductor Makers Concerned Price Cuts Could Hamper Growth," *Electronic News*, January 2, 1989, p. 22.

50. This value is roughly consistent with a report circulating in Japanese industry circles

These magnitudes are considerably larger—by a factor of roughly ten—than the increased profits for Japanese producers indicated in table 5-4. If accurate, these estimates would suggest that border controls were not the most costly aspect of STA-related policies for U.S. and European DRAM consumers, but rather that the underlying restraints on supply—production and investment—were the primary underlying cause of the financial distress experienced by chip buyers in 1988–89.

DRAMs versus EPROMs

Some have argued that the considerably greater market share of Japanese producers in DRAMs than in EPROMs explains the rather different behavior of these two products over the life of the STA.[51] One interesting question is why Japanese production of EPROMs plunged so precipitously, and imports into Japan surged, in dramatic contrast to the situation in DRAMs. The simplest possible answer is that foreign suppliers were more competitive, lower cost producers of what was clearly a commodity product, and that Japanese companies, unable to compete, exited the business. Since by most accounts, however, Japanese companies had a cost advantage in the manufacture of commodity semiconductors, this reasoning faces some obstacles.[52] Another possibility is that EPROMs were less profitable to manufacture than other products such as DRAMs. But when DRAM prices and production declined in 1989, EPROM manufacture continued to drop and imports to increase. Both of the above explanations are also undermined by the fact that, as the Japanese EPROM market tightened in 1987 and foreign manufacturers began to run into capacity constraints, the U.S. chip maker Intel contracted with the Japanese chip maker Mitsubishi to produce EPROMs

---

in the spring of 1989, according to which MITI had calculated that "bubble money" was running around 45 billion yen per month at that time.

51. Tyson argues that "In EPROMs, a product in which American companies still retained a significant share of the world market at the time the [STA] was signed, the outcome was very different. . . . American and other buyers did not report shortages in EPROM supplies, and American EPROM suppliers such as Intel asserted that EPROM prices never significantly rose above their marginal costs because of competitive market conditions." Laura D'Andrea Tyson, *Who's Bashing Whom? Trade Conflict in High-Technology Industries* (Washington: Institute for International Economics, 1992), p. 121.

52. Indeed, cost disadvantages faced by U.S. semiconductor manufacturers were one of the reasons why the Sematech manufacturing consortium was formed to improve the productivity and competitiveness of American chip producers.

sold in the Japanese market under the Intel brand name.[53] Mitsubishi, at least, could apparently manufacture and sell EPROMs at prevailing prices at a profit, and found it more attractive to do so than to increase production of DRAMs, which it also manufactured.

A third possible explanation is that the levels set for FMVs prevented Japanese producers from selling their output in foreign markets. But, as we have seen, these price floors were not binding in the U.S. market through 1988 and were never an obstacle to selling product in the Japanese market. Nevertheless, Japanese producers steadily cut back their output of EPROMs, and Japanese consumers increased their use of imported product in their stead.

This observation and the details of the Intel-Mitsubishi accord suggest an alternative explanation. Japanese producers as a group may have cut back production as a deliberate policy aimed at increasing imports and foreign market share, and assuaging friction with foreign chip makers, who had continued to market and sell EPROMs even after they dropped out of DRAMs. Although it is impossible to determine just how important political factors were in explaining all the unusual market developments observed after 1985, the review in chapter 4 of the details of the STA makes it clear that they were present and played some role. While EPROMs' role in the memory market declined precipitously in the 1990s (replaced by so-called flash memory chips, a superior electronically reprogrammable form of nonvolatile memory), EPROMs provided incumbent producers with a strong technology base for the new products.[54]

Import Promotion and the STA

Perhaps the most disputed element of the STA was the "secret" side letter specifying 20 percent as a reasonable expectation for foreign market share by 1991. After prices rose above FMVs in 1987, the focus for negotiations between the United States and Japan increasingly concerned the interpretation of this letter and the measurement and assessment of progress toward greater foreign participation in the Japanese semiconductor market.

53. Andrew Pollack, "Intel to Resell Japanese Chips," *New York Times*, July 30, 1987, p. D4.
54. MITI dropped EPROMs from its semiannual forecast in 1994.

As a blueprint for a voluntary import promotion measure, the details of the 1986 side letter left much to be desired. Even if the two sides could have agreed on whether or not the 20 percent "expectation" constituted a Japanese commitment, construction of a statistic measuring foreign market share in Japan soon became embroiled in yet another set of disputes. Details of a market share measure had not been specified in the STA or its side letters, and each side quickly adopted a definition furthering its own purposes. The U.S. government advocated the use of sales figures compiled by the World Semiconductor Trade Statistics (WSTS) program, a private data collection organization set up by chip producers, as the only accurate and internationally accepted figures on semiconductor sales. MITI, however, chose to construct its own ratio, measuring Japanese market size through a survey of sixty-three large Japanese chip consumers (including IBM Japan), and adding internal captive shipments by IBM to its Japanese subsidiary into its measure of foreign sales in Japan. The WSTS statistical system did not count IBM's output because IBM sold no product on the open market, and because IBM's products did not conform to industry standard designs and packaging. Because the MITI measure of market size in the denominator excluded a substantial number of Japanese consumers not included in the surveyed group of sixty-three large users, and because the measure of sales by foreign companies in the numerator included IBM's large internal shipments, it is not surprising that MITI's ratio was always considerably larger than the WSTS number cited by the American side.[55]

When the STA was renegotiated in 1991, the target of a 20 percent foreign market share had not been reached, by either the MITI or the WSTS calculation. The new arrangement raised market access issues to a new level of prominence and included within its main text the statement that "the Government of Japan recognizes that the U.S. semiconductor industry expects that the foreign market share will grow to more than 20 percent of the Japanese market by the end of 1992 and considers that this can be realized. The Government of Japan welcomes the realization of this expectation. The two governments agree that the above statements

---

55. See *Electronics Industry Almanac, 1992* (Tokyo: Dempa Publications, 1992), p. 909 (in Japanese). However, the WSTS statistics also exclude shipments by some small foreign producers that do not belong to the WSTS, as well as IBM, so that in theory (although almost certainly not in practice) it would be possible for the MITI number to exceed the WSTS-based measure of market share.

constitute neither a guarantee, a ceiling nor a floor on the foreign market share."[56]

How this new, formally established market share "expectation" was to be measured became an official issue. An appendix spelled out the details of a new statistical system and set out the principles for an official data collection program (DCP) that would define two distinct measures of foreign market share. The first formula (F1) basically took sales of semiconductors by DCP merchant market participants, added estimates of nonparticipant foreign sales, and divided the sum by an estimate of apparent Japanese consumption, based on official Japanese government production and trade statistics, less captive consumption (IBM Japan, for example). A second formula (F2) added in captive consumption to both numerator and denominator. The two formulas also differed in their definition of what constituted a "foreign" product: F1 assigned country origin based on the company doing "final assembly," except where "pure" assembly was done under contract; F2 assigned it by the nationality of the producer whose brand was stamped on the chip.[57]

Although the agreement itself specified neither formula as superior, predictably the governments of Japan and the United States quickly made it clear that each had a preferred choice. A press release from the Office of the U.S. Trade Representative set out its position that F1 "accurately describes changes in foreign market share," while the second formula was "not appropriate in measuring market access."[58] MITI took the opposite view, maintaining that the inclusion of both formulas in the pact made both equally useful measures of market share.

Figure 5-23 shows movements in all of these measures of foreign market share over time. Note that neither F1 nor F2 was officially calculated prior to the third quarter of 1991; the figures for F1 for earlier periods are SIA estimates. F1 appears generally to be slightly larger than

56. "Arrangement between the Government of Japan and the Government of the United States of America Concerning Trade in Semiconductor Products," June 11, 1991, sec. 2, para. 10.

57. "Arrangement," annex A., pp. 9–10. For production arrangements involving joint participation by both Japanese and U.S. firms (joint ventures, production under license, foundry arrangements), these definitions ultimately also involved considerable ambiguity, and the U.S. Department of Commerce ultimately created a detailed handbook defining on a case-by-case basis whether the output of a particular factory with such transnational links was to be considered "foreign."

58. U.S. Trade Representative, "Questions and Answers on the U.S.-Japan Semiconductor Arrangement," June 4, 1991.

Figure 5-23. *Alternative Measures of the Foreign Share of the Japanese Semiconductor Market, 1986–95*

Percent

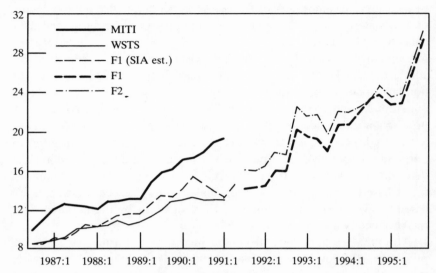

Sources: SIA, *Four Years of Experience*, app. 4; *Electronics Industry Almanac, 1992* (Tokyo: Dempa Publications, 1992), p. 910; "Trend in Foreign Market Share in Japan under the 1991 Semiconductor Arrangement," graph attached to Electronics Industry Association of Japan, "UCOM Chairman Comments on Foreign Semiconductor Market Share in Third Quarter of 1995," press release, December 14, 1995; and "Foreign Share of Japanese Chip Market Soars," *Computing Japan* (on line), March 21, 1996.

its intellectual precursor, the WSTS-based market share estimate. F2 is considerably smaller than its predecessor, MITI's survey-based market share estimate, but always substantially larger than F1, as might be expected.

Both F1 and F2 remained well below the 20 percent target through most of 1992, and bilateral trade talks in semiconductors seemed headed toward some tense moments as the end-1992 deadline approached. Since IBM had announced its entry into Japan's merchant semiconductor market in the summer of 1992,[59] MITI took the tack of pressing U.S. negotiators to include IBM's shipments to Japan in F1, in order to support its case that the goals of the 1991 arrangement had been met. Quite to

59. David Roman, "IBM Launches OEM Chip Push in Japan," *Electronic Buyers' News*, August 3, 1992, p. 3; Junko Yoshida, "IBM Japan's Efforts to Sell Chips Stalled: Distribs Will Sell Its 'High-Function' ICs," *Electronic Engineering Times*, August 10, 1992, p. 94.

everyone's surprise, however, F1 (as well as F2) jumped sharply up at the very end of 1992, just exceeding the 20 percent level. Even though F1 had dipped slightly below the magic 20 percent mark the following quarter, in early 1993, tensions were substantially relaxed by this development.[60]

Interestingly, the piercing of the 20 percent barrier for foreign market share in Japan in late 1992 came just as U.S. firms came roaring back in global chip markets. Extremely strong computer sales had buoyed markets for the complex, proprietary microprocessors in which American firms had a virtual monopoly, while markets for "commodity" memory chips, where Japanese producers were strongest, seemed mired in low margins and prices, amid a continuing glut of new capacity. For the first time since 1985, U.S. firms seemed to pass Japanese companies in sales.[61] Was U.S. trade policy finally achieving results, or was it, ironically, the fact that U.S. firms had been driven out of commodity memory, and forced by competitive forces into concentrating on the complex logic chips that were their true forte, that was responsible for both the global comeback and the Japanese success story? Put another way, was it the political pressure put on Japanese firms to buy American, or was it the shift in the market toward complex logic chips that accounted for the apparent success? At least some analysts argued that it was market forces, notably the shift in demand toward niches where technologically innovative American chip producers were strongest, and not bureaucratic pressure on Japanese chip consumers, that should be given the bulk of the credit.[62]

On the other hand, our discussion of EPROM market shares earlier in this chapter made it clear that there is at least some circumstantial evidence of a deliberate decision by Japanese producers to yield EPROM markets in Japan to foreign producers. Other prominent examples of substantial increases in U.S. market share in the face of formidable Japanese competitive capabilities may be found. A good example is the

60. The U.S. Trade Representative's June 1993 projections had suggested that F1 would drop to about 19.6 for the first quarter of 1993. David Lammers, "U.S. Uneasy about Japan Chip-Share Dip," *Electronic Engineering Times*, June 21, 1993, p. 4.

61. Bob Davis, "U.S. Chip Firms Seem to Lead Japanese Ones in Sales This Year, but Experts Aren't Sure Why," *Wall Street Journal*, December 24, 1992, p. 30; T. R. Reid, "U.S. Again Leads in Computer Chips," *Washington Post*, November 20, 1992, pp. A1, A42; and Ken Yamada, "Intel Is Ranked No. 1 Chip Maker for 1992 by Study," *Wall Street Journal*, January 5, 1993, p. B6.

62. William R. Cline, "Yen Appreciation Does Reduce Japan's Surplus," *Wall Street Journal*, May 20, 1993, p. A16.

Japanese market for application-specific integrated circuits (ASICs). The strongest U.S.-based producer in that market, Nihon LSI Logic, the Japanese subsidiary of LSI Logic, went from a 0.8 percent share of the Japanese market in 1988 to a 10.6 percent share in 1991.[63] Nihon LSI's management, when asked for a qualitative judgement, credited about half of its growth in the Japanese market to the effects of the STA's market access targets.[64] Other American semiconductor producers concur in their assessment of the important role of MITI in encouraging greater consumption of foreign semiconductors.[65] Finally, the frequency with which Japanese producers announced increases in the share of foreign products in their total chip procurement, and increasing "design-ins" of foreign product after reports of pressure from MITI, suggest at least some element of government intervention at work.

Available data shed some light on this question. If the aggregate share of foreign-based chips in the Japanese market at time $t1$ is S_{t1}, then

$$S_{t1} = \sum_i s_{it1} \, m_{it1},$$

where s_{it1} is the foreign share of product i at time $t1$, and m_{it1} is the share of product i in aggregate Japanese semiconductor sales at time $t1$. Then the change in foreign share from time $t1$ to time $t2$, $S_{t2} - S_{t1}$, can be decomposed into a portion due to increased foreign penetration of particular market segments, and a portion due to shifts in the composition of market demand. That is,

$$S_{t2} - S_{t1} = \sum_i s_{it2}m_{it2} - s_{it1}m_{it1}$$

$$= \sum_i m_{it1}(s_{it2} - s_{it1}) + \sum_i s_{it2}(m_{it2} - m_{it1}).$$

Or, in English, change in aggregate foreign market share from time $t1$ to time $t2$ $(S_{t2} - S_{t1})$ is the sum of changes in foreign penetration in market segment i $(s_{it2} - s_{it1})$ weighted by the share of segment i in total sales at time $t1$ (m_{it1}) plus the sum of changes in the share of market segment i in the total market $(m_{it2} - m_{it1})$ weighted by foreign penetration in that market segment at time $t2$ (s_{it2}).

63. Nihon LSI Logic's manufacturing facilities were a 50 percent joint venture with Kawasaki Steel. The market shares are calculated from data available on the *Nikkei Telecom* online computer database.
64. Author's interview with K. K. Yawata, Nihon LSI Logic, Tokyo, March 1991.
65. See Jacob M. Schlesinger, "U.S. Chip Makers Find 'Quotas' Help Them Crack Japan's Market," *Wall Street Journal,* December 20, 1990, pp. A1, A6.

The first terms summed after the second equality are the changes in market share due to increased foreign penetration, while the second set of terms are changes in market share due to changes in the overall composition of demand. So, if there are no changes in market composition ($m_{it2} - m_{it1} = 0$ for all i), all change in foreign market share is attributed to increased foreign penetration; conversely, if foreign penetration is unchanged in all market segments ($s_{it2} - s_{it1} = 0$ for all i), all change in foreign market share is attributed to changes in the aggregate composition of demand.

This is not the only possible decomposition of share change into a penetration component and a market composition component. The decomposition just described can be referred to as the "initial market composition" variant, since changes in foreign penetration within segments were weighted by the initial share of that segment in the aggregate market, while changes in market composition were weighted by final foreign market share in each segment. An alternative decomposition, to be referred to as the "initial foreign share" variant, can be given as

$$S_{t2} - S_{t1} = \sum_i m_{it2}(s_{it2} - s_{it1}) + \sum_i s_{it1}(m_{it2} - m_{it1}),$$

where the initial foreign share of each segment has been used to weight changes in market composition, and final market segment shares in total sales are the weights on changes in foreign share by segment. Still a third decomposition may be constructed that takes the mean of the previous two sets of weights on changes in market composition and foreign penetration, that is,

$$S_{t2} - S_{t1} = \sum_i \left(\frac{m_{it1} + m_{it2}}{2}\right)(s_{it2} - s_{it1})$$
$$+ \sum_i \left(\frac{s_{it1} + s_{it2}}{2}\right)(m_{it2} - m_{it1}).$$

This latter variant on the decomposition of changes in foreign market share has the virtue of using symmetric weights for both foreign market share and market composition changes.[66]

66. None of these decompositions is invariant to the level of disaggregation used. More or less disaggregated product categories will give somewhat different decompositions of the contribution of foreign penetration by segment, and market composition, to aggregate foreign market share change.

Table 5-5. *Foreign Share of the Japanese Semiconductor Market Due to Changes in Foreign Penetration and Composition, Selected Periods, 1986–94*[a]

Percent

Type of semiconductor	Period and type of device	Initial foreign penetration weights		Initial product composition weights		Mean of weights		Sum of effects	Shares
		Foreign penetration	Product composition	Foreign penetration	Product composition	Foreign penetration	Product composition		
U.S. companies 1986–90	*1986 U.S. share*								*1990 U.S. share*
Bipolar digital	24.5	-0.20	-1.10	-0.31	-0.99	-0.25	-1.05	-1.30	22.0
MOS memory	3.9	1.45	0.23	1.06	0.62	1.26	0.42	1.68	10.5
MOS micro	n.a.	n.a.	n.a.	n.a.	n.a.	n.a.	n.a.	n.a.	19.7
MOS logic	n.a.	n.a.	n.a.	n.a.	n.a.	n.a.	n.a.	n.a.	12.8
Total micro and logic	7.9	2.71	0.48	2.22	0.98	2.47	0.73	3.20	16.0
Analog	10.3	0.68	-0.46	0.89	-0.57	0.79	-0.56	0.22	14.8
Discrete	2.2	0.17	-0.07	0.20	-0.09	0.18	-0.08	0.10	3.0
Total	8.4	4.83	-0.92	4.06	-0.15	4.44	-0.54	3.90	12.3
1990–94	*1990 U.S. share*								*1994 U.S. share*
Bipolar digital	22.0	0.06	-0.92	0.14	-0.99	0.10	-0.95	-0.86	23.7
MOS memory	10.5	1.32	0.67	1.02	0.96	1.17	0.81	1.99	15.1
MOS micro	19.7	1.81	0.20	1.69	0.31	1.75	0.26	2.01	30.6
MOS logic	12.8	0.77	0.22	0.71	0.29	0.74	0.26	1.00	16.7
Total micro + logic	16.0	2.56	0.44	2.36	0.64	2.45	0.54	3.00	23.1
Analog	14.8	0.88	-0.26	1.00	-0.37	0.94	-0.31	0.63	21.3
Discrete	3.0	0.12	-0.09	0.14	-0.11	0.13	-0.10	0.02	3.6
Total	12.3	4.96	-0.18	4.69	0.09	4.83	-0.04	4.79	17.1

Non-Japanese companies

1990–94	1990 foreign share							1994 foreign share
Gate array, full custom, std cell	8.4	1.07	0.20	0.87	0.40	0.97	0.30	1.27	17.0
Microprocessor	72.2	0.38	0.81	0.27	0.92	0.33	0.86	1.19	82.2
Mask ROM	3.5	1.14	0.05	0.91	0.27	1.03	0.16	1.19	20.7
Industrial IC	22.0	0.53	0.41	0.38	0.56	0.45	0.48	0.94	30.3
Std logic, app specific std prod	16.7	0.58	0.08	0.56	0.10	0.57	0.09	0.66	21.9
Microperipherals	19.8	0.76	−0.23	1.10	−0.57	0.93	−0.40	0.53	50.1
DRAM	22.1	−0.11	0.58	−0.08	0.56	−0.09	0.57	0.48	21.1
EEPROM + flash	49.0	0.12	0.25	0.06	0.31	0.09	0.28	0.37	59.9
SRAM (>35ns)	11.9	0.52	−0.20	0.85	−0.53	0.69	−0.36	0.32	32.5
Analog/digital, D/A	31.2	0.10	0.16	0.08	0.18	0.09	0.17	0.26	35.6
Microcontroller (>4-bit)	12.9	−0.16	0.37	−0.11	0.32	−0.13	0.35	0.22	11.1
Prog logic, field prog gate array	92.2	−0.00	0.21	−0.00	0.21	−0.00	0.21	0.21	91.9
Other discrete	4.6	0.19	−0.02	0.21	−0.04	0.20	−0.30	0.17	8.6
SRAM (<35ns)	12.1	0.15	−0.02	0.18	−0.05	0.17	−0.04	0.13	30.4
OP AMP	28.6	0.16	−0.05	0.18	−0.06	0.17	−0.05	0.12	38.0
Microcontroller (4 bit)	2.5	0.13	−0.02	0.17	−0.06	0.15	−0.04	0.11	6.3
Transistors (<1W)	3.5	0.07	−0.00	0.07	−0.00	0.07	−0.00	0.07	5.3
Optoelectronics	3.5	0.07	−0.01	0.08	−0.02	0.08	−0.02	0.06	6.1
Consumer IC	9.1	0.28	−0.23	0.42	−0.36	0.35	−0.30	0.05	14.7
Transistors (>1W)	4.2	0.05	−0.02	0.06	−0.02	0.05	−0.02	0.03	6.3
Charge coupled devices	4.7	−0.02	−0.06	−0.03	−0.05	−0.02	−0.05	−0.07	3.9
EPROM	39.8	0.17	−0.27	0.26	−0.35	0.22	−0.31	−0.09	52.9
Other bipolar	13.0	0.12	−0.42	0.33	−0.62	0.23	−0.52	−0.29	19.5
Transistor-transistor logic	65.7	0.09	−0.64	0.17	−0.72	0.13	−0.68	−0.55	73.6
Total	15.7	6.41	0.95	7.00	0.36	6.71	0.65	7.36	23.1

Sources: Unpublished data from the Semiconductor Industry Association; MITI-Electronics Industry Association of Japan (EIAJ), "Present Condition of the Japanese Market," white paper, 1992; EIAJ, "Facts about the U.S.-Japan Semiconductor Arrangement," 1994, pp. 2-5, 2-6, 2-7, 2-17, 2-18, 3-7; and EIAJ, Semiconductor Facts, 1995, p. 3-7.

n.a. Not available.

Table 5-5 breaks out the contribution of increases in foreign market share by segment, and of changes in the share of different product segments in aggregate sales, to increased foreign market share. For given weights, adding across rows gives the total change in aggregate foreign market share due to changes in both foreign penetration and market composition in that product segment. Adding down columns gives the total contribution of changes in foreign penetration, or changes in market composition, respectively, to the change in aggregate foreign market share. The sum of the "foreign penetration" column may be interpreted as the increased share of foreign companies due solely to increased foreign market share in each product segment, taking the share of each product segment in overall sales as fixed. The sum of the "market composition" column can be interpreted as the increased share of foreign companies due solely to shifts in the share of each product in aggregate sales, taking the foreign market share in each product segment as fixed.

An immediate implication of this table is that shifts in the composition of demand in the Japanese market play only a very small role in explaining the overall increase in foreign market share over this period.

The calculations reported in the first part of this table, which analyzes SIA data on *U.S. company* sales in the Japanese market, are similar conceptually to the WSTS estimate of foreign market share, used over the five-year life of the first STA, described earlier. However, detailed product segment data are only available for U.S. firms participating in the WSTS program, and the share of these firms in the Japanese market was lower than the aggregate share of all foreign firms. (Also, the criterion for judging whether a chip produced through a joint arrangement between a Japanese and a foreign firm is Japanese or foreign is different from that used for the "official" market share statistics.)

The vast bulk of the aggregate change in foreign market share is attributable to increased foreign penetration of markets over either the 1986–90 or 1990–94 subperiod, not shifts in the composition of Japanese demand, which overall actually reduced market share. These tables (which, again, include only U.S. producers) show a large change in market share from 1986 to 1990 attributable to microcomponents and logic chips, followed by a lesser gain attributable to memory. This may reflect the fact that the principal non-American suppliers present in the Japanese market (for the most part, from Korea and Europe) mainly sell memory chips of various types. Interestingly, over the 1990–94 period, when sales by Intel, the U.S. microprocessor leader, soared and foreign market share shot up, increases in U.S. market share in Japan due to memory chip sales almost exactly equaled those due to microcomponents.

The second part of table 5-5 breaks out the sources of increased foreign market share in Japan during 1990–94 in much greater detail. This much more disaggregated set of data comes from a survey administered by the Nomura Research Institute (NRI) in collaboration with the Users' Committee of Foreign Semiconductors (UCOM), an organization of the sixty-two major Japanese semiconductor users set up in 1988 to encourage increased access by foreign semiconductor suppliers to the Japanese market.[67] The respondents to the NRI survey account for about two-thirds of Japanese semiconductor consumption.[68] The NRI survey uses a definition of "foreign" that appears to correspond to the F2 variant favored by the Japanese government (that is, a definition based on the label appearing on the finished chip).

The specific product areas accounting for the bulk of the foreign market share gain (each contributing half a percent or more and in the aggregate approaching 7 percent of a 7.4 percent increase in foreign market share) are ASICs (including gate array, full custom, and standard cell logic chip types), microprocessors, Mask ROMs and other memory (nonprogrammable and specialized memory chips), industrial ICs, standard logic and application-specific standard products, and microperipherals. The story told by these data is also one in which increased foreign penetration in specific product areas, rather than compositional shifts in the market, is the force driving increased foreign market share— 6.7 percentage points out of a total increase in foreign market share of 7.4 percent. The greater disaggregation adds more complexity to the picture, however. In both microprocessors and DRAMs, for example,

---

67. UCOM, affilated with the Electronics Industry Association of Japan, has sent trade missions overseas to stimulate chip purchases, held seminars on applications and design opportunities, and established a California liaison office to interface with smaller U.S. suppliers unable to set up Japan-based organizations.

68. Surveyed companies purchased $13.4 billion in semiconductors in 1990 (based on the NRI survey), compared with a total Japanese market of about $19.5 billion (based on WSTS statistics)—or 69 percent of the market. For the second and third quarters of 1994 (the period covered by the NRI survey in 1994), surveyed companies purchased $9.77 billion compared with a total Japanese semiconductor market (WSTS data) of $14.96 billion—or 65 percent of the market.

shifts in overall market demand toward these products were more important than increases in foreign market share in accounting for the net contribution of these products to the overall foreign share.

In fact, the detailed data shown here on the contribution of foreign memory chip sales in Japan reveal a bit of a puzzle and allow an important insight into the operation of the system for tracking foreign market share. Adding up market share changes of the various memory chips on the list of products in the NRI survey gives a net contribution of roughly 2 percent to foreign market share, about the same as the contribution that U.S. firms alone made to foreign market share in this segment, as shown in the analysis of WSTS data in the top part of table 5-5. At first glance, this is perplexing because it was widely reported that there were large increases in Korean DRAM shipments to Japan during this period, so one would have expected the increase in foreign share in memory chips to have been much larger than that corresponding to U.S. companies alone.[69]

A good part of the answer may lie in the NRI survey definitions, which, like the Japanese government's preferred F2 measure, apparently classified as foreign those products with non-Japanese labels. Newspaper reports suggest that with increasing frequency Korean DRAMs were being procured by Japanese companies for sale in Japan.[70] If shipped in assembled form to Japanese companies, then finished, stamped with a Japanese brand name, and sold in Japan, these products would be classified as Japanese, using F2 (and in the NRI survey would not be counted as foreign in table 5-5). They would, however, be counted as foreign by the U.S.-favored F1 formula.

Indeed, there is some good recent evidence that Japanese and Korean vendors were deliberately structuring deals designed to take advantage of the definitional complexities hovering over what was defined to be "foreign" and what "domestic." In March of 1995, NEC and Samsung announced a curious deal in which Samsung would ship Korean-made DRAMs to NEC in Japan, to be sold under NEC's brand name, while NEC's Scottish factory would ship an equal volume of DRAMs to Samsung's European assembly facility, where they would then be stamped

69. By 1994 imports from South Korea and Taiwan had risen to 15 percent of the Japanese memory market. "Memory Chip Imports from ROK, Taiwan Rise," *Nihon Keizai Shimbun*, March 11, 1995, p. 1.

70. See for example Tetsuya Iguchi, "Japanese, Korean Rivals Forging DRAM Alliances," *Nikkei Weekly*, October 18, 1993.

with Samsung's brand and sold in European markets. When investigated by a market analyst, NEC sources were forthright in describing the rationale for the transaction.[71]

The Samsung product sold by NEC in Japan would count toward NEC's purchases of foreign-made semiconductors under F1, scrutinized by the Americans, even though NEC's Japanese customers received the familiar product under the same familiar NEC logo stamped on product shipped to Japan from Scotland (which would not have counted as foreign in the Americans' eyes). From Samsung's perspective, the NEC product shipped by its European facility would be spared the stiff tariffs charged on DRAMs exported to Europe from Korea, and in addition would curry political favor with the European Community because of the "locally made" semiconductor product it had begun shipping. Yet nothing significant had fundamentally changed—NEC was producing the same quantities in its Scottish fab, and Samsung identical volumes on its Korean production lines.

Similar types of deals had been reported publicly in the past, but none so clearly targeting the market share calculations carried out under the STA.[72] This particular deal, to have begun in April 1995, is particularly intriguing because of the sharp jump in both F1 and F2 reported in the third quarter of 1995—more than 3 percent in just one quarter.[73] U.S. industry sources suggested that a significant part of this gain appeared to have been associated with a jump in Korean imports, as shown in the trade statistics. One may speculate that branding swaps of locally fabricated chips from Japanese companies operating in Europe and Japan to

71. Jim Handy, "NEC to Get 4MB DRAMs from Samsung Electronics," Dataquest, "DQ Monday Report" (on-line data service), March 13, 1995. See also, "Samsung, NEC in Euro DRAM Deal," *Electronic Engineering Times*, February 13, 1995, p. 16.

72. For example, Korea's Hyundai and Japan's Fujitsu announced a 1993 deal under which the Korean company would license its 4M DRAM chip to Fujitsu in exchange for the right to acquire DRAM chips made at a U.S. Fujitsu plant and sell them under its own brand name. Hyundai was thus able to avoid an 11 percent dumping tariff imposed by the United States on its Korean-made DRAMs. See Don Clark, "Fujitsu and Hyundai Electronics Form Unusual Joint Venture in Memory Chips," *Wall Street Journal*, October 7, 1993; and Jim DeTar, "Fujitsu, Hyundai In DRAM Talks," *Electronic News*, October 11, 1993, pp. 1–2. Korea's Goldstar also signed a 1993 pact with Hitachi to build chips under license from Hitachi's design that it would then supply to Hitachi in Japan. See David Lammers and Junko Yoshida, "Fujitsu and Hyundai in DRAM Pact," *Electronic Engineering Times*, October 11, 1993, p. 1.

73. As the result of this deal with NEC and of other export arrangements, Samsung was projecting a 50 percent increase in its exports to Japan in 1995, to reach a total of 130 billion yen. "Memory Chip Imports from ROK, Taiwan Rise," p. 1.

be sold with a Korean label for Korean-made product to be sold in the Japanese market with a Japanese label may have played some role in the marked 1995 rise in foreign market share in Japan.

To summarize, the available empirical evidence does not particularly favor the argument that shifts in the composition of Japanese demand, and a growing U.S. technical lead in microprocessors, were the principal sources of increased foreign share in the Japanese chip market. In fact, relatively little of the increase in foreign market share, over any period, was attributable to shifts in the composition of market demand. In every period examined, the bulk of the change in aggregate foreign market share was derived from increased foreign penetration of many individual product niches. Over 1990–94 the largest contributor to increased foreign market share seems to have been ASIC sales, an area where foreign firms had no gross technical advantage over Japanese producers. Thus, if anything, the available empirical evidence appears to support the anecdotal suggestions that Japanese consumers consciously substituted foreign products where they might previously have chosen offerings from domestic suppliers.

Conclusions

All the evidence reviewed here suggests that a variety of evolving policies associated with administration of the Semiconductor Trade Arrangement had important effects on global markets for DRAMs and EPROMs, the products subject to antidumping actions suspended as a result of the negotiation of this accord. In 1987 and 1988, government pressure on Japanese companies led to cutbacks in both production of and investment in these products. Output dropped and prices rose.

In response to foreign complaints about continued third-country dumping, a border control regime was established in an effort to reduce exports and increase export prices. Guidance was reportedly given to Japanese producers on desirable levels of exports to different foreign regions. Measures were taken to block flows from the secondary or gray market inside Japan to foreign customers. The largest producers' direct dealings with the largest foreign customers were more easily monitored and controlled.

Available data show that in the contract market, which accounts for the bulk of sales, and where dealings between a few large producers and a relatively small number of large customers account for most volume, a persistent differential between Japanese and foreign prices could be observed from 1987 through early 1989. In the gray market, on the other hand, in spot sales, which were inherently much more difficult to control, Japanese and foreign prices after some lag basically converged (but soared far above contract prices).

It is hard if not impossible to calculate the welfare impact of production and investment restraints, because it is difficult to be precise about what would have been produced absent these restrictions. It is easier to produce rough estimates of the welfare impacts of one isolated element of STA-related administrative measures, namely, border controls, and the resulting regional price differentials. Our results suggest a negative impact on welfare in the rest of the world amounting to perhaps 5 to 10 percent of regional consumption, over 1987–88, and a positive impact on Japanese welfare (with benefits derived from lower prices to Japanese consumers outweighing export profits forgone by producers).

Differences in the behavior of supply and demand between DRAM and EPROM markets differ mainly in the emergence of a significant price differential between Japan and the rest of the world in DRAMs (Japanese buyers paid less than buyers elsewhere for these products, but not for EPROMs). I have argued that this phenomenon is probably related to sharp declines in Japanese EPROM production and soaring imports of EPROMs into the Japanese market. Japanese producers may have held a lower market share in EPROMs than in DRAMs, but it is hard to see how this necessarily leads to reduced production and rising imports in Japan. The circumstances of this decline are consistent with the suggestion that the Japanese retreat in EPROMs may have been a deliberate political decision.

The analysis of available data on the growth of foreign companies' share of the Japanese semiconductor market supports the notion that pressures associated with the STA may have played an important role in increasing foreign sales. Only a small part of the increase in foreign market share could be attributed to shifts in the composition of Japanese demand toward products in which foreign firms already had a greater presence; most of the increase is clearly due to greater foreign penetration into individual product niches. Furthermore, a large part of the growth in foreign market share was associated with ASICs and memory chips, where foreign firms enjoyed no huge technical advantages relative to Japanese firms.

Appendix 5-A: The Economics of Contract Pricing

Given that perhaps 70 percent of DRAM sales are initially made as direct "contract" sales to large OEM users, it is useful to survey the nature of these contracts in some detail.

The Nature and Function of Long-Term Semiconductor Contracts

Typically "contract" sales are commitments to supply some quantity of parts, at some specified price, beginning on one specified future date and ending on another. At first glance, then, they might seem to be a special form of a continuing forward contract, that is, a legally binding promise to deliver product on some future schedule of dates at a single specified price. Further investigation suggests this is not the case, however.

Although these agreements are widely referred to within the industry as long-term contracts, they rarely seem to be legally binding commitments. The prices specified generally appear to hold when shipments under the contract begin, but often they do not persist over the life of the contract. Many contracts contain explicit provisions for renegotiation of the price downward, at the purchaser's option, in response to changing market conditions. In other cases, purchasers often successfully demand downward price adjustments even when the contract makes no explicit provision for such adjustments.[74]

Furthermore, because the system of price floors for DRAMs put into place by the U.S. Commerce Department in 1986 specified that the prevailing floor price at the time a legally binding contract is drawn up and signed shall remain in force throughout the life of the contract, there was an additional disincentive to producing a formal, legally binding document. To do so would have locked a purchaser into current prices, despite expected future declines in DRAM prices.

On the other hand, suppliers generally seem to respect contract prices as a de facto ceiling on prices charged their customers. This is generally irrelevant for products experiencing continuing price declines, such as DRAMs. But when the historically unprecedented increase in memory

---

74. An interesting compendium of DRAM contract "horror" stories—users and producers repudiating oral and written price commitments in response to changing market conditions—may be found in U.S. International Trade Commission, *64K Dynamic Random Access Memory Components* (June 1986) pp. A75–A88.

chip prices of 1987–88 occurred, preexisting contract prices were often respected and not raised (although some purchasers apparently did face cancellation or reduction of prescribed contract volumes at the negotiated price, after being informed of shortfalls in production). Thus, the long-term contract price may have served to constrain price increases over the life of preexisting contracts.

If contract prices are generally not legally or practically binding much beyond the original beginning date for the contract, what then is the purpose of entering into these informal, "handshake" commitments? Interviews with OEM purchasing managers suggest that ensuring the *quantity* of DRAMs to be purchased from suppliers is the major objective. In fact, purchasers frequently suggest that the critical issue in times of extreme shortage is not pricing, but getting adequate supplies. However, if one is willing to enter the spot market and pay the going price, one can always bid away scarce supplies from other users. But this may create other significant costs for the chip consumer that extend beyond the purchase price. Additional qualification costs or extensive additional testing may be required for purchases from new sources or gray market suppliers. Greatly increasing the number of sources of supply probably also greatly increases purchasing and qualification costs.

Producers of DRAMs face a different logic. Significant learning curve effects imply a sharp increase in the output of any given initial investment in DRAM production capacity over the product life cycle of that generation of DRAM, but there is considerable uncertainty about the exact timing and magnitude of yield improvements within the industry, and hence about aggregate supply. The producer must also be concerned about volatility in demand for the increasing quantities of DRAMs that will be flowing from existing fabrication lines in future periods. Quantity commitments lock purchasers into deliveries of a given producer's output and reduce the odds that large volumes of chips emerging from ever more productive factories will have to be liquidated in the gray market at a discount (in part reflecting new users' additional qualification and testing costs).

Empirical Evidence on DRAM Contracts

It is possible to examine empirically the nature of DRAM contracts and possible effects of the STA using a unique data source. Industry sources provided me with confidential data on OEM DRAM contracts covering the 1985–89 period. The data are drawn from contracts nego-

Table 5-A1. *Distribution of DRAM Contracts to Start (Lead) and Length, 1984–89*

Months lead/length	1M Contracts Frequency lead	Length	256K Contracts Frequency lead	Length	64K Contracts Frequency lead	Length
	July 1986 to Feb. 1989 (N = 128)		Oct. 1984 to Feb. 1989 (N = 174)		Jan. 1985 to Feb. 1989 (N = 83)	
0	0.29		0.41		0.43	
1	0.38		0.28		0.14	
2	0.04	0.01	0.08	0.01	0.04	0.01
3	0.09	0.30	0.12	0.14	0.19	0.10
4	0.08	0.02	0.09	0.03	0.10	
5	0.02			0.02	0.06	0.01
6	0.05	0.42		0.37	0.04	0.22
7	0.05	0.02	0.02	0.08		0.14
8		0.04		0.02		0.01
9				0.01		0.02
10	0.02	0.02		0.01		
11				0.03		0.01
12		0.16		0.24		0.46
13		0.01				
14						0.01
15				0.03		
17				0.01		
			Oct. 1984 to Aug. 1986 (N = 66)		Jan. 1985 to Aug. 1986 (N = 37)	
0			0.47		0.54	
1			0.29		0.16	
2			0.14			
3			0.09	0.09	0.08	0.19
4			0.02	0.06	0.11	
5				0.05	0.03	
6				0.55	0.08	0.24
7				0.06		0.14

tiated by a small number of European and North American OEMs; the bulk of the reported contracts refer to purchases by European users. Characteristics of the contracts that were collected include negotiation date, start date for shipments, period over which shipments are to be delivered, total quantity commitments over this period, contract price, nationalities of chip vendors and purchasers, chip organization and packaging, and chip speed (access time). After discarding contracts for which speed measures were unavailable (or that covered parts with a mixture of speeds); parts that used packages other than plastic dual in-line pin (DIP), plastic leaded chip carrier (PLCC, in the case of 256K DRAMs),

Months lead/length	1M Contracts		256K Contracts		64K Contracts	
	Frequency lead	Length	Frequency lead	Length	Frequency lead	Length
			Oct. 1984 to Aug. 1986 (N = 66)		*Jan. 1985 to Aug. 1986 (N = 37)*	
9				0.02		0.03
12				0.14		0.41
15				0.03		
17				0.02		
			Sept. 1986 to Feb. 1989 (N = 108)		*Sept. 1986 to Feb. 1989 (N = 46)*	
0			0.37		0.35	
1			0.28		0.13	
2			0.05	0.02	0.07	0.02
3			0.14	0.18	0.28	0.02
4			0.13	0.01	0.09	
5					0.09	0.02
6				0.27		0.20
7			0.04	0.09		0.15
8				0.03		0.02
9				0.01		0.02
10				0.02		
11				0.05		0.02
12				0.31		0.50
14						0.02
15				0.03		
Memorandum: Test for same distribution in both subperiods						
Chi-squared statistic			14.6	33	13.7	10.7
Degrees of freedom			5	13	6	9

Source: Author's calculations based on unpublished data provided by confidential industry sources.

and small outline cases (SO, in the case of 1M DRAMs);[75] and chips with relatively uncommon organizations,[76] a sample of 83 agreements for 64K DRAMs, 174 for 256K DRAMs, and 128 for 1M DRAMs remained.

The distribution of these contracts by lead time (months from negotiation date to start date) and duration (months from start date to end date) is shown in table 5-A1. For 64K and 256K DRAMs the contract

75. These involved a small number of observations divided among a relatively large group of other packages. Only 256K DRAMs with access times of 120 nanoseconds or faster were packaged in PLCC cases in the contracts in this sample.

76. Chip organizations other than 64K × 1, 256K × 1, 1M × 1, or 256K × 4 (the latter two are 1M DRAM types) appeared only in a relatively small number of contracts in my sample.

distributions are shown for periods before and after September 1986, when the STA went into force. (Only two of the contracts for 1M chips in this sample were negotiated before September 1986, and no such disaggregation is shown.) For 64K chips, all contracts in the sample were for parts with a 64K × 1 organization and plastic DIP packaging.[77] For 256K DRAMs, contracts covered both DIP and PLCC packages.[78] The 1M contracts include parts with both 256K × 4 and 1M × 1 organizations, and both DIP and SO cases, which were all found in abundant numbers. This proliferation of chip types demonstrates a trend toward increasing product variety in the DRAM market.

It is readily apparent that delivery on the vast bulk of these contracts begins within a very short period after their negotiation. The contract lengths cluster around three, six, and twelve months' duration. More than 40 percent of the contracts for 64K and 256K DRAMs could be considered "spot," in that shipments were scheduled to begin in the month they were negotiated; a smaller but still substantial fraction (28 percent of the 256K shipments; 14 percent of the 64Ks) were to begin in the month following the contract's negotiation. The great bulk of the remaining contracts for 64K and 256K parts began within two to four months, except for a small number of outliers clustered at five to seven months.

Contracts for 64K and 256K DRAMs negotiated after September 1986 tended to have longer lead times and to last longer than contracts before that date. A formal chi-square test comparing pre- and post-STA distributions generally confirms these casual impressions.[79] The period prior to the signing of the STA was one in which markets saw abundant supplies and generally declining prices, while 1987 and 1988 were generally marked by firm or rising prices and tightening supplies. Not surprisingly,

77. Four contracts for parts in ceramic cases were discarded from the sample; the rest of the discards were also in plastic DIP packaging.

78. A small but nontrivial number of contracts for parts with 64K × 4 organizations or for ZIP, ceramic, or SIMM packaging were dropped from the sample as well.

79. Both the lead times and lengths of contracts signed for the newer 256K DRAMs seem to have changed after the STA was signed, whereas lead times shortened but contract lengths did not seem to change in transactions involving the more mature 64K products. For 256K DRAMs we reject the hypothesis that these contracts were drawn from identical populations of lead times at the 5 percent significance level, and we reject a similar hypothesis on contract lengths at the 1 percent level. For 64K DRAMs we reject the hypothesis of a single distribution of lead times at the 5 percent level, but we are unable to reject the hypothesis of a common distribution of contract lengths at even the 10 percent significance level.

1M DRAM prices, almost all of which are based on negotiations after September 1986, more closely resemble the general patterns of distribution and clustering seen in contracts for 64K and 256K DRAMs over that same period rather than the earlier period. Thus DRAM contracts seem to have been drawn up earlier, and to have covered longer time spans, after the STA went into effect.

Elsewhere I have studied the prices embedded in these contracts using a simple econometric model in which these deals are viewed as forward contracts.[80] This econometric analysis of DRAM contract price data for three successive generations of memory chips supports several general propositions.

First, the simple model of the term structure of contract prices employed seemed quite consistent with these data: formal statistical tests did not lead to rejection, and estimated coefficients were largely unaffected by the term structure assumed. The point of departure was a model in which bias (deviation from the expected price in a future period) in contract prices serves to compensate purchasers for the transfer of risk to them by producers. In this model the impact of contract lead time on contract price reflects the "bias" in forward prices. The empirical results supported the presence of an expected negative bias ("normal backwardation," to use the term coined by John Maynard Keynes) in forward contract prices for DRAMs.

Second, the econometric results showed a rather dramatic decline of the bias term from the 64K generation of DRAMs, to the 256K generation, to the 1M generation, suggesting that the market viewed prices for later generation chips as considerably less volatile than previous generations of chips over the 1985–89 period. This, of course, was precisely what the administrative pricing guidelines and mechanisms imposed on the DRAM market with the advent of the STA would have been expected to accomplish, so these estimates suggest that the market believed that the STA actually did make DRAM prices less volatile.

Third, my suggestion that these contracts, beyond the initial purchase at the contracted price, mainly represent quantity commitments is supported by the generally small magnitudes and statistical insignificance of the effects of contract length as a determinant of contract pricing.

Fourth, analysis of price differentials faced by American and European purchasers of DRAMs suggested much smaller differentials than

80. See Flamm, "Measurement of DRAM Prices."

had been indicated elsewhere, and I could not reject the null hypothesis of no regional differences between the United States and Europe.

Fifth, estimated quantity discounts were very small and statistically insignificant. Other available data on contract pricing, disaggregated by size of transaction, also seem to show little systematic relationship between price and contract size.[81] Since interviews with purchasing managers at computer companies in 1989 frequently suggested that the largest companies often did pay lower contract prices, I interpret this to imply that the overall size of the account, rather than the size of one sales transaction for a single type of chip, determined any applicable discounts.[82]

Sixth, a general pattern of Korean vendors selling their product at somewhat higher prices emerged, consistent with anecdotal observations by market participants.[83] In effect, transactions with Korean producers probably more resembled a short-term, spot market–like exchange than the less volatile, long-term agreements generally characterizing contracts between large purchasers and producers. During a period of scarcity, the Korean producers appear to have charged a higher price, above the long-term contract price and approaching the spot gray market price. During a period of relative glut, on the other hand, we would expect this spot price to lie below the long-term contract price. This analysis is consistent with the reports in the trade press on Korean producer Samsung's dealings with its American distributors.[84] It is confirmed by a U.S. ITC

81. See U.S. International Trade Commission, *64K Dynamic Random Access Memory Components*, 1985, pp. A40–A45.

82. A 1992 U.S. International Trade Commission report suggests that contract sales may reasonably be divided into large, "Tier 1" premium OEM accounts, and a second tier of smaller accounts, including franchise distributors and value-added resellers. The report actually suggests that premium customers may pay higher prices, because of their more demanding and lengthy qualification requirements. See U.S. International Trade Commission, *DRAMs of One Megabit and Above*, p. A-43. Since this report covered a period of slack demand in which prices were generally declining quite sharply, one might interpret this to mean that second-tier contracts were shorter term, more like spot transactions.

83. In all likelihood a single Korean firm—Samsung—was responsible for all of the Korean product shipped in this sample of contracts.

84. At the peak of the DRAM shortage, in the summer of 1988, Samsung attempted to hike its prices to levels that its American distributors protested left them uncompetitive, and ended its ship-from-stock-and-debit policy, which effectively let distributors hold price-protected inventory. The distributors had thus been able to capture the benefit from rising prices by maintaining large inventories of product, without the downside risk that falling prices would pose with a large inventory. The practice was reinstated when the market slowed down in early 1989. See *Electronic News*, August 15, 1988, p. 47; February 27, 1989,

investigation, which found that all sales of Korean DRAMs to U.S. importers were made on a spot basis.[85]

Appendix 5-B: Construction of Estimated FMVs

This appendix briefly describes the relationship of foreign market values (FMVs) to cost concepts reported to the U.S. Commerce Department by Japanese producers. Producers were required to file quarterly submissions containing extensive cost data, part of which were reported in ranged form (reported value within plus or minus 20 percent of actual value) in public versions available for inspection at the Commerce Department.[86]

Cost of Manufacturing (COM). The first major subtotal reported was cost of manufacturing, equal to the direct materials, subcontract expense, direct labor cost, and production overhead expense accrued in the successive manufacturing stages of wafer fabrication, assembly, and final test.[87] When divided by number of good chips yielded by the manufacturing process, this generated a per unit manufacturing cost subtotal. To this was added an imputation for R&D cost calculated by taking the ratio of all semiconductor-related R&D in that quarter to the total cost of manufacturing all semiconductors that quarter (excluding R&D) and multiplying this ratio by the per unit manufacturing cost subtotal just

---

p. 27; April 3, 1989, p. 35. When prices turned down sharply in early 1990, Samsung shocked its American distributors by doing away with the customary "price protection" altogether, so that the effective prices paid by authorized distributors would more closely follow the ups and downs of the spot market (and more of the price risk would be transferred to the distributors). American distributors complained bitterly about Samsung's "broker mentality" (that it was behaving like a broker, passing on the full volatility of gray market spot pricing into its contracts with distributors, rather than moderating price swings with its contract pricing practices). See Bob Ferguson, "Samsung Notifies Distributors: Ship and Debit Will End Feb. 1," *Electronic News,* January 22, 1990, p. 34; Ferguson, "Samsung Terminates Six Distributors," *Electronic News,* February 5, 1990, p. 38; and Ferguson, "Brajdas Files Suit vs. Samsung over Inventory Returns, Credits," *Electronic News,* July 2, 1990, p. 32.

85. See U.S. International Trade Commission, *DRAMs of One Megabit and Above,* p. A-45.

86. This appendix is based on SIA, *Antidumping Law Reform*; and the author's interviews with Commerce Department officials and industry sources in 1990. The text describes the state of practice about 1990; in the first few quarters after the STA went into effect there was some variation as these procedures were developed.

87. Production overhead expense included depreciation of capital equipment, allocated across product lines according to generally accepted accounting practices.

Table 5-B1. *Foreign Market Value (FMV) Projected Cost Estimates for Japanese DRAM Manufacturers, by Company, Selected Quarters, 1986–89[a]*

Dollars per chip

Year and quarter	Company					Maximum + 25%[b]	Unweighted average
	Mitsubishi	NEC	Fujitsu	Hitachi	Toshiba		
256K DRAMs							
1987:1	1.86					2.33	1.86
1987:2	1.87		2.48		2.08	3.10	2.14
1987:3	2.65		1.76	2.57	2.45	3.31	2.36
1987:4	2.68		2.69	2.19	2.10	3.37	2.42
1988:1	2.25		1.49	1.89	1.36	2.81	1.75
1988:2	1.69		2.31	1.98	1.84	2.89	1.96
1988:3	1.72		2.50	1.89	2.19	3.12	2.07
1988:4	2.07		2.40	2.15	2.23	2.99	2.21
1989:1	2.47		2.10	1.95	2.15	3.09	2.17
1989:2	2.39		2.49	2.11	2.20	3.11	2.30
1989:3	2.34		2.31	2.30	2.02	2.92	2.24
1989:4	2.03		2.07	2.19	1.88	2.74	2.04
	1.97	1.58	2.06	2.02	2.74	3.43	2.07
	1.93	1.78	1.87		2.80	3.50	2.10
	2.90	2.14	1.75	2.16	2.03	3.62	2.20
1990:4	2.15	2.16	1.59	1.71		2.71	1.90
		2.51	1.78	2.34	1.89	3.14	2.13
			2.02	2.98	1.99	3.73	2.33
1991:3		2.13	1.92	4.39	1.80	5.49	2.56

described. The R&D counted by the Commerce Department was generally considerably greater than that reported to MITI in compiling Japanese statistics on IC R&D, including both corporate-level R&D and research grants to universities. The resulting number was per unit cost of manufacturing.

Sales, General, and Administrative Expenses (SG&A). The next category of expense was dealt with in a manner analogous to R&D. A ratio of SG&A to cost of manufacturing was calculated for all semiconductor sales, then multiplied by COM. This ratio was not permitted to fall below a minimum floor of 10 percent of COM.

Cost of Production (COP). Adding SG&A to COM yielded an estimate of cost of production.

Profit. A profit rate of 8 percent was applied to COP. This practice differed from typical antidumping cases (in which an 8 percent profit rate was generally taken as a minimum).

Year and quarter	Company					Maximum + 25^b%	Unweighted average
	Mitsubishi	NEC	Fujitsu	Hitachi	Toshiba		
1M DRAMs							
1987:1	23.75					29.69	23.75
1987:2			22.14		13.31	27.67	17.72
1987:3	15.67		15.96	11.34	9.47	19.95	13.11
1987:4	13.32		9.78	11.85	8.41	16.65	10.84
1988:1	10.21		5.92	13.98	6.81	17.48	9.23
1988:2	8.10		6.90	8.48	6.44	10.59	7.48
1988:3	9.22		7.51		8.20	11.52	8.31
1988:4	8.76		6.20	8.76		10.95	7.90
1989:1	9.45		5.44	6.33	7.10	11.82	7.08
1989:2	9.10		6.09	7.84	6.98	11.37	7.51
1989:3			4.54	6.61	5.06	8.27	5.40
1989:4			3.65	4.68	4.71	5.88	4.35
	6.33		3.71	5.62	5.62	7.91	5.32
	5.48	5.72	4.87		4.89	7.15	5.24
	4.70	5.22	4.89	4.14	2.93	6.52	4.38
1990:4	5.26	4.35	3.90	3.22		6.58	4.18
		3.31	3.97	3.79	4.06	5.08	3.78
	4.98	3.89	4.44	3.83	3.33	6.23	4.09
1991:3	4.87	4.35	3.95	4.86		6.09	4.51

Source: Based on data taken from unpublished company submissions to the U.S. Department of Commerce. "FMV" date is the quarter after the quarter to which the projected cost applies and two quarters after the quarter in which the projection was submitted.

n.a. Not available.

a. Cost concepts used are, by company:

256K DRAMs: Mitsubishi, export sales price (ESP) for M5M4256AP type chip; NEC, average of constructed value (CV) for 41256P, 41256CF, and 41256L type chips; Fujitsu, CV for MB81256/7P type chip; Hitachi, average of CV for 50256P NMOS and CV for 51258P CMOS type chips in dual in-line package; Toshiba, average of ESP for 41256AP and 41256P type chips for 1987:2 to 1989:1 and 41256AP only for 1989:2 to 1989:4 and ESP for 256K × 1 NMOS in dual in-line package for 1990:1 to 1991:3.

1M DRAMs: Mitsubishi, ESP for M5M41000P for 1987:1 and ESP for M5M41000AP for 1987:3 to 1991:3; NEC, average of CV for 421000C/421001C/421002C, 421000LA/421001LA/421002LA and 421000V/421001V/421002V for 1990:2 to 1991:3; Fujitsu, CV for MB81C1000/1P chip type; Hitachi, CV for 1Mx1 CMOS chip type in dual in-line package; Toshiba, ESP for TC511000P for 1987:2; average for TC511000P and TC511000AP for 1987:3 to 1988:3 and ESP for 1M CMOS DRAM in dual in-line package for 1989:1 to 1991:2.

b. If reported value is 20 percent less than true value, true value is 25 percent greater than reported value.

Constructed Value (CV). Profits were added to COP to produce a constructed value (CV). To a close approximation, constructed value was taken as an estimate of the fair value of the product. Minor adjustments (which could be positive or negative) for selling expenses were then made, depending on whether the product was sold to the customer after entering the United States or before. In the former case the adjusted CV was called exporters' sales price (ESP); in the latter, purchase price (PP).

Because the adjustments were generally small, ESP and PP were essentially the same value as CV.

Foreign Market Value (FMV). The floor price for a company's DRAM exports was related to these cost concepts in the following manner. Thirty days after the beginning of a quarter, companies were required to submit data showing actual costs in the previous quarter and projected costs for the current quarter. The de facto practice in the Commerce Department was to require the use of the previous quarter's actual costs as projected costs for the current quarter, because those were the only numbers that could be based on objective calculations using actual data and did not rely on hypothetical assumptions about cost declines and yield improvements.

Projected ESP and PP for the current quarter were used to set FMVs for sales delivered to a customer inside or outside the United States for the next quarter. Thus FMV in quarter t was approximately equal to projected cost in quarter $t - 1$. This in turn was approximately equal to actual reported cost in quarter $t - 2$.

Estimates of Company FMVs. Public submissions contained at least some ranged data on the above cost concepts for Mitsubishi, NEC, Fujitsu, Hitachi, and NEC. FMVs for Matsushita, Oki, and Texas Instruments Japan were calculated as a single weighted average of the FMVs for the five larger producers. Product types and reported data changed over time for many companies. Although some data were reported in each quarter for every company, reporting was not consistent. Table 5-B1 reports estimates of FMV based on projected costs for each of these companies where such a number could be constructed. In all cases the estimated FMVs are based on prior-quarter projections of CV or ESP. Where multiple versions of these chips were accounted for on separate cost tables, the numbers reported are an unweighted average of available data on unit costs for these different chip types.

CHAPTER SIX

Dumping in DRAMs

DUMPING has traditionally been defined as selling at a lower price in a foreign market than in one's home market. Since the mid-1970s, however, a different concept—that of sales at a cost less than a constructed "fair value"—has become an alternative standard, embodied in U.S. trade law, for finding that imports are being dumped in the U.S. market.[1] It has been estimated that, since 1980, about 60 percent of all dumping cases have been based on charges that foreign firms were selling at a price below some constructed average cost.[2] Perhaps the most widely publicized application of this standard has been in the case of imports of dynamic random access memory (DRAM) chips, the most important type of semiconductor in terms of dollar value of sales. Micron Technology's 1985 petition for relief from dumping of Japanese 64K DRAM imports specifically acknowledged that prices for these chips in the U.S. market may actually have been marginally higher than prevailing prices in the Japanese market.[3] Micron instead based its complaint on the charge that Japanese chips were being sold at prices that did not cover their full costs of production.

1. A brief history of the origins of this new standard may be found in Pietro Nivola, "Trade Policy: Refereeing the Playing Field," in Thomas E. Mann, ed., *A Question of Balance: The President, the Congress, and Foreign Policy* (Brookings, 1990), pp. 229–30; and Gary N. Horlick, "The United States Antidumping System," in John H. Jackson and Edwin A. Vermlost, eds., *Antidumping Law and Practice: A Comparative Study* (University of Michigan Press, 1989), pp. 133–34.
2. Horlick, "United States Antidumping System," p. 136.
3. See Micron Technology's petition to the U.S. Commerce Department, International Trade Administration, and the U.S. International Trade Commission, *Petition for the Imposition of Antidumping Duty* (June 1985), pp. 11–14.

As described in chapter 4, the U.S. government expanded the investigation of Japanese DRAM dumping to include 256K and 1M DRAMs. Folded into investigations of U.S. industry charges of dumping of erasable programmable read only memory (EPROM) chips, the matter culminated in the controversial U.S.-Japan Semiconductor Trade Arrangement (STA) of 1986. One of the outcomes of the STA was a system of floor prices for Japanese DRAM and EPROM imports administered by the U.S. Commerce Department, based on the calculation of a so-called foreign market value (FMV), which was derived from the "fair value" constructed cost comparisons enshrined in the dumping provisions of U.S. trade law.[4]

Although the FMV calculations were dropped from the 1991 successor agreement to the STA, Japanese producers were required to continue to collect the relevant data, to facilitate a "fast response" dumping investigation should the need arise. Thus, one may surmise that the implicit threat of a dumping investigation continued to give the FMV calculation a significant, if shadowy role in determining lower bounds on pricing of Japanese chip exports to U.S. (and possibly third-country) markets.

The idea of requiring producers to maintain prices at or above some concept of full long-run average cost is hard to defend, either as a positive description of what a profit-maximizing producer in a "competitive" market would choose to do, or as a normative guide for efficient resource allocation. It is, however, possible to construct an economically coherent argument that pricing below *marginal* cost can serve as a warning signal of "strategic" behavior by producers, which in some circumstances can

4. According to U.S. law, dumping is defined as sales of imports at "less than fair value" that cause "material injury to a U.S. domestic industry." To determine the fair value to which U.S. import prices are to be compared, the Department of Commerce is instructed by law to determine foreign market value (FMV) using sales prices in the exporter's home market. If the data are insufficient, Commerce must then use prices for sales to third countries. If these data too are inadequate, Commerce must calculate a "constructed value" to estimate the FMV. The statute says that actual foreign sales data may be disregarded when, first, they fall below total costs of production over an extended period of time and in substantial quantities, and second, they do not permit full recovery of all costs within a reasonable time and in the normal course of trade. The department is reported to take the "extended period" as its investigation period, "substantial quantities" as 10 percent of sales by volume, and generally to ignore the second requirement. See N. David Palmeter, "The Antidumping Law: A Legal and Administrative Nontariff Barrier," in Richard Boltuck and Robert E. Litan, eds., *Down in the Dumps: Administration of the Unfair Trade Laws* (Brookings, 1991), pp. 71–75. Also see Tracy Murray, "The Administration of the Antidumping Duty Law by the Department of Commerce," pp. 23–27 and 38–40, in this same volume.

justify policy intervention by government. However, in an industry subject to learning economies (where unit production cost falls with cumulative production experience), producers may rationally choose, for completely "competitive," nonstrategic reasons, to engage in "forward pricing," that is, to sell at a price that is below current marginal cost.

Is below-marginal-cost pricing for nonstrategic reasons empirically relevant in the semiconductor industry? Is it reasonable to defend even some revised version of a constructed cost test for dumping, based on *marginal* cost, as a reasonable trip wire for government scrutiny of possible strategic behavior by foreign producers? Perhaps most interesting, what can we deduce about the relationship between price and production costs using a minimally realistic model of the product life cycle when large upfront investments in capacity constrain output, large and relatively fixed investments in R&D create economies of scale, and learning economies are likely to be significant? How is an "FMV-like" system likely to constrain producer behavior in these circumstances? Because these characteristics are typical not just of semiconductor manufacture but of a broad range of high-technology products, the answers to these questions, and the methodology used in the inquiry, are of some importance.

This chapter lays out a simple analytical framework that can be used to compare the time path of output and prices in a nonstrategic, competitive (open-loop Cournot-Nash equilibrium) semiconductor industry with variants of constructed fair value that would be associated with the same path for output. The model is applied with empirically based parameters associated with 1M DRAM chip production, to explore how pricing of semiconductors is likely to be constrained, over the product life cycle, by constructed values in the form of FMV-type pricing rules. The basic model should also prove useful in analyzing many other interesting questions about the potential impact of public policies on high-technology industries subject to scale and learning economies.

The Economic Rationality of Below-Marginal-Cost Pricing

To a first approximation, stripped of a variety of practically important administrative and accounting issues, the FMVs constructed by the U.S. Commerce Department resemble an economist's concept of average cost of production, plus a fixed 8 percent markup that ostensibly reflects "normal" profit.[5] (This completely arbitrary markup is ignored in the rest of this discussion.)[6] A result taught in nearly any introductory economics course is that under some circumstances (for example, a downturn in demand) it can be economically rational for a producer in a competitive industry to sell at a price less than full average cost, as long as short-run marginal cost is covered. As long as a firm at least covers the variable costs of running a production line, and the cost of producing an incremental unit on that line, it makes economic sense to continue operating, even if revenues are insufficient to recover the full historical cost of the initial investment in developing the product and building the factory.

Thus most economists would find it entirely normal that, over at least some periods, observed prices fall short of full (long-run) average cost of production.[7] Any policy measure that prohibited marginal cost pricing by foreign exporters, while leaving domestic producers unaffected, would—if it actually affected market outcomes—increase domestic production at the expense of domestic consumers and (possibly) foreign producers. It would also arguably deny foreign producers national treatment, forbidding them the right to engage in economic behavior permitted of domestic firms.

If price falls short of the full average cost of production over a long period, of course, one may safely predict that some firms will exit the industry, and the industry will shrink to the point that the remaining firms at least recover full average costs over the life of their sunk investments. Thus, if sustained dumping according to constructed "fair values" (pricing below long-run average cost) is observed in a competitive industry, one may generally infer that excess capacity exists, and that exit will follow. But the observed pricing behavior may still reflect normal, competitive behavior on the part of those firms pricing below FMVs.

Is there ever any economic justification for remedial policies triggered by selling below a constructed FMV? The point at which many economists would agree that something other than "competitive," nonstrategic

5. See chapter 4, appendix B. The administrative and accounting issues so casually passed over here are widely believed to systematically bias the system toward findings of dumping and toward higher dumping margins. Richard Boltuck and Robert E. Litan, "America's 'Unfair' Trade Laws," in Boltuck and Litan, *Down in the Dumps*, pp. 13–14.

6. Also ignored is a requirement that overhead costs of no less than 10 percent be added to materials and fabrication costs in calculating cost of production.

7. See, for example, Alan V. Deardorff, "Economic Perspectives on Antidumping Law," in Jackson and Vermlost, *Antidumping Law*, pp. 30–33, for further elaboration on this point.

behavior might be going on occurs when a firm's price falls short of its
short-run marginal cost or, even more obviously, average variable cost
(which bounds short-run marginal cost from below over the relevant
range).[8]

In considering why a firm might rationally choose to produce and sell
at a price that fails to cover the current marginal cost of production, it is
helpful to distinguish between "strategic" and "nonstrategic" behavior.
I shall label a firm's behavior "strategic" when it explicitly takes account
of the impacts of its decisions on the behavior of other economic agents,
and "nonstrategic" when decisions consider the actions or choices of
other agents as fixed, unaffected by one's own actions. Predatory pricing,
as alleged by the U.S. semiconductor industry of its Japanese rivals
(described in chapter 3), can be classified as a form of strategic behavior.

One possible explanation, consistent with nonstrategic behavior, for
pricing below marginal cost is that increased production may lower a
firm's future production costs, through learning effects (discussed below). In this case, measured current marginal cost overstates "true"
marginal cost, which should take into account the cost-reducing effects
of current production on future output.[9]

But another possible explanation for behavior of this sort is a strategic
motive on the part of the dumper: either predation, defined here as
actions intended to encourage other firms to exit the industry; limit pricing, intended to discourage entry by other firms; or a defensive response
against predatory behavior by others.[10] In this case the rents from the

---

8. See Janusz A. Ordover and Garth Saloner, "Predation, Monopolization, and Antitrust," in Richard Schmalensee and Robert D. Willig, eds., *Handbook of Industrial Organization*, vol. 1 (North-Holland, 1989), pp. 579–90, for a detailed survey of the literature
on tests for predatory behavior.

9. Such learning economies can also be used as a strategic instrument, with a firm's
production decisions taking into account the impact of its learning on the actions of its
rivals. See Drew Fudenberg and Jean Tirole, "Learning by Doing and Market Performance," *Bell Journal of Economics*, vol. 14 (Autumn 1983), pp. 522–30, for such a model.

Deardorff, "Economic Perspectives on Antidumping Law," pp. 37–38, points out that
low-priced sales designed to build brand loyalty or otherwise alter consumer preferences
might also rationally lead a producer to sacrifice current profitability for future rents, and
to price below marginal cost. In effect, greater current output shifts future demand schedules, and current marginal revenue then understates "true" marginal revenue. Such
"demand-side learning effects," however, may be considered a form of strategic behavior,
since they are designed to alter the behavioral response to price of other economic agents,
in this case consumers rather than rival firms.

10. The modern rehabilitation of the theory of predation focuses on its impact on rival
firms' expectations about future profitability: as an exit-inducing investment in "disinfor-

exercise of monopoly power must be received later to justify absorption of a temporary loss on output shipped now.

Many forms of strategic behavior by firms, including predation, are regulated *within* national markets by antitrust laws. Thus a constructed cost test, used in the framework of the dumping laws, might be interpreted as a second-best attempt to remedy behavior by foreign firms that, if carried out on a purely domestic basis, would be considered the domain of antitrust policy. Unable to impose domestic policy standards on a foreign firm's behavior outside the national market, a national government can instead impose controls on the manifestation of that behavior—the pricing of sales to importers—in the domestic market.

Since, absent learning effects, pricing below short-run marginal cost is sufficient (but not necessary[11]) to conclude that a firm is acting strategically in its pricing policies, it may seem reasonable to review its activities when such behavior is observed, and to take corrective action if the intent is deemed to be predation and the potential impact is significant. If the predatory firm should succeed in its objective, it is at least possible that monopoly rents paid out later by national consumers, plus deadweight losses, could more than offset the windfall to national consumers from the temporary episode of low import prices now; it is these losses that justify some policy intervention.[12]

From this point of view, the economic problem with cost-based definitions of dumping is not that they are used at all, but that they use the wrong cost concept (long-run average cost instead of short-run marginal cost) as the prima facie trigger for considering intervention. This per-

mation" about the predator's cost structure, for example, or as the consequence of asymmetric financial constraints among competing firms created by imperfections in capital markets. The basic references are Paul Milgrom and John Roberts, "Limit Pricing and Entry under Incomplete Information: An Equilibrium Analysis," *Econometrica*, vol. 2 (March 1982), pp. 443–60; and David M. Kreps and Robert Wilson, "Reputation and Imperfect Information," *Journal of Economic Theory*, vol. 27 (August 1982), pp. 253–79. Useful interpretations are found in Paul Milgrom, "Predation," in John Eatwell, Murray Milgate, and Peter Newman, eds., *The New Palgrave: A Dictionary of Economics* (London: MacMillan, 1987), pp. 937–38; and Jean Tirole, *The Theory of Industrial Organization* (MIT Press, 1988), pp. 367–80.

11. Criticism of a short-run marginal cost test for predation generally argues that the rule is not stringent enough; prices above short-run marginal cost may still be associated with socially costly predatory activity. See Tirole, *Theory of Industrial Organization*, pp. 372–73; and Ordover and Saloner, "Predation, Monopolization, and Antitrust," pp. 579–80.

12. As, for example, Deardorff, "Economic Perspectives on Antidumping Law," pp. 35–36.

spective also leads one to focus on the close relationship between "fair trade" laws and competition and antitrust policy. It might be argued that some binding international standards for competitive business behavior (and their enforcement) might be offered as a constructive alternative to national "fair value" dumping tests based on constructed costs, as remedies for predation.

I will not attempt in this chapter to evaluate whether predation is a plausible description of what was going on in the DRAM marketplace in the 1980s; that question is explored—but not resolved—in the next chapter. I merely note that predatory behavior was one of the allegations made by the U.S. industry in pressing its case for protection. However, the modern theory of predation has been interpreted to suggest that high-technology industries are particularly important places to look for such behavior.[13]

Cost Structures and Dumping in the Semiconductor Industry

High-technology industries, which face large sunk costs in research and development (R&D) relative to sales, are by nature, along with highly capital-intensive industries, particularly prone to trade friction involving charges of dumping based on constructed cost tests. When fixed investments in R&D or factories are very large in relation to a firm's sales, a significant gap between average variable cost and long-run average cost will exist, and short-run marginal cost may fall significantly below long-run average cost for a substantial range of economically rational output levels. In such a case, perfectly competitive behavior may often trigger pricing below long-run average cost—and thus dumping charges— in a downturn.

High-tech industries are also particularly prone to dumping allegations because of the peculiar way in which R&D investments are treated by trade law (and many companies') accounting principles. An investment in a capital facility, for example, is not charged immediately against company revenues when construction is begun, or even when it is com-

13. Paul Milgrom argues that "policymakers should be especially sensitive to predatory pricing in growing, technologically advanced industries, where the temptation to discourage entry is large and the costs of curtailed entry even larger." (Milgrom, "Predation," p. 938.) See Kenneth Flamm, "Semiconductor Dependency and Strategic Trade Policy," *Brookings Papers on Economic Activity: Microeconomics* (1993), pp. 249–344, for further consideration of the plausibility of strategic behavior in semiconductor competition.

pleted; instead the cost is spread, by means of depreciation charges, over the period in which the facility is to be used. One may argue that accounting depreciation is at least an attempt to approximate the profile of true economic depreciation. The cost of an R&D investment, in contrast, is generally charged against revenues at the moment it is incurred, not spread over the investment's economically useful life.

This "front-loading" of R&D, it is sometimes argued, when processed through constructed cost calculations, leads to artificially high "constructed values" for high-tech imports—such as DRAMs—when initially shipped, in effect retarding technological progress. Defenders of this practice argue that since R&D charges are often allocated on the basis of sales, rather than identified with some particular product, the practical effect is to spread R&D charges over generations of products, through time (although it clearly remains true that a company just entering an industry after making a fixed R&D investment will necessarily have an initially high constructed cost).

The semiconductor industry is both technology intensive and capital intensive: it spends almost 15 percent of sales on R&D; it also typically spends an even larger fraction of sales—15 to 20 percent annually—on capital investments. Demand for semiconductors is also notoriously cyclical, so that it is not surprising to find that constructed cost tests for dumping were invoked in the 1985 industry downturn.

In addition, semiconductor production is believed to be characterized by learning economies. Unit production costs are believed to fall sharply with accumulated production experience. This further complicates our discussion of the borderline between nonstrategic pricing behavior and strategic activities. The key result, first elaborated by Michael L. Spence is that, with learning economies but without strategic interactions with rivals, a rational firm will generally equate marginal revenue to a value below its current short-run marginal cost of production, as it takes into account the cost-reducing effect of current production on future production costs.[14]

Although the Spence model is not directly applicable to the analysis of production decisions in the semiconductor industry, the point it makes greatly complicates the issue of whether or not constructed cost dumping tests—amended perhaps to use short-run marginal cost rather than long-run average cost as the indicator for possible intervention—can be jus-

---

14. "The Learning Curve and Competition," *Bell Journal of Economics*, vol. 12 (Spring 1981), pp. 49–70.

tified as a reasonable safeguard against predation by foreign producers. For in the Spence model, even with nonstrategic behavior, economically rational firms will engage in forward pricing, producing more even when marginal revenue falls short of marginal cost in anticipation of the cost-reducing effects of greater production.

Modeling the Semiconductor Product Life Cycle

The same considerations that make it likely that below-cost dumping will be observed in semiconductors also complicate any effort to build a tractable economic model of producer behavior. Any such model needs to address four distinctive features of semiconductor production:

—*learning economies,* the manner in which net output from a given semiconductor fabrication facility rises with accumulated production experience. This creates what amount to dynamic economies of scale.

—*capacity constraints,* due to the long lag between when facilities investments are started and when they are capable of mass production. It typically takes a year to a year and a half or more for a new facility to become operational. Once available for production, debugging manufacturing processes on "pilot production" can take another six months to a year.[15]

—*short product life cycles,* due to accelerated technological change. In DRAMs, a new, higher density memory chip has been introduced approximately every three years, rendering older chips obsolete. The newer chips use technologically more advanced manufacturing processes, requiring new production facilities (or, at a minimum, extensive retrofitting of older facilities). Coupled with the long gestation periods for new investments, this effectively means that capacity for a given generation of chip is locked in at the very beginning of the product life cycle. By the time the product is entering large-scale mass production, it is essentially too late to enter the market.

—*large R&D investments.* The sunk investment required for product and process R&D for a new generation of DRAM has typically been roughly equal to the cost of a high-volume manufacturing facility, and

15. For example, a new state-of-the-art fabrication facility in Taiwan reported in 1992 that it took four months to qualify production processes, followed by five months of further work to raise production from 1,000 to 10,000 wafers a month, at a facility with a current production rate of 15,000 wafers per month. See Klaus C. Wiemer and James R. Burnett, "The Fab of the Future: Concept and Reality," *Semiconductor International* (July 1992), pp. 96–98.

has been increasing rapidly. For the upcoming 256M DRAM, for example, industry sources have estimated both the R&D and plant investment required to be about $1 billion each.[16]

In my somewhat stylized depiction of the industry, a DRAM producer will be assumed to produce a homogeneous commodity, perfectly substitutable for that of other producers.[17] Difficult issues concerning the timing of the switchover from one generation of DRAM to another, and intergenerational externalities, are ignored by assuming that a DRAM producer faces a fixed period over which the product is sold, and that costs to develop and produce the product are relevant to that generation of DRAM alone. The product life cycle begins at time 0 and ends at time 1 (hence, the unit of time is the "product life cycle"). Every producer faces revenue function R, giving total revenues at any moment t as a function of its own production $y(t)$ and the aggregate output of all other producers $x(t)$. (All revenues and costs are measured in constant-dollar terms.)

Semiconductor production is believed to be characterized by strong learning economies. As was explained in chapter 5, output from a fully utilized chip fabrication facility is not fixed over time, but instead rises on an S-shaped profile (see figure 5-17 and the accompanying discussion). Although the Spence model of learning economies is rather unsuitable for analyzing production decisions in an industry where capacity constraints may be important, as is the case in the semiconductor industry, it shows, as we have seen, that even with nonstrategic behavior economically rational firms will engage in forward pricing. For simplicity in modeling producer behavior, I will, following Spence, ignore discounting over time on the grounds that product life cycles are short (typically, a new generation of DRAM is introduced every three years) and the additional complexity is substantial.[18]

---

16. Comments by an IBM executive at the Workshop on Government Roles in Commercial Technology, John F. Kennedy School of Government, Harvard University, September 14, 1992. "According to estimates from IBM, Siemens, and Toshiba, the cost of designing and qualifying a quarter-micron process is in excess of $1 billion." Adam Greenberg and J. Robert Lineback, "IBM, Toshiba, Siemens in 256M DRAM Alliance," *Electronic News*, July 20, 1992, p. 4.

17. This is not an unreasonable approximation. See Kenneth Flamm, "Measurement of DRAM Prices: Technology and Market Structure," in Murray F. Foss, Marilyn E. Manser, and Allan H. Young, eds., *Price Measurements and Their Uses* (University of Chicago Press, 1993), pp.157–206, for more detailed discussion of this issue in the context of semiconductor price indexes.

18. Also following Spence, I will measure all variables in *real* terms, taking out the

In semiconductor production, plant capacity may be measured in terms of "wafer starts," the number of slices of silicon, on which integrated circuits are etched, that can be processed per unit of time. At any moment t, $w[E(t)]$ functioning chips are yielded per wafer processed, where w is an increasing function of $E(t)$, "experience" through time t. How one defines relevant "experience" is a subject explored below. I will parametrize the impact of output y on relevant experience E as

$$\dot{E} = \frac{dE}{dt} = \frac{y}{K^\gamma},$$

where K is capacity and γ is a parameter taking on a value between zero and one. (For notational simplicity, time will sometimes be suppressed as an argument of time-varying variables.)

Some of the variable cost of producing a chip is incurred with every wafer processed, but some (the cost of assembly and final testing, for example) is incurred only with good, yielded chips. Let c be the variable cost incurred every time a wafer is processed, and d the variable cost incurred every time a good, yielded chip is taken through subsequent stages of production (assembly and final testing). If a wafer processing facility is utilized at rate $u(t)$ (where u is between zero and one), total variable costs at any moment are $dy + cuK$. Note that $y(t,K) = w[E(t)]\, u(t)\, K$.

Let up-front, sunk costs independent of output levels (such as R&D costs) be equal to F, and let fixed capital investment costs required for a facility processing K wafer starts be equal to r per wafer start. The producer's problem is to maximize (again, undiscounted) life cycle profits on this product,

$$\max_{u(t),K} \int_0^1 \{R[x(t),y(t)] - d\, y(t)$$

(6-1)
$$- cu(t)K - rK\} \, dt - F$$
$$\text{with } y(t) = w[E(t)]\, u(t)\, K$$
$$\text{s.t. } \dot{E} = \frac{y}{K^\gamma} = w[E(t)]\, u(t)\, K^{1-\gamma}.$$

---

effects of inflation, so that any applicable discount rate would be smaller than if measured in nominal terms. This, coupled with the shortness of the product life cycle for DRAMs, should make my undiscounted life cycle profit calculation a reasonably tolerable approximation to the more complex discounted flow.

316 DUMPING IN DRAMS

Assume that firms simultaneously choose initial capacity investments K and a time path for utilization rates, which give rise to a path for output over time, which they then proceed to follow. This assumption that capacity investments in DRAMs are committed at the beginning of the product cycle is not terribly unrealistic: as already noted, it typically takes a year or more to get a new fabrication facility up and running, and a new generation of DRAM is introduced roughly every three years.[19]

For the moment, take γ to equal zero (in other words, absolute cumulative production is the relevant measure of experience). As in the Spence model, I will assume a Nash equilibrium in output paths: given rivals' actual choices of capacity and a time profile for the utilization rate, equation 6-1 is maximized by every firm. For the purposes of this chapter, firms' behavior in this static game is assumed to be nonstrategic, since they take their rivals' capacity and output choices as unaffected by their own.[20]

Spence's Model

If wafer processing capacity K is not fixed over the life cycle but is continuously variable, as is implicit in Spence's formulation, then we have a special case of the above model in which r is zero (capital costs are included in wafer processing cost c and some arbitrary initial scale for capacity K is set), capital is a completely variable input, and a producer is free to choose any nonnegative u (that is, u is unbounded above, not bounded by one) and to produce any yielded chip output desired. Under these circumstances, as is easily shown in appendix 6-A, formal maxim-

19. The world record for bringing a new fabrication facility on line is said to be nine months. See Larry Waller, "DRAM Users and Makers: Shotgun Marriages Kick In," *Electronics*, November 1988, pp. 29–30.

20. An alternative, to be presented in the next chapter, is to set up a two-stage competition among rival firms, with capacity investment as the initial phase, followed by a second stage in which firms choose output paths subject to capacity constraints. The solution of the static game presented here corresponds to the open-loop (nonstrategic) equilibrium of this two-stage game, in which a firm's first-period choice of capacity takes its rivals' choices in both periods as given. The alternative equilibrium concept will assume second-period subgame perfectness, that is, that firms take into account the effect of their first-period capacity choices on their rivals' second-period output paths. This creates strategic interactions among firms. See Avinash Dixit, "Comparative Statics for Oligopoly," *International Economic Review*, vol. 27 (February 1986), p. 114; and Carl Shapiro, "Theories af Oligopoly Behavior," in Schmalensee and Willig, *Handbook of Industrial Organization*, vol. 1, pp. 383–86.

ization of the objective function in equation 6-1 yields the first-order condition

$$(6\text{-}2) \qquad\qquad R_y = d + \frac{c}{w} - \frac{\delta}{K^\gamma} ,$$

that is, u is chosen so that marginal revenue is set equal to current marginal cost $(d + c/w)$ less a term proportional to nonnegative adjoint variable δ, which captures the future cost-reducing effects of current production. Adjoint variable δ, in turn, is determined by the transversality condition

$$(6\text{-}3) \qquad\qquad \delta\,(1) = 0$$

and equation of motion

$$(6\text{-}4) \qquad\qquad \dot{\delta} = -\frac{c}{w}\, u\, K\, w_E .$$

By differentiating both sides of equation 6-2 with respect to time, we immediately see that marginal revenue R_y must be constant over time, and therefore, by equation 6-3, equal to current marginal cost at the end of the product cycle, $d + c/w[E(1)]$.

In short, with continuously variable capacity, a profit-maximizing producer will choose that level of output at which marginal revenue equals terminal (not current!) marginal cost. In industry parlance this is forward pricing. With a constant elasticity and autonomous (time-independent) demand a constant price proportional to terminal marginal cost will result.

Although this model provides an appealing explanation of the phenomenon of forward pricing, which is a notable empirical feature of business practice within the semiconductor industry, the actual trajectory of pricing suggested by this model (with a constant elasticity of demand, price is fixed at some constant level over the entire product cycle) is quite inconsistent with observed behavior.[21] Chip prices typically drop very quickly over the first part of the product cycle, drop less quickly as the product approaches maturity, and fall very slowly, if at all, at the end. As shall be seen in a moment, a more realistic treatment of capacity constraints yields a more plausible trajectory for prices.

21. Andrew R. Dick, "Learning by Doing and Dumping in the Semiconductor Industry," *Journal of Law and Economics*, vol. 34 (April 1991), pp. 133–59, for example, invokes the Spence model to motivate his assumptions about the time path of semiconductor prices over the product life cycle, but ignores the constant pricing prediction of the Spence model.

The Baldwin-Krugman Conundrum

The pioneering attempt to incorporate learning economies into a stylized, empirical model of the semiconductor industry is that of Baldwin and Krugman (B-K).[22] The B-K model focuses on regional segmentation of the U.S. and Japanese semiconductor markets, in order to simulate the impact of market closure policies, and takes an approach to producer behavior that differs significantly from that of Spence. B-K constrain firms to operate at full capacity over the entire product cycle; the choice variable for the firm is initial capacity, which once set, determines output levels over the entire product life cycle. The first-order condition for an optimum is that the life cycle revenue from the addition of a marginal unit of wafer processing capacity just equals the cost of building and operating that marginal unit of capacity (since all capacity is always fully utilized, the distinction my model draws between investment costs and wafer processing costs is immaterial).

Firms in the Spence model are never capacity constrained; firms in the B-K model always operate at their capacity constraint. The Spence model has firms forward pricing— maintaining marginal revenue constant over the life cycle, equal to their terminal marginal cost. The B-K model has marginal revenue—and price—falling smoothly over the life cycle. Thus, while the striking forward pricing behavior of the Spence model has disappeared (but not *all* forward pricing)[23], a more empirically plausible path for prices has replaced it.

22. Richard E. Baldwin and Paul R. Krugman, "Market Access and International Competition: A Simulation Study of 16K Random Access Memories," in Robert C. Feenstra, ed., *Empirical Methods for International Trade* (MIT Press, 1988), pp. 171–97. A somewhat different exposition of this model is given in Elhanan Helpman and Paul R. Krugman, *Trade Policy and Market Structure* (MIT Press, 1989), chap. 8. The later interpretation (henceforth referred to as H-K) differs in some significant respects from B-K. For example, the learning curve in B-K has yields improving with cumulative wafers processed (that is, faulty chips have the same yield-enhancing effects as good ones), whereas H-K presents a more conventional view of the learning curve, with yield rates rising with cumulative output of *yielded* (good) chips. The B-K assumption on yields, although not the accepted approach to modeling yield improvement within the industry, simplifies the mathematical structure of the model.

23. There is a great deal of confusion in the industry about what exactly forward pricing means. Does it mean that learning economies are to be taken into account when forecasting marginal costs, and prices, for future deliveries in forward contracts with large customers, or does it mean—as in the Spence model—going even further and producing quantities such that marginal revenue falls *below* current marginal cost? Note that even when output is capacity constrained one can have aggressive "forward pricing" in the sense of price

Unfortunately, when B-K actually calibrated their model to empirical data, their results indicated that, in the intuitively appealing case of a (Cournot-) Nash equilibrium in output, only *two* firms would populate the industry in zero profit equilibrium, in part because of the steep declines in unit cost due to learning economies assumed.[24] This gross deviation from actual market structure led them (despite their theoretical reservations about the approach) to specify firm behavior in terms of conjectural variations, and even this tactic yielded disturbingly high conjectural variations, fueling doubts about the underlying model.[25] As I point out below (see note 43), B-K actually *underestimated* learning economies as reported in the sources they cite, deepening the perplexity created by their results.

In addition, as Kala Krishna notes, the algebraic tractability of the B-K specification of firm behavior has been purchased by excluding the possibility of some interesting forms of strategic competition.[26] (Because B-K empirically calibrate conjectural variations, strategic interactions among firms exist.) Investments in capacity may be undertaken with strategic objectives, to convince rivals to reduce output or to exit, or to dissuade them from entering an industry, creating additional monopoly power which can then be exploited. Constraining firms to operate at full capacity over the entire product life cycle may restrict them to suboptimal output paths, where monopoly power is not fully exploited. It also hinders analysis of interesting policy questions regarding the potential welfare impact of strategic policies which may foster the creation and exercise of monopoly power (explored in chapter 7).

A variant of the B-K model can be fit into the framework outlined above for the Spence model, after suitable amendments. Utilization rate

---

(and, necessarily, marginal revenue) falling below current marginal cost: if one examines the simulations in this chapter, price is below both current marginal and average cost in the earliest portion of the product cycle (see table 6-4). Thus capacity-constrained output and aggressive forward pricing are not mutually inconsistent—the firm is producing as much as it can and reducing the price to whatever level is needed to sell it all.

24. See Baldwin and Krugman, "Market Access and International Competition," p. 185.

25. Baldwin and Krugman, "Market Acccess and International Competition," p. 195; and Helpman and Krugman, *Trade Policy and Market Structure*, p. 173. The conjectural variations approach has each firm assuming that if it increases its output by one unit, each of its competitors will change its output by v units, where v can take on a range of values. The case of $v = 0$ is the Cournot assumption.

26. "Comment on 'Market Access and International Competition,'" in Robert C. Feenstra, ed., *Empirical Methods for International Trade* (MIT Press, 1988), pp. 198–202.

u is constrained to equal one at all times, and the objective function in equation 6-1 is maximized with respect to K alone. The right-hand side of equation 6-4 is replaced by a more complex variant (corresponding to $u = 1$), and a new equation determining optimal capacity choice is added:

$$(6\text{-}5) \quad \int_0^1 \left[\left(R_y - d - \frac{c}{w} + \frac{\delta}{K^\gamma} \right) uw - \gamma \frac{\delta}{K^\gamma} uw - r \right] dt = 0,$$

where the B-K specification fixes u equal to one and γ equal to zero.

A More Realistic Model of the Semiconductor Product Cycle

It is possible to create a more realistic model of firm behavior, in which firms can continuously adjust output as in the Spence model, yet also face capacity constraints on output, as in the B-K model. I briefly summarize here the more detailed exposition laid out in the appendixes to this chapter. The firm's problem is to maximize equation 6-1 by choosing both an initial level of capacity K and a profile for time-varying utilization rate $u(t)$ for that capacity, which determines output at any moment in time. The optimal level of capacity chosen satisfies equation 6-5; the left-hand side of this equation can be interpreted as the net marginal return on additional investment in capacity. It also must be true that the optimal path must over some interval be capacity constrained; that is, $u(t) = 1$.

In general, the optimal path for $u(t)$ will be made up of three types of segments: interior segments, where $0 < u < 1$; lower boundary segments, where $u = 0$; and upper boundary segments, where $u = 1$. Within an interior segment, equations 6-2 and 6-4 will hold, as in the Spence model, as will a form of forward pricing: marginal revenue will be held constant, set equal to current marginal cost less δ/K^γ—the marginal cost-reducing value (over the remainder of the product life cycle) of an additional unit of output—at the endpoint of this interval.

With additional assumptions one can further sharpen the characterization of the optimal behavior of a profit-maximizing firm. I shall assume a symmetric-industry equilibrium with N identical firms, autonomous demand (that is, demand is not an explicit function of time), and concavity of total industry revenues in industry output (as would be the case, for example, with a constant-elasticity demand function and price elasticity exceeding unity). If we further assume that firm marginal revenue exceeds the initial value of current marginal cost as industry output approaches zero (as must be the case with a constant-elasticity demand, so some production will always be profitable), we can exclude the possibility of no firms entering the industry in a symmetric equilibrium. Although a Nash equilibrium in utilization rates is assumed for the moment, for expositional purposes I will parametrize a firm's perceptions of other firms' reactions to changes in its output in terms of a constant, non-negative conjectural variation. (The two interesting cases that motivate this parametrization are Cournot-Nash equilibrium—conjectural variation equal to zero—and a collusive, constant-market-share cartel—conjectural variation equal to $N - 1$.)[27]

Under these assumptions, optimal u must decline over an interior segment, and u must be continuous in time. Key features of the behavior of production and utilization rates over time are that y must be increasing when u is constant (because of learning economies); u must be decreasing when y is constant (for the same reason); and y is nondecreasing and u nonincreasing over time (a consequence of the additional assumptions). Therefore the optimal path of u must look like an upper boundary segment, possibly followed by an interior segment. The Spence forward-pricing result of marginal revenue being set equal to terminal marginal cost will hold whenever the firm is producing but is not capacity constrained (that is, when $0 < u < 1$, along an interior segment).

Note that nothing about the specific shape of the learning curve (function w) beyond the fact that it is increasing in experience ($w_E > 0$) has been assumed in arriving at this characterization of optimal policy. Let the time at which a firm switches from full-blast production to constant-output production be t_s (with full-blast production over the entire product life cycle an important possibility).

In short, with this simple description of the semiconductor product life cycle we derive a more realistic specification of firm behavior that captures both the importance of capacity investments and the ability of firms to exploit fully what monopoly power they enjoy by varying utili-

27. The major behavioral assumption excluded by a nonnegative conjectural variation is Bertrand competition in prices. Because DRAMs are essentially a homogeneous commodity sold in well-developed secondary spot markets, specifying that producers sell at a single market price and choose the quantities they will sell is the natural assumption. Moreover, David M. Kreps and Jose A. Scheinkman, "Quality Precommitment and Bertrand Competition Yield Cournot Outcomes," *Bell Journal of Economics*, vol. 14 (Autumn 1983), pp. 326–27, have shown that in a two-stage game, where first-stage capacity investments are followed by a second-stage Bertrand game in prices and a particular ("efficient") rationing rule, the outcome is a Cournot equilibrium in output.

zation rates over time. The model is simple enough to be empirically tractable. Firms will make some capacity investment, operate at that capacity for some period of time, and then possibly switch to a constant-output path (with constant marginal revenue, but decreasing utilization of capacity, as yields continue to rise) over the remainder of the product life cycle. (A period of full-blast production during which yields rise very sharply is in fact a pervasive feature of producer behavior in this industry, and has been given its own special name: "ramp-up".)

Preliminary Observations on Pricing over the Life Cycle

Even with its relatively general structure, the above analysis provides several insights into questions concerning pricing and dumping over the product life cycle. First, pricing below current marginal cost will *never* be observed near the end of the product cycle among competitive, non-strategic, profit-maximizing firms. This follows immediately from the fact that, if any output is being produced, marginal revenue will never be less than the right-hand side of equation 6-2 (see appendix 6-A), which at time 1—the end of the product cycle—equals current marginal cost $(d + c/w)$. Since price exceeds marginal revenue, price also must exceed current marginal cost in some neighborhood of time 1, the end of the product cycle.

Second, it is common practice in the semiconductor industry to note that chip *prices* seem to fall with cumulative experience, just as consideration of learning economies suggests that unit *costs* should. Indeed, analysts often identify average unit cost with price, then estimate an empirical "learning curve" by regressing the logarithm of market price against the logarithm of cumulative output for the industry. Slopes of this line typically are found to be around -0.3.[28]

The analysis just laid out, however, suggests that this implicit interpretation of prices as proportional to some variant of current unit costs is generally hard to justify. We have seen that firms may be forward pricing, with prices set below current marginal costs of production. In this case, to break even (or make a profit) over the product cycle, prices must, over at least some other period, be set above current marginal costs. Thus

28. See, for example, Integrated Circuit Economics, *Mid-Term 1989* (Scottsdale, Ariz., 1989); Boston Consulting Group, *Perspectives on Experience* (Boston, 1972); and Douglas Irwin and Peter J. Klenow, "Learning-by-Doing Spillovers in the Semiconductor Industry," *Journal of Political Economy*, vol. 102, no. 6 (1994), pp. 1200–27.

prices may in general be below current marginal costs over some period, and above current marginal costs at other times, and the practice of treating price as proportional to some definition of unit cost becomes questionable.[29]

However, in one important special case, namely, when production proceeds at full blast over the entire product life cycle, an explicit relationship between price and cumulative volume does emerge. In that case, output is fixed by initial capacity investments and the effects of learning economies on yields, while price is determined by the parameters of the demand curve, given whatever output is produced at full capacity. It can be shown that in this case

$$\frac{\partial \log P}{\partial \log E} = \frac{\partial P}{\partial E}\frac{E}{P} = \frac{\epsilon}{\beta},$$

where ϵ is the elasticity of yield with respect to cumulative output (the "learning curve" elasticity), and β the price elasticity of demand.[30] Thus, in this special case the slope of the relationship between the logarithm of price and the logarithm of industry cumulative output is the learning curve elasticity *divided* by the demand elasticity, and so reflects both supply and demand parameters.[31]

---

29. Needless to say, since average unit cost (in which fixed costs such as R&D and capacity investments are allocated to different time periods over the product cycle, and the average of this fixed cost per unit produced is added onto marginal cost) must lie above current marginal cost, the same criticism applies to identifying price with any definition of average unit cost.

30. To get this result, assume the industry is made up of N identical firms, each producing output y. Let industry demand be given by inverse demand function g, so that $P = g(Ny)$. With full-blast, capacity-constrained production, y at any moment in time is given by $w(e)K$, with K representing capacity (in wafer starts) per firm and w yielded (good) chips per wafer, a function of firm cumulative experience e. E ($= Ne$) is industry cumulative experience. Substituting for y in the expression for P, and differentiating with respect to E, we end up with

$$\frac{\partial P}{\partial E}\frac{E}{P} = \left[\frac{\partial g}{\partial y}\frac{y}{g}\right]\left[\frac{\partial w}{\partial e}\frac{e}{w}\right],$$

where the first expression in brackets is $1/\beta$, and the second ϵ.

31. In the special case of $\epsilon = 0.5$, $\beta = -1.5$ (which I shall argue is roughly what real-world parameters suggest), we then have an elasticity of price with respect to cumulative volume of about -0.3. With $\epsilon = 0.5$, we also have what is commonly called a "70 percent learning curve" (referring to the fact that unit cost declines by 30 percent as experience is doubled; see note 43 in this chapter). The empirical coincidence of a -0.3 elasticity of price with respect to cumulative output, and a 30 percent decline in cost with a doubling of cumulative output, has led to enormous confusion, since these two parameters will generally

Closing the Model

Can we say anything about the relationship between price and long-run average cost? The model sketched out thus far takes the number of firms in the industry—which will affect profitability and pricing—as given. One "natural" way to close the model is to specify that firms enter the industry until rents earned by producers—that is, the integrand in equation 6-1—just equal zero. The zero profit condition then determines N, the number of firms entering the industry (we will ignore the difficulties created if one insists that N be an integer). Zero profits mean that total life cycle revenues just equal total life cycle costs. Therefore (after dividing both concepts by total output over the product life cycle) average life cycle price must equal average life cycle cost per unit.

But is there any clear relationship between current price and current "fully allocated" average cost at any given moment? Current short-run marginal cost is a relatively clear concept: it is the additional current cost

---

not be identical. Indeed ϵ with a 70 percent learning curve is actually equal to 0.5, not 0.3.

Irwin and Klenow's 1994 study, "Learning-by-Doing Spillovers," which like all other studies regressing price on cumulative output finds an elasticity of price with respect to cumulative output of about $-.3$, interprets this to imply an 80 percent learning curve. For the reasons just given, this parameter can actually be interpreted as supporting a much larger, 70 percent learning curve in a capacity-constrained operating environment.

Irwin and Klenow recognize that if firms are capacity constrained, their methodology is suspect. They argue that an aggregate semiconductor industry capacity utilization measure produced by the SIA varied between 43 and 78 percent over 1978–92: "if these capacity figures apply to DRAM production, then the assumption that capacity constraints do not bind seems appropriate" (p. 1214). This assumption is unwarranted, however, since the SIA survey also includes older, economically marginal facilities that produce specialized, older-vintage, and custom chips on fully depreciated, obsolete lines as demand warrants. Current-generation DRAMs are generally produced in leading-edge facilities with new equipment that are basically run at full capacity around the clock for at least the first years of their lives. Stories in the business press (see chapters 3 and 4) suggest that, at the very minimum, during periods of cyclical boom in chip demand aggregate DRAM output has been constrained by available capacity.

The only available data on plant-level capacity utilization for current-generation semiconductor fabrication lines appear to support the notion of a capacity-constrained operating environment. In their detailed, international study of sixteen semiconductor lines (twelve of which used current generation 6-inch wafers and equipment, four of which used slightly older 5-inch wafers and equipment, and one of which used older 4-inch wafers), the Competitive Semiconductor Manufacturing survey of the University of California, Berkeley, found that "most but not all of the fabs in our sample were fully loaded [relative to their capacity] throughout their four-year period of observation." See Robert C. Leachman, "Introduction," in Robert C. Leachman, ed., *The Competitive Semiconductor Manufacturing Survey: Second Report on Results of the Main Phase*, report CSM-08 (Engineering Systems Research Center, University of California, Berkeley, 1994), pp. 17–18.

saved by producing one less unit at any given moment. This is the incremental cost saved when output is reduced by one unit.[32] In my model, current short-run marginal cost—$d + c/w$—is constant at any moment and equal to current average variable cost.

To define a fully allocated current average cost, however, it is necessary first to define an intertemporal cost allocation rule to spread fixed entry costs F and capital costs rK over the product life cycle. Dividing the capital and entry cost allocated to some instant in time by output produced at that moment yields a current average fixed cost per unit produced. If this current average fixed cost is added to current average variable cost (identical to short-run marginal cost in my model), we have a long-run average cost (LRAC) concept that satisfies the basic requirements of a long-run average cost: when multiplied by output at that moment, and when all such products are summed over all moments, total costs of production over the entire product life cycle are given.

In one special case—when d, the yielded chip output–sensitive part of variable cost, is "small" (in a sense to be defined in a moment), and it is assumed that free entry by other competing firms into the industry forces life cycle profits to zero—a clear pattern in the relationship between price and both marginal and average cost over the product life cycle exists. Differentiating our industry demand function with respect to time (and assuming determinants of demand other than price remain constant), we must have

$$\frac{\dot{P}}{P} = \frac{1}{\beta}\frac{\dot{z}}{z},$$

linking rates of change in price over time with growth rates in industry output, where β is the price elasticity of demand and z is industry output.

Now, recall that we have defined $LRAC$ as

$$LRAC = \frac{cuK + rK + F}{y} + d,$$

while marginal cost MC is just

---

32. When a firm operates at less than full capacity, this is identical to the increased cost incurred in producing one more unit. When operating at full capacity, the incremental cost of an additional unit is effectively infinite; the marginal cost curve is L-shaped, with a kink at full-capacity output.

$$MC = \frac{cuK}{y} + d,$$

and average cost always exceeds marginal cost by the average fixed cost. Differentiating with respect to time, then, gives[33]

$$\frac{LR\dot{A}C}{LRAC} = -\left[1 - \frac{d}{LRAC}\right]\frac{\dot{y}}{y} + \left[1 - \frac{\frac{rK + F}{y} + d}{LRAC}\right]\frac{\dot{u}}{u} < 0$$

$$\frac{\dot{M}C}{MC} = -\left[1 - \frac{d}{MC}\right]\left(\frac{\dot{y}}{y} - \frac{\dot{u}}{u}\right) < 0$$

and both marginal and average cost must be falling over the entire product life cycle.[34] More interestingly, because demand is elastic ($\beta < -1$) and the industry is populated by identical firms (so $\dot{z}/z = \dot{y}/y$), we must then have

$$\frac{\dot{P}}{P} - \frac{LR\dot{A}C}{LRAC} = \left(\frac{1}{\beta} + 1 - \frac{d}{LRAC}\right)\frac{\dot{y}}{y} - \left[1 - \left(\frac{\frac{rK + F}{y} + d}{LRAC}\right)\right]\frac{\dot{u}}{u} > 0,$$

if output-sensitive variable cost d is "small" in the sense of other costs' share in average cost (AC), $1 - d/LRAC$, exceeding $1/\beta$ in absolute value.

Because our assumption about entry means that total life cycle costs are exactly equal to total life cycle revenues, price less the fully allocated long-run average cost as defined above (that is, profit per unit), multiplied by output and summed over every moment of the product cycle, must be exactly equal to zero. Thus, if price exceeds the fully allocated average cost concept at any instant, it must fall below fully allocated average cost at some other instant over the product cycle, and vice versa. Under the assumption of "small" output-sensitive variable costs, price must fall more slowly than average cost, and average cost falls over the entire product life cycle. Therefore, for there to be zero profit over the

33. Since (see appendix 6A)

$$\frac{\dot{y}}{y} > \frac{\dot{u}}{u}, \left[1 - \frac{d}{LRAC}\right] > \left[1 - \frac{d}{LRAC} - \frac{\frac{rK+F}{y}}{LRAC}\right] > 0, \frac{\dot{y}}{y} \geq 0, \text{ and } \frac{\dot{u}}{u} \leq 0.$$

34. Since the expressions in brackets all are fractions lying between zero and one, y and u are both nonnegative, y is nondecreasing, u is nonincreasing, and y must be increasing when u is constant and constant when $\dot{u}$ is negative.

Figure 6-1. *Relationship between Semiconductor Price and Cost, Assuming "Small" Output-Sensitive Variable Costs*

Dollars

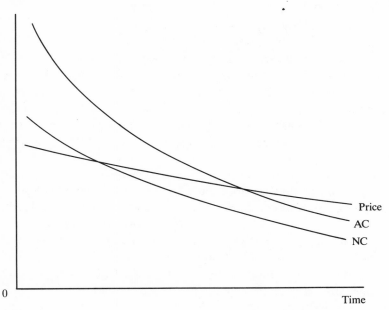

0

Time

Source: See text.

product life cycle, price must originally lie below average cost in the early part of the product cycle, then rise above it at the end. In this case the relationship among various cost concepts will appear as in figure 6-1, and we would expect to observe below-average-cost "dumping" in the early part of the product life cycle as the consequence of normal competitive behavior.[35]

The scenario sketched out in figure 6-1, however, depends critically on the assumption that costs incurred after wafer processing are a "small" component of cost throughout the product cycle.[36] As we shall see in a

35. This argument is made by example, in the special case of all firms engaging in full-blast production over the entire product life cycle, in Richard E. Baldwin, "The US-Japan Semiconductor Arrangement," discussion paper 387, Centre for Economic Policy Research, London, March 1990, pp. 17–19; and Baldwin, "The Impact of the 1986 US-Japan Semiconductor Agreement," *Japan and the World Economy*, vol. 6 (June 1994), pp. 129–52. Baldwin implicitly assumes that all variable costs are yield-sensitive (that is, that $d = 0$).

36. The behavior shown in figure 6-1 is also assumed, without explanation, in Dick, "Learning by Doing," pp. 133–59.

DUMPING IN DRAMS

moment, our best efforts to describe the semiconductor manufacturing process empirically suggest that this assumption is *incorrect,* and that the precise timing of episodes of below-cost pricing is considerably less predictable.

More generally, however, if our assumption that entry drives long-run profits to zero is a realistic one, below-LRAC "dumping" *must* be occurring at some point during the product cycle. This follows from the fact that any "standard" cost allocation rules for fixed costs will define an average fixed cost, which, when added to current average variable cost, cannot be exactly equal to actual price *at every moment* during the product cycle. For firms to earn zero profits, therefore, prices must sometimes be above average cost, and sometimes below—with below-cost dumping observed at the latter moments.

With learning economies present, a cost allocation rule yielding an average cost coinciding with price must generally be a function of *all* of the parameters of the control problem, and the allocated fixed cost will, in general, take on negative values as well as positive values. Since the cost allocation rules actually used to spread fixed costs over the product cycle—by firms or by the U.S. Commerce Department—are generally functions only of the size of the fixed costs, and of time, and produce only nonnegative values, it is essentially guaranteed that there will be an episode of below-LRAC dumping if learning economies are present and the industry is in a symmetric, zero-profit equilibrium.[37]

---

37. Define a cost allocation rule $g(z,t)$, where z is a vector of arguments and t is time, such that

$$\int_0^1 g(z,t) \, dt = F + rK.$$

Define fully allocated average cost (FAAC) by

$$FAAC = \frac{c}{w} + d + \frac{g}{y}.$$

that is, current average variable cost plus average fixed cost. We know that the optimal path must contain a capacity-constrained segment, and that along this portion of the optimal path

$$\dot{P} = P'N\dot{y} = P'NKw_E\dot{E}.$$

in symmetric industry equilibrium. If price P is to always equal FAAC along this segment, however, differentiating the expression for FAAC with respect to t, and setting this equal

Some Further Assumptions

If not these, then what paths over time for prices and costs *would* one expect to see in the simple model outlined here? Our next step is to take this simple control model and solve it to explicitly derive an individual firm's behavior over time. To sharpen our characterization of a profit-maximizing firm's optimal policy, we must address some additional issues.

LEARNING ECONOMIES. We shall approximate the learning curve by specifying that:

(6-6) $w(E) = \phi E^\epsilon$, with $E(0) = E_0, 0 \le \epsilon \le 1$.

This gives yielded chips per wafer as a function of experience E. This functional form is best regarded as an approximation: mass production typically starts at initially low yields; after a while yields start to rise quickly, then flatten out in a pattern closer to a logistic curve. (This function may be thought of as another view of the data shown in figure 5-17, where the x-axis now shows cumulative volume of good chips produced instead of time; good chips increase in a nonlinear way over time as the result of learning economies.[38]) Analytical tractability is the grounds for selecting this approximation. Note that a dummy experience value E_0 is used as an argument in the function to specify some initial

---

to the last expression, we must at every moment of this interval have

$$\frac{dg}{dt} = w_E \dot{E}\left(P'NK + \frac{c}{w^2} + \frac{g}{Kw^2}\right)y$$

$$= Kw_E \dot{E}[(1+\beta)P - d].$$

Since in equilibrium N, P, and other variables will generally be functions of all the parameters of the optimal control problem, a function g that satisfies this last equation must generally include all parameters of the control problem as arguments, unless $w_E = 0$, in which case g is constant. (In this case, I note in appendix 6-A that all capacity is utilized and output is constant over the entire product life cycle.) Thus, as long as there are learning economies (w_E does not equal zero), a cost allocation rule g varying only with F, r, K, and t cannot satisfy the requirement that $P = FAAC$, for arbitrary values of the parameters of the control problem, over this capacity-constrained interval.

Also, we have already noted that it is possible for P less than current marginal cost to be optimal in the presence of learning economies. (Indeed, the simulations reported below contain examples of such behavior.) Reexamining the definition of FAAC, it is clear that g must be negative for $P = FAAC$ to hold true over such an interval.

38. See also Integrated Circuit Engineering, *Mid-Term 1988* (Scottsdale, Ariz., 1988), p. 6-35.

Figure 6-2. *Empirical Approximation of the Learning Curve in Semiconductor Manufacture*

Yielded chips per wafer

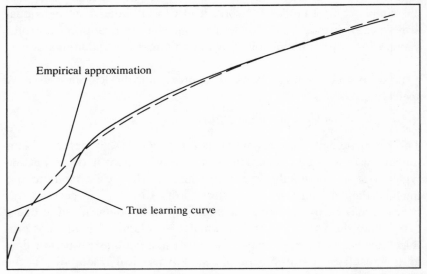

Experience

Source: Author's calculations.

nonzero yield—without this constant, yields would stay stuck at zero forever.[39] The assumed approximation to the "true" learning curve is depicted in figure 6-2. The approximation somewhat distorts yields, output, and pricing in the very earliest portion of the product cycle.

Defining "experience" raises additional issues. It is customary to use cumulative output as a proxy for experience in empirical studies, and most published empirical studies of learning economies have taken this approach. But using absolute, company-wide production experience as the determinant of any single facility's productivity implies that running, say, ten facilities in parallel produces the same yields at the end of a given period as running a single facility to produce the same total output over a much longer period. In the semiconductor industry, it is widely believed

39. B-K use the same functional form but do not face the stuck yield problem, because the argument in their learning curve is experience in processing gross wafers, not yielded good chips. The latter specification is generally industry practice in estimating learning curves.

that improved manufacturing yields come from two main sources: refinements of the operation of the production line (with each new refinement building on previous experience), and die shrinks (reductions in the feature size for chip designs, made possible by improved use of existing process equipment). These are iterative and sequential in nature. That is, lessons learned from running a line over some period of time are then applied to refine the operation of that line over a subsequent period.

However, by this logic, if numerous identical production lines are run in an identical fashion over the same period of time, then the same lessons are being learned, in parallel, on each line, and yields at the end of the period should be no higher than if only a single line were being run. Of course, if a new line—with no experience and lower yields—is put into operation after an older line has been running for some time, and it is possible to completely transfer the fruits of experience across facilities, then the maximum experience on any one line would be the experience variable determining production yields. Because all investment occurs at a single initial moment in my simple model, all lines will have identical amounts of production experience at any subsequent moment in time, and cumulative output per facility is the desired measure of experience.

It is possible that the lessons learned on different lines are not the same, if completely different "experiments" in production refinement are being conducted at every production facility. If, once again, experience can be completely transferred across facilities, and there is no duplication in lessons learned in different facilities, then it might be argued that company-wide, absolute cumulative output, rather than cumulative output per unit capacity, is the relevant experience variable.[40] Empirical discussion suggests, however, that the transfer of experience across facilities is quite costly.[41]

---

40. Or perhaps even industry-wide cumulative output, if complete cross-company diffusion of the lessons of production experience occurs.

41. C. H. Stapper and others, "Evolution and Accomplishments of VLSI Management at IBM," *IBM Journal of Research and Development*, vol. 26 (September 1982), p. 541, note that IBM, "with hard work on both sides of the ocean" has been able to adapt yield breakthroughs and transfer them between European and American plants within IBM. Noboru Ishihara and Hideki Wakabayashi, with Makoto Sumita, note, "Thus, much of the accumulated knowhow with respect to semiconductor processing technology comes from a section where it cannot be documented. For this reason, when engineers are unable to travel, the transfer of technology becomes difficult." See "Nomura Analyst Report: The Semiconductor Industry in the 1990s," Nomura Research Institute, Tokyo, 1991, p. 18. Yui Kimura, *The Japanese Semiconductor Industry: Structure, Competitive Strategies, and Performance* (Greenwich, Conn.: JAI, 1988), p. 50, comments that learning economies "are

A related issue is whether significant intercompany learning effects are significant, that is, whether yields are influenced by experience not just on a particular fab line or company, but by the experience of other companies. Such a specification would significantly alter the details of the model developed in this chapter.

One way to parametrize these differences in the conceptualization of how intrafirm learning economies work is to define experience as cumulative output divided by K^γ, where γ takes on a value of zero if absolute, company-wide cumulative output is the correct experience variable, and one if experience per facility (or unit capacity) is what is relevant.[42] This means

$$(6\text{-}7) \qquad \dot{E} = \frac{y}{K^\gamma} \text{ with some initial } E(0) = E_0$$

defines $E(t)$.[43] My approach will be an agnostic one: I will solve the model using both zero and one as possible values for γ, and then ask which seems to predict more empirically plausible behavior. While the "true" value almost certainly lies somewhere between these two extremes, it is my prior belief that it should be substantially closer to one. With plausible empirical assumptions, it turns out that parameter γ plays a critical role in defining the nature of an industry equilibrium and re-

---

only partially transferable across plants and across firms as the yields often depend on specific conditions of fabrication processes of a particular plant."

42. The relationship between this specification and the "per fab" and "per company" specifications of learning effects may be sketched out as follows. Let Y be total company output, q output per fab, K company capacity, f capacity per fab (plant size), and m the number of fabs per company. The basic hypothesis is that $E = q\, m^\rho$, with ρ an "appropriability" parameter taking on a value of zero if only the plant's own experience is relevant to yields, and one if all company-owned plants' experience is relevant. Any intermediate degree of appropriability can be assumed by choosing the appropriate value for ρ. Since $m = \frac{K}{f}$, and $q = \frac{Y}{K}$, then $\dot{E} = f^{\gamma-1}\frac{Y}{K^\gamma}$, $\gamma = 1 - \rho$. If capacity is measured in units such that f approximately equals one, then my assumed relationship holds as it stands. If not, rescale experience variable E as $E' = E\, f^{1-\gamma}$, and substitute for E in equation 6-6.

43. An alternative specification might make cumulative output, or cumulative output per unit capacity, the state variable, subject to some initial value; this alternative state variable times K to some power would then be the argument of w, the function giving yield per wafer. Such a specification, however, makes initial yield (with no experience) a function of the scale of capacity investment, which is undesirable. (In that case, increasing or decreasing capacity simply to raise initial yield on every line will play an entirely artificial role in determining optimal capacity.)

solves the B-K conundrum.[44] An exact solution for $E(t)$, assuming capacity-constrained output (see appendix 6-B), is given by

(6-8) $\qquad E(t,K) = [E_0^{1-\epsilon} + K^{1-\gamma}\phi t(1 - \epsilon)]^{\frac{1}{1-\epsilon}}.$

FINAL TEST AND ASSEMBLY YIELDS. A tested, just-fabricated, "good" die is not yet a finished integrated circuit. The dice produced on the wafer fabrication line must then be assembled into a sealed package, then subjected to a rigorous final testing process. Although yields of good, tested chips assembled from good dice may also show some evidence of a learning curve, the impact of learning in this stage of the integrated circuit production process is thought to be quite small relative to learning economies in the wafer fabrication phase of integrated circuit manufacturing.

I will model assembly and final test yields by assuming a fixed yield of final good chips from good dice produced on the wafer fab line, $v = \xi y$; v is "net" good, assembled and tested integrated circuits produced from quantity y of "gross" good dice yielded by wafer fabrication. Thus, after converting net, finished integrated circuit demand to a gross (defect-inclusive) demand for fabricated chips, we can pose the optimization problem in terms of choosing a time path for wafer fab output y (as opposed to net output ξy), and otherwise ignore the additional yield losses in the assembly and final test stages of production.[45] In interpreting

---

44. The existing empirical literature on learning curves gives us little help in deciding the correct specification. If data on cumulative output from a given facility, or aggregate data from a group of facilities with fixed capacities, are used to estimate the relationship $y = wK$ using equation 6-6, we get an equation like

$$Ln[y(t)] = a + \epsilon \, Ln[Q(t)],$$

that is, giving the natural logarithm of total output as a linear function of the natural logarithm of cumulative output Q, even if cumulative output per unit capacity is the relevant experience variable. The effects of capacity size K have been absorbed into constant a. Data from different facilities of varying size within a single company, or from different companies, along with an additional variable controlling for capacity size, are required to identify and estimate γ.

45. The critical assumption is that all good chips coming off the wafer fab line incur all the costs of assembly and final test before being culled.

If we denote DRAM consumers' inverse demand function for finished chips by $\bar{P}(\xi x, \xi y)$, then the producer's maximization problem, taking into account assembly and final test yield losses, is to

$$\max_{u(t),K} \int_0^1 \bar{P}[\xi x(t, \xi y(t)] \, \xi y(t) - \frac{d}{\xi}\xi y(t) - \frac{c}{w\xi}\xi y(t) - rK$$

the results, we must only remember to divide all gross per unit cost and revenue measures (price, marginal revenue, marginal cost, etc.) emerging from the optimization analysis by ξ, in order to get the net cost and revenue measures per good unit observed in the chip marketplace.

DRAM DEMAND. We must specify a demand function for DRAMs, and an industry structure, in order to calculate marginal revenue R_y. I shall assume a constant-elasticity demand function of the form

(6-9) $$z = \alpha P^\beta,$$

where z is aggregate demand for DRAMs, P is the price of DRAMs, and the industry is made up of N identical firms. With this specification, we have

(6-10) $$R_y = \left(\frac{N\,y}{\alpha}\right)^{\frac{1}{\beta}}\left(\frac{\sigma}{\beta} + 1\right),$$

where parameter σ equals the conjectural variation plus one, divided by N. With Cournot competition, σ is $1/N$; with a constant-market-share cartel, $\sigma = 1$.

Model Solution

Next, I briefly summarize the method used to solve numerically for an optimal policy. Full details may be found in appendixes 6-A and 6-B. It is useful to categorize optimal policies in terms of two possibilities. One possibility is that full-blast production is followed by an "interior segment" where a firm is producing at less than full capacity. In this case an

---

with $\xi y(t) = \xi uwK$

$$\dot{E} = uwK^{1-\gamma}.$$

Now, if we define an inverse demand function for "gross" chips (including product ruined in assembly and final test) by

$$P[x(t), y(t)] = \bar{P}[\xi x(t), \xi y(t)]\xi$$

and substitute, we get exactly the maximization problem given earlier in equation 6-1, where

$$R[x(t), y(t)] = P[x(t), y(t)]y(t).$$

optimal policy boils down to picking both an optimal capacity K and some optimal time t_s to switch from full-blast production to constant-output production. The other possibility is that the firm runs at full capacity throughout the product cycle. In this regime, necessary conditions for the firm are only used to solve for an optimal capacity.

Optimal Output Decisions: With Interior Segments

When the firm produces at less than full capacity, an optimal, profit-maximizing policy must set the difference between marginal cost and marginal revenue equal to δ/K^γ, the value of an additional unit of current production in reducing future production costs over the remainder of the product cycle. As in the Spence model, marginal revenue along this interior segment will be constant, equal to terminal marginal cost. We may solve a differential equation giving experience at time t_s, $E(t_s, K)$ after a period of full-blast production through optimal switchpoint t_s to an interior segment, and using this result, derive an equation giving t_s as a function of K, N, and other parameters of the control problem:

$$(6\text{-}11) \quad \left[\frac{N\phi E(t_s, K)^\epsilon\, K}{\alpha}\right]^{\frac{1}{\beta}} \left(\frac{\sigma}{\beta} + 1\right)$$

$$= \frac{c}{\phi[E(t_s, K) + \phi K^{1-\gamma}(1 - t_s)\, E(t_s, K)^\epsilon]^\epsilon} + d.$$

This is just the condition that marginal revenue at time t_s (on the left-hand side) equals current marginal cost at terminal time 1 (on the right-hand side).[46]

A second equation giving optimal capacity may be derived from equation 6-5. After solving for δ over both interior and boundary segments, and substituting into equation 6-5, we have an expression implicitly giving K as a function of optimal t_s, and N. Together with equation 6-11, for given N and various other parameters, we have two equations in two unknowns. An optimal t_s and K pair must solve these two equations.

---

46. The expression within the outermost parentheses in the denominator on the right-hand side of this equation is experience at terminal time 1, after producing from t_s to time 1 at a constant output level; the entire denominator is yield at time 1.

Optimal Capacity: With No Interior Segment

In many important cases the optimal path may not contain an interior segment, so we never switch from full-blast production, and $u(t)$ will always equal one. The transversality condition (equation 6-3) will still hold, however, and using this boundary value we can solve an equation of motion for δ, and derive a different version of equation 6-5, which implicitly determines K as a function of N and other parameters.

Since, however, this expression gives us a necessary condition for optimal K conditional on full capacity utilization over the entire product cycle, we must be careful to ensure that such a path is in fact a Cournot equilibrium. In searching for Cournot equilibria, then, attempts were made to solve both the two-equation system characterizing an optimal policy with interior segments, for a t_s and K pair, and the single equation giving optimal K assuming full capacity utilization throughout the product cycle. Solutions found were then checked as possible Cournot equilibria, by perturbing both firm capacity K and switching time t_s by 0.01 in all feasible directions, while maintaining the hypothesized equilibrium output path for all other firms, and calculating the impact on firm profitability (which should necessarily be negative if the perturbations are from a Cournot equilibrium). The procedure assures us that we have found a local maximum satisfying the first-order necessary conditions.

Plausible Parameter Values

The final step in the simulation of firm behavior is to decide on empirically plausible parameter values to be used in this model. A significant effort was devoted to constructing realistic and accurate parameter estimates.

LEARNING ECONOMIES. I am unaware of any published studies of experience curves in the semiconductor industry that control for the effects of varying facility capacities (that is, that estimate γ); however, there are numerous published estimates of learning curve elasticity ϵ based on the relationship between the logarithm of output (or of cost) and the logarithm of cumulative output. For DRAMs there are several published reports of an empirical 72 percent "learning curve," meaning that current unit cost drops by 28 percent with every doubling of output, corresponding to $\epsilon = 0.47$.[47]

To estimate the parameters of the learning curve for 1M DRAMs, estimates of "typical" wafer yields based on historical data and projections for the last four years of a five-year product life cycle were used to derive nonlinear least squares estimates of parameters corresponding to E_0, ϕ, and ϵ in equation 6-6.[48] Parameter γ was assumed to equal one; because the unconstrained estimate of E_0 was a small number very close to zero, I imposed a value of 0.01. This was the largest power of 10 that, when substituted into equation 6-8 to constrain parameter estimation, left other parameter estimates unchanged from values produced by the unconstrained estimation procedure. Learning elasticity ϵ was estimated to be 0.49, while ϕ had an estimated value of 31.[49]

47. With a constant wafer processing cost as the only cost element (the model that underlies these studies), we have

$$\text{unit cost} = \frac{c}{w} = \left(\frac{c}{\phi}\right) E^{-\epsilon}.$$

A learning elasticity ϵ equal to 0.47 is solved from the 72 percent learning curve, since $2^{-\epsilon} = .72$. See Robert N. Noyce, "Microelectronics," *Scientific American*, September 1977, pp. 62–69; and Office of Technology Assessment, *International Competitiveness in Electronics*, OTA-ISC-200 (1983), p. 76. Engineers at IBM, on the basis of studies of production costs for IBM bipolar integrated circuits in the 1960s and 1970s, derived a virtually identical 71 percent learning curve. See William E. Harding, "Semiconductor Manufacturing in IBM, 1957 to the Present: A Perspective," *IBM Journal of Research and Development*, vol. 25 (September 1981), p. 652. Douglas W. Webbink's 1977 survey of the integrated circuit industry notes that interviewed companies believed ϵ to generally lie in the 0.32 to 0.52 range, depending on the type of device. *Staff Report on the Semiconductor Industry* (Federal Trade Commission, 1977), p. 52.

Note that B-K appear to have erred in interpreting the number reported in the 1983 Office of Technology Assessment study's report of a 72 percent learning curve—their basis for assuming that $\epsilon = 0.28$, when it actually corresponds to $\epsilon = 0.47$!

Irwin and Klenow, "Learning-by-Doing Spillovers," interprets their results as reflecting an 80 percent learning curve, but this depends critically on the assumption that producers are not capacity constrained over their product life cycle. In a capacity-constrained environment, their results are consistent with a 70 percent learning curve (see note 34).

48. Since the unit of time is the (assumed five-year) product cycle, yields after two years correspond to time .4, after three years to .6, and so on. The data are given in VLSI Research, *Depreciation Schedules: Their Impact on the Competitiveness of Semiconductor Manufacturers*, San Jose, Calif., 1990, addendum 2. The data in this addendum correspond to a "typical" wafer fab running 2,500 wafer starts per week, at full capacity over the product life of the 1M DRAM. Conversation with Dan Hutcheson, VLSI Research, August 19, 1991.

49. If instead γ were set equal to zero, the estimate of ϕ would have risen from 31 to 36, but the estimated ϵ would not have changed.

To further check whether this critical parameter seems to accurately reflect the reality of 1M DRAM production, actual company-specific quarterly production estimates for the six largest 1M DRAM manufacturers were used to estimate learning elasticity ϵ. "Experience" at time t is given by

$$E(t) = E_0 + \frac{Q(t)}{K^\gamma},$$

where $Q(t)$ is cumulative production through time t and K is capacity. Recall that company output at any instant (y) is just the product of yield (w, described in equation 6-6) times utilized wafer capacity (uK). By choosing a period of time over which capacity is approximately constant and fully utilized ($u = 1$), the last term in this product will just equal a company's level of capacity. If E_0 (the initial starting value of experience) is small relative to $Q(t)/K^\gamma$ (current accumulated experience) over the period analyzed, then

$$\ln[y(t)] = \ln(\phi K^{1-\gamma\epsilon}) + \epsilon\ln[Q(t)] + (\epsilon E_0 K^\gamma)\frac{1}{Q(t)}$$

must hold.[50] The expressions in K in the above equation may be regarded as part of firm-specific coefficients on two variables—a constant and the inverse of cumulative output—and ϵ the coefficient of the logarithm of cumulative output in a regression equation.

We may wish to entertain the possibility of *interfirm* experience effects (that is, that $w = \phi\, E^\epsilon\, G^\zeta$, where G is the cumulative experience of the *rest* of the industry, and ζ is a parameter reflecting both the potential effect of others' experience on one's own cumulative output and the degree of appropriability of the experience of others. In this case we can develop an augmented relationship between firm i's output, and own and rest-of-the-industry experience, through reasoning analogous to that sketched out above, given by:

$$\ln[y_i(t)] = b_{0i} + b_1\ln[Q_i(t)] + b_2\ln[\overline{Q}_i(t)] + b_{3i}\frac{1}{Q(t)} + b_{4i}\frac{1}{\overline{Q}_i(t)},$$

where b_1 is an estimate of ϵ, b_2 is an estimate of ζ, overbarred variables refer to the rest of the industry, and other coefficients (b_0, b_3, b_4) vary

50. This makes use of the fact that, approximately, $\ln(1 + x) = x$, for small values of x.

across firms. If interfirm learning effects on yields are important, we would expect b_2 to be positive and the coefficients of inverse rest-of-the-industry cumulative output to be nonzero.

The equation also provides a simple test for the hypothesis that γ equals zero, since in that case the coefficient of inverse cumulative output should be constant across firms. We would not expect this test to be a particularly powerful one against the alternative that γ equals one, however. The coefficients on inverse cumulative output would also be approximately constant if firms had roughly similar levels of capacity, or if the starting value of experience E_0 was very small.

The above equation was estimated using data on cumulative output and current production for the six largest 1M DRAM producers (Toshiba, Hitachi, Fujitsu, NEC, Mitsubishi, and Samsung) from the second quarter of 1988 through the first quarter of 1989, a period of booming demand when trade press accounts suggest that DRAM output was capacity constrained.[51] Results are shown in table 6-1.

The estimated coefficients provide no evidence that interfirm learning was significant.[52] All of the rest-of-the-industry coefficients were statistically indistinguishable from zero. The point estimate of ζ, the elasticity of yield with respect to others' experience, was actually negative—a finding which, if accepted at face value, would indicate that other firms' experience *reduces* one's own yield. Again, however, none of these inter-

---

51. The data are Dataquest estimates of quarterly output. Reported shipments by Motorola have been added to Toshiba's output, and reported shipments by Intel to Samsung's output (since it is believed that most Motorola chips were fabricated by Toshiba, and Intel chips were "private labeled" Samsung output during this period).

52. Irwin and Klenow, "Learning-by-Doing Spillovers," find some evidence of weak interfirm spillovers (an additional unit of a firm's own output contributes three times as much relevant experience as an additional unit produced by a competitor). Their study is critically dependent on the assumption of no capacity constraints (see note 34) and the assumption that a firm's marginal revenue—proportional to industry price, with a constraint of proportionality dependent only on firm market share—is set equal to marginal cost. In their study, however, price is measured using Dataquest's estimates of average selling price, which average together long-term contract prices with spot market prices actually observed in the market during any quarter. Since marginal revenue is defined as the revenue from an incremental unit produced and sold into the market under prevailing conditions, the relevant price for marginal revenue calculations is arguably the spot market price. Thus, their marginal revenue measure almost certainly underestimates true marginal revenue (and marginal cost, assumed to be set equal by profit-maximizing producer behavior absent capacity constraints) in tight markets, and overestimates marginal revenue and cost in slack markets.

Table 6-1. *Regression Results from Analysis of Firm-Level Learning Curves*[a]

Variable	Regression 1			Regression 2			Regression 3		
	Estimated coefficient	Standard error	t-statistic	Estimated coefficient	Standard error	t-statistic	Estimated coefficient	Standard error	t-statistic
Log of cumulative output									
Own	0.51	1.86	0.27	0.67	0.17	4.05	0.52	0.07	7.86
Rest of industry	-0.79	2.19	-0.36						
Firm constants									
Toshiba	15.40	10.65	1.45	2.69	2.04	1.32	4.33	0.75	5.78
Hitachi	14.29	11.67	1.22	2.21	1.83	1.21	3.94	0.68	5.78
Fujitsu	14.92	11.94	1.25	2.40	1.79	1.34	4.06	0.67	6.05
NEC	14.84	11.36	1.31	2.42	1.85	1.31	4.12	0.69	5.99
Mitsubishi	14.15	11.27	1.26	2.02	1.84	1.10	3.96	0.68	5.78
Samsung	16.12	11.89	1.36	2.94	1.62	1.81	4.42	0.66	6.66
Own inverse cumulative output									
Toshiba	-16.09	19.22	-0.84	-0.77	1.62	-0.48			
Hitachi	4.87	11.47	0.42	0.27	0.42	0.63			
Fujitsu	3.79	5.25	0.72	0.13	0.30	0.44			
NEC	12.14	12.83	0.95	0.21	0.46	0.45			
Mitsubishi	6.15	10.61	0.58	0.75	0.44	1.71			
Samsung	0.55	0.52	1.04	-0.06	0.09	-0.70			
All companies							-0.13	0.05	-2.56
Rest-of-industry inverse cumulative output									
Toshiba	5.08	11.32	0.45						
Hitachi	-50.67	77.41	-0.65						
Fujitsu	-63.01	74.92	-0.84						
NEC	-102.84	101.43	-1.01						
Mitsubishi	-54.81	86.55	-0.63						
Samsung	-61.73	63.55	-0.97						
Memoranda									
R^2	0.997			0.99			0.98		
Std. error of regression	0.08			0.083			0.0938		
Degrees of freedom	4			11			16		

Source: Author's calculations.

a. Results of test for absence of interfirm effects: *F*- statistic with 7, 4 d.f. = 1.12; Wald chi-square with 7 d.f. = 7.84. Test for common inverse cumulative output coefficient: *F*- statistic with 5, 11 d.f. = 1.89; Wald chi-square with 5 d.f. = 9.45.

firm effects were statistically significant, and a joint hypothesis that they were all zero could not be rejected at any reasonable significance level.[53]

The point estimate of ϵ in the full model and in the model with a common coefficient on inverse cumulative output was barely over 0.5 (corresponding to a 30 percent learning curve), and 0.67 with no interfirm effects (corresponding to a 37 percent learning curve). This confirms other evidence suggesting substantial learning economies. A formal statistical test for a common coefficient on inverse cumulative output (corresponding to the hypothesis that $\gamma = 0$) was not entirely conclusive.[54] In summary, all available data seem to point to a large yield elasticity with respect to production experience, close to 0.5, and at least some evidence may suggest that absolute cumulative output is not an appropriate choice as the "experience" variable.[55] I shall use .49 as my estimate of ϵ, and 31 as my estimate of ϕ (the constant that determines chips initially yielded per wafer).

THE DEMAND FOR 1M DRAMS. There is little reliable information on the price elasticity of demand for DRAMs. Robert Wilson, Peter Ashton, and Thomas Egan estimate this price elasticity to range between -1.8 and -2.3 on the basis of a graph of the logarithm of bit price versus the logarithm of bits sold; William Finan and Chris Amundsen report a price elasticity of -1.8 on the basis of a simple regression of the logarithm of bit price on the logarithm of bits sold worldwide.[56] Neither of these

---

53. The F-statistic for this hypothesis was 1.12 with 7, 4 degrees of freedom; the Wald chi-square statistic was 7.84 with 7 degrees of freedom.

54. The F-statistic leads us not to reject the hypothesis at either the 5 or the 10 percent significance level, while the Wald chi-square statistic leads us to reject the hypothesis at the 10 percent, but not at the 5 percent significance level.

55. Estimation of a learning elasticity requires data on either current and cumulative output, or current average variable cost and cumulative output. The dubious practice of using price as a proxy for current unit cost will almost certainly lead to incorrect results, since the simple models of pricing behavior reviewed above suggest that market prices will diverge from either current average or marginal cost.

56. Robert W. Wilson, Peter K. Ashton, and Thomas P. Egan, *Innovation, Competition, and Government Policy in the Semiconductor Industry* (Lexington, Mass.: Lexington Books, 1980), pp. 126–27; William F. Finan and Chris B. Amundsen, "An Analysis of the Effects of Targeting on the Competitiveness of the U.S. Semiconductor Industry," study prepared for the Office of the U.S. Trade Representative, the Department of Commerce, and the Department of Labor, Quick, Finan and Associates (Washington, 1986), p. C-18; and Finan and Amundsen, "Modeling US-Japan Competition in Semiconductors," *Journal of Policy Modeling*, vol. 8 (Fall 1986), p. 321.

estimates makes any attempt to control for the effect of variation in the overall level of economic activity on chip demand. In previous work I estimated an overall price elasticity of demand for semiconductors used in the computer industry of -1.6, assuming a quality adjustment equivalent to the improvement in bit density observed in DRAMs, and chip use in computers fixed in proportion to computer output.[57]

To get as reliable an estimate as possible for 1M DRAM demand, I estimated a demand function giving the logarithm of quantity shipped of 1M DRAMs as a linear function of the logarithm of real 1M DRAM price, logarithms of real prices for 64K and 256K DRAMs (as possible substitutes), the logarithm of real U.S. GNP, and a linear trend.[58] The last two variables were included to capture the effect of fluctuations in overall economic activity and intergenerational "transition" effects.[59] The implicit GNP price deflator, rebased so that the fourth quarter of 1989 was equal to one, was used to deflate all monetary values to "real" fourth-quarter 1989 levels. Deflated GNP and substitute DRAM prices were converted to indexes with a value of one in the fourth quarter of 1989; as a result, the constant in a regression equation may be interpreted as the level of DRAM demand corresponding to fourth-quarter 1989 values for these variables.

In principle, because of the long time lag in the DRAM production process (almost a full quarter), we would not expect disturbances in the demand for DRAMs to have a significant effect on production within

57. See Joseph Grunwald and Kenneth Flamm, *The Global Factory: Foreign Assembly in International Trade* (Brookings, 1985), pp. 130–32.

58. This demand equation should be interpreted as a "final" demand for DRAMs. That is, demand for semiconductor inputs can be derived from a cost function C for semiconductor-using products of the form $C(\mathbf{p}, Z)$, where $\mathbf{p}$ is a vector of input prices (including semiconductors) and Z is some given level of electronic system output making use of semiconductor inputs. Demand for DRAMs is simply the partial derivative of C with respect to the price of DRAMs, $C_{Pdram}(\mathbf{p}, Z)$. The demand for Z is a function of electronic system prices (P_Z), the prices of other finished goods (P_O), and the level of aggregate economic activity (X); that is, $Z = f(P_Z, P_O, X)$. Electronics systems prices in turn are related to their unit cost of production $C(p, Z)/Z$ in competitive markets, so $P_Z = g[C(\mathbf{p}, Z)/Z]$. We then have $Z = f\{g[C(\mathbf{p}, Z)/Z], P_O, X\}$ implicitly giving Z as a function of $\mathbf{p}$, P_O, and X. Substituting this implicit function for Z in derived demand for DRAMs $C_{Pdram}(\mathbf{p}, Z)$, we then have a "final" demand for DRAMs as a function of input prices, prices of other finished goods, and the overall level of economic activity.

59. The data on quantity are Dataquest estimates and cover quarterly worldwide shipments from the second quarter of 1985 to the fourth quarter of 1989 by merchant producers. Data for DRAM prices are also quarterly Dataquest estimates of average sales price over this same period. Real (deflated) GNP and the implicit GNP price deflator are taken from *Economic Report of the President*, various years.

that same quarter, and therefore we would not expect simultaneity between price and quantity within any given quarter to be an important complication in estimating this demand function. Nonetheless, in addition to ordinary least squares, I used an instrumental variables estimator to estimate the coefficients of this equation.[60] The estimated regression equations, using both methods, are shown in table 6-2.

All fully specified regressions produced an estimated price elasticity near −1.5. Dropping either the linear time trend variable or the logarithm of real GNP as a measure of aggregate economic trends affecting demand had little effect on the estimated own-price elasticity. Interestingly, dropping both GNP and the time trend substantially raised the estimated price elasticity, to −2.1, near the range where other estimates of DRAM price elasticities—which, as already noted, ignore variation in aggregate economic activity—have clustered.

On the basis of these results, −1.5 was used as an estimate of 1M DRAM own-price elasticity β, and the value 190,000 as an estimate of product life cycle demand "level" α.[61] To transform this demand function into a demand for "gross" fabricated dice (prior to test and assembly losses), it was assumed that net output of tested and finished chips equals 0.9 times the number of good dice produced in wafer fab.[62] With the functional form assumed, a simple transformation of α is merely substituted for its original value in order to derive the appropriate inverse demand function.[63]

COST PARAMETERS. Using estimated 1989 values found in the 1990 VLSI Research study, I estimated r (capital cost per unit of product cycle wafer capacity) to be $240. Variable cost per wafer processed (including

60. The logarithms of cumulative output through the previous quarter of 1M, 256K, and 64K DRAMs were used—in addition to the other exogenous variables included in demand—to instrument DRAM price variables. Cumulative output would be expected to affect supply of DRAMs, via learning economies, but have no direct effect on demand. Values lagged one quarter were used as additional instruments.

61. Exp(22.97) multiplied by 20 (= 190,000 million) gives the demand that would be observed at a 1M DRAM price of $1 over a twenty-quarter (five-year) product cycle, given real output and substitute price levels prevailing in the fourth quarter of 1989.

62. For estimated test and assembly yields in this general neighborhood, see VLSI Research, *Depreciation Schedules*, addendum A; and ICE, *Mid-Term 1988*, pp. 7-16 to 7-17.

63. That is, $P(\xi z)\xi = (\xi Ny/\alpha)^{1/\beta}\xi = (Ny/\alpha')^{1/\beta}$, where $\alpha' = \alpha\xi^{-(1+\beta)}$.

Table 6-2. *Regression Results from Analysis of 1M DRAM Demand*[a]

Variable	Ordinary least-squares			Two-stage least-squares		
	Estimated coefficient	Standard error	t-statistic	Estimated coefficient	Standard error	t-statistic
Log of price						
64K	1.23	2.43	0.51	0.78	2.67	0.29
256K	0.63	1.68	0.37	−0.01	1.86	−0.01
1M	−1.47	0.49	−2.99	−1.54	0.51	−3.01
Trend	0.29	0.27	1.06	0.20	0.29	0.67
Log real GNP	−1.76	36.26	−0.05	10.77	39.67	0.27
Constant	22.97	1.02	22.54	23.24	1.09	21.40
R^2			0.95			0.949
Std. error of regression			0.726			0.736
Degrees of freedom			12			12
Log of price						
64K	1.17	2.02	0.58	1.17	2.14	0.55
256K	0.59	1.45	0.41	0.23	1.54	0.15
1M	−1.47	0.47	−3.11	−1.53	0.49	−3.14
Trend	0.28	0.10	2.81	0.27	0.10	2.70
Constant	22.97	0.98	23.51	23.22	1.04	22.43
R^2			0.9503			0.95
Std. error of regression			0.698			0.702
Degrees of freedom			13			13
Log of price						
64K	0.66	2.38	0.28	0.27	2.57	0.10
256K	−0.07	1.56	−0.04	−0.58	1.66	−0.35
1M	−1.55	0.49	−3.19	−1.61	0.50	−3.21
Log real GNP	33.99	13.77	2.47	35.70	14.32	2.49
Constant	23.01	1.02	22.47	23.33	1.08	21.56
R^2			0.946			0.944
Std. error of regression			0.929			0.739
Degrees of freedom			13			13
Log of price						
64K	5.40	1.65	3.27	5.46	1.74	3.14
256K	0.87	1.77	0.49	0.36	1.88	0.19
1M	−2.17	0.49	−4.44	−2.19	0.52	−4.23
Constant	23.67	1.15	20.51	23.87	1.23	19.42
R^2			0.92			0.919
Std. error of regression			0.852			0.857
Degrees of freedom			14			14

Source: Author's calculations.

a. Prices instrumented with constant, trend, cumulative output of 64K, 256K, and 1M DRAMs, and LRGNP; LRGNP and cumulative outputs lagged one year.

materials, labor, and wafer probe testing) was estimated to be $390.[64] Test and assembly costs were assumed to equal 23 cents for the integrated circuit package, and about 52 cents for assembly and final testing, for a total of 75 cents per device produced.[65]

Overhead is normally a significant part of semiconductor cost. From aggregate historical data for the 1981–87 period, I have assumed 36 cents in general, administrative, and selling costs for every dollar of direct manufacturing cost.[66] Thus, estimates for c, d, and r given above were marked up an additional 36 percent. Table 6-3 shows the assumed empirical parameter values used.

Baseline Simulations

Table 6-4 gives the optimal values of t_s and K derived from numerical solution of the optimal control problem described above. The roots of a system of two nonlinear equations in two unknowns (equations B-1 and B-2 in appendix 6-B), or one equation in one unknown (in the case where full-blast production over the entire product life cycle is the optimal policy, equation B-2' in appendix 6-B) were sought. Table 6-4 also shows a "gross rent," defined as profits net of all costs other than fixed entry cost F, received by each producer. The columns of table 6-4 correspond to different assumed numbers of firms in the industry, and the rows to differing assumptions about parameter γ, which defines the experience variable relevant to learning economies.

Since identical firms are assumed to make up the industry in equilibrium, one may "close" the model by assuming free entry, that is, that firms enter the industry up to the point where gross rent per firm just covers the fixed cost of entry F. Because we are restricted to an integer number of firms, we define the equilibrium number of firms as that where one more entrant reduces rent per firm below the entry cost F. As a

64. VLSI Research, *Depreciation Schedules*, addendum 2, puts material and labor cost at $381 per wafer processed in 1989; a $10 wafer probe test cost is added based on ICE, *Mid-Term 1988*, p. 7-9.

65. The package cost comes from conversations with Dan Hutcheson of VLSI Research, and the assembly cost is estimated to range from 7 to 20 cents offshore, or from 10 to 50 cents per device onshore (in the United States, Europe, and Japan) according to ICE, *Mid-Term 1988*, pp. 7-16 to 7-18. I have used a "typical" value of 32 cents. Final testing cost is estimated in ICE, p. 7-18 to be 20 cents per unit, for a grand total of 75 cents for package, assembly, and final testing.

66. The data on which this calculation is based are found in ICE, *Mid-Term 1988*, p. 7-20. I have excluded R&D and interest expense as elements of overhead.

Table 6-3. *Assumed Values for Empirical Parameters*

Symbol	Parameter	Assumed value
α^0	"Level" of life-cycle demand for assembled and tested units at \$1/chip	190 billion units
β	Price elasticity of demand	-1.5
ξ	Share of good, yielded chips as a fraction of good dice after assembly and final test	0.9
α	Level of demand for "gross" fabricated dice (including units rejected at final test)	$\alpha_0\,\xi^{-(1+\beta)}$
ϕ	Learning curve wafer fab yield "level" parameter	31
E_0	Initial "experience" at time 0	0.01
ϵ	Experience elasticity of wafer fab yield	0.49
γ	(Gamma) parameter determining experience variable	0–1
m	Overhead expense per dollar of direct manufacturing cost	0.36
d	Cost of packaging, assembly, and final testing per fabricated unit	\$0.75 $(1+m)$
c	Fabrication cost per processed wafer	\$390 $(1+m)$
r	Capital cost per unit life cycle wafer processing capacity	\$240 $(1+m)$

Source: Author's calculations.

consequence of the integer number of firms, the symmetric equilibrium so defined will generally be characterized by some small, positive rent (net of F).

I shall assume that the fixed entry cost (primarily total R&D costs for the 1M DRAM) that must be invested before mass production of the 1M DRAM can occur is between roughly \$250 million and \$500 million. Thus, for $\gamma = 1$, if F amounts to \$250 million, we would expect to find fourteen identical firms in the industry, each with facilities capable of producing 4.66 million wafer starts over a five-year product life cycle. With F at \$500 million we would expect nine producers, each with capacity to produce 6.94 million wafer starts over the product cycle. In either case the optimal policy would involve full-blast production over the entire life cycle. Thus one observation that emerges immediately from table 6-4 is that, with γ equal to one (which I argued earlier is a heuristically appealing specification), small differences in fixed entry costs can make a large difference in the industrial structure of the industry (number of firms observed). The same cannot be said for γ much less than one.

Table 6-5 summarizes some characteristics of industry equilibria derived from table 6-4 under differing assumptions about F. I have taken F as either $500 million or $250 million; these values are best interpreted as bracketing a range of feasible values. Alongside the equilibrium number of firms, the Hirschman-Herfindahl index of concentration is also shown.[67]

To get at the issue of whether or not dumping is observed, I have calculated observed prices and various cost concepts at 100 equally spaced points over the product life cycle. One useful cost concept is current short-run marginal cost, which in my model happens to be constant at any moment in time, coincides with average variable cost, and is equal to $d + c/w$. This is the incremental cost saved when output is reduced by one unit. Another important cost concept is fully allocated long-run average cost (LRAC). To define this concept I have assumed straight line depreciation in spreading capital and fixed entry costs over the product life cycle: an equal amount of these fixed costs is allocated to every moment in time. Capital and fixed entry cost per unit is then calculated by dividing fixed costs corresponding to time t by the number of units $y(t)$ produced at that moment. Adding average variable cost to average fixed cost gives LRAC $= d + c/w + F/y + r/uw$. Multiplying LRAC by output at any instant, and summing these costs for every instant over the product cycle, gives the total cost of producing some time-varying path of output over the entire product cycle.

Table 6-5 shows that, assuming $\gamma = 1$, price falls short of short-run marginal cost over the first 3 percent of the product life cycle, and falls short of average cost over roughly the first third of the cycle. With $\gamma = 0$, in contrast, price is less than marginal cost only at the very beginning of the product cycle; price is less than average cost over two distinct periods: at the very beginning, and over roughly the second quarter of

67. This index is defined as

$$HHI = \sum_{i=1}^{n} s_i^2,$$

where s_i is the market share of company i. The index ranges in value from one, with monopoly, to zero, with a competitive industry composed of an infinite number of equally sized firms. In the special case of N identical firms, this index is just equal to $1/N$.

Table 6-4. Solution of the Optimal Control Problem, Nonstrategic Baseline Simulation

	Number of firms															
	1	2	3	4	5	6	7	8	9	10	11	12	13	14	15	16
Gamma = 1																
Wafer-processing capacity (K)[a]	13.49	19.08	16.03	13.34	11.31	9.79	8.62	7.69	6.94	6.32	5.8	5.36	4.98	4.66	4.37	4.11
End, full-blast output (t_s)	1	1	1	1	1	1	1	1	1	1	1	1	1	1	1	1
Gross rent[b]	29,370	10,380	4,985	2,902	1,894	1,332	988	761	604	491	407	343	293	253	221	195
Gamma = 0.9																
Wafer-processing capacity (K)[a]	15.46	21.92	18.4	15.73	12.95	11.19	9.84	8.77	7.91	7.2	6.6	6.09				
End, full-blast output (t_s)	1	1	1	1	1	1	1	1	1	1	1	1				
Gross rent[b]	32,360	10,370	4,360	2,180	1,191	676	382	204	90	15	−36	−70				
Gamma = 0.8																
Wafer-processing capacity (K)[a]	16.96	23.91	20.12	16.76	14.23	12.32	10.86	9.68	8.74	7.96	7.3	6.74				
End, full-blast output (t_s)	1	1	1	1	1	1	1	1	1	1	1	1				
Gross rent[b]	35,570	10,410	3,752	1,455	477	5	−240	−371	−442	−479	−496	−500				
Gamma = 0.7																
Wafer-processing capacity (K)[a]	17.683	24.59	20.84	17.48	14.94	13	11.492	10.291								
End, full-blast output (t_s)	1	1	1	1	1	1	1	1								
Gross rent[b]	38,840	10,630	3,296	846	−151	−601	−811	−907								

Gamma = 0.68

Wafer-processing capacity (K)[a]	17.73	24.56	20.86	17.53	15	13.07
End, full-blast output (t_s)	0.98	0.95	0.96	0.98	1	1
Gross rent[b]	39,480	10,700	3,232	747	−258	−707

Gamma = 0.6

Wafer-processing capacity (K)[a]	17.77	24.4	20.81	17.57	15.11	13.22
End, full-blast output (t_s)	0.86	0.82	0.84	0.86	0.88	0.9
Gross rent[b]	41,950	11,050	3,061	428	−619	−1,074

Gamma = 0.5

Wafer-processing capacity (K)[a]	17.57	23.82	20.45	17.39	15.04	13.23
End, full-blast output (t_s)	0.72	0.68	0.7	0.72	0.75	0.77
Gross rent[b]	44,820	11,630	3,029	197	−924	−1,406

Gamma = 0.3

Wafer-processing capacity (K)[a]	16.69	22.06	19.18	16.53	14.46	12.86
End, full-blast output (t_s)	0.51	0.47	0.49	0.51	0.53	0.55
Gross rent[b]	49,660	13,020	3,414	187	−1,120	−1,697

Gamma = 0

Wafer-processing capacity (K)[a]	14.88	19.02	16.82	14.76	13.12	11.83
End, full-blast output (t_s)	0.31	0.28	0.29	0.31	0.33	0.34
Gross rent[b]	54,880	15,140	4,492	782	−796	−1,543

Source: Author's calculations.
a. Millions of product-cycle wafer starts.
b. Millions of dollars.

Table 6-5. Semiconductor Industry Characteristics under Alternative Symmetric Equilibrium Conditions[a]

Gamma	Number of firms	Hirschman-Herfindahl index	Final wafer yield[b] $w(E(1))$	Initial company output rate[c] $y(0)$	Final company output rate[d] $y(1)$	Segments of product cycle[e] Price is less than SRMC	Price is less than LRAC
				$F = \$500$ million[f]			
1.0	9	0.1111	442.4	22.5	3,070.1	0–0.03	0–0.32
0.9	6	0.1667	557.2	36.3	6,235.2	0–0.03	0–0.29
0.8	4	0.2500	758.5	54.4	12,712.6	0–0.02	0–0.19
0.7	4	0.2500	1,005.8	56.7	17,580.4	0–0.01	0–0.17 0.93–1.00
0.68	4	0.2500	1,063.2	56.9	18,283.4	0–0.01	0–0.16 0.87–1.00
0.6	3	0.3333	1,397.8	67.6	24,906.8	0–0.01	0–0.07
0.5	3	0.3333	1,796.8	66.4	27,263.7	0–0	0–0.05
0.3	3	0.3333	2,789.3	62.3	31,038.4	0–0	0–0.02 0.48–0.50
0	4	0.2500	4,321.3	47.9	28,019.6	0–0	0–0.01 0.24–0.63

$F = 250$ million

1.0	14	0.0714	442.4	15.1	2,061.5	0–0.03	0.–035
0.9	7	0.1429	550.4	31.9	5,416.0	0–0.03	0–0.29
0.8	5	0.2000	735.1	46.2	10,460.6	0–0.02	0–0.24
0.7	4	0.2500	1,005.8	56.7	17,580.4	0–0.01	0–0.15 / 0.97–1.00
0.68	4	0.2500	1,063.2	56.9	18,283.4	0–0.01	0–0.15 / 0.91–1.00
0.6	4	0.2500	1,313.7	57.0	20,157.1	0–0.01	0–0.12 / 0.71–1.00
0.5	3	0.3333	1,796.8	66.4	27,263.7	0–0	0–0.05
0.3	3	0.3333	2,789.3	62.3	31,038.4	0–0	0–0.02 / 0.49–0.49
0	4	0.2500	4,321.3	47.9	28,019.6	0–0	0–0.01 / 0.25–0.58

Source: Author's calculations.

SRMC = short-run marginal cost; LRAC = long-run average cost

a. $w(E(0))$ is yield of good chips per wafer fabrication process at beginning of product cycle; $w(E(0))$ is 3.25 in all cases.

b. Yield of good chips per wafer at end of product cycle.

c. Instantaneous company production rate at beginning of product cycle. In millions of chips per product life cycle.

d. Instantaneous company production rate at end of product cycle. In millions of chips per product life cycle.

e. Entire product cycle is from time 0 to time 1.

f. F is up-front sunk costs independent of output levels.

the cycle. Indeed, for all values of γ, given my assumptions about other parameter values, price falls short of marginal cost only at the very beginning of the product cycle. Further perusal of this table makes clear, however, that the timing of periods of sales at less than average cost is quite sensitive to the specification of the experience variable—depending on γ, such episodes can occur at the beginning of the product cycle, the middle, or the end, or in some combination.

Table 6-5 also shows that the value of γ makes a big difference in the structure of a symmetric-industry equilibrium. With cumulative output per facility ($\gamma = 1$) the relevant experience variable, a relatively large number of firms (nine to fourteen) populate the industry. With γ much below .9, no more than three or four firms make up the industry.

Figure 6-3 shows the path of price, marginal revenue, marginal cost, and average cost over time in the case where entry costs are $250 million and γ equals one. Figure 6-4 shows the time path for these variables over the product cycle when γ is instead equal to zero.

Ironically, the specification of firm behavior in the B-K model—full-blast production over the entire product cycle—turns out to be optimal if parameter γ is close to one (see table 6-4). The irony arises because the B-K model also specifies absolute cumulative output ($\gamma = 0$) as the experience variable, and given realistic choices for other parameters, optimal behavior would then require cutting back production to levels below capacity after about the first third of the product cycle.

REALITY CHECKS. How plausible are these simulations, and do they suggest anything about the realism of various assumptions about parameters? One straightforward way to evaluate the model is to compare the predicted industry structure with observed industry structure. Recall that figure 5-15 showed Hirschman-Herfindahl concentration indexes constructed from Dataquest estimates of annual producer shipments of various generations of DRAMs.[68] For virtually all generations of DRAM

68. These indexes are calculated from unpublished Dataquest estimates of DRAMs shipped from 1974 through the end of 1989. Note that there were two distinct varieties of 16K DRAM, one with a single voltage power source, the other requiring dual voltages; each is treated as a separate product in this figure. In calculating concentration indexes for 1M DRAMs, I have allocated Motorola-labeled product to Toshiba (since virtually all of Motorola's product over this period is believed to have been assembled from Toshiba-fabricated dice, or produced by a Toshiba-Motorola joint venture); 1M DRAMs bearing the Intel label have been assigned to Samsung, since it is believed that virtually all of Intel's sales over this period were "private labeled" Samsung product. Neither of these adjustments has a particularly significant effect on the pattern of concentration.

Figure 6-3. *Time Profile of 1M DRAM Costs and Prices in Simulated Nonstrategic Equilibrium with Gamma = 1*[a]

Dollars per chip

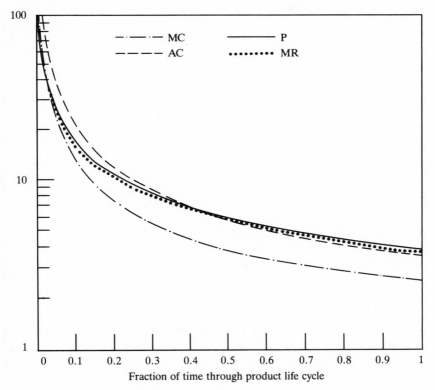

Fraction of time through product life cycle

Source: Author's calculations.
a. MC = short-run marginal cost; AC = long-run average cost; P = price; MR = marginal revenue.

the concentration index declines sharply from an very high initial level, as one producer after another comes on line with volume production. The index then levels off near .1, rising sharply at the end of the product cycle as producers drop the product line one after another. Although the early phases of the 256K and 1M DRAM may have been somewhat more concentrated than in earlier generations' life cycles, they too seem destined to eventually follow this pattern.

If one compares the Hirschman-Herfindahl indexes associated with my simulations with the pattern depicted in figure 5-15, only the results associated with the specification of cumulative output per facility

Figure 6-4. *Time Profile of Semiconductor Costs and Prices in Simulated Nonstrategic Equilibrium with Gamma = 0*[a]

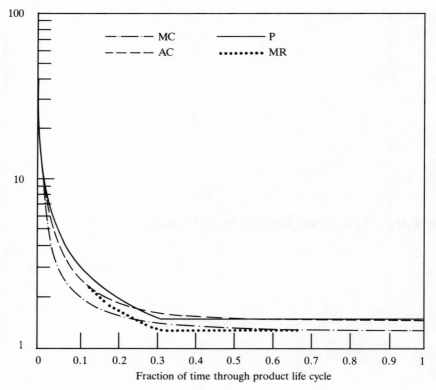

Dollars per chip

Fraction of time through product life cycle

Source: Author's calculations.
a. MC = short-run marginal cost; AC = long-run average cost; P = price; MR = marginal revenue.

($\gamma = 1$) as the experience variable fit reasonably closely. Note that my assumption of symmetric firms means that the associated Hirschman-Herfindahl index of concentration must be constant over time. While conceding that my model is at best an approximation to reality, I conclude that only a value of γ close to one yields predicted behavior that is reasonably close to industrial reality.

Another cut at this question may be had by comparing predicted with actual paths for DRAM prices over time. To do so, I have assumed that a five-year product cycle for the 1M DRAM effectively began in 1988 (although small quantities were produced as far back as late 1985, quan-

Figure 6-5. *Historical Prices for 1M DRAMs Compared with Simulated Nonstrategic Equilibrium Time Profiles*

Dollars per chip

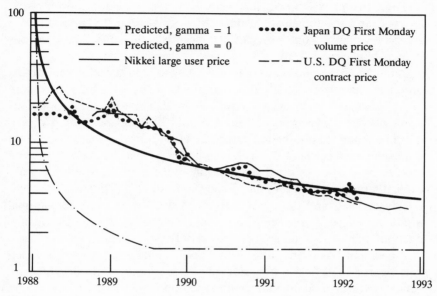

Source: Author's calculations.

tity production did not really ramp up until 1988). Figure 6-5 charts the actual behavior of one set of estimates of large-volume contract prices for 1M DRAMs in the U.S. and Japanese markets through autumn 1993, along with simulated 1M DRAM price levels associated with assumed γ equal to one and zero, respectively.[69] The period from the first quarter of 1988 through the first quarter of 1989 was a period of extreme shortage in DRAM markets, whereas the period after late 1989 was marked by lackluster demand. Given that the early portion of my empirical approximation to the learning curve is probably poorer than in later periods (see the discussion above), and that my assumption of symmetric firms is probably least appropriate in the early stages of the product cycle, I am not surprised to find that the very earliest part of the predicted time

69. These data are monthly averages of Nikkei and Dataquest estimates of average contract prices in these markets. The data are reported in *Computer Reseller News* and *Nihon Keizai Shimbun*, various issues. For more on the strengths and weaknesses of these data, and a thorough discussion of the segmented spot and contract markets in which DRAMs are sold, see Flamm, "Measurement of DRAM Prices," pp. 157–97.

path for prices seems the least accurate. All things considered, the simulation with $\gamma = 1$ seems to do a reasonable job of tracking real 1M DRAM prices. The simulation with $\gamma = 0$ clearly does not.

Thus, two pieces of evidence—observed and predicted concentration indexes, and the time path of DRAM prices—seem to suggest that a value of γ close to one provides significantly more realistic predictions than a value close to zero.

A final point to consider is that, according to industry folklore, DRAM producers have traditionally run their plants at full blast when they were in operation at all, a behavior that is consistent with the simulations presented here. However, beginning in mid-1989 Japanese DRAM producers announced production cutbacks. This raises three issues that I do not explore further here. First, DRAM capacity may be shifted, at some cost, to production of other types of integrated circuits, a possibility not explicitly incorporated into my model. Second, DRAM demand is notoriously cyclical, and the consequences of shifts in demand for optimal producer behavior are, again, not explicitly explored here. Third, production of DRAMs after the conclusion of the 1986 Semiconductor Trade Arrangement was clearly affected by political constraints, may have led to a degree of collusive behavior among producers, and otherwise involved political economic factors not incorporated into my model.

THE DUMPING ISSUE. Given empirical values deemed to be plausible in the case of 1M DRAMs, the exercises portrayed in tables 6-4 and 6-5 suggest that use of a short-run marginal cost test for dumping as a screen for potentially predatory behavior is only likely to give false positive results (pricing below current marginal cost absent strategic behavior) in the very earliest stages of the product cycle. One might interpret this to mean that a marginal cost–based dumping test might be defensible if some sort of exception to a marginal cost–based pricing standard is granted when a new product is first introduced. But it is not clear how robust this conclusion is to changes in the empirical parameters used in my simulations; further sensitivity analysis might shed greater light on this question.

The same cannot be said for an average cost test for predation. Depending on parameter values, episodes of below-average-cost pricing can occur in virtually any part of the product life cycle, even when producer behavior is entirely nonstrategic.

Figure 6-6. *Time Profile of Semiconductor Costs and Prices in Simulation with Very High Initial Yields*[a]

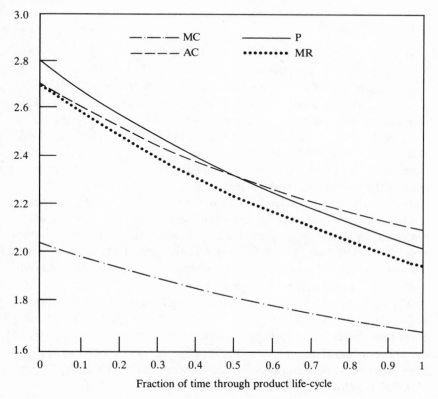

Dollars per chip

Fraction of time through product life-cycle

Source: Author's calculations.
a. MC = short-run marginal cost; AC = long-run average cost; P = price; MR = marginal revenue.

Indeed, while the simulations depicted in tables 6-4 and 6-5 all show an episode of below-average-cost pricing at the beginning of the product cycle, possibly followed by a later episode, it would be incorrect to assert that below-average-cost pricing will always necessarily be observed at the beginning of the product life cycle.[70] Figure 6-6 shows that by artfully changing a single parameter (in this case, by greatly raising initial yields, making $E_0 = 500$), assuming $\gamma = 1$ and $F = \$250$ million, one arrives at a symmetric-industry equilibrium where price *never* falls below mar-

70. Dick, "Learning by Doing," pp. 144–46, proposes this behavior.

ginal cost, and price falls below average cost only during the latter half of the product cycle.[71]

Conclusions

In recent years, pricing below a constructed long-run average cost has become the principal grounds for applying the U.S. dumping laws to U.S. imports of some foreign products. Although this practice has little obvious economic defense, it is possible to argue that a test based on marginal cost might serve as a useful screen for potentially predatory behavior by foreign exporters. However, in the presence of learning economies, such as are thought to be present in many high-tech industries, including semiconductors, below-marginal-cost pricing can be rational even in the absence of strategic behavior such as predation.

In this chapter I have developed a more realistic model of pricing over the life cycle of a product in which both fixed costs and learning economies are significant. Using empirically plausible parameters for production of 1M DRAMs, and assuming nonstrategic producer behavior, I have found that below-marginal-cost pricing is likely to be observed only in the very earliest stages of the product cycle.

The analysis has also shed considerable light on other facets of pricing and production over the product life cycle. A specification of learning economies based on cumulative output per facility as the "experience" variable was found to yield results that were considerably more realistic than other possible—and widely used—specifications. Contrary to popular belief, below-average-cost pricing does not necessarily have to occur near the beginning of the product cycle.

The model presented here appears to produce fairly realistic predictions of industry structure and pricing behavior when used with empirically plausible parameters. Extensions of this work to consider the possibility of strategic, noncooperative behavior on the part of producers, as well as cooperative or collusive behavior, are explored in the next chapter. Policy issues to be explored within this framework will focus on the impact of government policies on industry structure and aggregate national welfare.

71. The symmetric equilibrium depicted in this figure corresponds to an industry with eighteen producers, each with a product life cycle capacity of 3.88 million wafer starts, producing full blast over the entire cycle.

Appendix 6-A: A General Solution to the Problem of Optimal Capacity and Production Choice

Let:

$y(t) =$	company's output at time t
$x(t) =$	output of other companies
$F =$	fixed cost of entry
$R[y(t),x(t)] =$	company's revenues at time t
$d =$	assembly and test cost per chip
$c =$	wafer processing cost
$u(t) =$	utilization rate at time t
$K =$	fixed capacity
$r =$	capital cost per unit capacity
$w =$	yielded good chips per wafer
$Q =$	cumulative company output
$\gamma =$	the experience specification parameter, $0 \leq \gamma \leq 1$
$E =$	"yield relevant" experience $(E_0 + Q/K^\gamma)$.

The general problem described in the text is:

$$\max_{u(t),K} \int_0^1 \{R[y(t), x(t)] - F - dy(t) - cu(t)K - rK\}dt,$$

where $y(t) = u(t) \, w(E) \, K$,

$$\text{s.t. } \dot{E} = \frac{y(t)}{K^\gamma} = u(t) \, w(E) \, K^{1-\gamma} \quad w_E > 0,$$

$$u\epsilon[0, 1], \qquad\qquad\qquad w > 0.$$

Form the Hamiltonian (suppressing the arguments of functions for notational simplicity):

$$H = R - F - dy - cuK - rK + \delta \frac{y}{K^\gamma}$$

$$\text{with } \dot{\delta} = -\frac{\partial H}{\partial E} = -\frac{\partial H}{\partial y}\frac{\partial y}{\partial E} = -\left(R_y - d + \frac{\delta}{K^\gamma}\right) uw_E K,$$

where R_y denotes $\dfrac{\partial R}{\partial y}$, marginal revenue.

By the Maximum Principle:

choose $u(t)$ to maximize H, given $u \in [0,1]$, with $\delta(1) = 0$ (transversality condition).

There are three possible cases to consider:

$$0 < u < 1; \frac{\partial H}{\partial u} = \left(R_y - d + \frac{\delta}{K^\gamma} \right) \frac{\partial y}{\partial u} - cK = 0$$

(A-1.a)
$$\left(R_y - d + \frac{\delta}{K^\gamma} - \frac{c}{w} \right) Kw = 0, \dot{\delta} = -\frac{c}{w} u w_E K < 0,$$

$$R_y - d - \frac{c}{w} + \frac{\delta}{K^\gamma} = 0.$$

At an interior maximum, we also have a second-order necessary condition:

$$\frac{\partial^2 H}{\partial u^2} = R_{yy} (Kw)^2 \le 0.$$

(A-1.b)
$$u = 1 \, R_y - d - \frac{c}{w} + \frac{\delta}{K^\gamma} \ge 0$$

$$\dot{\delta} = -\left(R_y - d + \frac{\delta}{K^\gamma} \right) w_E K < 0.$$

(A-1.c)
$$u = 0; R_y - d - \frac{c}{w} + \frac{\delta}{K^\gamma} \le 0 \quad \dot{\delta} = 0.$$

Together with the transversality condition, this implies that $\delta \ge 0$ everywhere.

Along an interior segment, that is, possibility A.1.a,

$$R_y = d + \frac{c}{w} - \frac{\delta}{K^\gamma}$$

(A-1.d)
$$(\dot{R}_y) = \left(-\frac{c}{w^2} w_E \frac{y}{K^\gamma} \right) - \frac{\dot{\delta}}{K^\gamma}$$

$$= -c \frac{w_E}{w^2} u w K^{1-\gamma} - \left(-cu \frac{w_E K^{1-\gamma}}{w} \right) = 0,$$

which means R_y is constant along an interior segment.

The above analysis holds for any given K. Full optimization requires that optimal parameter K must be chosen to satisfy:

$$\int\limits_0^1 \frac{\partial H[\delta(t), K, u^*(t), E^*(t)]}{\partial K} \, dt = 0,$$

where $u^*(t)$ is the optimal utilization rate and E^* the trajectory of experience variable E corresponding to $u^*(t)$.[72] Then

$$\int\limits_0^1 \left[\left(R_y - d + \frac{\delta}{K^\gamma} \right) \frac{\partial y}{\partial K} + R_x \frac{\partial x}{\partial K} - \gamma \frac{\delta}{K^{1-\gamma}} y - cu - r \right] dt = 0.$$

For the moment, I will assume nonstrategic behavior—the firm perceives $\frac{\partial x}{\partial K} = 0$, so,

$$\int\limits_0^1 \left[\left(R_y - d + \frac{\delta}{K^\gamma} \right) uw - \gamma \frac{\delta}{K^\gamma} uw - cu - r \right] dt = 0.$$

Thus,

$$\int\limits_0^1 \left[\left(R_y - d - \frac{c}{w} + \frac{\delta}{K^\gamma} \right) uw - \gamma \frac{\delta}{K^\gamma} uw - r \right] dt = 0.$$

The expression in parentheses above:

$= 0$ if u is interior;
≥ 0 if $u = 1$;
≤ 0 if $u = 0$;

while the remaining expressions in brackets sum to a negative number.

This shows that the expression in parentheses must be positive, and $u = 1$, over some interval if optimal $K > 0$ (and any output is produced). The identical trajectory of output (and variable profit) could otherwise be produced at lower cost by choosing some smaller $\overline{K}$, then choosing utilization rate $\overline{u}(t) = K/\overline{K} \, u(t) < 1$.

The result just obtained is perfectly general. To explicitly solve for an optimal path, we add additional structure and make the following further assumptions:

72. See George Leitman, *An Introduction to Optimal Control* (McGraw-Hill, 1966), pp. 98–100.

—Industry inverse demand (and firm revenue function R) and learning function w are twice continuously differentiable functions in all their arguments.

—There are N symmetric firms.

—Industry revenues are an autonomous, strictly concave function of industry output.

Let $\overline{R}(z)$ be industry revenues and $z(t)$ industry output, such that $z(t) = x(t) + y(t)$; then $\overline{R}(z) = P(z)z$, where $P(z)$ is industry inverse demand.

I will assume $\overline{R}_{zz} = P''z + 2P' < 0$, for $z \geq 0$, (that is, $\overline{R}$ is strictly concave). So

$$\text{(A-2.a)} \qquad\qquad P'' < -\frac{2P'}{z} \quad \text{for } z \geq 0.$$

I have also assumed that $R[y(t),x(t)]$ is autonomous, not a function of time other than through $x(t)$ and $y(t)$. Since output is a perfectly homogeneous commodity, with a single market price assumed,

$$R[y(t),x(t)] = P(x + y)y,$$

$$R_y = P' \frac{dz}{dy}y + P = P'(1 + \lambda)y + P,$$

with λ ($= dx/dy$) the conjectural variation perceived by the firm. I shall regard λ as a constant varying between 0 and $N - 1$.

These limits parametrize λ as lying between two useful limiting cases of industrial organization:

$\lambda = 0$ with Cournot competition,

$\lambda = N - 1$ with a constant-market-share cartel made up of N identical firms.

I will be assuming $\lambda = 0$ for the moment, but I will develop my analysis of optimal utilization rates including the case of a constant-market-share cartel.

Note that

$$R_{yy} = P'' \left(\frac{dz}{dy}\right)^2 y + 2P' \frac{dz}{dy} = (1 + \lambda)[P''(1 + \lambda)y + 2P']$$

$$< (1 + \lambda)2P'\left[1 - \frac{y}{z}(1 + \lambda)\right] \text{ using (A-2.a).}$$

Now, consider two cases. First, under Cournot competition, $\lambda = 0$. Since

$$\frac{y}{z} \leq 1,\ 1 - \frac{y}{z}(1 + \lambda) > 0,\text{ and } R_{yy} < 0.$$

Second, with N identical firms in a collusive, fixed-market-share cartel, $\lambda = N - 1$,

$$\frac{y}{z} = \frac{1}{N},\ 1 - \frac{y}{z}(1 + \lambda) = 0,\text{ and } R_{yy} < 0.$$

In these two cases, firm revenue R is strictly concave in y, with $R_{yy} < 0$ everywhere. Now, since functions R and w are assumed twice continuously differentiable in their arguments, so too will be the Hamiltonian H, with $H_{uu} = R_{yy}(Kw)^2 < 0$. H is strictly concave in u. For given K, then, the necessary conditions for optimal u are also sufficient to guarantee maximization of H. More important, the strict concavity of H in u, and the constant bounds constraining feasible u, mean that we may invoke an appropriate theorem to conclude that optimal u is continuous in other arguments of H.[73] Since the other arguments of H are continuous functions of time, u must be as well.

Note also that $u(t)$ is a continuously differentiable function of time within an interior segment. This is a consequence of the linearity of $\dot{E}$ in u, and the strict concavity of the Hamiltonian.[74]

73. For example, a result proven by Debreu, found in Kevin Lancaster, *Mathematical Economics* (Dover, 1987), pp. 349–50, or a theorem due to Fiacco cited in Garth P. McCormick, *Nonlinear Programming: Theory, Algorithms, and Applications* (Wiley, 1983), pp. 245–46.
More directly, we may note that the strict concavity of H in u means that, at every moment, the $u(t)$ that maximizes H is unique. Since the set of feasible values from which $u(t)$ is chosen is compact, we may invoke theorem 6.1 from Wendell H. Fleming and Raymond W. Rishel, *Deterministic and Stochastic Optimal Control* (Springer-Verlag, 1975), p. 75, to conclude that $u(t)$ is a continuous function of time.
74. Consider optimal control $u^*(t)$ over some interior segment $t_1 < t < t_2$. Because of the strict concavity of the Hamiltonian, a local interior maximum of H must also be a global maximum of H with no constraints on control $u(t)$. Optimal control $u^*(t)$ over this interval must also be the optimal control for the problem

$$\max_{u(t)} \int_{t_1}^{t_2} [R(x,y) - F - dy - cuK - rK]dt,$$

subject to the initial and terminal conditions that $E(t)$ take on the values at times t_1 and t_2 associated with optimal control $u^*(t)$ in the original problem, but with no bounds on control $u(t)$. As Fleming and Rishel, *Deterministic and Stochastic Optimal Control* (corollary 6.1,

Next I restrict discussion to symmetric-industry equilibria, in which the industry is made up of identical firms. In this case, I will show that along an interior segment of such an equilibrium, $\dot{u} < 0$.

We already know (see possibility A.1.d) that along an interior segment,

$$(\dot{R}_y) = 0$$

$$= \frac{d}{dt}[P'(1 + \lambda)y + P]$$

$$= P''\dot{z}(1 + \lambda)y + P'(1 + \lambda)\dot{y} + P'\dot{z}$$

$$= P''N\dot{y}(1 + \lambda)y + P'(1 + \lambda)\dot{y} + P'N\dot{y} = 0$$

$$= \dot{y}[P''(1+ \lambda)Ny + P'(1 + \lambda) + P'N]$$

Is it possible that the expression in brackets equals zero?

If $P''(1 + \lambda)Ny + P'(1 + \lambda) + P'N = 0$

$$P'' = -\frac{P'(1 + \lambda + N)}{z(1 + \lambda)}$$

$$P'' = -\frac{P'\left(1 + \dfrac{N}{1 + \lambda}\right)}{z} > -\frac{P'}{z}(2)$$

$$\text{since } 1 + \frac{N}{1 + \lambda} \geq 2.$$

But this contradicts (A.2.a) and our assumption that $\overline{R}_{zz} < 0$. So we must have $\dot{y} = 0$.

Now $y = uKw$

$$\dot{y} = \dot{u}Kw + uKw_E \frac{y}{K^\gamma} = 0, \text{ which can only be true if}$$

$$\therefore \quad \dot{u} < 0.$$

Since u is continuous over time, when $u = 0$, it cannot jump to 1. Indeed, u cannot even become positive since $\dot{u} < 0$ as soon as $u > 0$. So, when $u = 0$, the optimal policy must remain $u = 0$.

p. 77), note, the linearity of the equations of motion in the control variable, and the concavity of the Hamiltonian, are sufficient to guarantee that the optimal control is a continuously differentiable function of time for this subproblem, with no constraints on $u(t)$.

If we add the further assumption that $\overline{R}_z(0)$ exceeds current marginal cost at time 0 (as must be true with constant elasticity < -1), then because $\dfrac{y}{z} = \dfrac{1}{N}$, $R_y = \overline{R}_z - P'z\left(1 - \dfrac{y}{z}(1 + \lambda)\right) \geq \overline{R}_z$, and possibility A-1.c cannot hold, it can never be optimal for $u = 0$ and nothing ever produced.

Note the role of learning economies in this model. If $w_E = 0$ everywhere (there is no learning), δ must be constant and equal to zero, and optimal u must be constant over the product cycle. Optimal capacity choice implies $R_y - d - c/w$ must be greater than zero, and therefore $u = 1$. The first-order conditions then require that K be chosen so that variable profit generated by a marginal unit of capital $(R_y - d - c/w)w$ just covers the cost of capital (r) and all available capacity is fully utilized over the entire product life cycle.

Appendix 6-B: Detailed Solution with Specific Demand and Learning Curve Assumptions

Specification of the Demand Curve

Let inverse demand be given by $P = \left(\dfrac{z}{\alpha}\right)^{\frac{1}{\beta}}$ with $z = x + y$, total industry output.

Consider industry sales, given by

$$\overline{R} = \left(\frac{z}{\alpha}\right)^{\frac{1}{\beta}} z \quad -1 < \frac{1}{\beta} < 0$$

$$\overline{R}_z = \left(\frac{1}{\beta} + 1\right)\left(\frac{z}{\alpha}\right)^{\frac{1}{\beta}} > 0$$

$$\overline{R}_{zz} = \left(\frac{1}{\beta} + 1\right)\frac{1}{\beta}\, \alpha^{-\frac{1}{\beta}} z^{\frac{1}{\beta}-1} < 0,$$

and note that $\lim_{z \to 0} \overline{R}_z = \infty$.

Now, marginal revenue for any individual firm is given by

$$R_y = P'\left(1 + \frac{dx}{dy}\right)y + P = P\left[\frac{P'zy}{Pz}\left(1 + \frac{dx}{dy}\right) + 1\right]$$

$$= P\left[\frac{1}{\beta}\frac{y}{z}(1 + \lambda) + 1\right],$$

where $\lambda = \dfrac{dx}{dy}$, the conjectural variation:

$\lambda = N - 1$ for a constant-market-share cartel;

$\lambda = 0$ under Cournot competition.

In addition, I assume that the industry is made up of N identical firms. In symmetric-industry equilibrium, each firm has market share $\dfrac{y}{z} = \dfrac{1}{N}$.

Let $\sigma = \dfrac{1 + \lambda}{N}$.

$$\text{So } R_y = P\left(\frac{\sigma}{\beta} + 1\right)$$

$$\text{where } \sigma = \begin{cases} 1, & \text{cartel;} \\ \dfrac{1}{N}, & \text{Cournot.} \end{cases}$$

Thus, in an industry made up of N identical firms,

$$R_y = \left(\frac{N\,y}{\alpha}\right)^{\frac{1}{\beta}}\left(\frac{\sigma}{\beta} + 1\right).$$

Specification of Learning Economies and Output

Let $w = \phi E^\epsilon$, with $\dot{E} = y/K^\gamma$, $E(0) = E_0$, $0 \le \epsilon \le 1$, $0 \le \gamma \le 1$. Initial yield (ϕE_0^ϵ) is independent of capacity choice. We are mainly interested in two specific cases: $\gamma = 0$ (learning depends on absolute production experience), and $\gamma = 1$ (learning depends on experience per unit capacity, or facility).

Then $y = \phi E^\epsilon u K$.

Over the interval from zero to t_s, where $u = 1$,

$$\dot{E} = \frac{y}{K^\gamma} = \phi E^\epsilon K^{1-\gamma}.$$

Solving this differential equation, we have

$$E(t,K) = [E_0^{1-\epsilon} + K^{1-\gamma} \, \phi t(1 - \epsilon)]^{\frac{1}{1-\epsilon}}, \text{ and } y(t,K) = \phi \, E(t,K)^\epsilon K.$$

We also know

$$\dot{\delta} = -\frac{\partial H}{\partial E} = -\left(R_y - d + \frac{\delta}{K^\gamma}\right) w_E K$$

$$= -\left(R_y - d + \frac{\delta}{K^\gamma}\right) \epsilon \, \phi \, E^{\epsilon-1} K,$$

which can be rewritten as the linear monic differential equation $\dot{\delta} + f_1(t,K) \, \delta = f_2(t,K)$, with

$$f_1(t,K) = \epsilon \, \phi \, E(t,K)^{\epsilon-1} K^{1-\gamma}, \quad f_2(t,K) = -[R_y(t,K) - d] \, f_1(t,K) K^\gamma.$$

Solving for δ, for some t in the interval $(0, t_s)$, given boundary value $\delta(t_s) = \delta_{ts}$, we have

$$\delta(t) = \frac{\delta_{ts} + \displaystyle\int_{t_s}^{t} f_2(\tau) e^{\int_{t_s}^{\tau} f_1(\mu) \, d\mu} \, d\tau}{e^{\int_{t_s}^{t} f_1(\tau) \, d\tau}}.$$

Define function $\delta_B(t,t_s,K,\delta_{ts})$ by the right-hand side of this equation.

With some difficulty, and a great deal of tedious algebra, we can then integrate this expression, and have

$$\delta_B(t,t_s,K,\delta_{ts}) = \left[\delta_{ts} + \frac{\beta + \sigma}{\beta + 1}\left(\frac{\phi K N}{\alpha}\right)^{\frac{1}{\beta}} E(t_s,K)^{\frac{\epsilon}{\beta}}\left\{1 - \left[\frac{E(t,K)}{E(t_s,K)}\right]^{\frac{\epsilon}{\beta}+\epsilon}\right\} K^\gamma \right.$$

$$\left. - d\left\{1 - \left[\frac{E(t,K)}{E(t_s,K)}\right]^\epsilon\right\} K^\gamma\right] \Big/ \left[\frac{E(t,K)}{E(t_s,K)}\right]^\epsilon.$$

Over the interval from t_s to one, optimal u is set such that $y(t,k) = y(t_s,k)$—in other words, constant output. $E(t_s,k)$, $y(t_s,k)$ are given by expressions in the last section. Over the interval from t_s to one, then,

$$\dot{E} = \frac{y(t_s,K)}{K^\gamma},$$

$$\text{so } E(t,k) = (t - t_s)\frac{y(t_s,K)}{K^\gamma} + E(t_s,K);$$

$$E(t,K) = (t - t_s)\, \phi E(t_s,K)^\epsilon\, K^{1-\gamma} + E(t_s,K)$$

$$= E(t_s,K)[1 + \Theta(t,t_s,K)],$$

where $\Theta(t,t_s,K) = \phi E(t_s,K)^{\epsilon-1}\, K^{1-\gamma}(t - t_s)$.

Θ is the ratio of incremental experience from time t_s to time t, to experience at t_s;

$$\dot\delta = -\frac{c\, u\, w_E K}{w} = -\frac{c\, y\, w_E}{w^2}$$

$$= -c\,\frac{\phi^2 E(t_s,K)^\epsilon\, K\epsilon E(t,K)^{\epsilon-1}}{\phi^2 E(t,K)^{2\epsilon}} = f_3(t\,;\,t_s,K);$$

where

$$f_3(t\,;\,t_s,K) = \frac{-c\epsilon E(t_s,K)^\epsilon\, K}{E(t,K)^{\epsilon+1}}\,.$$

Conditional on assumed switchpoint t_s, the solution to this differential equation can be written as

$$\delta(t) - \delta(1) = \int_1^t f_3(\tau\,;\,t_s,K)\, d\,\tau.$$

or, making use of the fact that $\delta(1) = 0$ (the transversality condition), define function δ_E as

$$\delta_E(t\,;\,t_s,K) = \int_1^t f_3(\tau\,;\,t_s,K)\, d\,\tau$$

$$= \frac{cK^\gamma}{\phi[E(t_s,K) + \phi K^{1-\gamma}(t-t_s)\, E(t_s,K)^\epsilon]^\epsilon}$$

$$- \frac{cK^\gamma}{\phi[E(t_s,K) + \phi K^{1-\gamma}(1-t_s)\, E(t_s,K)^\epsilon]^\epsilon}$$

$$= \frac{cK^\gamma}{\phi E(t_s,K)^\epsilon}\left\{\frac{1}{[1 + \Theta(t,t_s,K)]^\epsilon} - \frac{1}{[1 + \Theta(1,t_s,K)]^\epsilon}\right\}.$$

In particular, we can solve for $\delta_E(t_s\,;\,t_s,k)$, that is, the value of δ at t_s which, given some assumed t_s, solves the equation of motion over an interior segment and the transversality condition at time 1. In this special case, we have

$$\delta_E(t_s;t_s,K) = \frac{cK^\gamma}{\phi E(t_s,K)^\epsilon}\left\{1 - \frac{1}{[1 + \Theta(1,t_s,K)]^\epsilon}\right\}.$$

Solution of the Model

Our specification has ruled out the possibility that optimal $u = 0$. The optimal path for u will consist of a capacity-constrained ($u = 1$) segment through time t_s, possibly followed by an interior segment. For the moment, assume that $t_s < 1$.

Case 1. $t_s < 1$. At t_s, we also know that $u(t)$ will be entering an interior segment, so

$$\frac{\delta}{K^\gamma} = -\left[R_y(t_s,K) - d - \frac{c}{\phi\,E(t_s,K)^\epsilon}\right] \text{ must be true, at } t_s.$$

Thus, for given K, optimal t_s must satisfy

$$\frac{\delta_E\,(t_s;t_s,K)}{K^\gamma} + \left[R_y\,(t_s,K) - d - \frac{c}{\phi E(t_s,K)^\epsilon}\right] = 0.$$

or, substituting,

$$\text{(B-1)} \quad \frac{-c}{\phi[E(t_s,K) + \phi K^{1-\gamma}(1-t_s)E(t_s,K)^\epsilon]^\epsilon}$$
$$+ \left[\frac{N\phi E(t_s,K)^\epsilon K}{\alpha}\right]^{\frac{1}{\beta}}\left(\frac{\sigma}{\beta} + 1\right) - d = 0.$$

This is just the condition that marginal revenue equal current marginal cost at terminal time 1. If K is chosen nonstrategically to maximize profits, optimal K must satisfy

$$\int_0^1 \left\{\left[R_y(t,K) - d - \frac{c}{\phi E(t,K)^\epsilon} + \frac{\delta}{K^\gamma}\right]u\phi E(t,K)^\epsilon\right.$$
$$\left. - \gamma\frac{\delta}{K^\gamma}\,u\phi E(t,K)^\epsilon\right\}dt - r = 0.$$

Since the expression in brackets is zero after t_s, and u is one before t_s, this can be written as

$$\int_{ts}^{1} -\gamma \frac{\delta}{K^\gamma} u\phi E(t,K)^\epsilon \, dt + \int_{0}^{ts} \left\{ [R_y(t,K) - d]\phi E(t,K)^\epsilon \right.$$

$$\left. + \frac{\delta}{K^\gamma} \phi E(t,K)^\epsilon (1 - \gamma) \right\} dt - ct_s - r = 0.$$

We may substitute for δ with the functions δ_B and δ_E, described earlier. Noting that

$$u(t) = \frac{y(t_s,K)}{K\phi E(t,K)^\epsilon} = \frac{\phi E(t_s,K)^\epsilon K}{K\phi E(t,K)^\epsilon} = \frac{E(t_s,K)^\epsilon}{E(t,K)^\epsilon} = \frac{1}{[1 + \Theta(t,t_s,K)]^\epsilon}$$

over the interval from t_s to one, we can solve analytically for the first integral above, so, substituting,

$$-\frac{c\gamma(1-t_s)}{(1-\epsilon)} \left\{ \frac{\frac{1}{\Theta(1,t_s,K)} + \epsilon}{[1 + \Theta(1,t_s,K)]^\epsilon} - \frac{1}{\Theta(1,t_s,K)} \right\} + \int_{0}^{t_s} \left\{ [R_y,(t,K) - d] \right.$$

(B-2) $\cdot \phi E(t,K)^\epsilon + \delta_B\left(t,t_s,K,-\left[R_y(t_s,K) + d + \frac{c}{\phi E(t_s,K)^\epsilon}\right]K^\gamma\right)$

$$\left. \cdot (1-\gamma) \frac{\phi E(t,K)^\epsilon}{K^\gamma} \right\} dt - ct_s - r = 0.$$

An optimal choice of t_s and K, then, must solve equations B-1 and B-2.

Case 2. $t_s = 1$. The other possibility is that $t_s = 1$, and producers fully utilize available capacity throughout the product life cycle. Since there is no interior segment, equation B-1 does not have to hold at t_s. Instead, the transversality condition means that $\delta(t_s) = 0$, so $\delta_B(t, 1, K, 0)$ gives the value of $\delta(t)$ at any time t. Incorporating this into the first-order condition for optimal K, I then have

$$\int_{0}^{1} \left\{ [R_y(t,K) - d] \cdot \phi E(t,K)^\epsilon + \delta_B(t,1,K,0) \right.$$

(B-2′) $\left. \cdot (1-\gamma) \phi \frac{E(t,K)^\epsilon}{K} \right\} dt - c - r = 0,$

which can be solved for optimal K.

In searching for Cournot equilibria, then, attempts were made to solve equations B-1 and B-2 for optimal (t_s, K), and equation B-2' for optimal K (assuming $t_s = 1$).

Industry Profits

Total rents per firm, gross of fixed entry cost F, earned in an industry made up of N identical firms can be calculated as (for given optimal t_s and K):

$$\int_0^1 \left\{ \left[\left(\frac{Ny(t,K)}{\alpha} \right)^{\frac{1}{\beta}} - d \right] y(t,K) - cu(t)K - rK \right\} dt, \text{ or}$$

$$\int_0^{t_s} \left\{ \left[\left(\frac{Ny(t,K)}{\alpha} \right)^{\frac{1}{\beta}} - d \right] y(t,K) \right\} dt + \left[\left(\frac{Ny(t_s,K)}{\alpha} \right)^{\frac{1}{\beta}} - d \right] y(t_s,K)(1-t_s)$$

$$- cKt_s - \frac{cK(1-t_s)}{(1-\epsilon)} \left[\frac{\frac{1}{\Theta(1,t_s,K)} + 1}{[1+\Theta(1,t_s,K)]^\epsilon} - \frac{1}{\Theta(1,t_s,K)} \right] - rK.$$

CHAPTER SEVEN

Strategic Issues

TODAY it is frequently asserted in the United States that the semiconductor industry is a "strategic" or "critical" industry. Indeed the National Advisory Committee on Semiconductors (NACS), formed by congressional mandate, titled its first report to the president *A Strategic Industry at Risk*.[1]

Conceptions of "Strategic"

What precisely is meant by this use of the word "strategic"? There are at least three senses in which the word might apply to the semiconductor industry.

National Security

The NACS clearly had several concepts of "strategic" in mind. First, and most simply, the committee asserted that America's national security depended on the capabilities of its domestic semiconductor industry. The NACS's first report repeated the conclusions of a 1987 Defense Science Board task force's report on defense semiconductor dependency, arguing that U.S. and NATO forces "rely on a technolog-

1. National Advisory Committee on Semiconductors (NACS), *A Strategic Industry at Risk: A Report to the President and the Congress* (Washington, 1989). The NACS issued its last report in 1992, before disbanding.

ical advantage ultimately traceable to semiconductors to offset the numerical superiority of our adversaries."[2] It is hard to deny the force of this point, particularly in light of the historical role that the U.S. military played in driving the initial development and growth of the American semiconductor industry.

But the national security argument also raises nettlesome questions. The serious competition for American semiconductor companies comes from allied nations, not adversaries. Is there a serious threat that imports from our allies will be denied us?[3] Even if they were, the United States clearly does not lack the technical capability to develop the same products at home. Certainly, in that event, the United States would be stimulated to develop domestic sources of supply, and it is hard to envision a sudden military threat that would last long enough to make the time lag a serious issue. What is in doubt is not our absolute technical capacity to manufacture this or that semiconductor product, but our ability to manufacture semiconductors cheaply enough to be competitive in commercial markets. Maintaining a captive, technically advanced domestic capability to produce products for the military, albeit at high cost, rather than solving a military challenge, might worsen what is essentially an economic problem.

Whatever the abstract merits of the national security argument, the defense semiconductor dependency issue has been closely tied to more purely commercial concerns. The Defense Science Board's 1987 argument for intervention to strengthen the U.S. semiconductor industry explicitly as follows: because "*competitive, high-volume production* is the key to leadership in semiconductors," and "high-volume production is supported by the *commercial* market," and "*leadership* in commercial volume production is being *lost* by the U.S. semiconductor industry," "U.S. Defense will soon *depend on foreign sources* for state-of-the-art technology in semiconductors. The Task Force views this as an unacceptable situation."[4]

2. NACS, *Strategic Industry at Risk*, p. 6.
3. Although Shintaro Ishihara, a prominent right-wing politician in Japan, was to fuel this concern by suggesting in *The Japan That Can Say No*, trans. Frank Baldwin (Simon & Schuster, 1991), written with Sony's chief executive Akio Morita, that Japan had the power to withhold advanced chips, or make them available to the Soviet Union, and so alter the global balance of power.
4. *Report of the Defense Science Board Task Force on Defense Semiconductor Dependency* (Office of the Under Secretary of Defense for Acquisition, February 1987), pp. 1–2.

Clearly, however, the same logic (because of the shortcomings of domestic producers in commercial markets, U.S. defense will soon depend on foreign sources for state-of-the-art technology in product X) could be used to argue for intervention to maintain commercial preeminence in any or all high-technology industries (substitute for product X chemicals, biotechnology, communications, computers, aircraft, or instruments), virtually all of which have ties to some set of defense systems. Thus, pushed to its logical extreme, the national security argument calls for across-the-board market leadership in virtually all high-technology fields, and an end to the notion that competitive outcomes—on a global scale—should be left to impersonal market forces in this large and important set of industries.

The national security argument, if accepted and pushed to its logical limits, means that competitive market outcomes in high-technology products are acceptable only if they leave domestic producers (excluding even those of our allies and trading partners) firmly in command of global markets. This clearly is a recipe for the breakdown of an open international trading system in high-technology goods. If all nations were to take the argument seriously, high-tech autarky would be the inevitable result. If this attitude were to prevail in its most extreme form, there would be little need for further analysis, and the economic content of discussions of national policies toward high-technology industries would simply be about the least costly and most efficient means of attaining national self-sufficiency in these industries.

Economic realities, however, seem destined to doom this approach. The range of industries that could be labeled as high technology is so great, and growing so rapidly, that any policy of broad self-sufficiency would appear to be insupportably costly, no matter what its perceived merits from a strictly defense perspective. Furthermore, the revolution in transportation and communications technologies has pulled the world together in a way that makes it increasingly difficult to talk about purely *national* technological capabilities. Multinational firms based in different parts of the world link up with one another in a variety of types of alliances, and pull together specialized technical resources scattered about the globe, in order to create leading edge, high-tech products. However militarily appealing in the abstract, high-tech self-sufficiency is likely to be neither affordable nor feasible. Thus we turn to other meanings of the word "strategic" for the analytical purposes of this chapter.

Figure 7-1. *"Food Chain" Concept in Semiconductor Industry*

	United States 1990	World 1990	World 2000
Total manufacturing and service economy	$5.4 trillion	$20 trillion	$40 trillion
Electronics products and services	$384 billion	$751 billion	$2 trillion
Semiconductor manufacturing	$21 billion	$63 billion	$200 billion
Semiconductor materials and equipment	$9 billion	$20 billion	$60 billion
Fundamental sciences and processes	Software		
	Gases and Chemicals		
	Optics, materials, robotics		

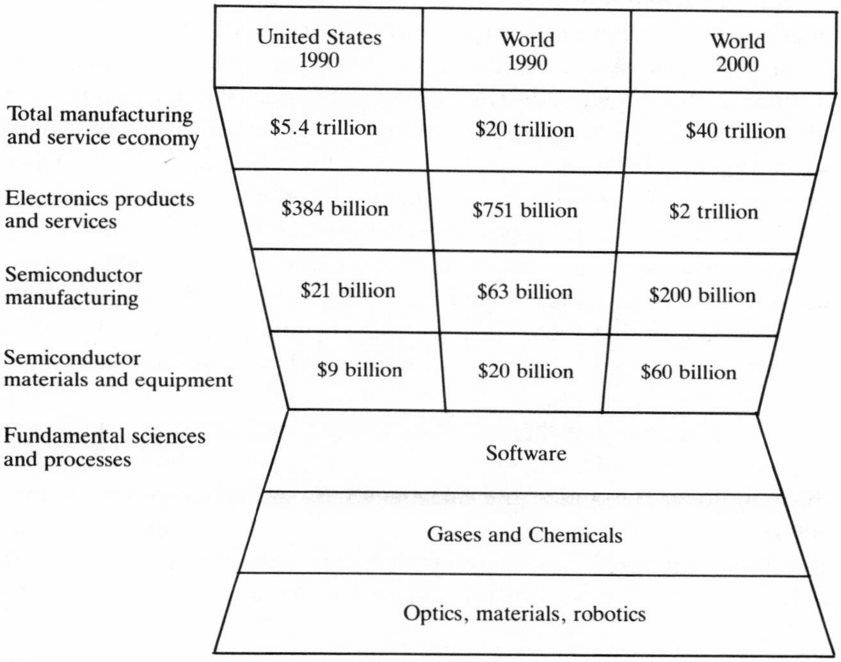

Source: National Advisory Committee on Semiconductors, *Attaining Preeminence in Semiconductors* (Arlington, Va.: February 1992), p. 9.

Strategic Sectors

Another set of arguments advanced by the NACS and others holds that certain industries are "strategic" because of their linkages to the rest of the economy. The simplest of these arguments is the "food chain" theory, often illustrated by something like figure 7-1. Upstream and downstream industries' competitive fortunes are interlinked in a complex ecological system that makes each dependent on the health of the others. As a report by the Semiconductor Industry Association put it, "The elements of the electronics industry are analogous to a 'food chain,' in which each component level, from silicon wafers up to finished electronic products, is dependent on the others. If one link is damaged, the others are automatically injured."[5]

5. Semiconductor Industry Association (SIA), *Key Facts and Issues* (1989), p. 24. The

The difficulty with this argument in its simplest form may be discerned by glancing at the input-output table for any major economy. Virtually all goods (other than final consumer goods) produced by any industrial sector also serve as inputs into most other sectors. Thus any policy of labeling as "strategic" industries those that supply significant inputs to other industries ends up declaring virtually all industries strategic. A policy arguing for support of "strategic" domestic industries then becomes a recipe for intervening to support virtually anything.

Clearly, if competitive products are available from foreign suppliers at a competitive price, no harm is necessarily done to other industries up and down the food chain. Indeed, a favorite response of skeptical economists to claims that the U.S. semiconductor industry is in some sense "strategic" has been to pose the question, "How are computer chips different from potato chips?"[6]

One might point out three important differences. First, a potato chip is a final consumption good, so that poor-quality, high-priced potato chips, although onerous to the consumer, have no direct impact on other industries. Semiconductors, on the other hand, are an input to a large and important array of user industries, for which access to the latest technology at a competitive price is critical. Second, entry into potato

---

"food chain" metaphor was widely applied in Washington during the policy debates over high-definition television of 1988–89: "Loss of consumer electronics has affected today's U.S. competitiveness. . . . The U.S. electronics infrastructure is like a biological food chain with many interdependent segments–when one link is weakened or destroyed, other links feel injury. Many believe competitive problems experienced today by the U.S. semiconductor industry can be traced to this lost segment." American Electronics Association, ATV Task Force Economic Impact Team, *High Definition Television (HDTV): Economic Analysis of Impact* (Washington, November 1988), p. i.

"When a personal computer adopts the kind of flat display screen associated with HDTV technology, it will not only create additional demand for the products associated with this new industry, but it will also require a large number of semiconductors to support the new projector and display technology that will be incorporated in the screen itself. The use of these inputs will give a boost to demand for products from these related industries, contributing to increases in their output and improving the chances that they will be profitable and have the funds to support further innovation.

These linkages create a kind of feed-back mechanism that provides its greatest advantages to our industrial base when the entire 'food chain' of the electronics industry is improved." Testimony submitted by Robert Cohen in *High Definition Television,* Hearings before the Subcommittee on Telecommunications and Finance of the House Committee on Energy and Commerce, 101 Cong. 1 sess. (Government Printing Office, 1989), p. 226.

6. The comment has been attributed to Michael Boskin, who was chairman of the Council of Economic Advisers in the Bush administration, but firm evidence that he actually said it appears to be lacking.

chip making is relatively easy, whereas entry into computer chip manufacture may involve large sunk investments and technology that is difficult to obtain. Third, computer chips are high-tech goods, requiring large, continuing investments in R&D; potato chips do not. An empirical literature tells us that much R&D is likely to have significant economic impacts that are not fully captured by the high-tech firm making the R&D investment.

These circumstances make it credible that one or both of two things might occur in the computer chip industry that would be unlikely to happen in the potato chip industry. Barriers to entry may make the exercise of monopoly power possible, allowing computer chip companies to earn monopoly rents. Spillovers of knowledge from chip R&D, as well as production, to other firms and industries may create externalities that result in economic benefits to firms other than the firm making the investment in R&D.

Many economists would agree that these two possibilities create conditions under which intervention to sustain a "strategic" industry might be economically justifiable.[7] First, if there are significant technological externalities in semiconductor production, so that social returns exceed private returns, policies to correct for presumed underinvestment in these activities may be appropriate. There is some empirical tradition of economists arguing that particular capital goods industries have been strategic because of informational spillovers between producers and users of these goods.[8]

In semiconductors it is in design, not manufacturing, that it seems most plausible to argue for technological externalities—that is, that semiconductor users are using the same technologies as semiconductor producers. One is hard pressed to think of examples of other industries that use the same manufacturing technologies as the semiconductor industry.[9] Cross-industry spillovers are more obvious in design. As the density of

7. Paul R. Krugman, for example, identifies these two criteria for "strategic" sectors as academically respectable economic arguments for selectively promoting specific industries. See his "Introduction: New Thinking about Trade Policy," in Krugman, ed., *Strategic Trade Policy and the New International Economics* (MIT Press, 1986), pp. 12–15.

8. See, for example, Nathan Rosenberg, "Technological Change in the Machine Tool Industry, 1840–1910," *Journal of Economic History*, vol. 23, no. 4 (1963), pp. 414–43; and Bo Carlsson and Staffan Jacobsson, "What Makes the Automation Industry Strategic?" *Economics of Innovation and New Technology*, vol. 1, no. 4 (1991), pp. 257–69.

9. Prominent exceptions include flat panel displays (used for the screens of laptop computers and similar products), which use a similar manufacturing technology, and the read-write heads used in computer disk drives.

circuit elements that can be inscribed on a silicon chip has increased, more and more of the proprietary design of an electronic system has been placed on a single chip, where before systems were built up from non-proprietary building block components. As a result, today the border between chip design and system design is exceedingly blurry. But this is an argument that systems design and chip design are essentially converging into a single industrial activity, undertaken by the same firm, and not necessarily an argument that there are obstacles that prevent spillovers from being captured within private firms, which might justify some corrective policy intervention.

Another kind of technological externality in semiconductor manufacturing might prove a more convincing concern. This involves the transfer of proprietary information between system designers and semiconductor manufacturers. With increasing levels of chip density has come the requirement that system designers transfer proprietary design information to chip manufacturers. Because Japanese chip manufacturers, in particular, are often arms of large conglomerates also involved in the design and sales of systems, reliance on external chip manufacturers has thus increasingly meant transferring sensitive design information to one's own competitors. In this case, however, the perils of "deindustrialization" in semiconductor manufacturing boil down to the absence of sufficient competition in chip manufacturing (forcing one to rely on one's own competitors to manufacture needed chips) and the absence of sufficient protection for intellectual property.

A similar argument might apply in the more specialized area of semiconductor production equipment. Eric von Hippel, for example, has argued that roughly two-thirds of the ideas for innovations in the machines used to manufacture semiconductors came from chip makers, through their close interaction with the machinery makers.[10] Conversely, one might argue that early access to the latest semiconductor manufacturing technology requires close interaction with semiconductor equipment makers, and that therefore maintenance of a competitive chip manufacturing industry demands a healthy local semiconductor production equipment sector. The argument has been made that these sorts of informational externalities link the semiconductor equipment industry to

10. Eric von Hippel, *The Sources of Innovation* (Oxford University Press, 1988), pp. 19–26.

the semiconductor industry downstream.[11] Indeed, this view was reflected in the decision of the Sematech, the U.S. semiconductor industry's R&D consortium, to focus its resources on improving the technical and economic health of its domestic equipment and materials suppliers.

On the other hand, in an age of drastically cheapened global transport and communications, and extensive overseas investment, it is not necessarily the case that a U.S.-based chip company is precluded by distance from establishing a close relationship with a foreign equipment manufacturer. To make the case for domestic industry support on these grounds, a further argument is required, namely, the assertion that American producers are unlikely to be able to obtain timely and competitive access to foreign chip making technology (this assertion is typically accompanied by the observation that Japanese chip making equipment companies have close ties to Japanese chip makers competing with American semiconductor firms).

This idea, in fact, is the application to semiconductor equipment of a second argument that is sometimes made as to why the semiconductor industry itself is strategic. This second argument hinges on the fact that monopoly profits may be earned in high-tech sectors. The role of semiconductors as an important input to many other sectors makes the potential exercise of monopoly power an extremely important concern, since market power in such an input may be extended downstream into user industries, by acquisition or vertical integration, allowing even greater monopoly rents to be collected.[12] In particular, if the recipients of monopoly rents are foreign nationals, then policies intended to create a domestic industry capable of seizing a share of these rents, or reducing their extraction from domestic consumers, may be desirable, even (in the extreme case) if domestic producers are less efficient.[13] As just observed, this argument is in essence made when contending that a domestic equip-

---

11. See for example, National Advisory Committee on Semiconductors, *Preserving the Vital Base: America's Semiconductor Materials and Equipment Industry* (GPO, 1990).

12. Otherwise the downstream user would normally substitute other inputs for the monopolized input as price is raised and dissipate some of the monopolist's potential rent in inefficient production. John M. Vernon and Daniel A. Graham, "Profitability of Monopolization by Vertical Integration," *Journal of Political Economy*, vol. 79 (July–August 1971), pp. 924–25.

13. The argument is presented in a simple, diagrammatic form in Kenneth Flamm, "Semiconductors," in Gary Clyde Hufbauer, ed., *Europe 1992: An American Perspective* (Brookings, 1990), pp. 260–63.

ment industry ought be preserved to limit the extension of foreign monopoly power into semiconductors.

Essentially both arguments—the argument from externalities and the argument from monopoly power—have typically been used to support the need to sustain the U.S. semiconductor industry as a strategic sector. On the monopoly power issue, the NACS wrote in 1989, "Our major competitors have a demonstrated capability—compelled either by policy or their own strategy and organization—to act in concert and exercise market power sufficient to control access to technology and its price. Compelling examples include the DRAM shortage, the inability of U.S. supercomputer companies to purchase the fastest foreign chips, and the inability of most U.S. chip producers to gain timely access to the most advanced foreign semiconductor manufacturing equipment and materials."[14] On the externalities question, the NACS wrote in 1992, "The close linkages between semiconductor and electronics manufacturers are important to spur innovations in semiconductors, and to incorporate quickly new technologies into successful products and services. These strong linkages help to create external economies—economic benefits that flow between semiconductor firms and their customers and suppliers, and also between competing semiconductor firms."[15]

Strategic Behavior and Strategic Policies

Another use of the word "strategic" comes from game theory, where strategic actions are defined as those intended to influence the moves of other players.[16] Firms act strategically when they take actions intended to alter the behavior of their rivals. Governments, too, can make strategic moves, when they implement policies intended to alter the behavior of the foreign rivals of their domestic firms, or even other foreign governments.

For example, suppose a government grants a subsidy, linked to value or volume of production, to a domestic firm selling against a single foreign rival in a foreign market. That subsidy will have the direct effect of

14. NACS, *Strategic Industry at Risk,* p. 14.
15. National Advisory Committee on Semiconductors, *Attaining Preeminence in Semiconductors, Third Annual Report to the President and Congress* (Washington, February 1992), p. 8.
16. This definition comes from Thomas C. Schelling, *The Strategy of Conflict* (Harvard University Press, 1960).

encouraging the domestic firm to increase output and increase its profit. By itself (taking the output of the foreign firm as given), this does not mean that domestic society as a whole benefits from the subsidy, since the increased profits for the domestic producer will generally be more than offset by the cost of the subsidy to national taxpayers. But it may also encourage the foreign rival to cut back its output, as news of the subsidy makes the rival realize that the domestic firm will definitely increase its output. Given this reality, the foreign rival may find that producing less output maximizes its own profits, allowing the domestic producer to expand its output even further and realize some additional profit. This indirect effect (working through the foreign producer's output decision) of the subsidy is the so-called strategic effect of the government policy, and is what makes it possible that such a *strategic trade policy* can work to benefit the interests of the subsidizing nation as a whole.[17]

Of course, the domestic welfare gains come at the expense of the foreign country, since the additional profit earned domestically is accompanied by a decline in the foreign company's profit. The foreign government may then be tempted to intervene to convince the domestic government to abandon its intervention, or to create an outcome more amenable to its own company's interest. The result is a strategic interaction between the two governments' policies.

The logical link with the concept of a strategic sector, discussed earlier, is that the literature on strategic trade policy (defined as policy that works by altering the behavior of foreign companies or governments) has focused on cases where benefits to a nation as a whole exist as a consequence of monopoly profits captured from foreign interests. The existence of significant monopoly power was one of the two economically defensible reasons why a sector might merit labeling as strategic.

But the other rationale, the presence of externalities, might also credibly inspire a strategic government policy. Policies used to force domestic investments by foreign companies, for example, in order to foster technological spillovers to domestic companies, might arguably be labeled a strategic trade policy.

17. This example is the classic model developed in James A. Brander and Barbara J. Spencer, "Export Subsidies and International Market Share Rivalry," *Journal of International Economics*, vol. 18 (February 1985), pp. 83–100. See also James A. Brander, "Rationales for Strategic Trade and Industrial Policy," in Krugman, ed., *Strategic Trade Policy*, pp. 23–46.

Nonstrategic policies (that is, policies that work only through their effects on domestic economic actors) might also be attractive in strategic sectors. Even if an economy or sector were entirely cut off from international trade, the government might wish to subsidize investments thought to be characterized by significant externalities. Or, in a strategic sector in which domestic firms with a unique competitive advantage face no foreign rivals, official encouragement of the formation of an export cartel may further national welfare at the expense of foreign consumers. In a real world linked by pervasive trade, however, virtually any policy affecting a strategic industry is also likely to have some effect on actual or potential foreign competitors' behavior, and hence also qualify as a strategic policy.

In principle, however, "strategic trade policy" is not conceptually equivalent to "government policy in strategic sectors." Strategic trade policy requires the existence of actual or potential foreign competitors, operating in a strategic sector, whereas economic justification for policies to promote strategic sectors, when motivated by externalities, does not necessarily require foreign competition. As a practical matter, though, the two ideas are closely identified with one another.

Discussion of government policies toward the semiconductor industry typically captures basic elements of all three distinct uses of the word "strategic." Semiconductors will be a strategic sector, meriting policy intervention, if the exploitation of monopoly power in a key semiconductor input is a credible likelihood, with potentially significant effects on downstream user industries, or if externalities are important. Strategic behavior by semiconductor producers (such as preemptive capacity investments or predatory pricing) may justify intervention by government. Strategic policies undertaken by governments (for example, protection, subsidy, or the toleration or facilitation of cartels) might also inspire a response by others.

Semiconductor Dependency and Strategic Trade Policy

It is quite remarkable that some key elements of the earliest episode of U.S.-Japan semiconductor trade friction, described in chapter 3, were to reappear a quarter of a century later, at the epicenter of controversy over the U.S.-Japan Semiconductor Trade Arrangement (STA) of 1986, described in chapter 4. In 1959, as in 1986, the semiconductor industry based its appeal to the federal government for protection on assertions about the industry's strategic nature, and the Japanese response was the formation of an export cartel.

Over the intervening two and a half decades, as a consequence of rapid technical innovation within the U.S. semiconductor industry, the semiconductor threat from Japan at first receded. But in the very late 1970s, as integrated circuits drawing on the large-scale Japanese R&D effort of the mid-1970s began to be exported in significant quantities, American producers once again began to raise alarms about their Japanese competitors. The market for memory chips—specifically, dynamic random access memory chips (DRAMs), the single most important type of semiconductor in terms of volumes produced, and erasable programmable read only memory chips (EPROMs)—was the reentry point for Japanese producers into global semiconductor competition, and the focus for trade frictions between the U.S. and Japanese semiconductor industries. And once more, strategic issues were placed at the center of debate.

The balance of this chapter examines the principal argument for government intervention in semiconductor trade as it has come to be articulated as a *strategic* economic issue for the United States: is dependency on foreign suppliers of semiconductors likely to be an empirically persuasive motivation for public policy?[18] The analysis will focus on the core arguments for activist, strategic policy interventions by the U.S. government: that the "dumping" of Japanese chips in the U.S. market in the early 1980s constituted predatory behavior on the part of Japanese chip producers, designed to secure monopoly power in key semiconductor markets, and that collusive behavior designed to exploit that monopoly power, once secured, could create significant costs for the U.S. economy, therefore justifying the investment of resources in defensive countermeasures.

The essentials of this argument are spelled out in figure 7-2. For simplicity I have assumed constant returns to scale in chip production and have assumed that technological investments and past learning have

18. By focusing on monopoly power as the motive for strategic behavior, I ignore the other possible reason for thinking semiconductors might be a strategic sector, namely, technological externalities. There is little compelling empirical evidence on this issue, particularly on aspects critical to the design of policy: whether spillovers cross industry boundaries as well as firm boundaries, whether they are inherently confined to a local geographic region, and whether alternative means (such as revised norms for intellectual property rights, or design and technical standards) can serve to internalize what might otherwise be externalities for an individual firm.

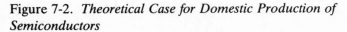

Figure 7-2. *Theoretical Case for Domestic Production of Semiconductors*

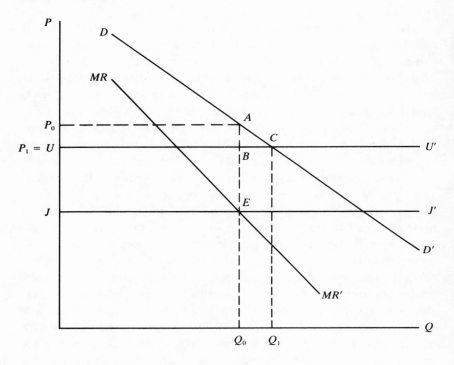

created a cost advantage for Japanese chip producers, so that their constant cost of production is *J* dollars (and their marginal and average cost schedule is represented by line *JJ'*). Domestic companies have higher unit costs of *U* dollars (and their cost schedule is given by line *UU'*). Foreign demand is given by demand curve *DD'*, and marginal revenue by line *MR-MR'*. If Japanese producers are then able to act as a profit-maximizing cartel, without fear of foreign entry, they will produce output Q_0, charge price P_0, and collect monopoly profit P_0AEJ. If entry and exit in the industry are costless, however, foreign producers can effectively contest the market, and this threat of entry places a cap on price at *U*. One may argue, however, that sunk costs—in the form of specialized and short-lived capital investments—are substantial in this industry. No individual domestic firm, then, will be willing to invest in a high-cost production facility, because in the event of a price war the cartel can lower its price below U and still fully cover costs, while the high-cost foreign producer will be forced to produce at a loss or to shut down and lose an amount equal to its sunk costs.

If the alternative is to face a cartel charging price P_0, however, it is clearly advantageous from the viewpoint of the foreign country to subsidize high-cost domestic production of output Q_1 at cost level UU'. Monopoly profits equal to P_0ABP_1 that would otherwise have been paid to the cartel are saved, and, in addition, consumer surplus equal to triangle ABC is gained. Subsidy of high-cost domestic production is superior to passive acceptance of uncontested, cartelized imports.

One should note that fear of dependency on foreign suppliers is not an exclusively American preoccupation. Chapter 2 of this book described an episode in which a similar fear appears to have motivated the Japanese government to invest in augmenting the domestic capability to supply the materials used by its semiconductor industry.

The U.S. semiconductor industry's analysis of its Japanese rivals' behavior, documented in chapters 3 and 4, has evolved historically. Through 1980 the story told was what might now be described as the "conventional" account of strategic trade policy: formal and informal barriers protecting Japan's domestic semiconductor market against American imports worked to promote the development of the Japanese industry and its global market share, to the detriment of the sales (and profits) of U.S. producers.[19] After 1980, however, as it became clear that prices in the American and Japanese markets were essentially identical, allegations of dumping—that Japanese producers were pricing exports below cost—necessarily and explicitly began to include an element of predatory behavior on the part of these companies, and an element of collusion. A more unconventional story of strategic behavior surfaced: below-cost pricing, it was asserted, was calculated to induce exit by American producers, after which Japanese producers would jointly raise prices and extract monopoly rents, which would provide a return on their investment in predation.

The history of this discussion and the American trade policy response it provoked could be interpreted skeptically as a self-fulfilling prophecy.

19. This is the "import protection as export promotion" policy described by Paul R. Krugman, "Import Protection as Export Promotion: International Competition in the Presence of Oligopoly and Economics of Scale," in Henryk Kierzkowski, ed., *Monopolistic Competition and International Trade* (Clarendon Press, 1984), pp. 180–93.

Clearly, in response to increasing trade frictions with the United States in the early 1980s, the Japanese government pressured its semiconductor industry to reduce exports, in essence sanctioning an export cartel. On the other hand, the U.S. industry has interpreted this same behavior as the normal second stage of a successful predatory campaign—rent extraction—thus validating the original assertions about Japanese intentions. The extent to which creation of export cartel–like market structures is the handmaiden of a misguided trade policy, or the fruit of successful predation, is the key issue.

The Semiconductor Trade Arrangement of 1986 drove this debate to new extremes. The complex evolution of the administrative mechanisms created under the auspices of this agreement was described in chapter 4. Perhaps the most interesting element of this most recent episode of U.S.-Japanese semiconductor trade friction is some evidence suggesting that after 1988, for perhaps the first time in their history, major Japanese semiconductor producers were able to maintain a significant degree of cooperation (or, less charitably, collusion) in a key product market despite the absence of explicit, overt regulatory pressures from the Japanese political system. The long-feared predatory threat might finally have surfaced, albeit with considerable support from policies put into place from 1986 to 1988.

In this chapter the model of semiconductor producer behavior developed in chapter 6 is modified and applied in order to assess the potential empirical significance of the threat of collusive behavior that has been the intellectual underpinning of an increasingly activist U.S. policy in semiconductors since the early 1980s. Putting aside the issue of whether collusive behavior was the proximate cause or the unforeseen effect of American trade policy, just how large an economic threat might it represent in the worst case? How much might the United States reasonably be willing to spend on "anticartel insurance"?

To examine these issues analytically, I have developed what I have argued is an at least minimally realistic model of pricing and production over the life cycle of a high-technology industry such as the semiconductor industry, in which both learning and scale economies and capacity constraints are important. I will apply this model using empirical parameters relevant to the production of 1M DRAMs, a product where actual historical behavior has been interpreted as a real-life episode to which both claims—first of predatory behavior, and then of collusive organization of foreign production to increase monopoly rent extraction—are relevant.

The DRAM Crisis of 1988

To begin, the most interesting data suggesting the proposition that the behavior of Japanese DRAM producers has not always reflected normal competitive market forces are drawn from observation of the market for DRAMs in 1987 through 1990. The relationships between prices and the foreign market values (FMVs) constructed by the U.S. Commerce Department raise interesting questions.

Figure 5-8 showed that, by late 1987 and continuing through late 1989, U.S. contract prices for 256K DRAMs had risen substantially above FMVs, and therefore that the FMVs were not constraining U.S. DRAM import prices over this period. From 1990 on, however, at least some Japanese DRAM imports appear to have been priced out of the U.S. market, as 256K DRAM prices fell below average FMVs. Indeed, Japanese 256K DRAM production fell quite sharply from 1990 on. Despite rapid and deep cuts in Japanese production, U.S. prices dropped below the FMV levels.

Figure 5-9 showed a similar pattern in 1M DRAMs. From late 1987 through early 1990, U.S. contract prices stayed above average (and over most of this period, above even the highest-cost Japanese company's individual) FMV. Unlike in 256K DRAMs, however, U.S. prices roughly tracked average FMV over the remainder of the STA's lifetime, and Japanese companies cut neither production nor exports of 1M DRAMs to the degree visible in 256K parts. Moderate cuts in Japanese output were apparently successful in boosting prices to levels at or above the Commerce Department FMVs.

By early 1989 semiconductor demand had begun to weaken. In response to a downward drift in DRAM prices, Japanese manufacturers began to cut back output in order to maintain high prices. Japanese newspapers talked of "coordination structures" among Japanese companies being used to achieve "high price stability." This was a distinctly different situation from that in 1988, when prices greatly exceeded price floors as the industry was at full capacity utilization. With all producers operating close to capacity, it was more difficult to argue that the industry's own restraint, rather than an earlier history of politically mandated restraints on capacity expansion and production, must be causing high prices. With idle capacity, it became more plausible to argue (as did the U.S. industry) that collusive firm behavior must be the principal cause of "abnormally" high prices exceeding FMVs.[20]

Causes of the Crisis

A more neutral assessment would begin by noting that at least three explanations for the historically unprecedented run-up in chip prices in 1988 have been seriously discussed. Logically, the 1988 crisis could be explained by any possible combination of these three distinct arguments:

—Innocent miscalculation by producers (on the demand side, under-estimation of the recovery in chip demand in 1988; on the supply side, unexpectedly slow growth in yield rates in semiconductor production in 1987 and 1988) could explain a sustained shortfall in supply.

—The implementation of the STA in 1987 (when Japanese producers were being "guided" by their government to reduce both output and investment) might be expected to have had an effect on chip supply in 1988. That is, we might imagine a purely exogenous, "political" shock to Japanese suppliers' production and investment decisions.

—We might imagine a deliberate, collective decision by a group of suppliers accounting for most of global production to exploit their monopoly power in order to seek greater monopoly rents.

Two distinct variants of the third story exist. The first argues that such organized rent collecting was largely opportunistic, facilitated by the creation, with the implementation of the STA, of a joint information gathering and price monitoring framework for Japanese chip supply. The second argues that the exploitation of monopoly power essentially reflects the private decisions of a collusive grouping of predatory producers (perhaps aided or abetted by the state), but basically independent of the evolving resolution of semiconductor trade frictions.

Some version of the latter variant has been an element of the story of Japanese predation told by U.S. chip producers since 1982. The factual case for this version was reasserted most coherently in a study produced

20. See Semiconductor Industry Association, *Four Years of Experience under the U.S.-Japan Semiconductor Trade Agreement: "A Deal is a Deal"* (San Jose, Calif., November 1990), p. 65.

for the SIA and given wide circulation by others citing it.[21] The argument starts from the circumstances of the "production coordination" by Japanese suppliers in late 1985 and early 1986 (described above), which attracted little contemporary public comment outside Japan (unlike in 1982, when the U.S. Justice Department became involved). Arguing that moves to reduce DRAM exports and production predated the STA and therefore must have been independent of government policy, proponents of this view instead attribute "production coordination" to Japanese chip makers passing over some threshold of monopoly power as American producers withdrew:

> the move toward production regulation by the Japanese producers' group began in 1985, well before the Semiconductor Arrangement had even been conceptualized, much less actually put in place. . . . Thus, by the third and fourth quarters of 1985, Japanese DRAM producers had few competitors left except each other. It was at this precise moment—in late 1985— that reports began to appear of joint actions by the Japanese DRAM producers to stabilize price competition by coordinated curtailments in output.[22]

In fact, a careful reading of the historical record refutes the specifics of these claims. Cuts in exports by the big Japanese producers occurred *after* political pressure had been brought to bear on the industry by top politicians in July 1985. Furthermore, most U.S. firms dropped out of DRAMs *after* the initial cuts in semiconductor exports by Hitachi, NEC, and Toshiba in the summer of 1985. And rather than being "conceptualized" as a strategy for the first time in 1985, Japanese chip exporters had previously jointly cut semiconductor exports—as a response to both political pressure and government "guidance"—back in 1982, and before that in 1981, possibly in 1979, and definitely in 1959, when Japanese market share was very much smaller than in 1985.

On the other hand, business practices that would in the United States have been characterized as anticompetitive clearly have been observed with some frequency in Japanese industry. Mark Tilton has recently pro-

---

21. This analysis is repeated uncritically in Office of Technology Assessment, *Competing Economies: America, Europe, and the Pacific Region* (1991), pp. 11–12; and in Laura D'Andrea Tyson and David B. Yoffie, "Managing Trade and Competition in the Semiconductor Industry," in Laura D'Andrea Tyson, ed., *Who's Bashing Whom? Trade Conflict in High-Technology Industries* (Washington: Institute for International Economics, 1992), pp. 117–18.

22. Thomas R. Howell, Brent L. Bartlett, and Warren Davis, *Creating Advantage: Semiconductors and Government Industrial Policy in the 1990s* (Washington: SIA and Dewey Ballantine, 1992), p. 117.

vided persuasive documentation of the continued toleration (and in some cases active encouragement) of cartels in Japan's basic materials industries by MITI and the Japan Fair Trade Commission, even after such activities were explicitly rendered of questionable legality by changes to laws passed in 1987.[23] And from time to time, details have emerged in U.S. legal proceedings that suggest that Japanese industry associations continue to serve as the venue for attempts to restrict or limit competition. Chapter 2 described how groupings within the Electronic Industries Association of Japan worked to organize both domestic and export cartels for televisions in the 1960s and 1970s. In the mid-1980s, a more recent court case showed, the Japanese Camera Manufacturers Association contained both a Design Committee and a Price Committee to which manufacturers were required to submit plans for features and pricing for new cameras.[24]

The traces of similar behavior are sometimes visible in semiconductor trade. In chapter 2, I documented a failed attempt in the 1970s by some elements in Japanese government and industry to reorganize semiconductor producers into a smaller number of firms. Designed as a countermeasure to the imminent liberalization of foreign imports and investment, the proposal had each reorganized firm specializing in particular chip types. Though some joint tie-ups in R&D ultimately emerged from the IC Liberalization Countermeasure program, inability to agree on who got what among Japanese chip companies appeared to prevent any industrywide cartel-like structure from emerging.

In 1981, an attempt to organize production cutbacks among major Japanese memory chip makers was reported in the Japanese press. This

23. Tilton studied cartels in the aluminum, cement, petrochemical, and steel industries. See *Restrained Trade: Cartels in Japan's Basic Materials Industries* (Cornell University Press, 1996). As Tilton notes, the Fair Trade Commission is far from independent of other Japanese government bureaucracies. By informal custom the head commissioner is a former Ministry of Finance official, while the other four commissioner slots are reserved for a retiree from the Finance Ministry, MITI, the Justice Ministry, and for a former official inside the FTC itself (pp. 48–49). On restraints in the Japanese steel industry, see also Thomas R. Howell and others, *Steel and the State: Government Intervention and Steel's Structural Crisis* (Boulder, Colo.: Westview, 1988), pp. 202–07.

24. See *In the United States District Court for the District of New Jersey, Honeywell, Inc., a Delaware corporation, Plaintiff, v. Minolta Camera Co. Ltd., a Japanese Corporation, and Minolta Corporation, a New York Corporation, Defendants,* Civil Action 87-4847, Transcript of Trial Proceedings, vol. 16 (Newark, N.J., October 11, 1991), pp. 2274–2295; vol. 33 (Newark, N.J., November 6, 1991), pp. 4960–4966; and vol. 63 (Newark, N.J., January 2, 1992), pp. 9727–9734. See also Barnaby J. Feder, "Honeywell-Minolta Dispute Teaches Conflicting Lessons," *New York Times,* February 12, 1992, p. D1.

episode is notable in that several Japanese producers appear to have attempted to coordinate production of semiconductors among themselves in the absence of any political initiative to deal with trade friction. But the attempt apparently failed. In September 1981, as prices for the newly introduced 64K DRAM plummeted, NEC, Hitachi, and Fujitsu each announced a freeze on increased production. When producers Oki Electric and Mitsubishi broke ranks and announced output increases late that year, however, "this, together with the rapidly rising demand [for 64K DRAMs], broke down the tacit agreement among NEC, Hitachi, and Fujitsu to hold back production."[25]

As 64K prices continued to fall in late 1981, trade frictions did intensify. As recounted in chapter 2, MITI intervened months later, in February 1982, counseling restraint in its "guidance" to Japanese producers. (Japanese market share in 64K DRAMs actually peaked in the last quarter of 1981, then went into a sustained decline.) Restraints on exports seemed to stick only after bureaucratic intervention to resolve a deepening external political crisis.

Chapter 3 described how revelation of the existence of a shadowy Council of Nine organized to set prices and allocate markets for discrete semiconductors in Japan played some role in the U.S. government's decision to pursue a section 301 complaint against Japan in 1985. What is unknown, however, is how effective this organization was in accomplishing its objectives. On the one hand, both this episode and the 1981 attempt at coordinating production cutbacks suggest that in the first half of the 1980s, at least, collusive and (at least formally) legally questionable behavior was not unknown in the Japanese semiconductor industry. On the other hand, the 1981 restraint attempt failed, and the Council of Nine's influence does not seem to have pushed Japanese semiconductor prices by 1985 to notably extreme levels relative to world prices.

On balance, then, frameworks imposed or invented in collaboration with politicians and bureaucrats in order to deal with trade friction, rather than sudden changes in market power, seem the more compelling explanation for historical episodes of successful restrictions on supply. In all cases prior to 1989, successful producer restraint coincided with active government initiatives.[26]

25. Yui Kimura, *The Japanese Semiconductor Industry* (Greenwich, Conn.: JAI Press, 1988), p. 66.
26. Indeed, Tilton notes that even in highly concentrated basic materials industries, active MITI involvement was needed in order for cartels to succeed. "In aluminum, cement,

Finally, there is one additional area where continuing American industry complaints might be characterized as a story about a more subtle form of private industrial collusion (fostered historically, to be sure, by Japanese government industrial policies—formal and informal—of the sort described in chapter 2). An alleged propensity for Japanese companies to buy Japanese, even when foreign companies offer equivalent products at lower prices, has continued to raise hackles through the present day, and in fact still fuels American arguments that continuing pressure from government trade negotiations is needed to offer them a fair shot at the Japanese market.

Such a propensity has sometimes been openly acknowledged within the Japanese semiconductor industry. A 1985 industry roundtable discussion of silicon wafer market conditions, for example, noted that Japanese buyers would not touch foreign wafers at the same price as domestic products, and would consider them only if prices were 10–15 percent below the price for Japanese-made product.[27] Although one might argue that delivery problems and quality differentials could account for a continuing reluctance to buy foreign inputs, other evidence would seem to contradict this interpretation. By 1987, for example, Toshiba was saying that "imported silicon wafers have almost totally satisfied Japanese users in terms of quality and shipping date, which used to be major problems of those imported wafers. We will keep the import ratio at 30 percent this year, considering our subsidiaries' policy and the effort of domestic

---

and petrochemicals the policymaking process followed this pattern: MITI pushed industry to go along with cartels that individual firms privately believed were necessary but that they could not coordinate on their own." Tilton, *Restrained Trade*, p. 207.

27. As anonymous industry executive "B" noted, at a time when the industry was in a period of peak demand, "Japanese corporations usually don't buy imported and marketed wafers even if they are of the same quality and a little cheaper. They may have tried to obtain ones in the latest shortage, but corporations overseas, then, might well have had no surplus of supply. Even if Japanese corporations want to buy overseas products on the basis of quality, ignoring the somewhat different yield involved, the commercial deal may not be worked out unless the prices are lower than those produced in Japan by 10 to 15 percent; and even if the prices so favor them, they will never buy very complex super LSIs. Although the wafers are flown in in only one day, a communications gap still remains a solid barrier between nations. Overseas corporations, after all, have to build their own plants in Japan if they are to succeed in selling wafers in Japan, I presume, as the Monsanto Corporation is going to." Monsanto, after building such a plant in the late 1980s, was still unable to significantly increase its share of the Japanese silicon wafer market, and ultimately exited from the silicon wafer business. "Silicon Semiconductors: Point in Question and Prospects; A Roundtable Talk With the Names of Speakers Withheld," *Kinzoku Jihyo*, no. 1125, April 5, 1985, p. 171.

silicon producers, but we will request even tougher price reductions of silicon producers in 1988."[28] Indeed, Japanese silicon wafer producers refused in 1987 to permit their Japanese customers to import wafers from the Japanese wafer makers' U.S. subsidiaries and take advantage of significant price differentials between the United States and Japan.[29] As the Japanese industry coped with the entry of new producers (Japanese steel companies) and a stronger yen, wafer makers complained that price leadership was not being properly maintained by the large producers and the "orderliness" of the market was disappearing.[30]

28. "Toshiba Trying to Reach 30% Materials Import Target This Year," *Rare Metal News*, no. 1396, April 1, 1987.

29. "According to users, there are even some American [epitaxial] wafers whose prices are half of Japanese products, including exchange rate differentials and with American standards. Based on this fact, it is very likely in the future that epi wafers from Japanese U.S.-based silicon producers, such as Shinetsu Semiconductor, Mitsubishi Metal, and Osaka Titanium, will be reimported.

Silicon producers point out that 'the cost of domestic silicon wafers increases in order to meet the strict Japanese standard, so it is irrational to compare price per surface area without taking quality into account.' In addition, Japanese silicon producers operating in the U.S. also mention that 'we are currently operating near full capacity in the U.S., and we cannot sacrifice U.S. users by reexporting the products to Japan.' " "Toshiba Trying to Reach 30% Materials Import Target," p. 1.

30. At a 1987 anonymous roundtable discussion, when the new fiscal year price negotiations came up for discussion, executive "E" complained:

> The persons in charge of ordering materials for the device manufacturers are trying to use the high yen value as a weapon. The industries budgeted at 1 dollar equal to 150 yen, but the yen has appreciated to 130 yen per dollar, and they are looking at the fact that the costs do not match up.
>
> In the past, domestically produced materials and parts have been of high quality. There had been a long history of transactions, so they were willing to purchase at somewhat high prices. Recently, however, the struggle to survive is forcing a change in thinking to move toward the less expensive imported products. There was one company that was rejoicing at having saved 100 million yen due to such a switchover. There is less of a consciousness in the industry of taking care of the parts and raw materials manufacturers.

Executive "D" added:

> Up until about the summer of last year [1986], there was quite a bit of difference in purchase price among different customers for material of the same specifications. If price cuts were undertaken, they were done while considering the health of the overall industry; they had a price-leader-type effect.
>
> However, beginning around the fall of last year, the orderliness of the situation disappeared. The main suppliers were not necessarily the price leaders. Other companies would cut prices on their own initiative, and a great deal of confusion was created in the market. It was difficult to make a distinction between the price of

Mark Tilton described a similar willingness to pay a high premium for domestic inputs in Japanese electronics industries purchases of polypropylene. According to Tilton's Japanese sources,

> Domestic polypropylene prices in recent years have run from a third to half again as high as the price of imports. . . . This price disparity is real and does not merely represent quality differences . . . 90 percent of imported polypropylene is standard-grade and completely identical to standard-grade domestic polypropylene. . . . Despite this disparity, imports never accounted for more than 3.1 percent of domestic market share in 1982–92. . . . The reporter said that the reason is security of supply and reliability of delivery. Electronics makers generally have a long-term relationship with a single petrochemical supplier, although not usually because of keiretsu ties. Thus, electronics firms buy domestic materials from a single supplier as a strategy to maximize their own security and efficiency."[31]

Even today, there seems to be a similar propensity by Japanese electronic equipment makers to favor procurement of domestically made chips. A 1994 Japanese electronics industry survey found the value of semiconductors directly procured by major users from overseas sources to be quite low, but noted that "many users are annoyed by the domestic overseas price differential for certain products. Thus direct overseas procurement will likely increase in cases where there is only one source in Japan, or the price difference is 50–60% for certain products (high performance linear and operating amps, etc.), or where the difference is around 20% and the Japanese supplier fails to make efforts to narrow the gap."[32]

Similarly, Mark Tilton notes that an executive in a large Japanese electronics firm reported to him in 1994 that while his firm continued to be reluctant to buy foreign steel, it now permitted some of its small subcontractors to buy imported steel. Although it still did not directly purchase imports, by threatening to buy foreign steel and giving domestic steelmakers the opportunity to discount prices to match import prices, the electronics maker could persuade domestic steel producers to lower

---

monitor wafers and dummy wafers. Everything was confused.

"Anonymous Roundtable Discussion on Silicon: Japan Consumes Half of the World's Wafers; Forecasts of the Outcome of U.S.-Japan Semiconductor Trade Friction," *Kinzoku Jihyo*, no. 1303, June 5, 1987, pp. 188–89.

31. Tilton, *Restrained Trade*, pp. 162–63.

32. See Osamu Otake, "High-Gear Overseas Production and Semiconductor Demand," *Tokyo Denshi* (February 1995), pp. 8–18, in *Foreign Broadcast Information Service Daily Report: Science and Technology*, August 30, 1995, electronic version.

prices from the standard "large user" price levels on that portion of their demand that they could buy from overseas sources.[33]

In short, evidence of a residual "buy Japanese" preference can still be discerned within some corners of the Japanese electronics industry. In many cases—particularly in semiconductors—quality levels for products produced by major global suppliers, whether Japanese or foreign, no longer seem to vary significantly aross companies and therefore do not provide much of an explanation for such a preference.[34]

There is another set of economic reasons why companies might prefer to buy from domestic sources at higher prices linked to "relational contracting." A long-term relationship with a preferred supplier may be fostered to obtain superior service and delivery, or partnerships created with suppliers to develop technologically superior new inputs, at the price of commitments to maintain procurement even when input prices are somewhat above market levels. There is also some evidence of a long-standing concern in Japan over the vulnerability to external disruption of supplies procured from overseas, although the overtones of economic warfare wrapped into this worry make it more a national security issue than an industrial concern (and it is hard to see how Japan's vulnerability in its dependence on foreign industrial inputs—other than natural resources—is any more severe than that faced by most other industrial nations). Finally—and often explicitly linked to government concerns about the security of supplies for key inputs—there is a long history of government policy formally and informally encouraging "buy Japanese" attitudes within the industrial sector, although most of the means of formal encouragement have now been dismantled.[35] Separating out what portion of the explanation for a domestic preference is economic rather than the result of politics, history, informal government policy, and the prevailing business culture is likely to remain an impossible task.

Somewhat less hopeless is the analysis of less subtle forms of industrial collusion. One of the most provocative issues raised by the operation of the STA is the extent to which cartel-like private behavior was observed

33. Tilton, *Restrained Trade*, pp. 185–86.
34. Quality differentials are not typically raised as an issue by Japanese companies today in purchasing foreign semiconductors. Virtually all the major Japanese semiconductor producers market foreign-manufactured chips in Japan with their own company's Japanese brand name stamped on them.
35. Tilton makes a powerful case that informal encouragement remains in the basic materials industries he studied, primarily because of a continuing government strategic interest in maintaining the security of supply. See *Restrained Trade*, chap. 7.

after the arrangement went into operation and what the potential consequences of such behavior are. I turn next to these issues.

Anecdotal Evidence on Private Collusion

The period from early 1989 through early 1990 is the most interesting
part of the five-year history of the first STA from the standpoint of
allegations of collusive behavior by Japanese companies. Market DRAM
prices in the United States and Japan remained at levels considerably
above U.S. FMVs, at a time of weakening demand, while companies
were reducing their output levels well below capacity. By this time the
Japanese government was under considerable foreign pressure *not* to
intervene directly and order domestic chip producers to reduce output,
so the manner in which Japanese DRAM producers were reducing their
output prompted intense foreign interest.

It would be helpful to use actual data on pricing and costs to examine
empirically the credibility of allegations of collusive behavior, but unfortunately no data on price-cost margins for DRAMs are available. Some
noisy information on average cost is available in public submissions to
the Commerce Department filed by companies as part of the process of
setting FMVs, but some possibly contentious assumptions are required
to use these data.[36] The only direct evidence on the question, then, is
necessarily anecdotal.

In one widely publicized episode, the three largest Japanese DRAM
producers were reported to have cut output sharply, in rapid succession,
on the day after the failure of an attempt by U.S. chip consumers and
producers to organize a joint DRAM manufacturing venture (U.S. Memories) was announced. Reports in the U.S. press fueled assertions that
the Japanese were "acting much like a cartel." Some analysts close to the
U.S. industry even charged that Japanese companies had flooded the
market and forced prices down in a deliberate effort to torpedo U.S.
Memories (see notes 180 and 181 in chapter 4).

Actually, major Japanese companies had cut shipments and reduced
production long before U.S. Memories failed, in an effort to stabilize
prices going back to the second quarter of 1989. Available data suggest
that, after increasing production of 1M DRAMs during the first quarter
of 1989, domestic producers began to cut back production in the late

36. These data are analyzed later in this chapter.

spring, sliced shipments even more, and increased their stocks of parts held in inventories.[37] In the early fall of 1989, most major Japanese chip producers announced large production cuts. At the end of July NEC announced that it had scaled back its plans to increase production of 1M DRAMs.[38] In September, Toshiba, Hitachi, and Mitsubishi each reported that they were cutting back their production levels by about 10 percent.[39] In late October, NEC announced that it too would actually cut current production levels in the first quarter of 1990.[40]

Contemporary Japanese press accounts asserted that Japanese chip makers were collectively cutting back production to slow the decline in prices, to achieve "high price stabilization." Talk of a "coordination structure" also reappeared in the trade press (see chapter 4).

In interviews in November and December of 1989, semiconductor executives at several Japanese companies were quite frank with me about their intention to continue to cut back DRAM production in order to stabilize prices. Asked why, given current prices that were certainly far above his company's production costs, he did not cut prices in order to stimulate sales, one top executive motioned toward a skyscraper visible through the window, the headquarters of a rival electronics giant, and remarked that if he cut prices, his neighbor would as well, setting off a round of continual price cutting. Asked how he could know how much to cut production in order to stabilize prices, without knowing the plans of other companies, his colleague responded that the matter was complicated, with "many aspects," and declined to elaborate further.[41]

The subject of coordination among companies is obviously a delicate one, and while rumors of golf course meetings are a staple of conversation

---

37. See chapter 4. The basic chronology of events in 1989 was set out as follows in a presentation by Hitachi's H. Nakagawa: supply exceeded demand in the second quarter of 1989; booking adjustment by users in the second to fourth quarters of 1989; and production adjustment by manufacturers after the third quarter of 1989. H. Nakagawa, "Semiconductor and EDP Market," slides from a presentation given circa late 1989 (given to the author in December 1989).

38. NEC withdrew its plan to increase monthly production of 1M DRAMs from 6 million to 8 million chips. *Nikkan Kogyo*, July 29, 1989, p. 1; *Nihon Kogyo*, July 29, 1989, p. 1.

39. "Major Semi-Conductor Manufacturers to Reduce 1M-DRAM Production by 10% from September," *Nihon Keizai Shimbun*, September 14, 1989, p. 10.

40. The cut was from 6 million units per month to 5 million units. *Dempa Shimbun*, October 31, 1989, p. 1; *Nikkan Kogyo*, October 31, 1989, p. 1; *Nihon Kogyo*, October 31, 1989, p. 5.

41. Author's interview in Tokyo, December 1989.

inside the industry, it is difficult to find someone with first-hand information who is willing to discuss it, even on an unattributed basis. Nonetheless, I did manage to interview one reliable source who had been present at a number of meetings in Japan where information on individual companies' production and capacity was discussed. According to his account, these informal and unofficial meetings were attended by managers from the semiconductor divisions of major Japanese companies. Each company had available to it computer printouts of other companies' output and capacity by product, and all companies had the same data in their printouts. No government officials were present at these meetings.[42]

After 1989, observers writing in the Japanese trade press noted the trend toward increasingly tight oligopoly control of Japanese DRAM production. As one chip executive remarked to an American government official in early 1990, "Since the Semiconductor Agreement, we [Japanese DRAM manufacturers] have moved from competing for market share to market sharing."[43]

Empirical Tests for Competition

Press reports, industry gossip, and the occasional first-hand account are suggestive and perhaps even believable, but can be made more credible by checking the observed empirical behavior of prices for 1M DRAMs over this period for a certain degree of consistency with the story. The basic idea is this: profit maximization requires that marginal revenue, as perceived by the firm, be set equal to marginal cost. With "competition," a firm's marginal revenue curve will lie above that corresponding to a collusive grouping of firms in a constant-market-share cartel. If we can somehow obtain an upper bound on marginal cost, then marginal revenue computed under competitive assumptions should always lie at or below that upper bound. If marginal revenue calculated under competitive assumptions lies well above the bound on marginal cost, it suggests that those competitive assumptions are incorrect, and is consistent instead with alternative stories, including collusive behavior.

---

42. The last meeting attended by my source, as of the date of my interview, was in 1990.

43. Personal communication with the author.

Because DRAMs are essentially a homogeneous commodity sold in well-developed secondary spot markets, the natural assumption is that producers sell at a single market price and choose the quantities they will sell. The benchmark adopted here for competitive behavior will be "Cournot competition," the assumption that producers take the output of other firms as unaffected by their own output decisions, at any moment in time.

For example, in the last quarter of 1989 Toshiba—the largest and almost certainly the lowest-cost maker of 1M DRAMs—was cutting production of these chips at a time when market prices in the United States and Japan hovered between $9 and $10 per chip. Its FMV was widely believed in Japanese industry circles to be about $4.50 at the time.[44] Indeed, during the shortage of the summer of 1988, Toshiba's production costs had been widely believed in Japan to be in the neighborhood of 500 yen (under $4),[45] and its costs in 1989 must surely have declined even further, as the result of both learning economies and less intensive capacity utilization. With Toshiba's world market share at well under one-third,[46] and a price elasticity of about −1.5, Toshiba's marginal revenue would have to have greatly exceeded 80 percent of price with Cournot competition. With production at levels well below capacity, it is difficult to believe that marginal cost could have been more than 50 percent greater than the Commerce Department's already more than generous estimate of average cost, as embedded in the FMV. Therefore, unless the various parameters just used are significantly different from their true values, Cournot competition must have been a rather poor description of Toshiba's competitive environment at that moment.

---

44. Fourth-quarter 1989 FMVs for 1M DRAMs were reported by one Japanese industry source to have been set as follows: Toshiba, $4.50; Hitachi, $5.20; Fujitsu, $5.50; Mitsubishi, $7.00; NEC, $7.50. Author's interview in Tokyo, December 1989.

45. See Stefan Wagstyl, "Winning When the Chips Were Down," *Financial Times,* May 10, 1988, p. 33; Stefan Wagstyl, "Turning Hard Times Into Good Times," *Financial Times,* June 14, 1988, p. 22; and Terry Dodsworth, Louise Kehoe, and Stefan Wagstyl, "The Chase to Catch the Japanese," *Financial Times,* July 25, 1988, p. 18. At the time, 1M DRAMs were selling for about four times Toshiba's estimated cost of production, or 2,000 yen. Toshiba claimed to be achieving 65 percent yields over this period, at a time when the industry as a whole averaged under 40 percent.

46. Dataquest estimates show Toshiba with under 20 percent of world 1M DRAM production in the last quarter of 1989, and under 30 percent even if Motorola's sales are counted in the Toshiba market share (Motorola sold Toshiba-fabricated chips under its own name and operated a DRAM manufacturing joint venture with Toshiba at the time).

More precisely, we have perceived marginal revenue of firm i, MR_i, given by

$$(7\text{-}1) \qquad MR_i = P\left[\frac{s_i}{\beta}(\lambda_i + 1) + 1\right],$$

where λ_i is the so-called conjectural variation, giving the perceived response of other producers' output to a one-unit increase in the firm i's output; s_i is the market share of firm i; and β is the industry's price elasticity of demand. In the case of Cournot competition we have λ_i equal to zero; with a constant-market-share cartel we have λ_i equal to the inverse of the firm's share of cartel output less one. Because price elasticity β is negative, the firm's perceived marginal revenue in a cartel will always be less than perceived marginal revenue with Cournot competition. Heuristically, we may wish to think of conjectural variation λ_i as varying between zero and a positive number equal to the inverse of the firm's share of cartel output less one, with the parameter value reflecting the "degree of competition."

Before proceeding further, it is necessary to discuss briefly some additional complications peculiar to the economics of semiconductor production. The key issue is learning economies.

Learning Economies

Semiconductor production is believed to be characterized by learning economies, in which unit production costs fall sharply with accumulated production experience. A key result due to Spence and generalized by Fudenberg and Tirole is that, with learning economies, a rational firm will generally equate current marginal revenue to a value below its current short-run marginal cost of production, as it takes into account the effect of current production on reducing future production costs. That is, a profit-maximizing firm sets

$$MR(t) = MC(t) - \delta(t) = \hat{MC}(t),$$

where δ is a positive "shadow value" of experience, which declines to exactly zero at the end of the product life cycle (designated as time 1). Alternatively, the right-hand side of this equation may be labeled $\hat{MC}$, or "true marginal cost," equal to current marginal cost MC, adjusted for the future cost–reducing effects of current output. This analysis assumes that any capacity constraints are nonbinding.[47] (With capacity constraints, MR may exceed $\hat{MC}$.)

This relationship allows us to characterize, at least crudely, the relationship between prices and costs in an industry with learning economies. We must have

(7-2) $0 = MR - \hat{MC} \geq MR - MC > MR - \overline{MC},$

where MC is current marginal cost and $\overline{MC}$ is an upper bound exceeding MC. Under the null hypothesis of quantity-taking Cournot competition, we may substitute $P(1 + s_i/\beta)$ for MR in equation 7-2, and after deducting off our upper bound on MC we should have a quantity that is unequivocally nonpositive.

Under the alternative hypothesis of cartel behavior, perceived marginal revenue will be less than under Cournot competition, and

$$MR_{Cournot} - \hat{MC} > MR_{cartel} - \hat{MC} = 0,$$

so that Cournot marginal revenue less "true" marginal cost will be positive (although $MR_{Cournot} - \overline{MC}$ may be negative if upper-bound $\overline{MC}$ exceeds MC by enough). Our basic procedure for testing for competitive behavior, then, will be to construct an upper bound on marginal cost, then deduct it from estimated marginal revenue maintaining competitive Cournot assumptions. Under competition, the result should necessarily be less than or equal to zero. Under the alternative hypothesis of collusion ($\lambda_i > 0$), marginal revenue less an upper bound on current marginal cost may be positive (although it need not be).

Bounding Marginal Cost

The most difficult issue in an exercise like this is always to find data that allow one to say something reasonable about marginal cost. For two

47. It does, however, permit discounting of future revenues and costs by the firm when it makes its output choices. An equation such as this (with current MR set equal to current MC less a positive term) can easily be shown to hold in asymmetric industries (with firms of different sizes), as firms continuously make production choices over time in the case of nonstrategic behavior (in which rivals are assumed to have precommitted output paths). A similar result has also been shown to hold in a symmetric-industry equilibrium (identical firms) when firms behave strategically in their choice of current output levels (that is, where firms manipulate rivals' output levels in future periods, by signaling lower future costs through increases in their own current output, via learning effects). See Drew Fudenberg and Jean Tirole, "Learning-by-Doing and Market Performance," *Bell Journal of Economics*, vol. 14 (Autumn 1983), p. 527.

Japanese companies—Mitsubishi and NEC—the public cost submissions discussed in chapter 5 contain a reasonable amount of data on a cost concept known as cost of manufacture (see appendix 5-B). COM is essentially equal to average variable cost, plus some allocation of depreciation charges for the manufacturing facility in which 1M DRAMs are produced, plus some allocation for R&D costs.

Variable costs of DRAM production essentially have two components. Briefly, silicon wafers are processed into semifinished chips, then tested. The good "yielded" chips are then assembled and packaged, then tested again. The yield of good chips is what is believed to increase significantly with experience as the result of learning economies. Some portion of variable cost is sensitive to the processing cost for the silicon wafer starting down the production line, while another portion varies with the number of good, yielded chips that are then assembled and tested. If the variable processing cost per wafer is c, and the cost per yielded chip is d, then total variable cost is given by $y(c/w + d)$, where y is finished chips and w is yield of good chips per wafer. Yield w, in turn, may be expected to be an increasing function of experience, as measured by cumulative output Q.

For given experience (cumulative output), then, yield levels are fixed, and marginal cost is likely to be approximately constant and equal to average variable cost. This is not an exact description of the cost structure, but it is likely to be a decent approximation. Current-generation 1M DRAMs were generally produced in large-scale factories of recent vintage that were producing a relatively small number of products.[48] Marginal decisions to increase output of DRAMs basically require the producer to cut back on production of some other product (such as SRAMs or microprocessors), but this is unlikely to have any significant impact on the variable cost of processing a wafer containing DRAMs or of testing or assembling a finished DRAM. In effect, the typical short-run marginal cost curve is likely to be L-shaped: approximately constant

48. See for example, W. C. Holton and others, *JTECH Panel Report on Computer Integrated Manufacturing (CIM) and Computer Assisted Design (CAD) for the Semiconductor Industry in Japan* (McLean, Va.: Science Applications International Corporation, 1988). "The large semiconductor wafer fabrication plants visited by the JTECH panelists are engaged in high-volume production of VLSI standard products such as dynamic and static memories and microprocessors. A single wafer fabrication line is typically used to produce perhaps five to fifteen different products (for example, static and dynamic random access memory (RAM) devices of different bit capacity, several different microprocessors) using one or two standard processes" (pp. 14–15).

marginal cost as output of DRAMs is increased, then soars up to infinity as plant capacity is reached.

Now, the Commerce Department's measure of the cost of manufacture, COM, is defined as

$$COM = \frac{[TVC + D(t)] \, [1 + \rho(t)]}{Y},$$

where TVC is total variable cost (assumed proportional to output Y, given experience Q and a corresponding yield rate), $D(t)$ is the depreciation cost allocated to DRAM manufacture at time t, and $\rho(t)$ is the ratio of R&D to semiconductor sales for the company as a whole. By the above reasoning, we then have

$$COM = \left[MC + \frac{D(t)}{Y} \right] [1 + \rho(t)],$$

where MC is current marginal cost. Since depreciation cost per finished chip, $D(t)/Y$, is a large positive number (see chapter 6), and ρ is positive, we may consider COM an upper bound on current marginal cost.

In fact, COM is likely to be significantly above MC. Depreciation costs are a significant share of total production cost (likely to exceed 20 percent),[49] and fragmentary data suggest that the estimate of ρ produced by the Commerce Department's procedures was also high (about 0.36 for Mitsubishi).[50] Although the data for Mitsubishi and NEC also suggested that there was a lot of variance across firms in these ratios,[51] it seems

49. A rule of thumb in the semiconductor business is that there is a 1:1 overall ratio between the capital cost of new capacity and the volume of annual sales generated by that capacity. With a five-year life for a new production line and straight-line depreciation, this suggests that depreciation amounts to roughly 20 percent of sales. For a product such as DRAMs, whose manufacturing costs account for relatively more of the value of sales than the overall semiconductor industry average, one would expect this ratio to be even greater.

50. For fifteen data points describing Mitsubishi's 256K and 1M DRAM costs at various times from 1987 through 1991, the average estimate of ρ—estimated as COM divided by the total cost of wafer fabrication, assembly, and testing—was .36. This corresponds to a cost structure with R&D amounting to 21 percent of constructed value: $\rho / [(1 + \rho)(1.08)(1 + s)] = .21$, with s an estimate of sales, general, and administrative expenses as a share of COM. A single companywide estimate of ρ was applied by the Department of Commerce to all Mitsubishi semiconductor products. See appendix 5-B.

51. It was also possible to estimate ratio s—sales, general, and administrative expenses (SG&A)—as a share of COM using various data from Mitsubishi and NEC. Using ninety-two observations for various models of DRAMs over the 1987–91 period, and either the relationship COP = $(1 + s)$ COM or Constructed Value = $(1.08) (1 + s)$ COM, s for NEC averaged 0.39. This corresponds to a cost structure with SG&A amounting to 26 percent

reasonable to conclude that COM is likely to be well above MC, and serve as quite a high upper bound on marginal cost, when output is not constrained by capacity limitations.

Both marginal revenue and this upper bound on marginal cost for 1M DRAMs were then calculated for Mitsubishi and NEC over the period from the third quarter of 1989 through the fourth quarter of 1990. Since (see the discussion above) Japanese firms are known to have started to cut back production of 1M DRAMs in the third quarter of 1989, we can safely assert that after that quarter production was not capacity constrained. Therefore, profit-maximizing producers at that point would have been setting perceived marginal revenue equal to marginal cost (and not exceeding marginal cost as might be the case with output constrained by capacity). And because DRAM production fell short of capacity, we would also expect to be on the flat part of an L-shaped marginal cost curve, with marginal cost approximately constant.

Figures 7-3 and 7-4 plot marginal revenue under the assumption of Cournot "competition" for Mitsubishi and NEC using several different sets of price data (a large-user price in Japan and an estimated Japanese spot price, as reported by *Nihon Keizai Shimbun*, a U.S. contract price as estimated by Dataquest, and a Japanese contract price estimated by Dataquest), and quarterly market share estimates from Dataquest.[52] Under competitive assumptions, these estimates of marginal revenue should be well below the generous upper bound on marginal cost that COM should provide.[53]

They are *not* in 1990—in either U.S. or Japanese markets, using either spot or contract prices as the measure of revenues generated by a marginal sale—and exceed this upper bound on marginal cost by a substantial margin. I conclude that the (admittedly imperfect) cost data that are

---

of constructed value ($= s/[(1.08)(1+s)]$). For Mitsubishi, using thirty-one observations on various types of DRAMs, s averaged just 0.165, corresponding to a cost structure with SG&A equal to 13 percent of constructed value.

52. The NEC and Mitsubishi fourth-quarter 1990 market shares were used in the first-quarter 1991 estimates of their respective marginal revenues, since market share estimates for that quarter were unavailable. A price elasticity of demand of -1.5 was used in constructing these marginal revenue estimates.

53. Actually, the upper bound on COM shown is 1.25 times COM (in the case of NEC, 1.25 times the maximum of the estimates of COM reported for three chip variants). This is because with these ranged data, reported COM may be up to 20 percent less than the actual value (and the actual value therefore up to 25 percent greater than the reported value).

Figure 7-3. *Mitsubishi Competitive Marginal Revenue versus Cost of Manufacture (COM) Reported to Commerce Department, 1989–91*

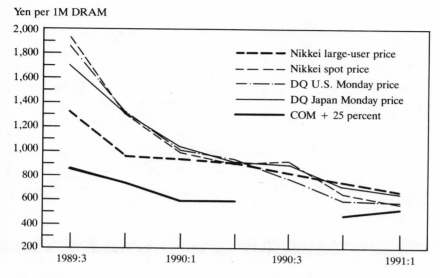

Yen per 1M DRAM

Source: Author's calculations and Dataquest and Nikkei data. Calculations explained in the text. Marginal revenue based on use of alternative price series is graphed.

available strongly support the contemporary accounts of collusive market-sharing behavior by Japanese producers.

The Costs of Facing a Cartel

Although the STA appears to have played—at the very least—a catalytic role in the move toward market sharing by Japanese producers, one can entertain the thought that this particular Pandora, having emerged from her box, is probably the most persuasive argument to be mustered for the proposition that declines in U.S. semiconductor manufacturing capability pose strategic issues for the U.S. economy. Put most simply (and abstracting from the many nuances that muddy discussion of what actually has occurred behind the scenes within the Japanese semiconductor industry since the 1986 accord), I wish to ask what the cost to U.S. chip consumers might be if they faced a hypothetical foreign DRAM cartel, and therefore what they might reasonably be willing to spend (for example, by subsidizing entry by U.S. firms) to ensure that such a cartel could not be formed.

Figure 7-4. *NEC Competitive Marginal Revenue versus Cost of Manufacture (COM) Reported to Commerce Department, 1989–91*

Yen per 1M DRAM

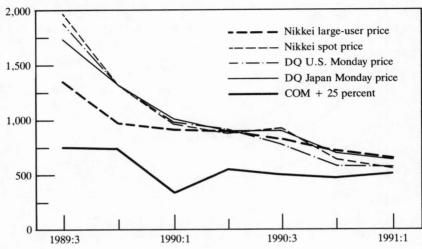

Source: See figure 7-3.

I propose to frame this argument in three stages. First I will review a simple model of the semiconductor product life cycle that assumes non-strategic, noncooperative behavior on the part of producers. I will then show how strategic, but noncooperative behavior might be expected to change producer decisions. Finally, I will sketch out a particular scenario for cartel formation and simulate its impact on American chip consumers.

Modeling the Impact of Cartelization

My modeling efforts will focus on the 1M DRAM for two reasons: first, it was the most recent generation of DRAM chip for which relatively reliable data were available at the time the research was begun; and second, its product life cycle largely corresponded to the period of the STA. This second consideration was particularly important, since I wished to approximate the potential impact of collusive behavior over a period when it was alleged to have occurred.

The argument that collusive behavior could have significant effects in DRAMs is not inherently unreasonable. In chapter 5 it was observed that during the early years of 256K and 1M DRAM production the industry was somewhat more concentrated than the historical norm, but by 1989 concentration in both these products looked similar to earlier historical levels. Levels of geographic concentration offered greater promise as a reason for concern. At the time the STA was signed, reductions in production or exports by Japanese producers could have had potentially great impacts on aggregate worldwide supply of DRAMs.

Strategic Behavior

Next I modify the model developed in the last chapter in order to permit noncooperative, strategic behavior on the part of producers. As in the previous chapter, competition will be thought of as occurring in two periods. In the first period firms sink resources into capacity investments. In the second period firms select a time profile for capacity utilization rates and output, given the available capacity resulting from their first-period investment decisions. Given first-period capacities, the second-stage game in utilization rates results in a static Cournot-Nash equilibrium. Now, however, I assume that this two-stage game is characterized by subgame perfection. When firms select first-period capacities, they now correctly anticipate the resulting second-stage equilibrium. (Though in the first period rivals' capacity choices continue to be perceived as unaffected by one's own capacity choice.) Strategic interactions arise because first-period capacity choices correctly anticipate effects on one's own and competitors' second-period utilization decisions.[54]

Since the second-period subgame remains identical to that analyzed in the nonstrategic case (that is, taking capacities as given, optimal utilization rates are exactly as analyzed earlier), we need only consider how the determination of optimal capacity is changed. As before, I assume a symmetric equilibrium of identical firms.[55]

To do so, we must add an additional term to equation 6-5 which reflects the anticipated effects of increasing one's own capacity on second-stage outputs. Given the optimal path for the utilization rate $u^*(t)$ and the

54. Another possible approach to strategic behavior, not pursued here, would be to make capacity investments sequential and permit "first-moving" firms to anticipate and preempt capacity investments by "follower" firms, in a von Stackelberg–type model. A major component of such an approach would be to describe the economic basis for asymmetries among firms.

55. Formal analysis of how strategic competition in capacity investments alters the model developed in the last chapter is presented in appendix 7-A.

corresponding trajectory $E^*(t)$ for the experience variable, the necessary condition for optimal capacity choice is that

$$(7\text{-}3) \quad \int_0^1 \left[\left(R_y - d + \frac{\delta}{K^\gamma} \right) uw + R_x \frac{\partial x}{\partial K} - \gamma \frac{\delta}{K^\gamma} uw - cu - r \right] dt = 0$$

must hold (see appendix 7-A). Note that this partial derivative is with respect to K alone, evaluated with $u(t)$ set equal to optimal control $u^*(t)$. This differs from the expression developed from equation 6-5 because, in the nonstrategic case, initial capacity investments were not viewed by the firm as affecting rivals' second-period utilization decisions; that is, $\partial x/\partial K = 0$ was assumed. With more than two firms in the industry ($N > 2$), and with symmetric equilibrium, concavity of the industry revenue function is sufficient to guarantee that rivals will reduce their output along an interior segment. Since R_x is negative, the additional effect added on to equation 6-5 must be positive, which means that the marginal return on additional capacity will be positive at the level of investment corresponding to the nonstrategic equilibrium. With strategic competition, additional capacity will seem attractive.

Along a "boundary" segment (when available capacity actually constrains production), however, the response by competitors to a marginal increase in one's own capacity will be zero. Intuitively, when a firm's marginal revenue already exceeds its true (corrected for learning economies) marginal cost, so that it chooses to operate at maximum capacity, a small increase in a competitor's output—which reduces the firm's marginal revenue (with a concave industry revenue function)—will still leave the firm wishing to produce at maximum output. Thus, if the symmetric industry equilibrium is one in which firms are everywhere capacity constrained, producing at full blast throughout the product cycle, the strategic equilibrium will coincide with a nonstrategic equilibrium. Strategic behavior will have no effect on the industry equilibrium.

Note that I am allowing only partially strategic behavior with this modeling strategy. There are two possible instruments of strategic behavior: capacity choice and output (capacity utilization) choice. Output choice has potential strategic effects via learning economies, insofar as output at any moment in time affects future cost structure. (Without learning economies, current output choice has no effect on future cost structure, and therefore no strategic value.) A firm might choose current output taking into account the impact of its own current output choice on rivals' future output choices, rather than taking the rivals' future output path as given. To reduce complexity, I do *not* allow strategic output choice, and output paths are chosen as open loop, in which firms precommit to an output path at the beginning of the second stage of the game.[56]

However, I *do* permit strategic competition in capacity. Firms *do* use capacity choice strategically, in order to affect rivals' choice of (open-loop) output path in the second stage of the game. Thus strategic competition is permitted in capacity choice, but not in output choice (via learning effects).

Table 7-1 reworks table 6-4 to reflect strategic behavior, with equation 7-3 used instead of equation 6-5. If no solution for a system based on equations 6-9 and 7-3 was found, the alternative, corresponding to full-blast production throughout the product cycle, is identical to (unmodified) equation 6-5, since $\partial x/\partial K = 0$ holds along a capacity-constrained segment. In what I earlier tentatively concluded was the most realistic set of assumptions about how learning economies work (learning occurs at the plant level, with parameter $\gamma = 1$), strategic behavior appears to have no impact on the symmetric-industry equilibrium, and full-blast production throughout the product cycle still prevails. In the case where learning occurs at the company level ($\gamma = 0$, which had a substantial interior segment with nonstrategic behavior), capacity per firm increases by about 50 percent relative to the nonstrategic case, and the equilibrium number of firms drops from four to three, for the range of fixed R&D costs considered plausible. The resulting path of DRAM price is shown in figure 7-5. Strategic behavior thus has no impact with the "per fab line" specification of learning, but a large impact in the case of companywide learning.

For the reasons outlined earlier, the $\gamma = 1$ case continues to appear most plausible. Thus I conclude that—within the framework developed here—my best efforts at discovering realistic values for empirical parameters relevant to 1M DRAM production suggest that strategic competition in capacity investments had little potential for shaping market outcomes in the industry.

---

56. One possible (but only partial) empirical justification for this specification is that there are very long lags between capacity utilization decisions and the resulting output. The production process for a chip takes almost a full quarter year, so any change in output rates is observable by one's rivals only after a substantial lag.

Table 7-1. Solution of the Optimal Control Problem, Strategic Simulation

	Number of firms															
Simulation	1	2	3	4	5	6	7	8	9	10	11	12	13	14	15	16
Gamma = 1																
Wafer-processing capacity (K)[a]	13.49	19.08	16.03	13.34	11.31	9.79	8.62	7.69	6.94	6.32	5.8	5.36	4.98	4.66	4.37	4.11
End, full-blast output (t_s)	1	1	1	1	1	1	1	1	1	1	1	1	1	1	1	1
Gross rent[b]	29,370	10,380	4,985	2,902	1,894	1,332	988	761	604	491	407	343	293	253	221	195
Gamma = 0.9																
Wafer-processing capacity (K)[a]	15.46	21.92	18.4	15.73	12.95	11.19	9.84	8.77	7.91	7.2	6.6	6.09				
End, full-blast output (t_s)	1	1	1	1	1	1	1	1	1	1	1	1				
Gross rent[b]	32,360	10,370	4,360	2,180	1,192	676	382	204	90	15	−36	−70				
Gamma = 0.8																
Wafer-processing capacity (K)[a]	16.96	23.91	20.12	16.76	14.23	12.32	10.86	9.68	8.74	7.96	7.3	6.74				
End, full-blast output (t_s)	1	1	1	1	1	1	1	1	1	1	1	1				
Gross rent[b]	35,570	10,410	3,752	1,455	477	5	−240	−371	−442	−479	−496	−500				
Gamma = 0.7																
Wafer-processing capacity (K)[a]	17.683	24.65	20.86	17.48	14.94	13	11,492	10.291								
End, full-blast output (t_s)	1	0.98	0.99	1	1	1	1	1								
Gross rent[b]	38,840	10,590	3,277	846	−151	−601	−811	−907								

Gamma = 0.68

Wafer-processing capacity $(K)^a$	17.73	24.76	21.07	17.61	15	13.07
End, full-blast output (t_s)	0.98	0.94	0.95	0.98	0.997	1
Gross rent^b	39,480	10,580	3,056	676	−264	−707

Gamma = 0.6

Wafer-processing capacity $(K)^a$	17.77	24.97	21.73	18.19	15.48	13.46
End, full-blast output (t_s)	0.86	0.79	0.79	0.82	0.85	0.88
Gross rent^b	41,950	10,630	2,289	−113	−963	−1,294

Gamma = 0.5

Wafer-processing capacity $(K)^a$	17.57	24.9	22.29	18.69	15.89	13.8
End, full-blast output (t_s)	0.72	0.64	0.62	0.65	0.69	0.72
Gross rent^b	44,820	10,880	1,614	−858	−1,643	−1,902

Gamma = 0.3

Wafer-processing capacity $(K)^a$	16.69	24.07	22.8	19.15	16.23	14.08
End, full-blast output (t_s)	0.51	0.4	0.36	0.4	0.44	0.47
Gross rent^b	49,660	11,840	1,108	−1,593	−2,399	−2,600

Gamma = 0

Wafer-processing capacity $(K)^a$	14.88	22.11	22.65	19.03	16.04	13.87
End, full-blast output (t_s)	0.31	0.2	0.16	0.19	0.22	0.25
Gross rent^b	54,880	13,650	1,569	−1,508	−2,445	−2,746

Source: Author's calculations.
a. Millions of product-cycle wafer starts.
b. Millions of dollars.

Figure 7-5. *Historical Prices for 1M DRAMs Compared with Simulated*
Strategic Equilibrium Time Profiles

Dollars per chip

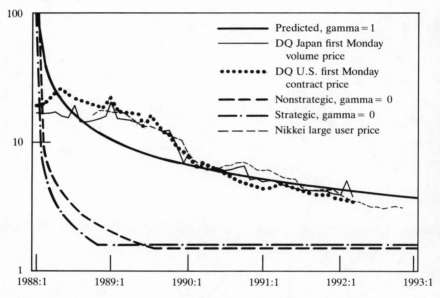

Source: Author's calculation and Dataquest and Nikkei data. Calculations explained in the text.

Collusion and Cartel

Now it is possible to attempt to answer the question posed at the
beginning of this chapter. Relative to a benchmark of strategic or nonstrategic—but *competitive*—behavior, can the impact on U.S. consumers
of the systematic exploitation of monopoly power by a foreign DRAM
cartel, that is, of *collusive* behavior, be estimated?

The simplest variant of the story to be told to motivate this calculation
runs along the following lines. Suppose, for one reason or another, that
firms in country U believe that the fixed sunk cost F required for them
to enter the industry is higher than for firms in country J (that is, that
the firms in country U believe themselves to be less efficient producers).
In a competitive equilibrium where entry ensures zero economic profits,
there will then be no entry by country U firms, and only (more efficient)
country J firms will populate the industry. Country J firms alone then

Table 7-2. *Simulated Welfare Effects of Cartelization on the DRAM Market*[a]

| Simulation | Strategic competition in capacity with: | | Change |
	Competitive output choice	Second-period cartel	
Gamma = 1[b]			
t_s, end full-blast output	1	0.237	. . .
$P(1)$, final price (dollars)	3.77	9.37	. . .
Consumer surplus (billions of dollars)	148.9	117.4	−31.5
Industry net profits (billions of dollars)	0.045	4.564	4.5
Gamma = 0[c]			
t_s, end full-blast output	0.16047	0.037	. . .
$P(1)$, final price (dollars)	1.61	4.08	. . .
Consumer surplus (billions of dollars)	287.9	186.4	−101.5
Industry net profits (billions of dollars)	3.957	34.57	30.61

Source: Author's calculations.
a. Symmetric industry equilibrium assumed.
b. F equals $250 million; N equals fourteen firms; and K equals 4.655 million product-cycle wafer starts.
c. F equals $250 million; N equals three firms; and K equals 22.648 million product-cycle wafer starts.

play the two-stage game described above and select sunk capacity investments corresponding to the strategic (or the nonstrategic) case.

Now suppose that, after the first (capacity selection) stage but before the second (production) stage, an external force intervenes and changes the rules. The country J firms are to be permitted—perhaps even encouraged—to form a cartel structure, to reduce output and raise prices. How much will consumers of DRAMs lose? How much will producers of DRAMs gain?

Carry the argument a step further. Suppose country J firms, if left to their own devices, are perfectly prepared to form a cartel on their own, but that to do so requires that some "critical mass" of existing DRAM producers be country J firms. Then country U can prevent cartel formation by subsidizing entry by its own firms (which are assumed to be behaviorally, or for legal reasons, indisposed toward collusion). What level of subsidies should country U be willing to pay to ensure against cartel formation?

Table 7-2 shows the loss to consumers and increased producer profits associated with a second-stage switch from a noncooperative Cournot-Nash equilibrium to cartel formation. Initial capacities are assumed to have been chosen as the first stage of the competitive game described above, with an anticipated competitive second stage. Now, however, the conjectural variation parameter σ has been reset from a competitive, Cournot value of 1/N to the collusive, cartel value of one, in calculating the second-stage output path.

It is striking that the lost consumer welfare is so very much larger than the profits gained by producers.[57] There are two reasons for this. First, fixed cost per unit produced rises considerably when output is cut back sharply in the second stage relative to the competitive output levels for which capacity was initially chosen. Second, the existence of learning economies means that, when producers cut back output, they also raise their variable costs per unit over the remainder of the product cycle. Thus, much of what might have been additional monopoly profit, without scale and learning economies, is chewed up by higher unit costs incurred when output is reduced.

Thus, for example, in what I have argued is the more realistic case of $\gamma = 1$, a switch from competition to collusion in 1M DRAMs costs over $30 billion in forgone consumer welfare, over a five-year product cycle, while producing monopoly profits of only about $4.5 billion over the same period. (Note that at the height of the real-world 1M DRAM shortage of 1988–89 some analysts' estimates put "excess" industry profits at levels of $1 billion to $2 billion per year.)[58] Clearly, if these calculations are correct, even if country U accounts for only 40 percent of final DRAM consumption, its consumers should be willing to pay $2.4 billion or so annually to prevent the formation of a cartel (40 percent of $30 billion is $12 billion, or $2.4 billion annually). Of course, there may be other, less costly ways to inhibit cartel formation by foreign producers: if governments can agree to enforce some common international antitrust standards, for example, and foreign governments apply such norms to their producers, the benefits of competition might be available to U.S. consumers at no cost.

---

57. In the case of a constant-elasticity demand curve, it can be shown that the total consumer surplus CS is given by:

$$CS = -\left(\frac{\alpha}{1 + \beta}\right)\left[\int_0^{ts} P(s)^{1+\beta}ds + (1 - t_s)P(t_s)^{1+\beta}\right].$$

58. See Kenneth Flamm, "Internationalization in the Semiconductor Industry," in *Semiconductor Materials and Equipment Institute* (Newport Beach, Calif., January 1989). These numbers cover both 256K and 1M DRAMs.

More Complex Stories about Cartels

The models developed in this chapter and the previous one, despite their many simplifications, seem to do a tolerable job in tracking reality in at least some respects. The extreme sensitivity of my simulations of symmetric equilibria to the specification of the experience variable relevant to learning economies was, to me at least, quite unexpected. Gratifyingly, the "per fab" specification of learning economies appears to solve the somewhat puzzling results of earlier research: Cournot behavior, coupled with a reasonable empirical approximation to the costs of real-life DRAM production, then gives much more realistic predictions about industrial structure.

The story told in the last section raises some interesting questions, not all of which are easily answerable. First, how realistic is this story? Is the cartel threat—and the resort to anticartel insurance—or some lesser degree of collusion a serious possibility? Were the DRAM price and output manipulations of 1988–89 a purely transitory phenomenon, or has some threshold been crossed into a less competitive, more cooperative world of "market sharing"? Continued production cutbacks by Japanese producers after 1989 may suggest a break with pre-1986 patterns of firm behavior. Also, efficiency differences between groups of firms could be tied to a story about intergenerational externalities, where incumbents have lower costs because they are incumbents.

Second, if country J firms can form a cartel, why not do so in stage one, and choose more profitable levels of capacity investments? My answer is that, at stage one, entry by others will be stimulated if reduced capacity investments imply that positive rents will be earned, while at stage two new entry is no longer possible.

More important is the issue of whether cartel pricing would attract additional entrants in later plays of this game. This is a complicated matter, but one can construct at least the germ of an argument as to why entry by noncountry J firms might not occur. Let us imagine that nonincumbent country U firms operate at a cost disadvantage (say in fixed entry cost F) relative to incumbent country J firms, and that entry by a single nonincumbent is enough to completely disrupt the cartel. Let us also imagine that capacity investment is decided as the first stage of a two-stage competitive game (if it were apparent that the number of firms and levels of capacity set in the first stage were other than at competitive levels, such as to imply rents in the second stage, further entry by other firms would presumably be stimulated). Then we have a terrible dilemma for the potential country U entrant: if it does not enter, the country J firms may form a cartel in the second stage, in which case it must sit and watch as others earn rents. If it does enter, however, it disrupts the cartel, guarantees competition in the second stage, and thus guarantees itself a loss in a zero-profit equilibrium (since it produces at a cost disadvantage relative to the rest of the population of firms).

Third, my consumer welfare calculation assumes that monopoly rents are extracted by a cartel at arm's length from a competitive user industry, which then passes on cost increases directly to final consumers. It is well known that if an input is not fixed in proportion to output, a monopolist controlling the pricing of an input maximizes profit by integrating forward into the user industry that purchases that input, in order to avoid substitution away from the input.[59] Since chip-producing firms in country J may also be systems producers, there would appear to be few barriers to this occurring in the long run. The welfare cost to final consumers may increase if this were to occur.

Fourth, what if a domestic chip-consuming industry is imperfectly competitive and extracts rents from sales to foreign consumers? Can this not increase the welfare loss from a domestic standpoint, particularly if forward integration by foreign chip producers and the exit of domestic equipment producers accompany monopoly rent extraction?

Fifth, the logic of anticartel insurance suggests that chip consumers as a group ought to be willing to subsidize entry by cartel-busting entrants. (The ill-fated U.S. Memories consortium, in which chip users proposed to fund entry into the DRAM business, comes to mind.) However, there will be a free rider problem. All consumers gain from the entry of a cartel buster, whether or not they have actually helped pay for it. A government role would therefore seem to be needed to handle this "public good" aspect of entry.

Sixth, little attention has been paid to intergenerational externalities. The fact that several firms—particularly Korean companies and Japan's NMB Semiconductor—were able to enter the DRAM business in the mid-1980s with virtually no prior experience suggests that such externalities, if they exist, are not so significant as to be an insuperable barrier to entry.

59. See Vernon and Graham, "Profitability of Monopolization," pp. 924–25.

Finally, the degree of collusion and, indeed, all the details in the policy-related story I am telling are extreme. The costs calculated here probably should best be regarded as an upper bound on the costs corresponding to a more realistic scenario. Is there a convenient way to parameterize less stark forms of behavior? In particular, one might want to imagine a scenario in which most producers participate in a cartel, but one or two "outsiders" form a competitive fringe and do not cooperate with the cartel (but do not break it either). With asymmetric firms, however, characterization of equilibrium paths becomes much more difficult.

Conclusion

So-called strategic arguments for an activist U.S. policy in semiconductors have focused on two issues: the possible existence of technological externalities—spillovers—between semiconductor production and other types of industrial activity that are not captured by producers, and the extent to which monopoly profits can be extracted from downstream chip users by an upstream cartel. This latter issue has been particularly important in shaping the debate over trade policy in semiconductors.

There is plenty of general anecdotal evidence of Japanese industry activities that probably would have violated antitrust laws had they ocurred in the United States (and at least some of which were arguably also illegal in Japan) in past decades, and continuing up to the present day. There are even some specific episodes that suggest similar behavior in the Japanese semiconductor industry. However, in all cases where it appears that such activities were successful in creating cartel-like industrial structures, some degree of active or tacit government support for such organization can also be discerned.

The latter point clearly holds true in semiconductors. Before 1986, episodes in which Japanese semiconductor producers successfully managed a reduction in their collective output or exports were in all cases undertaken with the support of MITI and generally were undertaken as a response to acute, external trade friction. From 1986 through 1989 the organization by MITI of formal structures designed to control output and prices was overt and explicit. Restraint in production and exports was clearly linked to a deliberate government policy. It is in some sense circular to argue that the policy was justified by the predatory intent of restraints that were implemented as part of the policy.

After 1989, however, the visible hand of MITI as organizer of restraints had disappeared, and it is at least possible to argue that evidence of collusive behavior reflected wholly private behavior. Available data—anecdotal and statistical—in fact, seem to show Japanese DRAM producers behaving in a way that does not appear very competitive in 1990. One is left to wonder, however, whether behind the scenes there was tacit government support for a policy of private market sharing, to replace the cutthroat competition sometimes practiced by the Japanese industry, and minimize the potential for renewed trade friction.

My simulation results, however, made it clear that a foreign chip cartel, at least in principle, can create a substantial net welfare loss for an economy making significant use of the cartelized input. Under the right empirical circumstances, a serious argument can be made for subsidizing domestic capabilities as a form of anticartel insurance. Whether those circumstances existed—or continue to exist—in semiconductors is likely to remain the subject of continued debate.

Appendix 7-A: Solution of the Model with Strategic Behavior

The basic framework is to conceive of competition as occurring in two distinct stages. In the initial period firms make capacity investment decisions. In the second period firms decide what utilization rates to apply to their sunk investments in capacity. I assume that the two-period equilibrium is characterized by subgame perfection. Given their first-period capacity decisions, the second-period game in utilization rates is assumed to be characterized by a static Nash equilibrium in utilization rates. When firms select their first-period capacities, they are assumed to be able to anticipate correctly the effects of their choices on the second-period equilibrium. In the first period, however, they take the capacity choices of their rivals as unaffected by their own. The strategic interactions come from the anticipated effects on their rivals' *second*-period output decisions of their own first-period capacity investments. This contrasts with the *nonstrategic* model outlined in appendixes 6-A and 6-B, where the firm viewed both the first- and second-period actions of competitors as unaffected by its own actions in either period.

Given first-period capacity investments, the solution for optimal utilization rates is identical to that given in appendixes 6-A and 6-B. As before, we restrict our attention to symmetric equilibria of identical firms.

Strategic effects enter the analysis when we consider optimal capacity choice. Along the optimal trajectory for the utilization rate $u^*(t)$ and the corresponding trajectory $E^*(t)$ for the experience variable, a necessary condition for an optimum choice of capacity K is that

$$\int_0^1 \frac{\partial H[u^*(t), E^*(t), K, \delta(t)]}{\partial K} dt = 0$$

$$\int_0^1 \left[\left(R_y - d + \frac{\delta}{K^\gamma} \right) \frac{\partial y}{\partial K} + R_x \frac{\partial x}{\partial K} - \gamma \frac{\delta}{K^{1+\gamma}} y - cu - r \right] dt = 0.$$

Note that this partial derivative is, with respect to K alone, evaluated with $u(t)$ set equal to optimal control $u^*(t)$. This differs from the expression developed in appendix 6-B (equation B-2) because of an additional term,

(7A-1) $$\int_1^0 R_x \frac{\partial x}{\partial K} dt,$$

added onto the left-hand side. In the nonstrategic case, initial capacity investments were not viewed by the firm as affecting rivals' second-period output decisions, i.e., $\partial x / \partial K = 0$ was assumed. The remainder of this appendix develops term C-1 explicitly.

First I adopt some additional notational conventions: variables subscripted with y refer to the firm making a strategic calculation, and variables subscripted with i to its $N - 1$ rivals.

A firm sees each rival as maximizing its own Hamiltonian, H_i, exactly analogous to the Hamiltonian shown in appendix 6-A, subject to the constraint that firm i's utilization rate u_i be less than or equal to one. (I rule out the possibility of a symmetric equilibrium with no output produced [$u_i = 0$] for reasons described in appendix 6-A.) The firm takes its rivals' capacities as given and is interested in how its rivals' capacity utilization rates vary with its own capacity choice.

Form the Lagrangean,

$$L_i = H_i + \mu_i(1 - u_i)$$

$$= R_i(y, x_i, \Sigma_{j \neq i} x_j) - F - dx_i - cu_i K_i - rK_i + \frac{\delta_i x_i}{K_i^\gamma} + \mu_i(1 - u_i),$$

with $x_i = u_i w(E_i) K_i$, μ_i a nonnegative Lagrange multiplier, and

$$\dot{\delta}_i = -\left(R_{ix_i} - d + \frac{\delta_i}{K_i^\gamma}\right) u_i w_{Ei} K_i.$$

Applying the maximum principle using this Lagrangean, we have the following first-order condition:

$$\frac{\partial L_i}{\partial u_i} = \frac{\partial H_i}{\partial u_i} - \mu_i = 0.$$

Since

$$\frac{\partial H_i}{\partial u_i} = \left(P'x_i + P - d + \frac{\delta_i}{K_i^\gamma}\right) w(E_i) K_i - cK_i,$$

for each rival i, a firm knows that the first-order conditions determining u_i and μ_i are

(7A-2) $$\left(P'x_i + P - d + \frac{\delta_i}{K_i^\gamma}\right) w(E_i) K_i - cK_i - \mu_i = 0$$

$$\mu_i(1 - u_i) = 0,$$

$$\text{with} \quad \mu_i \geq 0.$$

The firm does comparative statics by taking derivatives with respect to K_y, which gives, in the case of the first equation of this two-equation system (in two unknowns):

$$w(E_i) K_i \left[P''\left(\Sigma_{j\neq i} \frac{\partial x_j}{\partial u_j}\frac{\partial u_j}{\partial K_y} + \frac{\partial x_i}{\partial u_i}\frac{\partial u_i}{\partial K_y} + \frac{\partial y}{\partial K_y}\right) x_i + P'\frac{\partial x_i}{\partial u_i}\frac{\partial u_i}{\partial K_y} \right.$$
$$\left. + P'\left(\Sigma_{j\neq i} \frac{\partial x_j}{\partial u_j}\frac{\partial u_j}{\partial K_y} + \frac{\partial x_i}{\partial u_i}\frac{\partial u_i}{\partial K_y} + \frac{\partial y}{\partial K_y}\right) \right] - \frac{\partial \mu_i}{\partial K_y} = 0.$$

Because the firm faces identical symmetric competitors, however,

$$w(E_i) = w(E_i), \quad K_i = K_j, \quad \frac{\partial u_i}{\partial K_y} = \frac{\partial u_j}{\partial K_y},$$

and we need solve only for $\dfrac{\partial u_i}{\partial K_y}$ and $\dfrac{\partial \mu_i}{\partial K_y}$ instead of $2(N-1)$ such terms.

Let $\dfrac{\partial y}{\partial K_y}$ be $u_y w(E_y)$, where subscript y denotes the values of these variables for the firm varying K_y.

Since $\dfrac{\partial x_i}{\partial u_i} = w(E_i) K_i$, the comparative statics are thus given by

$$
\begin{bmatrix}
[w(E_i)K_i]^2 \, [N - 1)(P''x_i + P') + P'] & -1 \\
-\mu_i & (1 - u_i)
\end{bmatrix}
\begin{bmatrix}
\dfrac{\partial u_i}{\partial K_y} \\[2ex]
\dfrac{\partial \mu_i}{\partial K_y}
\end{bmatrix}
=
$$

$$
\begin{bmatrix}
-[w(E_i)K_i]u_y w(E_y)(P''x_i + P') \\[3ex]
0
\end{bmatrix}
$$

Let M be the square matrix on the left-hand side of this equation, and $|M|$ its determinant.

Then,

$$
\begin{bmatrix}
\dfrac{\partial u_i}{\partial K_y} \\[2ex]
\dfrac{\partial \mu_i}{\partial K_y}
\end{bmatrix}
= \frac{1}{|M|}
\begin{bmatrix}
(1 - u_i) & \mu_i \\[2ex]
1 & [w(E_i)K_i]^2[(N - 1)(P''x_i + P') + P']
\end{bmatrix}
\cdot
$$

$$
\begin{bmatrix}
-w(E_i)K_i u_y w(E_y)(P''x_i + P') \\[3ex]
0
\end{bmatrix}
$$

where $|M| = [w(E_i)K_i]^2\{(N - 1)(P''x_i + P') + P'\}(1 - u_i) - \mu_i$.

There are two general cases to consider. First, along interior segments, where $\mu_i = 0$ and $u_i < 1$, $|M|$ will be negative in a noncooperative, symmetric Nash equilibrium when $N > 2$. In this case the term in braces must be negative.[60]

---

60. By the strict concavity of the revenue function, $zP'' < -2P'$.

Therefore $P''x_i < \dfrac{x_i}{z}(-2P')$

Second, along boundary segments, where $\mu_i > 0$ and $u_i = 1$, $|M| = -\mu_i < 0$.[61]

The general expression for $\dfrac{\partial u_i}{\partial K_y}$ is

$$(7A\text{-}3) \qquad \frac{\partial u_i}{\partial K_y} = \frac{-(1 - u_i)w(E_i)K_i u_y w(E_y)\{P''x_i + P'\}}{|M|},$$

which, in a symmetric Nash equilibrium with $N > 2$, will be negative when $u_i < 1$ (because the expression in braces will be negative; see note 60). Noting that, in a symmetric equilibrium, $u_i = u_y = u$, $E_i = E_y = E$, $K_i = K_y = K$, and so on, and $\dfrac{x_i}{z} = \dfrac{1}{N}$,

---

$$P''x_i + P' < \left[\frac{x_i}{z}(-2) + 1\right]P'.$$

Since $P' < 0$, the right-hand side of this inequality will be negative as long as

$$\frac{x_i}{z}(-2) + 1 > 0$$

$$\frac{x_i}{z} < \frac{1}{2}.$$

In a symmetric equilibrium, $\dfrac{x_i}{z} = \dfrac{1}{N}$.

Therefore, $P''x_i + P' < 0$ as long as $1 < \dfrac{N}{2}$. $N > 2$ is sufficient for the bracketed term to be negative.

61. Actually there is a third possibility: that $\mu_i = 0$ and $u_i = 1$, in which case M is singular and $\partial u_i/\partial K_y$ undefined. This can only occur at the transition from a boundary to an interior segment and can never last more than an instant. (With $u = 1$, industry output must be rising and marginal revenue R_y falling. But with $\mu = 0$ over some interval with $u = 1$,

$$R_y - d - \frac{c}{w} + \frac{\delta}{K^\gamma} = 0$$

$$(\dot{R}_y) = \frac{c}{w^2}w_E\frac{y}{K^\gamma} + \frac{\dot{\delta}}{K^\gamma} = \frac{c}{w^2}w_E\frac{wK}{K^\gamma} - \left(R_y - d + \frac{\delta}{K^\gamma}\right)w_E K^{1-\gamma}$$

$$= \frac{c}{w}w_E K^{1-\gamma} - \left(\frac{c}{w}\right)w_E K^{1-\gamma}$$

$$= 0,$$

which cannot be true in a symmetric-industry equilibrium with all firms at full capacity over some interval. Therefore, we cannot have $u = 1$ and $\mu = 0$ for more than an instant.)

$$(7A\text{-}4) \quad \frac{\partial u_i}{\partial K_y} = \begin{cases} 0, \; u_i = 1 \\[2mm] -\dfrac{u}{K} \dfrac{\left(\dfrac{P''}{P'}z\dfrac{1}{N} + 1\right)}{1 + (N - 1)\left(\dfrac{P''}{P'}z\dfrac{1}{N} + 1\right)}, \; < 0, \; u_i < 1. \end{cases}$$

A key point emerging from this analysis is that—in a symmetric equilibrium—over periods when output is capacity-constrained, a marginal increase in capacity will have no impact on rivals' output. Intuitively, when a firm's marginal revenue exceeds its marginal cost, a small increase in a competitor's output—which reduces marginal revenue—will still leave the firm wishing to produce at maximum output.

Returning to the task of simplifying expression C-1, we then have:

$$\int_0^1 R_x \frac{\partial x}{\partial K} dt = \int_0^1 P'y \frac{\partial x}{\partial K} dt = \int_0^1 P'y \sum_i w(E_i) K_i \frac{\partial u_i}{\partial K_y}$$

$$(7A\text{-}5) \qquad = \int_{t_s}^1 P'y(N - 1)wK\left[-\frac{u}{K} \frac{\left(\dfrac{P''}{P'}z\dfrac{1}{N} + 1\right)}{1 + (N - 1)\left(\dfrac{P''}{P'}z\dfrac{1}{N} + 1\right)} \right] dt,$$

since $\dfrac{\partial u_i}{\partial K_y} = 0$ when $u_i = 1$.

With a constant-elasticity demand, and β the price elasticity, this becomes:

$$(7A\text{-}6) \quad \int_{t_s}^1 \frac{-P(N - 1)y}{\beta N K} \left\{ \frac{1 + \left(\dfrac{1}{\beta} - 1\right)\dfrac{1}{N}}{1 + (N - 1)\left[1 + \left(\dfrac{1}{\beta} - 1\right)\dfrac{1}{N}\right]} \right\} dt =$$

$$-\frac{y(t_s, K)P(y(t_s, K))}{\beta NK}(N - 1)\left\{\frac{1 + \left(\frac{1}{\beta} - 1\right)\frac{1}{N}}{1 + (N-1)\left[1 + \left(\frac{1}{\beta} - 1\right)\frac{1}{N}\right]}\right\}(1 - t_s).$$

This term was added to the right-hand side of equation B-2 in solving for possible equilibria. Again, it is important to note that strategic investment behavior has no effect if the symmetric-industry equilibrium associated with the empirical parameters of the problem is capacity-constrained throughout the product life cycle.

CHAPTER EIGHT

Conclusion:
Mismanaged Trade?

SINCE 1959 the United States and Japan have been arguing over the rules for international industrial competition in semiconductors. In some respects, little has changed.

In 1959 massive Japanese exports of low-priced germanium transistors ignited concerns of American semiconductor manufacturers. The transistor's importance for computers, communications, and other electronic systems in an epoch when markets for advanced electronics were dominated by military applications led to appeals for protection from imports on grounds that national security was at stake. Japan's government and its transistor industry responded by constructing an export cartel designed to stabilize prices. Transistor consumers around the globe complained. The U.S. electronics industry was divided: semiconductor producers demanded protection, while systems producers fought policies that might raise the price of an important component and create a competitive disadvantage for them in commercial markets. The problem ultimately was to disappear because of the phenomenal rate of technological advance in microelectronics. As individual transistors were replaced by much more sophisticated integrated circuits, the American semiconductor companies pioneering these innovations created entirely new families of products and rendered the Japanese cost advantage in the older discrete semiconductors irrelevant.

Two and a half decades later a wave of low-priced Japanese memory chip exports again threatened a foundering U.S. semiconductor industry.

425

The crucial function of microelectronics in building high-technology equipment that provided the U.S. military with its overwhelming technological superiority prompted an influential 1987 Defense Science Board report calling for federal support of a government-industry consortium. The result was today's Semiconductor Manufacturing Technology (Sematech) consortium.

Concerns over national security also precipitated the U.S. government's pressure on Japan to negotiate the 1986 Semiconductor Trade Arrangement. The Japanese government worked with its industry to fix export prices and restrict domestic production and investment. As prices began to rise, semiconductor consumers around the world again loudly complained. Ultimately, the American chip industry surged back not because of an aggressive reentry into memory chip production, but because of rapidly improving technology in other areas. U.S. companies shifted into the design of more technically advanced chips that collapsed entire electronic systems onto a few chips—whether the microprocessors and logic chips used in ever more powerful personal computers and workstations or the specialized logic chips and processors used in telecommunications, disk controllers, and multimedia. Trailing behind was a Japanese semiconductor industry struggling to break out of what seemed to be an increasingly competitive and decreasingly profitable memory chip business.

Much has changed over these decades in other respects. In the 1960s American development of advanced semiconductor technology was dominated by a cold war defense budget that funded the necessary research. Over 70 percent of U.S. integrated circuits went into defense and space applications in 1965. Today the development of semiconductors is fueled by an enormous commercial market—the military market accounts for less than 2 percent of U.S. consumption. Equipment sold in the marketplace typically uses the latest components, and equipment shipped to military users (except for very specialized purposes) often lags behind technologically. What was once a major benefit conferred by a generous military budget is no longer a significant advantage for American industry.

In the early 1960s, too, Japanese demand for semiconductors was a tenth of sales in the U.S. market. Even though American industry understood that selling in foreign markets was critical to its success, penetration of the Japanese market was not crucial for an American electronics company (although access to the American market certainly was for Japanese producers). Indeed, U.S. companies often appeared to settle for unattractive deals in gaining very limited access to the Japanese market. By 1986, however, the Japanese semiconductor market had become critical: its size surpassed that of the United States for the first time.

In the 1960s, Japanese companies were in no sense the technological equals of American companies. Their success in producing inexpensive transistors was based on low wages, a relatively mature product, and a labor-intensive manufacturing process. By the late 1980s Japanese companies generally equaled or surpassed American companies in almost every area of the rapidly developing technology of semiconductor manufacture. Only in chip-level systems design and integration did American semiconductor manufacturers as a group seem to have a significant lead. And not only was the American monopoly on the technology no longer a monopoly, it was not even a duopoly. Korean and European competitors also stood at the frontiers of semiconductor manufacturing. Taiwan and other Asian countries poured resources into their own developing semiconductor industries and seemed poised to make a similar transition.

In one crucial respect, however, the semiconductor industry today plays exactly the role it did thirty years ago. In 1959 semiconductors became the focus for the first major episode of high-technology trade friction between the United States and Japan. In the early 1970s, semiconductors and computers became the focus for U.S. pressure on Japan to liberalize access to markets for high-technology products. In 1983 semiconductors constituted one of the first products of concern when the U.S.-Japan High Technology Working Group began to discuss sectoral trade and investment issues. In 1986 semiconductors became the first high-technology industry for which America and Japan agreed on a market share objective to create greater foreign access to the Japanese market. Today, as has been true for almost four decades, disputes over semiconductor trade continue to be the flash point for efforts to resolve international trade and investment disputes that ultimately will set precedents for other high-technology industries.

Why Semiconductors?

The economics of the semiconductor industry has caused it to be at the forefront of high-technology trade frictions. For one thing, the relation between product prices and costs of production is enormously complicated. Economies of scale are significant—there are large fixed investments in research and development for increasingly complex chips and in constructing high-volume production facilities. Learning economies are also very important; unit costs fall about 30 percent with every doubling of cumulative output. Thus the price of an integrated circuit has historically had only the loosest relationship with some notion of the average cost of manufacturing a chip at any given time. Companies make investment decisions based on projected revenues and costs over a five-year product cycle, knowing that when supplies are limited or demand is great, prices will soar and profit margins will be fat.

When chip demand falters or supplies are plentiful, prices may be pushed down toward the incremental cost of producing another chip, and revenues may not even cover the depreciation charges on the factories turning out the current generation (not to mention continual R&D expenses for the next generation). In fact, an economically rational manufacturer might be willing to sell chips for less than the current variable cost of production if the learning curve is steep enough—and predictable future decreases in manufacturing costs are large enough to justify continuing production through the money-losing times—in order to emerge a lower-cost producer in a more profitable environment.

These static and dynamic scale economies have also relentlessly driven manufacturers to sell as many chips as possible wherever possible to maximize the return on relatively fixed investments in R&D and production capacity and acquire the production experience needed to ride quickly down their learning curve. From the beginning, then, they have competed in an international marketplace.

The drive to sell in global markets has inevitably meant that laws prohibiting dumping can be used to attack foreign imports in times of slack demand, even though the exporters' behavior may be economically rational and no different from the practices of domestic producers. Behavior that might be construed to be a signal of anticompetitive, predatory intent in an industry lacking these characteristics is a predictable outcome of normal market forces in semiconductor production. In every economic downturn since significant U.S. imports of foreign ICs began in the late 1970s, dumping has been an issue. And it will resurface in future downturns as long as the current antidumping standards are in force.

Large R&D investments (historically, 10 to 15 percent of sales) in semiconductors have made government involvement with the industry inevitable in another way. Historically, the U.S. government paid a large part of America's semiconductor R&D bill on the grounds of protecting national security. Even without this rationale, many economists would agree that government subsidies are legitimate and desirable. As in other high-technology industries dependent on large R&D expenditures (and with society's return on these investments typically much greater than what a private investor can fence off and take claim to), governments can and should stimulate greater investment in developing new technology. But research subsidies can also be explicitly designed to serve strategic economic ends and help domestic manufacturers defeat foreign competitors in world markets. There is no way to distinguish between government support for R&D to make a nation's producers more productive and innovative contributors to the domestic economy, and support primarily intended to capture profits from foreign producers.

Although other high-technology industries may also have significant economies of scale associated with steep R&D requirements, large sunk investments required to enter production, or steep learning curves in manufacturing, few have them all simultaneously. Furthermore, semiconductors have experienced one of the steepest sustained rates of technological advance ever recorded for an important economic commodity (with prices for a bit of memory falling continuously for decades at an average rate of more than than 35 percent a year). This situation has been coupled with increasing importance in the economy (and interest on the part of governments) and a boom and bust cyclicality of demand that is extreme even by the standards of the rest of the electronics business. As a result, the semiconductor industry has seemed destined to be the pioneer in working out trade rules for all high-technology industries. Other high-technology industries emerging in the global market in the second half of this century often have had some of these characteristics, but rarely all, and never to the extent seen in the chip business.

Strategic Policy in Semiconductors

None of the factors shaping today's competitive environment was very clear before the late 1970s. The United States effectively dominated world semiconductor markets, primarily becaused of its unchallenged technological strength. Public policy during the cold war helped build the foundations for this supremacy, but it had never explicitly grappled with the economic implications of this advanced infrastructure, nor was economic benefit ever the purpose of its construction. U.S. investment in developing semiconductor technology was strategic only in the very narrow sense of the pursuit of qualitative military advantage.

The development of the Japanese semiconductor industry, however, was shaped by very different policies. Initially, semiconductors were merely an incidental appendage on a different beast. Through the 1960s the primary focus for Japanese industrial policy in electronics was the domestic computer industry. To sell in the Japanese market, American computer companies were forced into joint ventures with Japanese counterparts, endured sales ceilings, or accepted other restrictions. By the early 1970s, however, several U.S. partners of Japanese electronics companies had dropped out of the computer business. It had become clear that a frontal assault on the established mainframe computer market dominated by IBM was not going to work unless Japanese companies could offer lower prices to offset the American advantages in virtually all other aspects of the computer business.

By then the general design principles for a commercial business computer were well understood. The extraordinary decrease in the cost of computing power continued, primarily due to ever cheaper and more powerful components rather than design innovation.[1] So Japanese producers realized that by manufacturing state-of-the-art chips at lower cost and incorporating them into their machines, they might be able to compete in established markets for computer systems.

Meanwhile, the semiconductor manufacturing divisions of these same Japanese companies faced equally rocky prospects. In the mid-1960s, when American companies first began to ship computers based on advanced chips, the Ministry of International Trade and Industry undertook a small program to improve Japanese semiconductor technology. Japan's first semiconductor memory chips were built as part of a related MITI project, the development of a high-speed mainframe computer.

But these efforts had not closed the gap with American companies. Cheap American chips flooded world markets, including Japan, and Japanese producers complained loudly about American dumping. Despite

---

1. See Kenneth Flamm, "Technological Advance and Costs: Computers versus Communications," in Robert W. Crandall and Kenneth Flamm, eds., *Changing the Rules: Technological Change, International Competition, and Regulation in Communications* (Brookings, 1989), pp. 16–19.

stiff trade and investment barriers, including the refusal to allow foreign producers to establish Japanese subsidiaries, the share of U.S. imports in Japanese integrated circuit consumption steadily rose, peaking at 35 percent of the market in 1971.

As a further attempt to combat this trend, MITI introduced more informal means to discourage the use of foreign chips. Imports of advanced American calculator and memory chips—to be incorporated into, and often reexported as, Japanese calculators—were increasing quickly. Calculator manufacturer Sharp had started the rush to American chips, and MITI responded by making an example of the company, denying necessary licenses to allow it to use imported chips. American companies attempting to recruit Japanese semiconductor executives to help build their Japanese sales even found MITI lobbying the executives to dissuade them from breaking ranks.

MITI's tactics proved successful. Although Japan abolished trade and investment restrictions in the mid-1970s, the import share of Japanese chip consumption continued to shrink. Meanwhile, government agencies funded much larger programs than they had previously to improve Japanese semiconductor technology. From 1975 to 1980 the government-owned telephone company sponsored two consecutive development programs for advanced semiconductor equipment and devices, the so-called Very Large Scale Integration (VLSI) projects. In 1976 MITI began its own four-year VLSI project. After organizing Japanese companies into research consortia, these projects pumped large amounts of money into semiconductor R&D. Japanese and American semiconductor companies spent comparable shares of their sales revenues on R&D between 1976 and 1985, but from 1976 to 1979 the VLSI subsidies added resources half again the size of the Japanese industry's own funding of integrated circuit R&D.

The VLSI projects helped Japanese companies develop expertise in a new generation of semiconductor production technologies. The key to American success in semiconductors had been Silicon Valley and its American counterparts, dense clusters of technical expertise where chip manufacturers could buy the latest capital goods developed for use in production systems. Japanese manufacturers hoping to produce better and cheaper chips than their American competition would need production machinery superior to that used by American producers. When products developed by the VLSI projects came to market in the 1980s, Japanese companies began shipping the most advanced and reliable systems in key areas of the chip manufacturing process. With first access to this equipment, they would henceforth have a head start toward producing the best chips at the lowest cost.

The next logical step was to invest aggressively in production facilities that could use the new systems. Through the 1980s, Japanese semiconductor companies' ratio of capital spending to revenues far outpaced that of American companies. Together, the investment in new manufacturing technology and the capital spending blitz that followed made Japan the world leader in semiconductor manufacturing. U.S. companies' share of world sales peaked in 1975 then plunged while the Japanese share soared. And Japanese imports, which constituted one-third of Japan's apparent integrated circuit consumption in 1974, fell to 12 percent in 1986.

Alarmed at the earliest of these developments, U.S. chip producers formed an industry association in 1977 and began to complain to their government about their inability to secure access to Japanese markets and about Japanese government industrial policies. A few years later they began making frequent allegations of Japanese dumping in the U.S. market.

In response to American complaints, MITI reacted in two ways. It adopted a decidedly lower public profile. For example, the VLSI program was abruptly shut down in 1980 (although a private sector version continued through 1986). This lower profile was aided by the success of the Japanese effort: producers could stand on their own and government funds for R&D became relatively less important as Japanese chip production soared on a self-sustaining trajectory. At the same time, however, the U.S. complaints gave the ministry ample justification for continuing to provide "guidance" to Japanese chip makers as it attempted to manage semiconductor trade frictions with America in the 1980s.

In late 1985, as the semiconductor industry sank into a frighteningly deep downturn, American chip producers urged their government to respond to the perceived threat from Japan. The government reacted in two ways. With financial support from the Department of Defense, Sematech was formed to improve U.S. chip manufacturing. Unlike previous Pentagon investments in semiconductor production, Sematech focused on the technological and economic health of the infrastructure of the entire industry, not just on specialized producers and products catering to unique needs of the military. Meanwhile, the U.S. government pressed suits to stop dumping and other trade violations. The result was ultimately the Semiconductor Trade Arrangement of 1986.

Effects of the Semiconductor Trade Arrangement

The trade arrangement had three major features. First, Japan agreed to a set of pricing floors administered by the U.S. Department of Commerce on exports of DRAMs and EPROMs, and set up strict border controls to ensure that exports were not being priced under the floors. Second, Japan agreed to "monitor" an even broader variety of chips, based on company-specific cost and price data, and "take appropriate actions" to prevent dumping in export markets. Because of the inherent difficulty of controlling flows of such an easily transportable commodity as semiconductors, the appropriate actions evolved into administrative guidance to companies on production and investment levels. Third, MITI (in a "secret" side letter) agreed to a numerical benchmark for foreign market share, and took actions to encourage Japanese electronics manufacturers to buy more foreign semiconductors.

FMVs and Border Controls

The price floors, called foreign market values (FMVs), set by the Commerce Department and the border controls constructed by MITI to make them stick hit hard. Significant memory chip price differentials existed between the Japanese and American markets from 1987 through 1989. These differentials put U.S. computer and electronic equipment producers at a disadvantage because they had to compete against Japanese companies purchasing identical memory chips at considerably lower prices. From the standpoint of the aggregate welfare of the world outside Japan's borders, the impact of the regional price differentials was noticeable but not staggeringly large. In DRAMs, for example, I estimate that the net welfare impact of the FMVs and border controls was 5 to 8 percent of what would have been the value of rest-of-world DRAM consumption in the absence of border controls (see chapter 5).

Moreover, there were clear limits to how tightly markets could be controlled. Large regional price differentials existed during these years in the large user "contract" market, which amounted to 70 to 80 percent of consumption. In the much more freewheeling and entrepeneurial grey or spot market, however, regional price differentials were transitory (see chapter 5). The conclusion, perhaps, is that the more diffused, decentralized, and competitive the market, the tougher it is to swim upstream against fundamental economic forces, even for the most powerful of government bureaucracies.

In large part because of computer industry objections, the FMV system was dropped when a second semiconductor trade arrangement was signed in 1991. Instead, Japanese companies were required to collect and maintain the same data used for FMV calculations so as to create a faster antidumping procedure. In effect, a self-policing shadow FMV system was established (see chapter 4). MITI continued to collect data on export prices from companies and to issue semiannual forecasts of production and exports of monitored products. Whether this new framework would have maintained relatively high prices without additional government intervention during an economic downturn was never tested: from 1992 to 1995 the semiconductor industry enjoyed a sustained expansion of historic proportions.[2]

Production and Investment Controls

The effect of the system of restraints on production and capital investment that MITI created in 1987–88 is the most difficult outcome of the STA to assess. Any analysis would require hypotheses about levels of production and investment in the absence of restraints, and there is little data that can be used to construct a credible alternative to observed history. Under restraints, Japanese semiconductor investments surpassed 20 percent of sales in 1989 and 1990 (see figure 1-3). It is not unreasonable to suspect that cutbacks in production and investments in 1987–89 were the most economically significant element associated with implementation of the STA and clearly an important factor in the unprecedented increases in memory chip prices in 1988.

More indirectly, the Japanese government seems to have facilitated some "privatized" restraint by Japanese producers in 1989 and 1990 (chapter 7), after the official constraints were removed. This facilitation resulted from the government's sometimes open and sometimes tacit approval of "coordination" of Japanese production and investment, and

---

2. Driven by annual growth rates in PC shipments that averaged more than 23 percent from 1993 to 1995 and rapidly rising semiconductor content in electronic equipment, PC circuit boards alone accounted for one-third of world consumption in 1994. Lee Gomes, "Growth in Computer Shipments Slowing," *San Jose Mercury*, January 30, 1996; and Ronald A. Bohn and Mary A. Olsson, "Semiconductors: An Industry in Transition," *Red Herring*, no. 23 (September 1995).

its creation of an official apparatus to monitor production and investment through supply-demand forecasts, surveys of company prices, and questionnaires on investment plans. As in the past, an attitude of tolerance—and possibly indirect support—from the bureaucracy seems to have been necessary for the successful organization of coordinated action within Japanese industry.

In the end, however, this attempt to organize the Japanese industry may have backfired. High prices for memory chips and restraints on production and investment allowed South Korea's young semiconductor industry to charge into the global chip market. From 1987 to 1994 South Korea's three major chip producers quintupled their output of DRAMs, producing more than a quarter of world output by 1994. Built largely on memory chip sales, their market share for all semiconductors went from 1 percent in 1987 to 8 percent in 1994.[3] When the semiconductor boom started in 1992 and U.S. and South Korean companies responded with massive investments in a new generation of fabrication lines, Japanese companies held back and invested a considerably smaller share of their revenues in new facilities (see figure 1-3).[4] This inertia may also be explained by the collapse of Japan's speculative economic bubble and the damping effects on the Japanese economy that also occurred over roughly this same period.

Increased Foreign Market Share

In many respects the least ambiguous outcome of ten years of semiconductor trade arrangements was the effects on foreign sales of semiconductors to Japanese customers. The record, laid out in chapters 2 and 3, makes it clear that through the mid-1970s there was a determined and systematic, officially sanctioned effort to block, disrupt, and slow down foreign chip sales in Japan. Yielding to strong foreign pressure, formal trade barriers were removed by the mid-1970s and investment controls brought into conformity with OECD standards by the early 1980s. But from the mid-1970s through the mid-1980s, there remained some informal

3. Dataquest, "South Korea's Semiconductor Giants Increase DRAM Share Fivefold in 1994," online research highlight, November 27, 1995.

4. In 1995 VLSI Research estimated that Korean firms were reinvesting 30 to 55 percent of their semiconductor revenues in new plant and equipment, compared with 22 percent for U.S. firms and 15 percent for Japanese firms. Julie Chao and David P. Hamilton, "Bad Times are Just a Memory for DRAM Chip Makers," *Wall Street Journal*, August 28, 1995, p. B4.

actions and attitudes that hindered foreign sales. The sanction or toleration of dubious (from an antitrust perspective) private activities also made it more difficult for most foreign firms to break into Japanese markets.

With the implementation of the STA, the Japanese official attitude toward semiconductor imports publicly reversed course. Instead of removing obvious obstacles only after foreign pressure had been brought to bear, MITI became a promoter of foreign imports. And the change extended far beyond the "import now" placards (in English) attached to luggage carts at Tokyo's international airport. MITI officials visited users of chips to spread the new gospel. It supported industry overseas trade missions, and set up organizations and infrastructure to make it easier for foreign producers to reach Japanese chip consumers. Many large companies set, under government pressure, "voluntary" targets for increasing their use of imported chips. As industrial giant Matsushita's executives noted in interviews with the *New York Times*, "the 1986 trade agreement between the United States and Japan was instrumental in prompting the company to consider American chips."[5]

Other factors, of course, also explain the increase in the foreign share of the Japanese chip market from 9 percent in 1986 to almost 30 percent in 1991. An appreciating yen made foreign chips considerably cheaper. The quality of American chips improved: by the late 1980s defect and failure rates for both American and Japanese chips met the standards set by demanding multinational customers. And (as noted in chapter 5) Japanese companies seemed to learn how to cut deals with foreign companies that helped meet their "import share" targets without significantly changing their production and sourcing patterns.

But some attributions of the causes of the increasing foreign share of the Japanese market are wrong. It is not true that Japanese companies' increasing tendancy to use chips that U.S. manufacturers tended to specialize in was responsible for the bulk of the increase. An analysis (see chapter 5) of two different sets of data concluded that increases in the rate of foreign penetration in different market segments, as opposed to

---

5. Andrew Pollack, "The Surplus That Wouldn't Die," *New York Times*, February 27, 1996, p. D8. "'In 1988 we imported virtually nothing except some semiconductors,' said Toshihiko Murota, managing director of Matsushita Communication Industrial Company. . . . 'Even the semiconductors,' he conceded, 'were imported only because of the 1986 semiconductor trade agreement. Purchasing managers were comfortable buying from Japanese companies. So Matsushita went over their heads, requiring its division chiefs to come up with annual import targets.'"

shifts in Japanese demand between segments, accounted for most of the change.

And contrary to some assertions (particularly by European critics of the STA), the increases in foreign market share were not limited to American producers.[6] In 1986, when the first STA took effect, U.S. producers accounted for 93 percent of foreign sales of semiconductors in Japan (of a total foreign market share of 8.5 percent). By 1994 they accounted for 78 percent of a much larger (20.2 percent) foreign share. U.S. producers thus captured two-thirds of the expansion, European producers 5 percent, and Korean and Taiwanese companies 28 percent.[7] It is not surprising that some Asian producers are considerably less negative than European critics are about U.S. pressure to open the Japanese semiconductor market.[8]

Where Do We Stand?

The concerns that drove U.S. policy when the STA was negotiated in 1986 confront very different realities a decade later, in 1996.

Technological Competition

It is a much more competitive American industry that faces its international competitors today. One crude measure of this renaissance is the

6. Douglas Dunn, the president and CEO of Philips Electronics N.V. of the Netherlands, was reported to have told a conference in Tokyo that although the agreement was "aimed at boosting the share of foreign products in general, Japanese customers are pressured to buy U.S.-made products. Dunn said the share of European semiconductors in Japan remains at a mere 1 percent [which] shows that the pact has functioned to shut out third-country suppliers." AP wire report, February 2, 1996. See also "SIA to Japan: Let's Renew Pact," *Electronic Buyers' News* (February 12, 1996), p. 70.

7. These statements are based on unpublished Dataquest estimates of foreign companies' sales in the Japanese market. From 1986 to 1994 U.S. companies' share rose from 7.9 percent to 15.7 percent, European companies' from 0.5 percent to 1.1 percent, and Korean and Taiwanese companies' from 0.1 percent to 3.4 percent. Dataquest assigns sales to companies based on the brand name stamped on the chip.

8. For example, when Miin Wu, President of Macronix International, Taiwan's largest supplier of nonvolatile memory chips, was asked in late 1995 if the STA should be allowed to lapse, he replied: "If every country opens up its market for people to freely compete, there's no need for any agreements. Right now, the Japanese believe they are buying enough from overseas, so there's no need for an agreement. We should look at the facts, and whether or not Japan is really doing that. If things go the wrong way and Japan closes its door, then I'm sure the U.S. government will come back again." "Plain Talk On What's Ahead for Taiwan's Chip Business," *Electronic Buyers' News* (January 2, 1996), p. T2.

share of the world market held by the U.S. semiconductor equipment and materials industry. After years of steady decline to less than 45 percent in 1990 (slightly smaller than the share held by Japanese producers that year), the U.S. industry rebounded and exceeded 50 percent in 1992.[9] There is widespread agreement among Sematech member companies that the organization has been instrumental in improving the U.S. industry's manufacturing infrastructure and stimulating productive cooperation between equipment and materials suppliers and chip producers. In 1997 federal funding of Sematech will end, but member companies will continue to fund its activities at or above their current contribution.

A less persuasive but equally interesting measure of Sematech's performance is the perceptions of others. In the fall of 1995, MITI requested funds for the largest government semiconductor R&D effort since its 1976 VLSI project. Modeling the cooperative program on Sematech, the ministry organized Japanese semiconductor makers and equipment and material manufacturers into the Association of Super Advanced Electronic Technologies (ASET). MITI was to begin funding with a fiscal year 1996 subsidy of $100 million (10 billion yen) a year. Expected to run for 5 years, the program's goal is to develop basic processing technologies needed for 1 gigabit DRAMs. Japanese subsidiaries of three U.S. firms (IBM, Merck, and Texas Instruments) are among the twenty-one firms taking part.[10] Japanese wafer producers and MITI have banded together in an associated effort, the Super Silicon Crystal Research Institute (with a $70 million tab split equally between MITI and the companies), to develop ultralarge 16-inch (400 mm) silicon wafers for the next century.[11]

9. VLSI Research data, as cited in William J. Spencer and Peter Grindley, "SEMA-TECH after Five Years: High Technology Consortia and U.S. Competitiveness," *California Management Review*, vol. 35 (Summer 1993), pp. 13, 20. In the narrower category of wafer fabrication equipment, Dataquest data show U.S. producers' share steadily declining to a low approaching 35 percent in 1991, then increasing after 1992 (although continuing to fall short of Japanese producers' market share). See Clark Fuhs, "Worldwide Wafer Fab Equipment: Status, Tends and Forecast," presentation to the SEMI-Industry Strategy Symposium, Pebble Beach, California, January 1994.

10. "Ministry to Promote Advanced Electronics R&D," *Asahi Shimbun*, August 29, 1995, p. 11; "Unity of 10 Semiconductor Companies," *Asahi Shimbun*, October 13, 1995, p. 1; David Lammers, "Japanese Team Up for Semiconductor R&D," *Electronic Engineering Times*, October 30, 1995, p. 1; and Yoshiko Hara, "Japanese Set 12-inch, 16-Mbit Research Research Projects," *Electronic Engineering Times*, February 19, 1996, p. 10.

11. Unpublished photographic slides furnished to the author by the Semiconductor Equipment and Materials Institute, April 1996.

The research initiative was one of a number of new cooperative efforts to "strengthen Japan's inferior design power while further raising process technology strength in which Japan is already strong."[12] Japanese companies had become anxious after 1993, when their share of world output dropped below the U.S. share for the first time since 1985.[13] According to one Japanese analysis of the new initiatives, Sematech's establishment helped U.S. manufacturers to regain power. "Also, South Korean and Taiwanese manufacturers, who have become active in the market and are rapidly closing in on Japanese manufacturers, owe their rapid growth to their governments' assistance. . . . It is obvious that the Japanese semiconductor industry is preparing to launch a counterattack against foreign manufacturers in full cooperation with the government."[14]

In addition to the Super Advanced Electronics project, in 1995 MITI began a ten-year national R&D project on femtosecond technology— electronic devices that generate and manipulate ultrafast pulses and the basic science, technology, and materials needed to develop them. The Ministry of Education has also begun to try to remedy some long-standing problems in coupling academic science and education to industrial technology. In 1996 the ministry will set up at Tokyo University a center to design and evaluate complex chips; about one hundred other universities have requested similar centers.[15] More than twenty universities and ten large semiconductor companies have set up a four and one-half year, $10 million joint program to develop an advanced microprocessor design to be used in distributed and parallel computer systems, the architecture for such systems, and an operating system to be used with it.[16] This project is intended to be a prototype for future business-university cooperation.

The other key elements of the counterattack included establishment of the Semiconductor Industry Research Institute of Japan (SIRIJ) in 1994 and a cooperative Japanese industry effort to develop a new gener-

12. Taro Okabe, "Semiconductor Industry Research Institute Resolves Policy Framework," *Nikkei Microdevices* (December 1994), pp. 120–21.

13. "Ten Japanese Companies Together Try to Develop Semiconductor Production Equipment," *Nihon Keizai Shimbun*, July 8, 1995, pp. 1, 9.

14. Emi Yokota, "Chip Makers Preparing to Form United Front," *Ekonomisuto*, October 31, 1995, p. 42.

15. "Ministry of Education Seeks Practical Education in Semiconductors," *Nihon Keizai Shimbun*, September 4, 1995, p. 17.

16. "Research On State-of-the-Art Technology with the Cooperation of Businesses and Universities," *Nihon Keizai Shimbun*, October 2, 1995, p. 17.

ation of semiconductor production equipment using 12-inch (300 mm) diameter wafers. The SIRIJ was set up by the ten large Japanese chip producers as a think tank for "reactivating Japan's semiconductor industry."[17] One of institute's functions is to foster closer research ties with Japan's universities—it mimics the Semiconductor Research Consortium (SRC) administered by the U.S. Semiconductor Industry Association in this respect. The Semiconductor Technology Academia Research Center is pairing universities with companies on modestly funded research projects.[18]

What will probably become one of the SIRIJ's most controversial efforts is a joint Japanese industry project to develop next-generation production equipment and processes for 12-inch wafers, which was widely estimated to require $500 million to $1 billion in funding.[19] The MITI Super Advanced Electronic Technology project seems designed to provide funding equal to about half this amount for basic technology development. In early 1996 the ten large chip makers in SIRIJ announced the formation of Semiconductor Leading Edge Technologies Inc. (SELETE) to help fund and execute the development of 12-inch (300 mm) production equipment.[20] Funded by its membership at $350 million over five years, SELETE is oriented toward nearer-term equipment evaluation and development.[21]

An interesting feature of the Japanese 12-inch wafer effort is that it explicitly rejected a cooperative effort with American and other foreign companies. Sematech had begun organizing an international effort to accelerate 12-inch equipment development in 1994 and had invited Japanese companies to participate. After some discussion the Japanese de-

17. Okabe, "Semiconductor Industry Research Institute." The ten companies are NEC, Toshiba, Hitachi, Fujitsu, Mitsubishi, Matsushita, Oki, Sanyo, Sharp, and Sony. See also "Semiconductor Makers to Jointly Develop Advanced Equipment," August 23, 1995, in Foreign Broadcast Information Service, *Pacific Rim Economic Review: Japan*, August 24, 1995.

18. Okabe, "Semiconductor Industry Research Institute"; *Asahi Shimbun*, October 22, 1995, p. 3; and unpublished Semi photographic slides, April 1996.

19. David Lammers, "Japanese May Form Equipment Consortium," *Electronic Engineering Times*, July 17, 1995, p. 132; and "Ten Japanese Companies."

20. "Electronic Firms Form Alliance," AP wire story, February 13, 1996; Peter N. Dunn, "Japanese R&D Co-ops Proliferate; Some Confusion over Who Does What," *Solid State Technology*, April 1996, pp. 58, 61; and Hara, "Japanese Set 12-inch."

21. One Japanese executive has said that SELETE is "for tomorrow's real business while ASET is for the primary technology development needed in the next century." Hara, "Japanese Set 12-inch."

cided to go it alone, although some information will be exchanged between the two projects.[22] A clue to the logic behind this choice was contained in a report released by the Semiconductor Equipment Association of Japan in October 1995 that made recommendations on R&D, standards, and business practices, and ratified the emerging consensus for a major cooperative effort by government, industry, and universities to improve Japanese semiconductor competitiveness.[23] After noting the importance of further standardization and calling for creation of a permanent standards organization, as well as keeping "foreign semiconductor-related laws and regulations under observation," the report's recommendations also included "establishment of a Japanese-created world standard" as an industry objective.[24]

What all this shows is that despite the growth in strategic alliances among semiconductor manufacturers headquartered in different countries, the often fierce competition among Japanese chip producers in their home market, and a globalized market for semiconductor equipment and materials, at least some within government and industry continue to impute a national economic advantage to control over development of the most advanced semiconductor manufacturing equipment. They see enough of an advantage to turn down direct cooperation on R&D and thus saving real resources through cooperative efforts with rivals if those rivals are foreign. This view is by no means unique to Japan (Sematech's charter, for example, restricts membership to U.S.-based companies), but it contrasts markedly with much that is written about an increasingly borderless world of global alliances in high technology.[25]

22. See Naoyuki Mikami, "Next Generation Semiconductor Development Noted," *Shukan Toyo Keizai*, August 26, 1995, pp. 80–82; and author's interviews with Sematech officials, March 1996. Korean, Taiwanese, and European firms have joined the Sematech-sponsored international initiative.

23. The report called for the "establishment of a research and development organization composed of semiconductor, equipment, and materials manufacturers" and "the establishment of a government-academia-industry R&D organization for critical technologies." It also recommended changes in taxes on capital investment and revisions of some semiconductor and equipment business practices unique to Japan. "Proposal about Semiconductor-Manufacturing Equipment," *Nikkei Sangyo Shimbun*, October 3, 1995, p. 9.

24. See "MITI to Subsidize Semiconductor Research Association," *Pacific Rim Economic Review*, November 1, 1995, in FBIS, *Science and Technology*, November 2, 1995, cd-rom version, citing a *Nikkan Kogyo* press report.

25. Sematech sponsorship of an international 12-inch program is clearly a move toward greater international cooperation in technology development. And, with legislative constraints associated with federal funding disappearing (along with the funding), the organi-

Pricing and Dumping

The availability and pricing of semiconductors was an issue for two groups of American companies in the 1980s. Facing stiff competition from low-priced Japanese memory chips, semiconductor producers argued that Japan had organized a strategy to drive them from the market. Ultimately, they argued, dependence by U.S. users on a Japanese industry riddled with anticompetitive practices was dangerous. It would ultimately free Japanese suppliers to form a cartel and extract huge monopoly profits from U.S. electronic companies. Worse yet, dependence could create a competitive advantage for Japanese companies that would allow them to penetrate U.S.-dominated electronic equipment markets by delaying U.S. companies' access to advanced components or charging high prices for them.

These sorts of warnings were self-serving when they came from American chip producers seeking protection from Japanese imports. But they were not rejected out of hand by U.S. user industries, particularly the computer manufacturers, who might have been expected to be deeply interested in maintaining access to low-priced components. In the 1980s, U.S. supercomputer manufacturers had sometimes privately expressed concern about their dependence on a small group of Japanese suppliers for ultra-high-performance semiconductors, at a time when the supercomputer divisions of these suppliers were aggressively marketing new computer hardware based on superior performance against American systems.[26] (In addition, the U.S. computer industry had its own history of market access grievances in Japan.)

Concerns over access to components were reinforced by the "DRAM crisis" of 1988 (see chapter 4).[27] With only a minimal competing U.S. production capability for these memory chips remaining in place, U.S. computer makers saw prices for them soar and serious shortages develop.

zation may be reconsidering affiliations with foreign-based companies. The U.S. semiconductor industry, including Sematech, and the U.S. government supported a proposal for a joint venture between Japan's Canon and the U.S.-owned Silicon Valley Group to develop and produce semiconductor lithography equipment, but negotiations were not successful.

26. References to the occasional public discussion of this issue may be found in General Accounting Office, *International Trade: U.S. Business Access to Certain Foreign State-of-the-Art Technology*, GAO/NSIAD 91-278 (September 1991), p. 21.

27. A concise summary of U.S. computer industry thinking at the time may be found in National Research Council, Computer Science and Technology Board, *Keeping the U.S. Computer Industry Competitive: Defining the Agenda* (Washington: National Academy Press, 1990), chap. 2.

Compelled to cut back production and hold up introductions of new computers, American companies watched as Japanese competitors undercut their prices with access to cheaper chips, introduced new systems using the parts they were having trouble obtaining, and cut into their share of global markets. U.S. computer makers even went to the extraordinary length of sitting down with U.S. chip makers and organizing a crash joint venture to manufacture DRAMs.

This project never got off the ground, perhaps because of its sheer size and complexity. But it dramatized the tangibility of concerns over the potential impact of a semiconductor supplier cartel on user industries. The economic damage from a supplier cartel could be large (see the simulation models constructed in chapter 7), and if realized, would be an important concern for public policy.

Concerns about organized restrictions on supplies of semiconductors are not currently an issue for computer producers. New competition in DRAMs, stimulated by the large-scale market entry of Korean and Taiwanese producers and increased investments by the remaining U.S. and European producers, has created enough diversity in supplies to make a successful cartel impossible to construct within Japan and difficult even to visualize.[28] For other more specialized semiconductors the concerns have also receded, primarily because most U.S. producers of high-performance computers have adopted hardware strategies built around the high-volume commodity microprocessors in which U.S. companies are undisputed world leaders.[29]

Interestingly, many of the proposals to lessen U.S. companies' vulnerability from dependence on uncertain supplies of critical components, initiatives that were controversial when included in the proposal for a U.S. Memories venture, are today a routine feature of the semiconductor business. Long-term, five-year contracts between DRAM suppliers and

28. Although concentration in DRAM production remained high enough to prompt occasional thoughts that a small number of companies continued to have the ability to set prices in the short term, were they able to successfully coordinate their actions? In March 1996 Samsung, the world's largest DRAM producer, announced amidst plunging memory chip prices that it would be halving its production of 4 M DRAMs to fix its "price problem." Added to cuts announced weeks earlier by NEC and Hitachi, this action would cut monthly supplies from 50 million chips to 31 million by the end of 1996. See Newsbytes News Network, "Asia Technology Newsbriefs," on-line data, March 29, 1996; and Vision Multimedia Technologies, "Chip Talk," on-line data, April 15, 1996.

29. See Kenneth Flamm, "Controlling the Uncontrollable: Reforming U.S. Export Controls on Computers," *Brookings Review*, vol. 14 (Winter 1996).

computer firms are reported in the trade press.[30] Customers routinely buy equity stakes in chip suppliers' operations to ensure access to crucial supplies.[31] Joint ventures to share the enormous costs of developing and fabricating new chips are also common.[32]

What remains a concern for both U.S. chip producers and consumers—and a potent potential source of schisms between them—is the application of antidumping laws and price floors to semiconductor imports. From 1986 to 1989, Japanese government measures to implement the pricing provisions of the STA had the ironic effect of helping create the very cartel-like structure that had long been considered the main threat driving U.S. trade policy. The U.S. semiconductor industry argued that these measures had not been sought and strictly speaking were not part of the agreement. But without either the strictest of border controls or reductions in production, higher prices in world markets could not be sustained. Given the porousness of borders, the ingenuity of entrepeneurs, and the ease with which large volumes of chips can be transported, the Japanese government was probably correct in concluding that both sets of measures were needed to comply with the STA. The U.S. government, formally and informally, at first encouraged Japan to attempt to

30. See Mark LaPedus, "5-Year DRAM Deals," *Electronic Buyers' News*, December 18, 1995, p. 1.

31. See for example, Don Clark, "In the Hot World of Chips, Tradition Melts," *Wall Street Journal*, October 30, 1995, p. B4; and Mark LaPedus, "HP May Join Taiwan Venture," *Electronic Buyers' News*, January 8, 1996, p. 12. One pronounced trend in the semiconductor world has been for so-called fabless semiconductor companies—design houses that contract with outside foundries to manufacture their designs—to invest in the foundries to ensure the availability of capacity to manufacture their products. Examples in 1994–95 include disk controller producers Adaptec and Opti; graphics controller companies S3, Cirrus Logic, Trident, and Oak; specialized memory producers Alliance and Lanstar; and programmable logic makers Actel, Altera, Lattice, and Xilinx. See Mark LaPedus, "Alliance, S3 Sign Fab Deal—To Build Foundry in Taiwan with UMC," *Electronic Buyers' News*, July 17, 1995, p. 1; Ismini Scouras, "Lanstar Lines Up DRAM Fab," *Electronic Buyers' News*, February 12, 1996, p. 1; "Fabless Deals," Fabless Semiconductor Association, downloaded April 1996; and Dataquest, "The Fab Four," on-line abstract, Dataqauest, October 1995.

32. Recent announcements include joint ventures to build DRAM production facilities in the United States by partnerships between Siemens and Motorola, Toshiba and IBM, and Texas Instruments and Hitachi. Motorola is also considering joining an existing three-way IBM-Siemens-Toshiba R&D effort to work on technology for 1 gigabit DRAMs. See Clark, "In the Hot World of Chips"; Darrell Dunn and Mark Hachman, "Motorola Jumps into DRAM Fray," *Electronic Buyers' News*, October 30, 1995, p. 3; and Loring Wirbel, "TI, Hitachi Tip Twinstar Setup," *Electronic Engineering Times*, January 23, 1995, p. 92.

cut production and seal its borders. When protests erupted and the governments were forced to abandon production controls, sticking to the objectives of the arrangement meant driving the controls underground, in essence by tolerating private cartel-like behavior.

The contradictions built into these policies were ultimately resolved when buoyant demand and high chip prices between 1992 and 1995 made price floors irrelevant. Demand was so strong, in fact, that for the second time in the history of the industry the average cost of a bit of memory increased slightly in 1993. Memory costs on average seemed to be dropping at half the historical rate.

This is unlikely to continue. According to the U.S. semiconductor industry's projections for the next fifteen years, the cost of producing a bit of memory will decrease about 26 percent a year, about halfway between the earlier trend and the record of the early 1990s.[33] Furthermore, long-standing historical patterns are unlikely to have abruptly ended for the semiconductor industry. The sustained expansion of the early 1990s could not continue forever, and a cyclical plunge into weak demand, excess supply, and sharply dropping prices was virtually inevitable. Indeed, in early 1996 the industry seemed poised for just such a decline.

When downturns come, the rules for semiconductor pricing will again become an issue. American producers will likely urge limits on how low import prices can plunge and will launch antidumping cases to accomplish this end. Chip users will protest, with considerable justification, that the antidumping laws establish price floors that lack a sound economic basis and are biased against foreign producers (see chapter 6). Unless (as in 1986) some arrangement is negotiated that effectively sets a global floor on chip prices (rather than a U.S. floor), U.S. chip users will be tempted to move their manufacturing operations to take advantage of lower prices outside the United States. And if a worldwide undertaking to raise global chip prices uniformly is sanctioned by government actions, the paradox of U.S. trade policy facilitating the creation of the very cartel-like structure that has been the principal threat driving other trade, competition, and technology policies will unfold.

33. This is the implicit annual rate of decline calculated from Semiconductor Industry Association, *The National Technology Roadmap for Semiconductors* (San Jose, Calif.: 1994), p. 11.

Market Access

From the vantage point of early 1996, the pressure exerted to gain greater access to the Japanese market has paid off. The target of 20 percent market share written into the trade arrangements of 1986 and 1991 today seems of historical interest only as the foreign share of the Japanese market approaches 30 percent.

But there was a real history of Japanese discrimination against foreign products. When, despite considerable obstacles, foreigners established initial relationships with Japanese customers in the late 1960s and early 1970s, the Japanese government actively worked to disrupt them. The all-important long-term relationships were broken up. Thus, calls in recent years for "affirmative action" to assist foreign companies in gaining easier access to the Japanese market resonated in American ears: the barriers in Japan appear higher than elsewhere, and the ladders thrown up to scale them were deliberately smashed by the state.

Given the current situation, however, is it now time to declare what seems mainly historical to have been resolved? One is tempted to say yes. Today, all seems well. Under continuing political urging, a strong web of relationships between foreign and Japanese companies has been spun. It is hard to see how this network can be dissolved when, by most accounts, it seems to benefit both sides.

Still, some reservations must be expressed. First, much of the discussion of the continuing growth in strategic alliances between Japanese and foreign semiconductor firms seems to overstate the case. Precise and consistent definitions of what a strategic alliance is are critical in determining trends, and little such consistency exists. Collaborations between a foreign supplier and a Japanese customer resulting in the incoporation of a foreign chip into the design of a Japanese product certainly seem to have increased greatly in the past decade.[34] But because these collaborations may not extend beyond production of a very specific product, it is not clear whether they should be counted as strategic alliances.

Some data portray a volatile process of alliance formation. One source shows that strategic partnerships between U.S. firms and Japanese chip makers doubled in the years just before 1987 then decreased sharply in

34. See Electronics Industry Association of Japan, *Semiconductor Facts 1995* (Washington, 1995), p. 4-1.

1989 and 1990.[35] Other data show that after rising sharply (from 3 in 1980 to a peak of 120 in 1987) new Japanese semiconductor alliances have fluctuated considerably from year to year. From 1985 through 1994, according to industry consulting firm Dataquest, Japanese companies formed new semiconductor alliances with other companies an average of 86 times a year. In 1993 the number sank to 71, in 1994 it was 99, and in 1995 it dropped to 83.[36]

A second cause for concern is the potential impact of a sharp cyclical decline in the semiconductor market. The major Japanese producers would be faced with the choice of letting machines and men stand idle (or with even greater difficulty, reducing their workforce) or cutting back on purchases of semiconductors from outside vendors and using excess internal capacity to substitute for purchased parts. When demand is high and capacity fully utilized, as it has been for the past four years, this has not been an issue. In slack times there will be a strong economic incentive to switch to producing products internally. Although proprietary microprocessors and custom logic designs may not be easily switched to internal production lines, there are other chips procured from foreign companies—flash memories, static memories, DRAMs, and ASICs, for example—that are not going to be difficult to switch.

Thus it seems safe to predict that in a serious industry downturn, many alliances with outside vendors—strategic and nonstrategic—that are associated with purchases of foreign chips will be put under stress, perhaps to the breaking point. Although the economic logic driving such a contraction in foreign sales is readily apparent, foreign complaints and renewed trade friction are a very predictable consequence.

Third, the current absence of severe complaints over Japanese practices that impede foreign access to the Japanese market does not reflect a more general consensus that access as a whole is no longer an issue. Significant disputes exist in other parts of the Japanese economy. The key point for the semiconductor industry is that rather than pointing to formal trade barriers, most of the complaints by foreign businesses relate

35. See National Research Council, *U.S.-Japan Strategic Alliances in the Semiconductor Industry* (Washington: National Academy Press, 1992) p. 14.
36. Sheridan Tatsuno, "Beyond the Chip Wars: The Boom in Japanese Semiconductor Alliances," *Venture Japan*, vol. 1, no. 2 (1988), pp. 24–25; Dataquest, "1995 Japanese Semiconductor Alliances: A Slowdown," on-line abstract, Dataquest, January 1996; and "Japanese Semiconductor Alliances Taper Off in 1995," on-line research highlight, March 1996. Dataquest has revised the 1987 peak downward to 117.

to private anticompetitive practices, government competition policy, the continued use of informal administrative guidance by government officials, and the selective enforcement of antitrust laws.

Strictly speaking, competitive practices and antitrust matters are not trade issues. A playing field tilted against outsiders affects the would-be Japanese entrant as well as the foreigner. But insiders are likely to be Japanese, just as foreign entrants will almost certainly be outsiders. The effect of hidden barriers to competitive entry into the Japanese market will be to hinder foreigners and protect Japanese businesses.

Because there is no baseline for judging the adequacy of competitive conduct or antitrust policy that has been agreed to internationally and no responsibilities defined or enforced in the World Trade Organization, anticompetitive private behavior, arbitrary or selective enforcement of laws, and informal government guidance that impedes market access have no agreed international venue for settling disputes. Foreign companies can go to national authorities with complaints, but if anticompetitive behavior is tolerated by custom or law, or if national laws are selectively enforced by national authorities, or if bureaucrats issue undocumented guidance to manufacturers, there is no framework for resolving grievances except government-to-government negotiation.

In the Japanese semiconductor industry successful collusive behavior among firms seems to have occurred only with the explicit or implicit support of MITI.[37] Although government toleration of anticompetitive behavior in the semiconductor industry is not currently an issue, there is persuasive evidence that such behavior is sometimes tolerated and even encouraged in other areas of the Japanese economy, in some cases in apparent conflict with the letter of Japanese law (see chapter 7). Thus the fear sometimes expressed by American semiconductor producers that without some formal government-supported framework Japanese manufacturers will be tempted to slide back into the practices of the past is not without foundation. If the semiconductor industry were to face a serious downturn, the Japanese government might be tempted to permit (or even encourage) electronics producers to collude in order to keep

37. See chapter 7. In recent years, MITI has responded to foreign grievances about anticompetitive practices by referring the complainant to the Japan Fair Trade Commission, historically an ineffectual arm of the government. MITI recently displayed its power even within the JFTC's nominal sphere of antitrust enforcement by issuing an unprecedented notice to auto dealerships requesting "information about possible violations of the Antimonopoly Law." "Please Tell Us about Violation of an Antimonopoly Law," *Nikkei Sangyo Shimbun*, November 7, 1995, p. 14.

CONCLUSION   449

factories from closing in what has long been regarded as an economically
strategic industry. Is it unreasonable to worry when one can find evidence
that this happens elsewhere in the economy?

Where Do We Go from Here?

In the spring of 1996, after ten years of interaction through semiconductor trade arrangements, the American and Japanese chip industries
and their respective governments seemed again to be on a collision
course. The Americans seemed reasonably happy—although ready to
argue that market access in specific product sectors continued to be
restricted. The Japanese, however, were distinctly unhappy.

Their concerns were not hard to understand. With the STA apparently
succeeding in improving American chip firms' access to the Japanese
market, U.S. trade negotiators had since the closing days of the Bush
administration increasingly looked at its use of numerical benchmarks as
the proven formula for making headway on difficult matters of market
access. Japanese officials, aware that they had created a Frankenstein's
monster by agreeing to benchmarks in semiconductors and later to some
extent in autos, were adamant in their determination to erase numerical
targets or benchmarks from the trade policy map. Much of the conflict
in U.S.-Japan trade negotiations of the early and mid-1990s centered on
whether quantitative measures of market access were even to be collected, and if so, what purpose they would serve. Bilateral agreements
with the United States that contained specific targets were particularly
troublesome from Japan's perspective. Despite the World Trade Organization's procedures, put into place in 1994, for settling multilateral
disputes, the United States would have a legal basis for using its unilateral
powers to retaliate against Japan—section 301—if it could be argued that
a bilateral agreement had been broken.

In the fall of 1995, with Ryutaro Hashimoto serving as minister of
trade and industry (as well as deputy prime minister), MITI publicly
declared that "the agreement between the governments has ended its
historic mission." At a quadrilateral trade ministers' meeting in late October, then U.S. Trade Representative Mickey Kantor had called for its
extension. Hashimoto refused, declaring, "We do not need to go to the
trouble of signing a bilateral accord because the semiconductor market

has been rapidly turning international in recent years."[38] In 1996, with Hashimoto now prime minister and Trade Representative Kantor continuing to stress the need to negotiate a new agreement, both sides seemed locked into public positions that could only end in painful confrontation.

Both sides, it might be argued, were missing a unique opportunity to strengthen the institutional structures supporting international commerce in high-technology products. By recognizing the problems raised during ten years of semiconductor trade arrangements, and remedying them while reinforcing their contributions toward creating more open and competitive international markets, the two governments could strengthen the foundations of an open international trading system in high-technology products.

The most important criticism of the STA is of its implicit definition of a 20 percent market share as a floor for foreign sales in the Japanese market. Shares in an open, competitive market reflect not only the ability of firms to gain access to the market, but also the underlying competitiveness in price, quality, service, and support of the products being sold. Before the 1980s the Japanese government often took action—directly and indirectly, formally and informally—on behalf of its domestic producers to disrupt the ability of foreign firms to sell semiconductors in Japan. Unquestionably there were ample grounds for U.S. firms to seek compensatory measures. But after close to a decade of successfully pressuring the Japanese government to pursue aggressive affirmative action, the United States needs to ask whether it needs to return to what should be its long-term objectives for the international trading system. If open, transparent, competitive markets are a basic tenet of the structures we build to support global high-technology industry, a negotiated market share floor contradicts that goal. In fact, the U.S. government indicated in early 1996 that it was willing to drop this feature of the STA.[39]

A second criticism of the STA was that as a bilateral agreement it did not recognize the increasing globalization of semiconductor production. Asian producers, especially Korea and Taiwan, account for more than 10 percent of world chip output and European makers a little less than 10

---

38. "Collision Foreseen over Semiconductor Accord," *Nihon Keizai Shimbun*, November 3, 1995, p. 3.

39. In early 1996 chief U.S. trade negotiator Ira Shapiro declared to a San Francisco business conference that the United States would no longer seek numerical targets in semiconductors from Japan. Newsbyte News Network, "Japan Technology Newsbriefs 3/1/96" (on-line news feed), March 1, 1996.

percent. As China, Singapore, Thailand, and Malaysia enter the market, their contribution must be recognized and their importance embraced in defining the future rules of international competition. Multilateral discussions of semiconductor trade and investment issues are inevitable.

While the United States and Japan may no longer account for 90 percent of world output as in 1985, with 80 percent (and 60 percent of global consumption) they remain the giants of the industry.[40] If they can agree on a direction for semiconductor trade, they will have a powerful influence in shaping new multilateral institutions.

In a similar vein, it is sometimes argued that the web of strategic alliances and complex sourcing arrangements between international companies makes it increasingly difficult to define what is national and what is foreign. But this was also true in 1986 and in 1991. (Qualitatively, some of today's relationships may be more complex, but it is not clear that U.S.-Japan semiconductor alliances are quantitatively more significant than they were a decade ago.) As long as the words *foreign* and *national* continue to have operational significance for governments—as they do, for example, in technology, procurement, investment, and antitrust policies—in resolving disputes we will be compelled to struggle to define in a pragmatic way what is foreign and what is national.

A third critique of the STA is that with the creation of the World Trade Organization in 1994, a multilateral mechanism now exists to resolve the sorts of trade disputes that led to creation of the STA. The WTO may now be the proper mechanism to settle the semiconductor trade issues addressed by the STA. But although the WTO can deal with trade issues for which guiding principles have been agreed to in the General Agreement on Tariffs and Trade, many of the concerns addressed by the STA are not addressed in the GATT. Increasingly, anti-competitive practices (tolerated or even encouraged by government), informal and undocumented administrative guidance, and formal regulation that limits the ability of new companies to compete against those already in a market form the core of U.S. complaints about access to Japanese markets. There simply are no agreed upon international standards of conduct in these areas, and the suggestion that they be adjudi-

40. In some respects the industry has become *less* globalized. In 1970 U.S. and Japanese firms produced less than 75 percent of world output and European companies the balance. The sharp decline in the European industry is as notable as the rise of East Asia. For an analysis of European decline, see Kenneth Flamm, "Semiconductors," in Gary Clyde Hufbauer, ed., *Europe 1992: An American Perspective* (Brookings, 1990), pp. 225–92.

cated by the WTO begs the question of what standard is to be used in resolving them.

In the long run, international norms in competition and antitrust policy will be discussed and perhaps adopted.[41] But this will be a painstaking process that is likely to take decades given the varying approaches and attitudes among different nations. For the moment, to resolve basic problems, there would seem to be no alternative to direct discussions among governments.

The fact that the GATT-WTO structure does not address some important matters, particularly competitive practices and antitrust enforcement, that can create functional barriers to the entry of U.S. firms into foreign markets creates some serious tensions for U.S. policy. The GATT-WTO apparatus severely limits the extent to which trade policy can be used to apply pressure to resolve other issues. It is improbable to suppose that, stymied by the absence of effective trade policy leverage, the United States will accept that it can do nothing about privately administered (or informally constructed) barriers around foreign markets. What is more likely is the construction of new unilateral tools (for example, extraterritorial application of U.S. antitrust standards), which in the long run will be a step away from multilateral governance of an international trading system.

A final critique of the STA has been that the frictions it addressed are ancient history. With foreign chips (by whatever definition) taking 30 percent of the Japanese market, market access is no longer a major sore point. The worldwide industry is relatively healthy, and markets continue to grow rapidly. As a press release from MITI stated, "competitive foreign-based semiconductors have been firmly incorporated as indispensable products in the Japanese market. . . . Thus, the Japan-U.S. Semiconductor Arrangement has fully achieved its objectives and shall expire."[42]

If the semiconductor market slips into a major decline, however, this situation will change. Given the enormous increase in world chip production capacity expected in the next few years and the economic pres-

41. For a useful analysis, see F. M. Scherer, *Competition Policies for an Integrated World Economy* (Brookings, 1994). As Keio University professor Jiro Tamura comments in this book (after noting that Japanese antitrust law is modeled on U.S. law), Japan's "discrepancy in antimonopoly activity [with the United States and Europe] is based less on the law itself than on the weakness of the enforcement of the law" (pp. 112–13).

42. MITI, "Comment on the Announcement of the Market Share of Foreign-based Semiconductors," press release, Tokyo, March 19, 1996.

sures to keep the new factories running, prices are likely to come down steeply in a recession. There will almost certainly be accusations of unfair competition, acute trade frictions, and dumping cases. What are still largely nationally defined standards (subject to some loose guidance from the GATT) governing fair values for imports will be used to punish offending exporters. Within Japan at least some of the foreign semiconductors so "firmly incorporated as indispensable products" are likely to prove dispensable as electronics producers consider using idle factories to make products previously purchased from outside. If, along with the STA, reliable data on foreign chip purchases disappear, bitter disputes over what precisely is going on in the Japanese market are likely to erupt. In the absence of some stipulated facts, disputes over what is happening will add additional fuel to a furor over why it might be happening.

This brings up some of the potentially positive elements the STA has contributed to the dialogue on semiconductor trade issues. Gathering some agreed upon data on what has been a contentious issue for thirty years serves one very useful purpose. Facts prevent some unhappy companies from complaining to their government that their experiences are typical of a broader trend when they are not. Certainly, well understood data have permitted Japan to argue without dispute that what was once a problem is one no longer. Without such facts, additional disagreement is likely whenever trade issues periodically erupt—as they surely will in this boom and bust industry.

A second positive contribution of the STA is the establishment of a forum to facilitate speedy resolution of disputes. In the next downturn, both market access and dumping will certainly be issues. An agreed upon format for quickly talking about them can avoid time-consuming, acrimonious, and possibly inconclusive debates over how to structure talks and who is to sit at the table.

Third, from the U.S. perspective the STA has illuminated shadowy corners of the Japanese electronics industry. In an environment of mistrust built by decades of behind-the-scenes exclusionary maneuvering, the regular dialogue created by the STA has built confidence in the United States that attitudes have indeed changed. Questions can easily be asked and answers given. The private industrial relationships that have now been constructed may be able to sustain the atmosphere of openness and cooperation. But given the signs of continued official tolerance of legally dubious insider relationships in at least a few Japanese industries, there is some justification for U.S. chip companies' worries that if the lights go out, they will again be in the dark, on the outside. A continuing, regular channel for discussion and consultation could dispel much of the suspicion that might otherwise poison relations.

How, then, can we take the best features of a decade's experience with semiconductor trade policy, eliminate the negative elements, and build a world trading system in microelectronics that contributes to a prosperous, competitive, and technologically innovative global industry? One approach would be to extract bits and pieces of ideas already under discussion by government and industry and combine them in what might be called the Global Semiconductor Conference (GSC).

The Global Semiconductor Conference

The Uruguay Round and earlier rounds of trade talks pulled nations already committed to an open trading system and membership in the GATT into a regular forum that gradually enabled the world trading system to transform abstract principles into rules to deal with the real problems of a maintaining a living, open system. The GSC would have a similar purpose. It would create a forum that could gradually attempt to resolve some of the difficult policy issues that are an integral feature of international competition in semiconductors. In the process, if the solutions worked out in the rough-and-tumble laboratory of the semiconductor industry are useful, they might be incorporated into the more general GATT-WTO apparatus and address similar issues in other industries (particularly high-technology sectors with many of the same industrial characteristics).

In its simplest incarnation the GSC would consist of regularly scheduled multilateral meetings whose sole objective would be to work out the mechanisms needed to ensure a transparent and competitive international market for semiconductors. Just as the GATT makes adherence to some minimum set of commitments and responsibilities a condition for participation in decisions about further evolution of the system, membership in the GSC would also require adherence to a minimum set of ground rules. Those might reasonably be expected to include free trade in semiconductors, unrestricted foreign investment in semiconductor production and sales, a minimum set of guarantees for intellectual property, a commitment to collect common data needed to assess the transparency of markets, and a willingness to remove both formal and informal barriers to entry into national markets when they are identified in an agreed upon fashion.

The work of the GSC would be to add the detail needed to make these general rules into practical ones for competitive international semiconductor markets. In the future the agenda for the GSC might include other difficult issues. Given that national tests for dumping and other proscribed trade practices in the semiconductor industry vary considerably, as do remedies, and can still be manipulated to create disadvantages for foreign competitors selling in national markets, international harmonization of procedures is needed to define the rules of the road for competition. Current antidumping laws may be adequate for cement or pasta, but they do not address important features of the semiconductor industry and other high-technology industries. Procedures that take into account sunk R&D costs, economies of scale, and learning curves in an economically defensible way are needed.

Similarly, an R&D and technology group within the GSC could encourage greater cooperation on government-funded R&D programs to increase the productivity of global investments in basic microelectronics research and technology. A standards group of the GSC could define a framework to facilitate harmonization of international standards for design, manufacturing, equipment, and materials in the semiconductor industry, when there was broad agreement that uniform standards might be useful.

Finally, the difficult question of harmonization of antitrust rules and their enforcement is probably best addressed in small pieces. By working on this question within the restricted venue of the semiconductor industry, some experience useful to broader rules would likely be accumulated. In a sense, the semiconductor industry is already serving as such a laboratory, since the ongoing discussion of dumping and of antidumping remedies is inextricably linked to theories of predation and antitrust. A sensible way to deal with accusations of dumping in semiconductors will necessarily be part of a sensible approach to dealing with questions about anticompetitive practices and the role governments play in these practices.

From Here to There

In early 1996 the U.S. government and semiconductor industry had two key objectives as they negotiated with Japan: maintenance of consistent and reliable data that would permit analytical discussion of market access issues and some continuing framework that would permit resolution of trade and access issues as they arise. Both could contribute to the quest for an open, competitive world market in semiconductors.

Japan also seemed to have two primary demands in early 1996. It wished to replace a bilateral agreement with a multilateral process and wanted no numerical targets or benchmarks. These demands also seemed consistent with construction of an open, competitive trading system.

If Japan and the United States were to simply announce the formation of a multilateral Global Semiconductor Conference to begin operation just as the 1991 STA expired, all four objectives could be accomodated. The GSC would initially include a commitment to collect national data needed to evaluate market transparency and a framework for timely consultations on trade, investment, technology, or market access issues that are raised by GSC members. (As founding members and equal partners in this enterprise, Japan and the United States would be agreeing to provide data to each other and react to requests for consultation with equal alacrity.) The GSC would be open to membership by all nations that pledged to respect the ground rules established at its foundation.

To tackle an ambitious agenda that might include analyses of pricing practices, market access, technology programs, and technical standards clearly will require significant participation and support from industry. Just as clearly, it will require governments to participate. It is unthinkable that American semiconductor producers could sit down with foreign semiconductor producers and discuss pricing, markets, standards, and cooperative R&D efforts without the U.S. government present. From the perspective of a user industry, the meeting might seem to be the perfect cover for the creation of a supplier cartel. At a minimum some U.S. government oversight would be absolutely necessary to guarantee that user industry interests are respected (and antitrust suits against producers avoided).

In addition, most of the issues on the agenda would involve government responsibilities, and companies have no power to cut deals over these matters. Antitrust rules, dumping, and access to publicly funded technology programs are areas where governments make policy, not companies. A framework with industry-government meetings on technical matters that paralleled government-to-government negotiations over policy would need to be constructed.

Thus if the Global Semiconductor Conference or something like it is to be the way out of the rhetorical box that the United States and Japan find themselves in, it will involve both government and industry. It will not happen if the United States and Japan, whose companies control 80 percent of world semiconductor production, do not take the lead in building such an institution. It will be much harder to organize it later, as more countries become established in the industry, and ground rules become more complicated to negotiate. In 1996, then, an opportunity to resolve a political impasse and begin to explore solutions to some thorny policy issues involving global high-technology industries seems to have presented itself.

Mismanaged Trade?

If one looks only at the Semiconductor Trade Arrangements of 1986 and 1991, simple but contradictory answers to the question posed in the title to this book spring to mind. Overall, the system of pricing floors established under the STA probably worked to the net disadvantage of U.S. interests, and the cartel-like structures fostered by MITI in implementing what Japan perceived to be its objectives only made the situation more difficult. There was—and is—no way to set effective price floors for semiconductors that ultimately would not penalize U.S. user industries.

The market access the STA opened up has been a different story. There was a long history of formal and informal barriers that the STA helped tear down.

At a deeper level, two lessons emerge from trade frictions with Japan in semiconductors. One is that contradictions between tactical compromises and strategic, long-run principles in U.S. trade policy ultimately come home to roost. Having failed to achieve functional access to the Japanese market in the mid-1970s after formal trade barriers were removed, the United States essentially decided to use the informal system of MITI guidance and government collaboration with an industrial inner circle to achieve an outcome that at least resembled what it thought real systemic reform might have accomplished. Paradoxically, this decision probably strengthened what had been waning MITI influence in the Japanese semiconductor industry. It also enabled Japan to position itself as the champion of multilateralism in the international system and claim the high ground against government intervention and in favor of letting competitive markets determine economic outcomes. (This latter situation grated with particular intensity on American nerves, given Japan's history of government intervention in this and other high-technology industries, and official tolerance for collusion and regulation elsewhere in the Japanese economy.) Rather than supporting a system that would guarantee real competition in global markets for high-technology products, the United States has come to wear the mantle of defender of the status quo.

The second lesson from this history is the need to continue working to guarantee open, competitive markets for high-technology products. The economics of high-technology industries ensures that the trade regime in its current form is not going to do the job. Economies of scale will tempt governments to invent strategic policies that give their firms advantages against foreign competitors. Collusion that benefits domestic producers at the expense of foreign producers and consumers will remain at least theoretically possible. Dumping laws and antidumping procedures make it easy to further stack the deck against foreign manufacturers.

The large share of costs and revenues accounted for by R&D investments—the defining characteristic of a high-technology industry—virtually guarantees continued intervention by government. Even in an economy cut off from international trade, there is a large economic literature that argues that government support for research is desirable where it corrects for market failures (private investors' difficulties in capturing the full fruits of their R&D investments for their exclusive financial benefit) that cause private returns to investment in technology to fall short of the full benefit to society. Both case studies and statistical analysis have confirmed the empirical relevance of this argument.

International trade generates additional complexities. The possibility of shifting technology-created profits from foreign producers and consumers (or avoiding payment of returns to them) raises a second rationale for government intervention. If national policy can create situations in which technology-based returns can be captured overseas for national producers, national income and the standard of living are increased. This is the strategic trade rationale for intervention in high-technology industries.

The upshot is that, for both reasons, governments will be deeply involved in national investment in technology and technology-intensive industries. The challenge is to propose some way of neutralizing subsidies to R&D as a tool of profit-shifting trade competition, yet preserve the ability of governments to engage in socially beneficial public investment in R&D. Some move toward reciprocity in R&D—permitting companies from other countries to join some of one's subsidized research programs in exchange for one's own companies being permitted to join in a commensurate portion of another country's R&D projects—would seem a useful step in that direction.

Over the long haul the only certainty is that an expanded multilateral framework will be needed to guarantee the open, competitive trade in high-technology products that serves a common international interest. One can only hope that the international semiconductor industry, which pioneered public discussion of problems that are now obvious, will also lead the way in finding solutions that are not.

Index

Academic sector, 25, 41n7
Advanced Micro Devices (AMD), 89, 166, 217
Advanced Telecommunications Research Institute, 119
Advantest, 105, 144
AEA. *See* American Electronics Association
Akashi Seisaku, 110
Alps, 89
AMD. *See* Advanced Micro Devices
American Electronics Association (AEA), 216–17
American Telephone and Telegraph Company (AT&T): as captive or merchant producer, 21n21; patent licensing, 44, 49n30
AMI, 72, 89
Ampex, 129
Ando Electronics, 110
Antitrust issues, 150–51, 160, 195, 310, 436; Microelectronics and Computer Technology Corporation (*1983*), 148; Micron, 167; National Cooperative Research Act (*1984*), 148; nature of, 448; Union Carbide (*1988*), 123n254; Semiconductor Trade Arrangement and, 227; U.S. Memories, 217–18; Zenith Radio Corporation (*1974*), 134n23. *See also* Monopoly issues
Apple Computer, 218
Application-specific integrated circuits (ASICs). *See* Integrated circuits
ASET. *See* Association of Super Advanced Electronic Technologies
Asia Seisaku, 110
ASICs (Application-specific integrated circuits). *See* Integrated circuits

Association of Super Advanced Electronic Technologies (ASET), 438, 440n21
Automobile industry, 2, 166–67

Baldrige, Malcolm, 150
Bardeen, John, 30
Basic Technology Research Promotion Center, 118–19
Bell Telephone Laboratories: development of solid-state electronics, 8, 18, 30–32, 33; technology transfer, 41, 42, 43, 44, 49; VLSI technology, 90
Brattain, Walter, 30
"Bubble money." *See* Business sector
Bush (George) administration, 2n2, 449
Busicom, 76
Business sector: "bubble money," 277–78; development programs in semiconductors, 40–41; semiconductor consumption, 34n47, 35t; standards of competitive behavior, 311; U.S. company sales, 286t

Calculators: "calculator war," 75–77; integrated circuit supplies, 72, 74, 75–79; large-scale integration chips and, 71, 72, 73; mass production of, 68–69; miniaturization, 67; single-chip, 73; solid state, 60; transistorized, 68n86; U.S. chips in, 431
Canon: manufacturing processes, 105, 110–11, 144; purchase of U.S. integrated circuits, 72
Cartels. *See* Economic issues; Ministry of International Trade and Industry
CDL. *See* Computer Development Laboratories
Center for Integrated Systems, 147–48

462

China, 451
Clinton (Bill) administration, 2–3
Compaq, 218
Computer Development Laboratories (CDL), 96, 112–13
Computer industry: *1960s*, 430; *1980s*, 193, 201, 203, 213, 222; *1990s*, 283, 434n2; commercial, 48, 430; component purchases, 24; computer prices, 7; Dendenkosha Information-Processing System (DIPS–1), 93; IBM, 59, 60; integrated circuits in, 34–35, 61; Japanese industry, 39–40, 52, 66, 80–82, 125, 148, 430; Japanese regulation of, 53; Next Generation Basic Computer Technology program, 113n226; semiconductors in, 33–34, 48, 49, 63, 203, 220, 442; single-chip microcomputers, 73; U.S. industry, 442–43; Very High Speed Computer Systems (VHSCS), 61–62, 92, 93
Computer Technology Research Association, 59n61
Consumer interests, 272–78, 292, 293, 309n9
Contract sales. *See* Purchasing and sales
Corrigan, Wilfred, 211, 217, 244

Dainippon Printing, 105
Dataquest, 21, 236, 404
Dataquest *First Monday Report*, 244–47
Defense Advanced Research Project Agency (DARPA), 147–48
Defense issues: integrated circuits, 34–35; national security issues, 130–32, 150, 153, 372–82, 425–26, 429; purchases, 24; research and development funding, 29–38, 117n234, 373; transistors, 8, 30–34
Defense Science Board, 373
Dell Computer, 218
Demand. *See* Purchasing and sales
Denshinho (Electronics Industry Promotion Law). *See* Law on Temporary Measures for Promoting the Electronics Industry
Digital Equipment Corporation, 217
Distribution. *See* Purchasing and sales
DRAM. *See* Dynamic random access memory
Dumping: competitive behavior and, 317; constructed cost dumping tests, 312–13, 358; cost structures and, 311–34; definition, 305, 306n4, 308n5; effects, 211; economic analysis of, 305–71, 428, 445; European charges of, 173, 189, 191;

forward pricing and, 309–10, 314, 316–17, 318, 320, 321, 322; by Japan, 141–46, 149–50, 160, 162, 166, 167, 168–70, 172, 305, 383; by Korea, 224–25; litigation, 162, 168, 169–70, 172, 173, 176, 182, 189; prevention of, 187n92, 223–224, 227, 228, 292, 312–13, 428; sanctions, 187, 188, 211, 225, 230, 266; screening for, 356, 358; strategic and predatory, 140n41, 141–46, 150–51, 306–11, 322–28; third-country, 292; by the United States, 73–74, 136, 430–31. *See also* Purchasing and sales; Semiconductor Trade Arrangement
Dynamic random access memory (DRAM): *1988* shortage and crisis, 192–201, 203, 207–11, 235, 294–95, 300n84, 387–96, 442–43; costs, 402; design and development, 9–11, 82–83, 93–94, 98–99, 313–14; dumping cases, 173, 176, 182, 189; forecasting of production, 196–201, 222–23; product life cycle, 313, 354–55; quality, 145–46, 163, 232n9, 436; role in semiconductor industry, 8–9, 99, 383; Semiconductor Trade Arrangement and, 160–226; supply, 256–66, 443–45; trade issues, 138, 139, 143–44, 148–49; U.S.firms, 167n17. *See also* Dumping; Indexes; Manufacturing processes; Production processes; Purchasing and sales
Dynamic random access memory (DRAM), prices: below cost and quality dumping, 141–44, 145, 148–49, 168–69, 219–21, 355–58; competition and, 398–405; controls and guidance, 171–72, 175–76, 188, 189–90, 193, 198; cyclical effects, 162–63, 240–41; European markets, 188, 189–90; fair market values, 177n55, 183; Japanese, 184, 187, 265, 387–88; "pi" rule, 9–11, 240; regional differentials, 242–54, 433; two-tier, 138–41, 150–52; in the United States, 203, 210–12, 219–21; Semiconductor Trade Arrangement and, 237–38, 269, 297–301, 445
Dynamic random access memory (DRAM), production and production controls; *1980s*, 387, 390–91; *1990s*, 219–20; costs, 401–05; forecasting, 196–201; investment and, 202–04; learning economies, 336–37; memory chip supply, 254–66, 356; prices and, 323; Semiconductor Trade Arrangement, 180–86, 192–95, 201–08, 269–73; U.S. Memories and, 396–97

Dynamic random access memory (DRAM) products: *64K*: *64K* DRAM wars (*1981–83*), 148–58; controls and guidance, 160; development, 93–94, 99; dumping, 166, 167, 168–69, 172, 227; pricing and production, 160, 162, 168–69, 227, 235*f*, 247, 249*f*, 391; production, 196, 262, 391; Semiconductor Trade Arrangement and, 297–301; yield, 261*f*

Dynamic random access memory (DRAM) products: *256K*: controls and guidance, 175, 177, 187, 192, 198, 229–30; development, 99, 116; dumping, 160, 168, 169–70, 173, 176; prices, 169–70, 187, 203, 223, 244, 245–46, 247*f*, 250–51, 269, 387; production, 183–84, 192, 196, 198, 209*n*152, 262, 269, 352, 353; Semiconductor Trade Arrangement and, 160, 177, 269, 297–301; shortages, 203–04

Dynamic random access memory (DRAM) products: *1M*: controls and guidance, 182, 192, 193, 198, 215–16; costs, 401–05; demand, 341–43, 344; development, 96, 99, 116, 192; dumping, 356–58; learning curves and economies, 337–38, 400–401; prices, 212–16, 221–23, 245, 246*f*, 248*f*, 249*n*28, 251, 269, 270*f*, 354–56, 386–87, 398–99; production, 182, 192, 193, 198, 200, 212–13, 215, 220, 262, 269, 386–87, 396–97, 398–99; shortages, 200, 203–04, 208*n*148, 213; Strategic Trade Arrangement and, 298–301

Dynamic random access memory (DRAM) products: miscellaneous products: *1 gigabit*, 438; *1K*, 99*n*190; *4K*, 99*n*190; *16K*, 98–99, 139–40, 141, 142, 144, 232*n*8, 352*n*68; *4M*, 98*n*189, 217, 222–23; *16M*, 98*n*189; *256M*, 314

ECL. *See* Electrical Communications Laboratory; Emitter coupled logic
Economic issues: cartels, 382, 384–24, 405–17, 418, 443; contract pricing, 294–301; cost of manufacture, 402; dumping, 305–71; economies of scale, 7, 313; effects on demand, 342; employment, 23*f*, 24, 47–48; externalities, 380, 383*n*18; forward pricing, 309–10, 314, 316–17, 318, 320, 321, 322, 327–28; government policies and support, 3, 380–81; quality control, 145*n*59; high-technology industries, 458; recessions, 162, 167, 205; semiconductor and chip industry, 7, 17, 38; subsidies, 383–424;

value of the dollar, 247*n*27; value of the yen, 203, 393, 436; venture capital, 24–25; welfare issues, 381, 382, 412–14, 416, 418, 433, 458–49. *See also* Learning economies; Trade

EECA. *See* European Electronic Component Manufacturers Association
EEPROMs. *See* Electronically erasable programmable read-only memory chips
EIA. *See* Electronics Industries Association
Electrical Communications Laboratory (ECL), 40
Electronically erasable programmable read-only memory chips (EEPROMs), 176
Electronic Arrays, 144
Electronic components: boom of *1988*, 201; in Europe, 25; Japanese regulation, 53; ranking of industry, 13, 375–76; sales, 18*n*20; in the United States, 24, 129, 425
Electronics, consumer: development, 18; integrated chips in, 139; in Japan, 47, 49, 52–53, 54, 60, 68, 125, 133, 431; in the United States, 53, 129. *See also* Calculators; Television and radio
Electronics Industries Association (EIA: U.S.), 128, 129
Electronics Industries Association of Japan, 87, 135*n*25, 390
Electronics Industry Council, 53
Electronics Industry Promotion Law (Denshinho). *See* Law on Temporary Measures for Promoting the Electronics Industry
Electrotechnical Laboratory (ETL). *See* Ministry of International Trade and Industry
Emerson, 129
Emitter coupled logic (ECL), 176
Erasable programmable read only memory (EPROM): *128K*, 254, 255*f*; *256K*, 166, 254, 255*f*; *512K*, 269, 271*f*; *1M*, 269, 271*f*; dumping cases, 160, 162, 170, 173, 174, 176, 189; prices, 166, 174, 210–11, 239, 241, 253–55, 265–66, 269, 271, 278*n*51; production and production controls, 182, 198, 201, 239*f*, 262, 263*f*, 264*f*, 265–66; role in semiconductor industry, 383; supply and demand, 231, 256–66, 278–79, 283, 293
Esaki, Leo, 45*n*24
ETL. *See* Electrotechnical Laboratory
Europe: acquisition of solid-state electronics, 18; computer industry, 25–

26, 427; dumping negotiations, 225–26; Korean products, 234*n*13, 291; "logic wars," 73*n*105; products, 288; share of global chip market, 2, 23–27, 437, 450–51; Semiconductor Trade Arrangement, 173, 188–91; tariffs, 26; third-country trade, 160, 243. *See also* Semiconductor industry, Europe

European Electronic Component Manufacturers Association (EECA), 189

Everex Systems, 218

Export-Import Trade Law of *1952*, 133–34

Exports. *See* Trade

Fairchild Semiconductor, 35–36, 56, 58, 76, 88

FECL. *See* Foreign Exchange Control Law

FIL. *See* Foreign Investment Law

Fisher Ideal price index. *See* Indexes

FMVs (Foreign market values). *See* Trade, U.S.

FONTAC project, 59–60

Forecasting. *See* Ministry of International Trade and Industry; Semiconductor Trade Arragement

Foreign Exchange Control Law (FECL; *1949*), 55–56

Foreign Investment Law (FIL; *1950*), 55

Foreign market values (FMVs). *See* Trade, U.S.

Forward pricing. *See* Economic issues

Fransman, Martin, 220

Fuji Electronic Chemicals, 110

Fujitsu: *256K* EPROM, 166; Computer Development Laboratories, 96; DRAM production and trade, 116*n*232, 139, 142*n*45, 143, 192, 277, 291*n*72, 391; dumping, 170*n*28; joint ventures, 86, 95*n*180; subsidies, 66; supplier to NTT, 91; Ultra Advanced Computer Development Technology Association, 81, 85; U.S.facilities, 144, 167; Very High Speed Computer Systems, 61–62, 93; very large scale integration ICs, 96, 107, 110, 116

GATT. *See* General Agreement on Tariffs and Trade

GCA. *See* Geophysics Corporation of America

GE. *See* General Electric

General Agreement on Tariffs and Trade (GATT): government subsidies in, 3*n*4; Japanese integrated circuit imports, 79–80; Japanese tariffs, 147; MITI

monitoring system, 195–96; Semiconductor Trade Arrangement, 185, 188, 190, 230; third-country trade, 160, 190; World Trade Organization and, 451

General Electric (GE): joint ventures, 57; sale of Great Western Silicon, 122–23; Toshiba and, 129; trade adjustment assistance, 137; transister production, 33

Genkyoku, 53

Geophysics Corporation of America (GCA), 105*n*202, 111

Germanium. *See* Transistors

Gidwani, Ramesh, 184

Global Semiconductor Conference, 454–55

Goldstar Electron, 225, 291*n*72

Gray market. *See* Purchasing and sales

Great Western Silicon, 122–23

Hashimoto, Ryutaro, 449–50

Hayakawa Electric, 67, 68. *See also* Sharp

Hemlock, 122

Hewlett-Packard, 217, 218

HHI (Hirschman-Herfindahl index). *See* Indexes

High Technology Working Group (HTWG), 154–58

Hirschman-Herfindahl index (HHI). *See* Indexes

Hi-Silicon, 122

Hitachi: Computer Development Laboratories, 96; DRAM production and trade, 116*n*232, 139, 142*n*45, 143, 166–67, 169, 183, 192, 198, 200, 215, 291*n*72, 391, 397, 443*n*28; dumping, 170*n*28, 172; licensing agreements, 42, 43–44; market shares, 45; RCA and, 129; research and development, 41, 62–63, 82–83; subsidies, 66, 82–83; supplier to NTT, 91; transistors, 45, 48; Ultra Advanced Computer Development Technology Association, 81, 85, 95*n*180; U.S. facilities, 144, 167; Very High Speed Computer Systems, 61–62, 93; very large scale integration ICs, 96, 107, 110, 116

Hitachi America, 166

Hitachi Software Engineering, 110

Hoffman Electronics, 57–58

Home Electric Appliances Market Stabilization Council, 135*n*25

Honeywell, 129

Hong Kong, 131

HTWG. *See* High Technology Working Group

Hyundai Electronics, 225, 291*n*72

IBM (International Business Machines Corporation): antitrust case against, 90; as captive producer, 20–21; computers and products of, 59, 60, 61–62, 80, 81n128, 90, 280; entrance into Japanese market, 58, 282; integrated circuits and, 59; local foreign production, 20; memory products, 11, 261f; OS/2 operating system, 220; subsidiaries, 438; U.S. Memories, 217, 218
Ibuka, Masaru, 43
ICs. *See* Integrated circuits
IMF. *See* International Monetary Fund
Imports. *See* Trade
Indexes: aggregate price per bit of memory sold, 10; Fisher Ideal price index, 10, 11, 210–11, 237, 252–53; Hirschman-Herfindahl index (HHI), 254–59, 347, 352, 353–54
Integrated circuits (ICs): analog circuits, 66; application-specific, 173, 176, 192–93, 243–44, 283–84, 289, 292, 293; bipolar, 69n89; competition in, 73–77, 137; complementary metal oxide silicon (CMOS), 85; development and funding, 8, 34–38, 49, 56, 59–60, 96–98, 390; diffusion self-aligned (DSA), 62n75; electronic switching, 92; gate arrays, 192–93, 289; hybrid, 60n62; large-scale integration (LSI), 62–63, 67, 71–73, 85; Mask ROMs, 289; medium-scale integration (MSI), 62n74; metal oxide semiconductor (MOS), 62n75, 71; metal oxide semiconductor large-scale integration (MOS LSI), 71–72, 75, 82, 85, 86–87; negative metal oxide silicon (NMOS), 85, 86; packaging, 105; pricing, 138–39, 141–43; quality, 77–78, 92; random access memory (RAM), 62n75; silicon gate metal oxide semiconductor large-scale integration, 85; very large scale integration (VLSI), 62n74, 90, 93–113, 431, 432. *See also* Dynamic random access memory
Intel Corporation: DRAM production and sales, 98–99, 142n46, 288, 352n68; EPROM production and sales, 166, 170, 278–79; import contracts, 76–77; Japanese sales office, 89; production facilities, 167; U.S. Memories, 217
Intellectual property. *See* Patents and licensing
International Monetary Fund (IMF), 130
International Rectifier, 57
International Standard Electric (England), 41n7

International Trade Commission (ITC), 139–40
ITT Corporation, 41n7, 57, 89, 129

Jacobs, Bernard, 61
Japan, 18, 49–52, 56, 272–78. *See also* Semiconductor industry, Japan; Semicondutor Trade Arrangement; Trade, Japan
Japan Economic Journal, 205, 207–08, 245n25
Japan Electrical Manufacturers Association, 119
Japanese Camera Manufacturers Association, 390
Japanese Development Bank (JDB): investments in technological development, 52n33, 66; low-interest loans, 53–54, 60, 66, 121, 122
Japan Fair Trade Commission, 390, 448n37
Japan Radio, 57
Japan Silicon, 122
JEOL, 105
Josokuho, 53
Junkins, Jerry, 184, 208

Kane, Sanford, 217
Kantor, Mickey, 449–50
Kawasaki Steel, 122, 123
Kidenho. *See* Law on Temporary Measures for the Promotion of Specified Electronics Industries and Specified Machinery Industries
Kijoho. *See* Law on Temporary Measures for the Promotion of Specified Manufacturing Industries
Kobe Kogyo (Japan), 42–45, 47
Kokusai Electric, 105
Komatsu, 110
Komatsu Denshi, 123
Komatsu Hoffman, 57–58
Komiya, Ryutaro, 53
Korea: dumping, 224–26, 234n13; exports, 247n27, 290, 300–301; investment, 435n4; pricing, 162, 234n13, 300–301; products, 288; share of global chip market, 2, 23, 161, 219–22, 416, 426, 435, 437, 438, 443, 450. *See also* Samsung
Kyocera, 105
Kyodo Electronics Laboratories, 61, 86, 89n160, 95n180

Law on Temporary Measures for Promoting the Electronics Industry

(Electronics Industry Promotion Law or Denshinho; *1957*): passage and extension of, 52, 53*n*34, 63; "rationalization" cartels under, 54*n*39; research and development subsidies, 66; support classes, 53–54; trade protection, 54–55

Law on Temporary Measures for the Promotion of Specified Electronics Industries and Specified Machinery Industries (Kidenho; *1971*), 52*n*33, 53*n*34, 63

Law on Temporary Measures for the Promotion of Specified Manufacturing Industries (Kijoho; *1956*), 52*n*33

Learning curves and economies: costs and, 319, 323*n*31, 328, 383–84, 414; definition and role, 7, 17, 261; model of, 329–33, 336–41, 400–01, 407–09; pricing and, 307, 309, 318*n*23; semiconductor production, 295, 312, 313, 314–15, 318*n*22

Legislation, Japanese, 52–56

Letters, side: bilateral agreement on semiconductors of *1983*, 155, 157; Semiconductor Trade Arrangement, 170, 173–74, 176, 279–92, 433; Texas Instruments and, 70

Liberal Democratic party (Japan). *See* Political issues

Licensing. *See* Patents and licensing

LSI Logic, 122, 123, 217. *See also* Nihon LSI Logic

Machinery and Electronics Industry Law (*1971*), 52–53

Machinery and Information Processing Industries Law (*1978*), 52–53

Machinery industry, 63–64

Malaysia, 451

Manufacturing processes: advances, 8, 39, 103–06, 144, 377–78; capital-intensive nature, 15–16; costs, 16, 113, 232*n*9, 301–02; dynamic random access memory, 8–9, 98–99; electron beam production equipment, 62*n*75, 102; Japanese, 102–06, 125, 383–84, 432; metal oxide semiconductor technology, 71; learning economies, 17, 428, 429; local foreign production, 20; photolithographic processes, 30*n*33, 62*n*75, 102, 107; "steppers," 102, 110–11; techniques, 30*n*33, 377–78; timeframe, 16–17; transistors, 128; very large scale integration technology, 100–13; wafer

and silicon production, 106, 119–24, 377–78. *See also* Production processes

Matsushita, 45, 66, 129

Matsushita Electronics, 57

MCC. *See* Microelectronics and Computer Technology Corporation

Merck, 438

Michio Watanabe, 172

Microelectronics and Computer Technology Corporation (MCC), 148

Micron Technology, 162, 166, 167, 168–69, 183, 217, 222, 224, 226, 305

Military, U.S. *See* Defense issues

Ministry of International Trade and Industry (MITI): administrative guidance and control, 149, 155*n*97, 156–57, 175–201, 203, 229–30, 391, 395, 418, 431–32, 434–35, 444, 448, 457–58; authority and power, 43–44, 55–56, 63, 70, 119, 124–25, 134, 137; development of technology, 40, 70, 83, 106, 119, 430, 432, 438–41; dumping issues, 162; Electronics Industry Division, 53; Electrotechnical Laboratory (ETL), 40, 48, 60, 61, 82–83, 96, 109; monitoring and forecasting, 178–79, 192, 195–201, 202, 205*n*139, 209*n*152, 210, 216, 222–23, 224, 228, 229, 260; research and production cartels, 54, 59, 74, 80–81, 82*n*133, 83–86, 94–95, 111, 132, 134–36, 137, 150–51, 171–72; Semiconductor Trade Arrangement, 436; Sunshine Project (silicon), 119–24; technology contracts, grants, licenses, and subsidies, 43, 61–68, 76, 77, 79*n*123, 80–81, 83–86, 95–98, 113–19, 137–38, 438; Texas Instruments and, 58–59, 69–70. *See also* Semiconductor industry, Japan; Trade, Japan

Mitsubishi: calculator integrated circuits, 69; Computer Development Laboratories, 96; DRAM production and sales, 142*n*45, 215, 391, 397; dumping, 172; Intel and, 278–79; MITI subsidies, 67–68, 81; Texas Instruments and, 69; Ultra Advanced Computer Technology Research Association, 81, 86, 94; very large scale integration ICs, 96, 107, 110; Westinghouse and, 129

Mitsubishi Electric, 277

Mitsubishi Metals, 123

Mitsubishi TRW, 57–58

Models, statistical: Baldwin-Krugman (B-K) model of semiconductor industry, 318–20, 330*n*39, 332–33, 337*n*47, 352; Bertrand game, 321*n*27; capacity and production choice, 359–71, 407–11, 423–24; cartels, 406–17; competition, 398–405, 418–24; Cournot competition, 399, 404; Cournot-Nash equilibrium, 307, 309, 321, 334, 407, 413–14, 418; "food chain" theory, 375–77; game theory, 380, 407, 418; learning economies, 400–01, 408–09; marginal costs, 401–05; semiconductor industry, 312–58, 380, 383–424; strategic behavior, 407–11, 419–24

Momota, Tsuneo, 109

Monitoring. *See* Ministry of International Trade and Industry; Semiconductor Trade Arrangement

Monopoly issues: barriers to industries, 377; dumping and, 140n41; Japanese, 135, 170n30, 211; Ministry of International Trade and Industry, 54; monopoly power, 211, 319, 321–22, 379–80, 381, 383n18, 388; monopoly profits and rents, 379, 388, 414, 416. *See also* Antitrust issues

Monsanto Corporation, 392n27

Morita, Akio, 43, 70

Mostek, 167

Motorola: Hitachi and, 149n73; Japanese sales of *64K* DRAM, 149; joint ventures, 89; research and development funding, 33, 36; Toshiba and, 129, 352n68, 399n46; U.S. Memories, 217

Multinational companies, 18, 374

Multinational companies, U.S.: electronic component sales, 18; foreign affiliates, 18–19, 28n31; RCA, 42–43; research and development, 18–19; semiconductor assembly, 136–37

Nakasone, Yasuhiro, 187

National Advisory Committee on Semiconductors (NACS), 372

National Cooperative Research Act of *1984*, 148

National Science Foundation, 12

National security. *See* Defense issues

National Semiconductor, 88, 167, 217

NBK, 123

NCR, 218

NEC. *See* Nippon Electric Corporation

NEC Anelva, 110

NEC Toshiba Information Systems (NTIS), 81–82, 96, 111, 112

New Energy Development Organization (NEDO), 120, 123–24

New Japan Radio, 57

NGK Spark Plug, 105

Nihon Business Automation, 110

Nihon Kaizai Shimbun (Nikkei), 205, 214–15, 245–47, 250, 404

Nihon LSI Logic, 283–84

Nikon, 62n75, 105, 110–11

Nippon Columbia, 66

Nippon Electric Corporation (NEC): DRAM production and trade, 116n232, 139, 142n45, 143, 166–67, 169, 180, 183, 198, 200, 215, 277, 290–91, 391, 397, 443n28; dumping, 170n28, 172; Honeywell and, 129; integrated circuits, 60, 78n121, 92; joint ventures, 56–57, 81–82; market shares, 20–21, 45; New Computer Series Technology Research Association, 81–82, 85; research and development, 41, 62–63, 82–83; subsidies to, 66, 81–83; supplier to NTT, 91; Texas Instruments and, 58, 68, 69; transfer of technology, 56–57, 68; transistors, 45, 48; U.S. producers and manufacturers, 144; Very High Speed Computer Systems, 61–62, 92, 93; very large scale integration ICs, 96, 107, 110, 116

Nippon International Rectifier, 57

Nippon Kokkan, 122

Nippon Motorola, 89

Nippon Peripherals, 81, 95n180, 111

Nippon Steel, 122

Nippon Telegraph and Telephone Corporation (NTT): Basic Technology Research Promotion Center, 118–19; certification processes, 155–56; computers, 93; development of semiconductor technology, 40, 45, 90–94, 95, 106, 115–16; DEX 2 switching system, 92; patents, 116, 117t; privatization, 117–18; procurement practices, 137, 116n232; subsidies to semiconductor industry, 113, 117–18, 125–26, 137–38

Nittetsu Electronics, 122

NKK, 123

NMB Semiconductor, 17n18, 416

Nomura Research Institute (NRI), 289

North American Rockwell, 72. *See also* Rockwell International

Noyce, Robert N., 56, 138, 140, 142n46, 151

NTIS. *See* NEC Toshiba Information Systems

NTT. *See* Nippon Telegraph and Telephone Corporation

OCDM. *See* Office of Civil and Defense Mobilization

OECD. *See* Organization for Economic Cooperation and Development

Office of Civil and Defense Mobilization (OCDM), 128

Oki Electric, 144, 172, 183, 277, 391 computer manufacturing, 96; supplier to NTT, 91; Ultra Advanced Computer Technology Research Association, 81, 86, 94; Very High Speed Computer Systems, 61–62

Okinawa, 88, 131*n*10

Okura Group, 135

Organization for Economic Cooperation and Development (OECD), 80

Osaka Titanium Company (OTC), 110, 120–21, 122, 123

OTC. *See* Osaka Titanium Company

Pacific Semiconductor, 31, 57–58

Parkinson, Joseph, 181

Patents and licensing: licensing agreements, 129; "patent war" in Japan, 68–70; royalties, 71; semiconductor manufacture, 378; technology transfers, 42–44, 56–57, 92, 107, 109, 378; LVSI projects, 107, 109, 116, 117*t*. *See also* Semiconductor Trade Arrangement

Pattern Information Processing System (PIPS), 82

Philco, 33

Philips NV, 57, 129

"Pi" rule. *See* Dynamic random access memory

Political issues: Japanese semiconductor and computer industry, 27, 79–80, 84–85, 94, 167*n*17, 283, 284; Japanese trade policies, 130, 144, 166, 172*n*34, 373*n*3; Semiconductor Trade Arrangement, 230–31, 283, 284, 356; U.S. semiconductor and computer industry, 27–30, 283, 284

Polypropylene, 394

Prices. *See* Dynamic random access memory; Erasable programmable read only memories; Semiconductor Trade Arrangement; Semiconductor industry; Trade

Production processes: capacity constraints, 313, 318, 320, 321, 322, 323, 339; control of, 195–201, 203, 213, 434–35; cost, 9, 305, 308–09, 310, 312, 315, 319, 320–21, 322–28, 335, 344–49, 352, 402–05, 427–28, 445; labor in, 13; packaging, 296–301; plant capacity, 315, 317, 323*n*31,

325*n*32, 334–36, 337*n*48, 346–47, 356, 359–65, 407–09; pricing and profitability, 314–17, 318, 320–21, 322–28, 335, 345, 347, 352, 371; processes and timing, 13, 213, 231–32, 261–62, 316*n*19, 342–43, 409*n*56; production alliances, 446–47; product life cycle, 313–58; quantities produced, 238*f*, 239*f*, 260–66, 272–73; research and development in, 13, 14*t*, 31–32, 91–92, 200; silicon and memory chips, 62*n*75, 200; technological advances and transfers, 9, 91–92, 94, 331–32, 377, 431–32; testing, 232, 236, 295, 333–34, 345, 402; transistors, 45; very large scale integration equipment manufacturers, 96, 98*n*189, 431; yields, 198, 200, 258–61, 295, 314, 320, 322, 323, 329, 330–31, 333–34, 339, 341, 343*n*60, 402. *See also* Manufacturing processes

Purchasing and sales: authorized distributors, 233, 234; contract, spot, and gray market sales, 141–42, 143, 161, 180, 184, 209–10, 221, 230, 233–36, 243, 244, 248–52, 269, 292–301, 443–44; demand, 236, 240, 288, 289, 291–92, 293, 312, 334, 341–44, 365–71, 387, 388, 391, 426; domestic versus foreign products, 394–95; original equipment manufacturers (OEMs), 232–33, 235–36, 295–301; pricing, 233, 234, 236–54; "pulls," 233; quantity commitments, 235–36; regional and country differentials, 242–54, 259–60, 272, 293; semiconductor chips, 2, 8; supplies, 247*n*27, 254, 256–66, 272, 295; term of contracts, 294–301; tying arrangements, 243–44; volume prices, 244*n*22; warranting, 233. *See also* Dumping; Dynamic random access memory (DRAM) prices

Purdue University, 30*n*32

R&D. *See* Research and development

Radar, 29–30

Radio. *See* Television and radio

Rapoport, Carla, 176

Rashomon (film), 79

Raytheon, 33, 57

RCA: development of solid-state electronics, 18, 31, 33, 36; technology transfer, 42–43; trade policies, 129

RCA Victor, 129

Reagan (Ronald) administration, 153, 187, 188

Regency Corporation, 45

Remington Rand, 129
Research and development (R&D): cost accounting, 311–12; economies of scale, 14–17; General Agreement on Tariffs and Trade and, 3n4; government-industry collaborations, 59, 80–85, 428–29, 458; investment and costs, 7, 11–13, 14t, 25, 313–14, 315, 428–29, 458; semiconductor industry, 312
Research and development, Japan: 3.5 generation program, 80–81, 83, 93, 111; Dendenkosha Information-Processing System (DIPS), 93; electron-beam (E-beam) writing systems, 110, 111; femtosecond technology, 439; funding, 92–93, 102, 103f, 111–19; joint research labs, 113; as percent of sales, 12–13; Pattern Information Processing System project, 82–83; role of government, 45, 112n225; subsidies, funding, and incentives, 54, 60, 61–65, 66–68, 81, 96–98, 111–13, 116–17, 125–26, 152, 441; transfer of technology, 102, 103, 109–10, 116, 119; VLSI programs, 94–113, 116, 125, 431
Research and development, U.S.: funding, 27–30, 117n234, 147–48, 429; joint ventures, 147–48; tax credits for, 147; technology transfers, 18, 31–32, 41–44, 57. See also Semiconductor Manufacturing Technology consortium
Ricoh, 72
Rigaku Electronics, 110
Rogers, T. J., 217–188

Sadao Inoue, 186
Samsung: DRAM production and sales, 168, 212, 221, 226, 443n28; dumping and, 225, 226; growth, 222; Intel and, 352n68; NEC and, 290–91; U.S. distributors, 300–01
Sanyo, 129
SBFC. See Small Business Finance Corporation
SEH. See Shin Etsu Handotai
Selete. See Semiconductor Leading Edge Technologies Inc.
Sematech. See Semiconductor Manufacturing Technology consortium
Semiconductor industry: 1950s, 128–32; 1960s, 132–36, 382–83, 425; 1970s, 73, 77, 136–42; 1980s, 142–58, 157–61, 162–63, 192–201, 203, 207, 212, 222, 234–35, 234–54, 272, 292, 382–83, 425–26, 432; 1990s, 223, 289n68, 292, 427, 434, 435, 445; competition in, 124, 147, 148, 153,

158, 185–87, 223, 232, 254n33, 283, 316, 373–74, 380, 398–405, 441; cyclical nature, 16–17, 151–52, 157–58, 240–41, 312, 356, 382–83, 428, 429, 447; economic issues and, 427–29, 447–48, 452–53; foreign sales, 15; High Technology Working Group (HTWG), 153–58; history, 8, 18, 49, 429; investment in, 7, 376–77, 384–85; market shares, 20–21, 449; memory chips, 8; models, 307–71, 372–424; production, 7, 313; as a strategic industry, 372–82; technology and industrial structure, 7–17, 377; subsidies, 413–14, 428–29; supply and demand, 231–36; end users, 2. See also Dumping; Dynamic random access memory; Integrated circuits; Manufacturing processes; Research and Development; Semiconductor Trade Arrangement; Silicon chips; Trade; Transistors
Semiconductor Industry Association (SIA): formation, 138, 432; investigations and, 139, 140–41; responses to competition, 147, 164, 216–17, 432; section 301 complaint, 158, 165. See also Trade, U.S.
Semiconductor industry, Europe. See Europe
Semiconductor industry, Japan: 1950s, 40–49, 128; 1960s, 39, 49–72, 124, 125, 382–83, 426, 430; 1970s, 39–40, 72–116, 124–25, 137–40, 141, 143, 144, 145, 390, 435; 1980s, 116–24, 125, 146, 147–58, 162–226, 288, 382–83, 385, 387–98, 426–27, 435, 444; 1990s, 283, 286–87, 288–92, 438–54; competition and, 124–26, 153, 128–58, 383, 431; cooperation and collusion within the industry, 6, 107–13, 142, 163–65, 168, 186–87, 195, 204–08, 214–15, 219–20, 230, 279, 383, 385–86, 387–98, 417, 448–49; "Den-Den" suppliers, 91, 92; government policies and strategies, 5–6, 27, 39–40, 48, 52, 80–81, 113–19, 124, 125, 160–61, 385–86, 418, 430, 448; growth, 39–52, 99, 100, 113, 120–21; investment in, 15–16, 39, 57, 65, 80–81, 87, 122–23, 124, 128, 138, 163, 166, 193–95, 202–08, 215, 434–35; joint ventures, 56–57, 60–61, 68–70, 122; liberalization of, 73–90, 93, 94, 119, 124, 125, 130, 137, 390; market share, 20t, 21–24, 77–79, 148–49, 280, 451; production in U.S., 22, 144; products, 23, 45n24, 71, 99, 392; quality issues, 145; regulations, 53, 125;

technology transfer, 41–44, 124; transistors, 49–52. *See also* Dumping; Ministry of International Trade and Industry; Research and development, Japan; Semiconductor Trade Arrangement; Trade, Japan

Semiconductor Industry Research Institute of Japan (SIRIJ), 439–41

Semiconductor industry, U.S.: *1960s*, 132, 136–37, 382–83, 425, 426–27; *1970s*, 136–37, 141–46, 430; *1980s*, 146, 147–58, 162–226, 382–83, 425–26, 442; *1990s*, 283, 372, 437–54, 455–59; captive and merchant producers, 20–21, 24; competition, 6, 15, 88–90, 111, 128–58, 373–74, 383, 405–17, 429, 437, 442, 443, 451–52; fabless companies, 23, 444*n*31; foreign sales, 15; foreign subsidiaries and contractors, 136–37; government policies and strategies, 27–38, 153, 163, 373–74, 382–424, 429–30; investment in, 15–16, 144*n*51; joint ventures, 430, 444; market share, 20*t*, 21–24, 279–92, 437–38, 446–49, 451; products, 23, 71, 72–73, 278*n*51, 375*n*5, 389; quality issues, 145–46; research and development, 12–13; sales and trade issues, 2, 18*n*20, 73–74, 132, 145, 216; as strategic industry, 372–80, 405, 429–30; subsidies for, 380, 405–17, 429; transistors, 49–52; Very High Speed Integrated Circuit program, 37–38. *See also* Semiconductor Trade Arrangement

Semiconductor Leading Edge Technologies Inc. (Selete), 440

Semiconductor Manufacturing Technology consortium (Sematech), 38, 278*n*52, 379, 426, 432, 438–39, 440–41

Semiconductor Research Consortium (SRC), 440

Semiconductor Research Cooperative, 147

Semiconductors. *See* Dynamic random access memory; Integrated circuits; Transistors

Semiconductor Technology Academia Research Center, 440

Semiconductor Trade Arrangement (STA; *1986, 1991*): background, 127, 158–76, 227–29, 306, 426, 432; consequences of, 26, 160–61, 185–87, 210–11, 227–304, 386, 388, 395–98, 405, 433–37, 453–54, 457; cooperation and collusion within the industry, 163–65, 168, 186–87, 195, 204–08, 214–15, 219–20, 230, 279, 283, 388, 395–405, 453–54; criticisms of, 450–54, 457; European markets, 173,

188–91, 224; foreign market values, 177, 180, 191, 211, 213–14, 223–24, 228, 306, 307–08, 433–34, 450–51; general description, 1; Japanese views of, 175, 436; market shares, 170, 171, 172, 173–74, 175, 188, 206, 229, 279–92, 293, 435–37, 446–49, 450, 452, 457; monitoring and forecasting, 175–201, 203, 223–24, 228, 229–30, 243, 252, 272, 281–82, 283–84, 292, 306, 356, 388, 434, 444–45, 453, 457–58; patents, 228; prices, 171–72, 174, 175–80, 189–90, 241*n*19, 242–43, 297–304, 396; profitability, 204–08, 302, 307–08; public policy and, 1–6, 38; regional welfare, 272–78, 293; renewal and renegotiation, 223–26, 280–81, 306, 434, 449–50; side letters, 170, 173–74, 176, 178, 229, 279–92, 433; third-country markets, 160, 174, 176, 178–79, 180, 183, 185, 188, 190, 224, 229, 243, 292. *See also* Dumping

Sharp: calculators, 68*n*86, 72, 75, 431; DRAM production and sales, 142*n*45; joint ventures, 86, 95*n*180; North American Rockwell and, 72; Texas Instruments and, 69

Shiba Electric, 66

Shin Etsu, 110

Shin Etsu Handotai (SEH), 120, 123

Shockley, William, 30

SIA. *See* Semiconductor Industry Assoication

Side letters. *See* Letters, side

Siemens, 222

Signetics, 36, 89

Silicon and memory chips: "bi" rule, 240; costs, 9, 16, 50–51, 240, 337*n*47, 343, 345, 393*n*29, 402; crossover point, 223; die shrinks, 260*n*39, 331; economies of scale, 14–17; flash memory chips, 279; integrated circuits, 8; lifespan, 11; microprocessor performance, 8*n*7; prices of, 7, 317; production, 62*n*75, 200, 241, 260–61, 377–78, 402; product lives, 7; purchase and sales, 392; reduced instruction set computer chips, 36; research and development, 30, 49–52, 102, 105–06, 260*n*39, 438–39, 440–41; role in high-technology industries, 7; wafer starts, 315. *See also* Dynamic random access memory; Erasable programmable read only memory; Manufacturing processes; Production processes

Siltec, 123

Singapore, 451

SIRIJ. *See* Semiconductor Industry Research Institute of Japan
Small Business Finance Corporation (SBFC), 52*n*33
Sony Corporation, 43; market shares, 45; Texas Instruments and, 58, 68, 69–70; transistor products, 45, 48
Specified Manufacturing Industries Law. *See* Law on Temporary Measures for the Promotion of Specified Manufacturing Industries
Spence, Michael L., 312–13, 400
Spot sales. *See* Purchasing and sales
Sprague Electric, 137
SRAMs. *See* Static random access memory chips
SRC. *See* Semiconductor Research Consortium
STA. *See* Semiconductor Trade Arrangement
Standard Radio, 129
Stanford University, 147–48
Static random access memory chips (SRAMs), 176
Steel industry, 394–95
Strategic industries, 372–424
Strauss, Robert, 138*n*32
Sun Microsystems, 218
Super Silicon Crystal Research Institute, 438
Supply. *See* Purchasing and sales
Sylvania, 33, 129

Taiwan, 2, 23, 427, 437, 439, 443, 450
Takagi, Noboru, 109
Takeda Riken, 144
Tanahashi, 170
Tanaka, Kakuei, 94
Tandem Computers, 218
Tandy, 218
Tarui, Yasuo, 102, 109
Taxation incentives, 63, 65–66
TDK Electronics, 88
Technology transfers. *See* Research and development
Tektronix, 218
Television and radio: cartels, 390; controls in the *1960s*, 132–36; development, 45, 47; Japanese, 49, 66–67, 68, 131; licensing issues, 42–43
Telex, 90
Texas Instruments (TI): Canon and, 72; entrance into Japanese market, 57, 58–59, 60, 68–70, 77, 136; Japanese pricing, 142–43; Minuteman II guided missile, 35*n*50, 36; research and development

funding, 32*n*40, 33; role in semiconductor industry, 56; Sharp and, 69*t*; Sony and, 58, 68, 69–70; subsidiaries, 438; U.S. Memories, 217
Texas Instruments Japan, 77, 180, 183–84, 185*n*84
Textiles, 132
Thailand, 451
Third-country exports. *See* Trade
TI. *See* Texas Instruments
Tilton, Mark, 389–90, 394
Toko, 89
Tokyo Oka Kogyo, 110
Tokyo Telecommunications Engineering, 43. *See also* Sony Corporation
Tokuyama Soda, 120–21, 122
Toppan Printing, 105
Toray, 110
Toshiba: DRAM production and sales, 142*n*45, 143, 166–67, 192, 214, 215, 277, 352*n*68, 397, 399; dumping, 170*n*28; joint ventures, 57; licensing agreements, 42, 43–44; market shares, 20–21, 45; research and development, 41, 82–83, 105; silicon wavers, 392–93; subsidies, 66, 81–83; transistors, 45; New Computer Series Technology Research Association, 81–82, 85; U.S. facilities, 167; Very High Speed Computer Systems, 61–62; very large scale integration ICs, 96, 107, 110, 116
Toshiba Engineering, 110
Toyo Communication Equipment, 66
Trade: competition, 2, 3–5, 7, 17–27, 51–52, 73–77, 83–84, 86–87, 147; European, 26; exports in electronics, 49–52; frictions and negotiations, 3, 4, 7, 128–58, 383, 385–86, 391, 449, 451–53; intellectual property protection, 156; Semiconductor Trade Arrangement and, 2, 4, 446–49; strategic objectives and policies, 5, 21*n*21, 140, 155, 380–82, 426–27, 458–59; tariffs, 26*n*30, 155, 156; third-country issues, 131, 160, 172; World Trade Organization and, 451. *See also* Dumping; General Agreement on Tariffs and Trade; Semiconductor Trade Arrangement; *individual countries and products*
Trade and Exchange Liberalization Guidelines, 130
Trade, Japan: *1960s*, 73, 131–36; *1970s*, 73, 75–77, 83, 87, 88*t*, 136–46; *1980s*, 120–22, 123, 143, 148–58, 159–224, 389, 432; *1990s*, 162, 455–59; controls, 132, 149–50, 166; Council of Nine, 164–65, 391;

customs issues, 164n3; domestic, 135;
DRAM wars, 148–58; five-company
rule, 133–34; foreign sales, 2, 15–16,
75–76, 133, 262; imports and exports,
47, 49–52, 54–59, 60, 61, 62f, 70–88,
120–22, 123, 130, 160, 195, 209, 243,
262, 264, 389, 432; integrated circuits,
60, 73, 74–88; price- and quantity-fixing
agreements, 133–46, 149, 150–51, 162,
163–64; sales agents, 142n45, 143;
strategic and predatory sales policies,
152–53, 165; tariffs and quotas, 87, 131,
133, 147, 149–50, 152, 156; third-country
exports, 131, 179; two-tier pricing, 138,
140–41, 142, 143, 152; U.S.
interpretation of policies, 152–53, 216,
385–87, 392; voluntary impact expansion
(VIE) measures, 157. See also Dumping;
Semiconductor Trade Arrangement
Trade, U.S.: 1950s, 128; 1960s, 130–36;
1970s, 136–46, 430–31; 1980s, 148–58,
159–224; 1990s, 455–59; contract, spot,
gray market sales, 141; DRAM wars,
148–58; effects of, 127; foreign market
values (FMV), 177, 180, 191, 211, 213–14,
223–24, 266–72, 301–304, 387, 399;
government involvement in, 149–50, 153,
163; imports and exports, 128–29, 130–
32, 137, 141, 144, 156, 157, 163–64, 266–
72; national security and, 130, 132, 150;
section 301 issues, 158, 160, 164–66, 167,
171, 172, 176, 227, 391, 449; tariffs, 156.
See also Dumping; Semiconductor Trade
Arrangement
Transistors: consumer versus industrial,
50–51; development, 8, 30–33, 48, 49,
52; differences, 128; germanium, 49, 92,
425; Japanese, 40–42, 45–48, 128;
obsolesence, 11; technology transfers,
44; U.S., 48, 128. See also Silicon and
memory chips; Semiconductor industry
TRW Incorporated, 57–58

UCOM. See Users' Committee of Foreign
Semiconductors
Ulvac, 105
Union Carbide, 123n254
Unisys, 218

United States: cooperation with Japan,
130, 429–59; semiconductor industry
output, 1; Soviet Union and, 130; trade
policies, 4. See also Military, U.S.;
Multinational companies, U.S.;
Research and development, U.S.;
Semiconductor industry, U.S.;
Semiconductor Trade Arrangement;
Trade, U.S.
Users' Committee of Foreign
Semiconductors (UCOM), 289
Ushio Denki, 105
U.S. Memories, 214, 216–19, 244, 396, 416,
443
U.S. Semiconductor, 123

Very High Speed Computer Systems
(VHSCS). See Computers
Very High Speed Integrated Circuit
(VHSIC). See Integrated circuits;
Semiconductor industry, U.S.
Very Large Scale Integration (VLSI). See
Integrated circuits
VHSCS (Very High Speed Computer
Systems). See Computers
VHSIC (Very High Speed Integrated
Circuit). See Integrated circuits;
Semiconductor industry, U.S.
VLSI (Very Large Scale Integration). See
Integrated circuits
VLSI Technology Research Association,
100, 102, 107, 111–13
von Hippel, Eric, 378

Wafers. See Silicon and memory chips
Western Electric: research and
development, 31, 33, 736; technology
transfers, 41n7, 44
Westinghouse, 129
World Semiconductor Trade Statistics
(WSTS), 280
World Trade Organization (WTO), 449,
451–52

Yeutter, Clayton K., 172
Yukio Honda, 181, 205

Zenith Radio Corporation, 134n23
Zilog, 89

HD 9696 .S43 U466 1996

Flamm, Kenneth, 1951-

Mismanaged trade?